DATE DUE			
Unless Recalled Earlier			
OCT 1 5 1996			
OCT 1 6 1996			
MAY - 3 2000			

DEMCO 38-297

Antennas for Radar and Communications

WILEY SERIES IN MICROWAVE AND OPTICAL ENGINEERING

KAI CHANG, Editor
Texas A & M University

INTRODUCTION TO ELECTROMAGNETIC COMPATIBILITY
Clayton R. Paul

OPTICAL COMPUTING: AN INTRODUCTION
Mohammad A. Karim and Abdul A. S. Awwal

COMPUTATIONAL METHODS FOR ELECTROMAGNETICS AND MICROWAVES
Richard C. Booton

FIBER-OPTIC COMMUNICATION SYSTEMS
Govind P. Agrawal

ANTENNAS FOR RADAR AND COMMUNICATIONS: A POLARIMETRIC APPROACH
Harold Mott

Antennas for Radar and Communications: A Polarimetric Approach

HAROLD MOTT
Department of Electrical Engineering
University of Alabama

A WILEY-INTERSCIENCE PUBLICATION
JOHN WILEY & SONS
New York / Chichester / Brisbane / Toronto / Singapore

TK
6590
.A6
M68
1992

This text is printed on acid-free paper.

Copyright © 1992 by John Wiley & Sons, Inc.

All rights reserved. Published simultaneously in Canada.

Reproduction or translation of any part of this work beyond
that permitted by Section 107 or 108 of the 1976 United
States Copyright Act without the permission of the copyright
owner is unlawful. Requests for permission or further
information should be addressed to the Permissions Department,
John Wiley & Sons, Inc., 605 Third Avenue, New York, NY
10158-0012.

Library of Congress Cataloging in Publication Data:
Mott, Harold.
   Antennas for radar and communications: a polarimetric approach/
Harold Mott.
     p. cm. —(Wiley series in microwave and optical engineering)
   "A Wiley-Interscience publication."
   Includes bibliographical references.
   ISBN 0-471-57538-0
   1. Radar—Antennas. 2. Antennas (Electronics) 3. Electromagnetic
waves—Polarization. I. Title. II. Series.
TK6590.A6M68 1992
621.3848′3—dc20   92-9240

Printed in the United States of America

10 9 8 7 6 5 4 3 2 1

To Wolfgang-Martin Boerner

Preface

This book began as a revision of *Polarization in Antennas and Radar*, published by Wiley-Interscience in 1986, but the changes made were so extensive that it became inappropriate to consider it a second edition. The major part of the book is new, and most of the material that did come from the first text was rewritten. I had two goals in writing the book. One was to provide a polarimetrically correct introduction to antennas at the graduate and advanced undergraduate level. The other was to provide a consistent and thorough treatment of polarization theory as applied to antennas and targets for communications and radar.

The text is suitable for a one semester introductory course in antennas and covers all of the topics necessary for students who will use antennas for radar or communications. Chapters 1, 2, 3 (through Section 3.26), 4, and 5 are suggested for such a course. In addition to the standard definitions and developments of commonly used antenna texts, the book treats misaligned antennas by use of Euler angle matrices, antenna noise temperature, adaptive antenna arrays, polarization matching of antennas in a communication system, and power loss caused by radiation or reception of a quasimonochromatic signal by antennas that are frequency dependent. Many introductory texts do not lay a good foundation in polarization theory and consequently cannot properly treat such topics as polarization matching, scattering matrices, and radiation of nonmonochromatic signals. It is hoped that this book does so. If time is available after completion of the suggested chapters, the first eight sections in Chapter 6, which cover target scattering matrices and introduce the concept of optimal polarization in radar, can be included. Chapter 9 is concerned with target detection and recognition and offers an interesting way to complete a course in antennas.

The book is also extremely useful for those who have already taken a course in antenna theory from a standard text and wish to learn polarization theory. Chapters 1 and 2 can be omitted for those students and practicing engineers. An appropriate course of study, either formal or by informal reading, is Chapters 3, 4, 5, 6, 7, and 9, while Chapter 8 is useful for those interested in measurements. The background needed is a good understanding of calculus and vector analysis, some knowledge of matrices, and a knowledge of electromagnetics equivalent to that acquired by completing a good undergraduate text. The text treats matrices at a level somewhat above the most elementary, but the methods are covered in detail and should present no difficulty.

The value of the book as a reference for engineers and physicists, a use of particular interest to the author, is greatly enhanced by additions. Chapter 3 now includes stereographic, orthogonal, Lambert, Mercator, $\tau - \epsilon$, Aitoff–Hammer, and Mollweide polarization charts. Chapter 6 discusses change of polarization base, the Kennaugh process for choosing an optimum radar polarization, the Huynen fork, the Boerner extension of the Huynen fork, and the scattering matrix with geometric variables. The relationships among scattering matrix, scattering cross section, and radar cross section are developed. Scattering matrices and Huynen fork parameters are given for several targets. Chapter 7 on partial polarization, the longest chapter in the book, includes the Graves polarization power scattering matrix, the Mueller and Kennaugh matrices, the covariance and coherency matrices, and a discussion of the effect of antenna frequency dependence on transmitted-wave polarization and received power. Chapter 9 discusses the use of polarization as an aid in detecting targets in clutter and separating classes of targets.

Great care has been taken to identify the coordinate systems used in wave transmission and scattering from a target. Two target scattering matrices, Jones and Sinclair, are used for those targets that possess scattering matrices, depending on coordinates used for the scattered wave. Confusion is often caused by failure to be explicit in this matter. For the same reason, two target matrices are used for partially-polarized waves and for those targets that do not have a Jones or Sinclair matrix, namely the Mueller matrix and a similar one with reversed coordinates that is more useful to radar engineers. I have taken the initiative, after discussions with those who are familiar with the history of developments in polarization theory, of naming the second matrix after Edward Kennaugh, who first developed the matrix and whose contributions to radar polarimetry have not been surpassed.

I have been exceptionally fortunate in having two of the outstanding polarimetrists of this century, Professor Wolfgang-Martin Boerner and Dr. Jean Richard Huynen, interested in this project. Both provided notes, and both were most generous with their time and knowledge in making suggestions and clarifying points. Professor Boerner's students worked through the manuscript and provided useful feedback. Dr. Shane R. Cloude was also very kind in explaining certain points. I wish to thank them for their invaluable assistance. Professor Boerner's influence was critical. Without it, it is unlikely that I would have undertaken the task of writing another book. I am most grateful to Professor Boerner, and this book is dedicated to him.

HAROLD MOTT

Tuscaloosa, Alabama
July 1992

Contents

1 An Introduction to Antennas   **1**

    1.1 Introduction   1
    1.2 The Vector Potentials   2
    1.3 Integral Solutions for the Vector Potentials   7
    1.4 Approximations to the Potentials   9
    1.5 Far-Zone Fields   10
    1.6 Use of the Potential Integrals for Physical Structures   13
    1.7 Radiation Pattern   15
    1.8 Gain and Directivity   21
    1.9 The Dipole Antenna: Fields   25
    1.10 Reciprocity Theorem   29
    1.11 An Equivalence Theorem   30
    1.12 The Dipole Antenna: Input Impedance   32
    1.13 Waveguide Opening into Infinite Ground Plane   38
    1.14 The Receiving Antenna   42
    1.15 Transmission between Antennas   51
    1.16 The Radar Equation   52
    References   55
    Problems   55

2 Arrays, Broadband Antennas, and Noise   **58**

    2.1 Introduction   58
    2.2 Moment Method and Geometrical Theory of Diffraction   58
    2.3 Antenna Arrays   68
    2.4 Scanning with Planar Arrays   73
    2.5 Adaptive Arrays   79
    2.6 Broadband and Frequency-Independent Antennas   88
    2.7 Antenna Noise   98
    References   112
    Problems   113

x   CONTENTS

3 Representations of Wave Polarization   115

3.1 Introduction   115
3.2 The General Harmonic Wave   115
3.3 Polarization Ellipse for Plane Waves   118
3.4 Linear and Circular Polarization   125
3.5 Power Density   125
3.6 Rotation Rate of the Field Vector   126
3.7 Area Sweep Rate   128
3.8 Rotation of $\mathscr{E}$ with Distance   129
3.9 The Polarization Ratio   130
3.10 Circular Wave Components   131
3.11 Change of Polarization Base: Orthogonal Elliptical
     Components of a Wave   133
3.12 Relationship between P and q, and the Modified
     Polarization Ratio   135
3.13 Ellipse Characteristics in Terms of q 136
3.14 Ellipse Characteristics in Terms of P and p 138
3.15 Polarization Characteristics for Ranges of P, p, and q 138
3.16 The Transformations $p(q)$ and $q(p)$ 140
3.17 The Transformation for $u < 0$ 143
3.18 Polarization Chart as the p Plane   146
3.19 Coincident Points on the q and w Planes   147
3.20 Contours of Constant Axial Ratio and Tilt Angle   147
3.21 Contours of Constant $|p|$ 148
3.22 Contours of Constant ϕ 150
3.23 Stokes Parameters   155
3.24 The Poincaré Sphere   156
3.25 Special Points on the Poincaré Sphere   157
3.26 Other Relationships between the Variables   158
3.27 Mapping the Poincaré Sphere onto a Plane   159
3.28 Mapping onto the q and w Planes   163
3.29 Other Maps of the Poincaré Sphere   167
     References   179
     Problems   180

4 Polarization Matching of Antennas   182

4.1 Introduction   182
4.2 Effective Length of an Antenna   182
4.3 Relation between Effective Length and Gain   183
4.4 Received Voltage   185
4.5 Maximum Received Power   188
4.6 Polarization Efficiency—Match Factor   191
4.7 The Modified Friis Transmission Equation   194
4.8 Polarization Match Factor—Special Cases   196
4.9 Match Factor in Other Forms   201

4.10	Contours of Constant Match Factor	203
4.11	The Poincaré Sphere and Polarization Efficiency	210
4.12	Match Factor Using One Coordinate System	212
4.13	Polarization Match Factor—Misaligned Antennas	212
	References	219
	Problems	220

5 Polarization Characteristics of Some Antennas — 222

5.1	Introduction	222
5.2	Test Antennas for Determining Effect of Polarization	224
5.3	The Short Dipole	229
5.4	Crossed Dipoles (Turnstile Antenna)	232
5.5	Crossed Dipoles with Ground Plane	233
5.6	The Loop Antenna	235
5.7	Loop and Dipole	237
5.8	Waveguide Opening into Infinite Ground Plane	239
5.9	Horns	242
5.10	Paraboloidal Reflector	246
5.11	Narrow-Polarization-Beamwidth Array	261
5.12	Traveling-Wave Loop Antenna	264
5.13	The Axial-Mode Helix	267
5.14	Simple Waveguide System for Elliptical Polarization	273
5.15	Another Waveguide System	277
5.16	Lossless Power Combiner and Divider System	278
	References	287
	Problems	288

6 Polarization Changes by Reflection and Transmission — 290

6.1	Reflection at an Interface: Linear Polarization	290
6.2	Elliptical Waves	300
6.3	Reflection and Transmission Matrices	309
6.4	Scattering by a Target: Sinclair and Jones Matrices	311
6.5	Scattering with Circular Wave Components	316
6.6	Sinclair Matrix and Polarization Ratio	318
6.7	Change of Polarization Base: The Scattering Matrix	319
6.8	Polarization for Maximum and Minimum Power	322
6.9	The Scattering Matrix with Geometric Variables	336
6.10	Scattering Matrices and Huynen Fork Parameters for Some Common Scatterers	349
6.11	Additional Characteristic Polarizations	360
6.12	Characteristic Polarizations with Nonsymmetric Sinclair Matrix	361
	References	362
	Problems	362

xii CONTENTS

7 Partial Polarization 365

    7.1 Introduction 365
    7.2 Representations of the Fields 366
    7.3 The Coherency Matrix 373
    7.4 Stokes Vector of Partially Polarized Waves 391
    7.5 Polarization Ratio of Partially Polarized Waves 399
    7.6 Reception of Partially Polarized Waves 400
    7.7 Scattering by a Target: The Mueller Matrix 403
    7.8 The Kennaugh Matrix 415
    7.9 Power Spectral Densities of the Fields 419
    7.10 The Power Scattering Matrix 440
    7.11 Target Coherency and Covariance Matrices 451
    7.12 Characteristic Polarizations with Nonsymmetric
        Scattering Matrix 455
        References 459
        Problems 460

8 Polarization Measurements 463

    8.1 Introduction 463
    8.2 Monochromatic Polarization Parameters 463
    8.3 The Stokes Vector 470
    8.4 The Scattering Matrix 472
    8.5 Polarization Power Scattering Matrix 474
    8.6 Mueller and Kennaugh Matrices 476
        References 477
        Problems 478

9 Target Detection 479

    9.1 Introduction 479
    9.2 Rain Clutter 482
    9.3 Chaff 484
    9.4 Sea Clutter 485
    9.5 Target Glint 486
    9.6 A Matrix Target-Clutter Discriminant 489
    9.7 Separation of Constant-Null-Polarization Classes 492
    9.8 Polarization Diversity and Adaptation 497
        References 500

Appendix A The Mueller Matrix 502

Appendix B The Kennaugh Matrix 509

Index 513

Antennas for Radar and Communications

CHAPTER ONE

An Introduction to Antennas

1.1. INTRODUCTION

This chapter introduces the concepts of antenna pattern, antenna input impedance, gain, effective area of a receiving antenna, the equivalent circuit of an antenna, efficiency, the relationship between gain and effective area, the Friis transmission formula, and a simple form of the radar equation.

Antenna concepts cannot be introduced without illustrative examples, so a linear wire antenna, of which the half-wave dipole is the best-known example, is used to illustrate many of the ideas. Radiation from apertures is also discussed. Arrays and other types of antennas are considered in Chapter 2, and after the foundation of polarization theory is laid in Chapter 3, still other antennas are examined.

The task of designing an antenna system appears to be formidable, but fortunately it can be broken into simpler tasks that are more easily understood and carried out. The separation is convenient and enlightening. Consider the two antennas shown in Figure 1.1: a transmitting antenna (1) connected to a generator and a receiving antenna (2) connected to a receiver (which we treat as a load impedance).

1. To the generator, antenna 1 appears to be a load impedance or admittance (perhaps transformed by a connecting transmission line or waveguide). We wish to find this impedance for matching purposes and to determine the power accepted by the antenna.

2. A portion of the power accepted by the antenna is radiated and a portion is dissipated as heat. We need to find the total power radiated.

3. The transmitting antenna does not radiate equally in all directions. We must determine the directional characteristics of the antenna and find the power density (Poynting vector magnitude) at the receiving antenna.

4. To the receiver, antenna 2 appears to be a voltage or current source, with the source value partly determined by the incident power density.

2   AN INTRODUCTION TO ANTENNAS

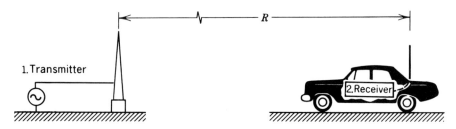

FIGURE 1.1   Transmitting and receiving antennas.

Antenna 2 also appears to have some internal impedance. We must determine the source value and the receiving antenna internal impedance so that power to the receiver load can be found.

5. The path from antenna 1 to antenna 2 may not be direct but may involve a reflection from ground (multipath) or a target (as in radar). Strictly speaking, these are not antenna problems, but since this book is concerned with radar applications, we are interested in appropriate descriptions of these phenomena.

6. The power to the receiver in Figure 1.1 depends on the polarization characteristics of both antennas and any change in polarization during propagation between the antennas. However, we will defer a discussion of polarization properties to later chapters. We must first establish a firm foundation for the development of these properties.

7. In general, the factors enumerated above depend on frequency, and antennas consequently have a finite bandwidth determined by impedance, radiation pattern, and so on. If all antenna, generator, target, and load parameters are known as functions of frequency, the bandwidth determination is straightforward, and we devote little attention to it in this chapter. In later chapters the frequency dependence of antennas is examined in more detail.

1.2. THE VECTOR POTENTIALS

If the sources of the electromagnetic field vary sinusoidally with time, so do the field quantities, and the Maxwell equations

$$\nabla \times \mathscr{H} = \mathscr{J} + \frac{\partial \mathscr{D}}{\partial t} \tag{1.1a}$$

$$\nabla \times \mathscr{E} = -\frac{\partial \mathscr{B}}{\partial t} \tag{1.1b}$$

$$\nabla \cdot \mathscr{B} = 0 \tag{1.1c}$$

$$\nabla \cdot \mathscr{D} = \hat{\rho} \tag{1.1d}$$

can be written in a more tractable form using one of the substitutions, shown here for the electric field intensity only, but applicable to all the field and source terms,

$$\mathscr{E} = \mathrm{Re}(\mathbf{E}e^{-i\omega t}) \tag{1.2a}$$

$$\mathscr{E} = \mathrm{Re}(\mathbf{E}e^{j\omega t}) \tag{1.2b}$$

If the form (1.2a) is used, the Maxwell equations become

$$\nabla \times \mathbf{H} = \mathbf{J} - i\omega \mathbf{D} \tag{1.3a}$$

$$\nabla \times \mathbf{E} = \mathbf{M} + i\omega \mathbf{B} \tag{1.3b}$$

$$\nabla \cdot \mathbf{B} = \rho_M \tag{1.3c}$$

$$\nabla \cdot \mathbf{D} = \rho \tag{1.3d}$$

while if form (1.2b) is used, they become

$$\nabla \times \mathbf{H} = \mathbf{J} + j\omega \mathbf{D} \tag{1.4a}$$

$$\nabla \times \mathbf{E} = -\mathbf{M} - j\omega \mathbf{B} \tag{1.4b}$$

$$\nabla \cdot \mathbf{B} = \rho_M \tag{1.4c}$$

$$\nabla \cdot \mathbf{D} = \rho \tag{1.4d}$$

These equation sets are called the Maxwell equations in time-invariant form, the time-invariant Maxwell equations, or the complex Maxwell equations. The choice of time variation has an effect on the form of solutions to the Maxwell equations. For example, the phase variation of a plane wave traveling in the direction of a propagation vector $\mathbf{k}$ in a homogeneous medium takes the form

$$e^{i\mathbf{k}\cdot\mathbf{r}} \qquad e^{-j\mathbf{k}\cdot\mathbf{r}}$$

depending on the choice of positive or negative sign in (1.2). It is desirable to use different symbols i or j in the exponent for the time variation, depending on which sign is adopted in (1.2). Both sign conventions are utilized, and confusion can be avoided by using i for the negative sign convention and j for the positive. In this text, $+j$ is used, together with the time-invariant Maxwell equations of (1.4).

In (1.3) and (1.4), a fictitious magnetic charge density ρ_M and magnetic current density $\mathbf{M}$ were added to the Maxwell equations. Physical quantities corresponding to these fictitious additions do not exist. It is convenient, however, when considering some antenna problems to replace the actual sources by equivalent sources on a surface. The magnetic current density and charge density account for discontinuities in field components across the boundary surface [1, 2].

4 AN INTRODUCTION TO ANTENNAS

For linear, isotropic media the field terms are related by the constitutive equations

$$\mathbf{D} = \epsilon \mathbf{E} \tag{1.5a}$$

$$\mathbf{B} = \mu \mathbf{H} \tag{1.5b}$$

$$\mathbf{J} = \sigma \mathbf{E} \tag{1.5c}$$

and the time-invariant Maxwell equations become

$$\nabla \times \mathbf{H} = \mathbf{J} + j\omega\epsilon\mathbf{E} \tag{1.6a}$$

$$\nabla \times \mathbf{E} = -\mathbf{M} - j\omega\mu\mathbf{H} \tag{1.6b}$$

$$\nabla \cdot \mathbf{H} = \frac{\rho_M}{\mu} \tag{1.6c}$$

$$\nabla \cdot \mathbf{E} = \frac{\rho}{\epsilon} \tag{1.6d}$$

Since the Maxwell equations are linear, we may use superposition to account for the two sets of sources and solve separately the equations for electric and magnetic sources:

ELECTRIC SOURCES

$$\nabla \times \mathbf{H}_J = \mathbf{J} + j\omega\mathbf{D}_J = \mathbf{J} + j\omega\epsilon\mathbf{E}_J \tag{1.7a}$$

$$\nabla \times \mathbf{E}_J = -j\omega\mathbf{B}_J = -j\omega\mu\mathbf{H}_J \tag{1.7b}$$

$$\nabla \cdot \mathbf{B}_J = \nabla \cdot \mathbf{H}_J = 0 \tag{1.7c}$$

$$\nabla \cdot \mathbf{D}_J = \epsilon \nabla \cdot \mathbf{E}_J = \rho \tag{1.7d}$$

where subscript J refers to the partial fields produced by electric current density $\mathbf{J}$ and electric charge density ρ;

MAGNETIC SOURCES

$$\nabla \times \mathbf{H}_M = j\omega\mathbf{D}_M = j\omega\epsilon\mathbf{E}_M \tag{1.8a}$$

$$\nabla \times \mathbf{E}_M = -\mathbf{M} - j\omega\mathbf{B}_M = -\mathbf{M} - j\omega\mu\mathbf{H}_M \tag{1.8b}$$

$$\nabla \cdot \mathbf{B}_M = \mu\nabla \cdot \mathbf{H}_M = \rho_M \tag{1.8c}$$

$$\nabla \cdot \mathbf{D}_M = \nabla \cdot \mathbf{E}_M = 0 \tag{1.8d}$$

where subscript M refers to the partial fields produced by magnetic current and charge densities $\mathbf{M}$ and ρ_M.

Consider first the real-time invariant Maxwell equations with electric sources. Since the divergence of $\mathbf{B}$ is zero, $\mathbf{B}$ can be represented as the curl of a vector potential **A**, commonly called the magnetic vector potential,

$$\mathbf{B}_J = \mu \mathbf{H}_J = \nabla \times \mathbf{A} \tag{1.9}$$

Substituting this equation into (1.7b) leads to

$$\nabla \times (\mathbf{E}_J + j\omega \mathbf{A}) = 0 \tag{1.10}$$

Since the curl of the gradient of a scalar function is identically zero, we may set

$$\mathbf{E}_J + j\omega \mathbf{A} = -\nabla \Phi_J \tag{1.11}$$

where Φ_J is a scalar potential. It is commonly called the electric scalar potential.

Equations may be developed for **A** and Φ_J, and from their solutions $\mathbf{D}_J$ and $\mathbf{B}_J$ may be found. If we take the curl of (1.9) and substitute it in (1.7a), we obtain

$$\nabla \times \nabla \times \mathbf{A} = \mu(\mathbf{J} + j\omega \epsilon \mathbf{E}_J) \tag{1.12}$$

Using a vector identity for the left side of this equation and substituting (1.11) into the right side gives

$$\nabla(\nabla \cdot \mathbf{A}) - \nabla^2 \mathbf{A} = \mu \mathbf{J} - j\omega\mu\epsilon \nabla \Phi_J + \omega^2 \mu\epsilon \mathbf{A} \tag{1.13}$$

At this point only $\nabla \times \mathbf{A}$ has been constrained ($= \mathbf{B}_J$). We are free to choose $\nabla \cdot \mathbf{A}$, according to Sommerfeld [3] and Panofsky and Phillips [4],

$$\nabla \cdot \mathbf{A} = -j\omega\mu\epsilon \Phi_J \tag{1.14}$$

With this choice, and with the definition

$$k^2 = \omega^2 \mu \epsilon \tag{1.15}$$

(1.13) becomes

$$\nabla^2 \mathbf{A} + k^2 \mathbf{A} = -\mu \mathbf{J} \tag{1.16}$$

The introduction of the relationship (1.14) between **A** and Φ_J makes it unnecessary to find Φ_J. The magnetic flux density $\mathbf{B}_J$ may be found from the vector potential **A** once (1.16) is solved. Then $\mathbf{E}_J$ may be found at points away from the sources by means of (1.7a), or alternatively we may find it from (1.11) and (1.14). Thus

$$\mathbf{E}_J = -j\omega \mathbf{A} - \frac{j}{\omega\mu\epsilon} \nabla(\nabla \cdot \mathbf{A}) \tag{1.17}$$

6 AN INTRODUCTION TO ANTENNAS

Potentials for use when only magnetic sources **M** and ρ_M are present may be developed from (1.8) by analogy with the process used with electric sources. Using (1.8d) allows us to define an electric vector potential **F** by

$$\mathbf{D}_M = \epsilon \mathbf{E}_M = -\nabla \times \mathbf{F} \tag{1.18}$$

where the negative sign is arbitrary.

Substituting this equation into (1.8a) gives

$$\nabla \times (\mathbf{H}_M + j\omega \mathbf{F}) = 0 \tag{1.19}$$

and the relationship to a magnetic scalar potential Φ_M becomes

$$\mathbf{H}_M + j\omega \mathbf{F} = -\nabla \Phi_M \tag{1.20}$$

Substituting the curl of (1.18) into (1.8b) we obtain

$$\nabla \times \nabla \times \mathbf{F} = \epsilon(\mathbf{M} + j\omega\mu \mathbf{H}_M) \tag{1.21}$$

and if we follow a process like that used previously and specify the divergence of **F** as

$$\nabla \cdot \mathbf{F} = -j\omega\mu\epsilon \Phi_M \tag{1.22}$$

we arrive at an equation for **F**:

$$\nabla^2 \mathbf{F} + k^2 \mathbf{F} = -\epsilon \mathbf{M} \tag{1.23}$$

Once **F** is obtained, $\mathbf{D}_M$ can be found from (1.18) and $\mathbf{H}_M$ either from the appropriate time-invariant Maxwell equation or from

$$\mathbf{H}_M = -j\omega \mathbf{F} - \frac{j}{\omega\mu\epsilon}\nabla(\nabla \cdot \mathbf{F}) \tag{1.24}$$

As a final step we superimpose the solutions and find for the total fields due to electric and magnetic sources

$$\mathbf{E} = -j\omega \mathbf{A} - \frac{j}{\omega\mu\epsilon}\nabla(\nabla \cdot \mathbf{A}) - \frac{1}{\epsilon}\nabla \times \mathbf{F} \tag{1.25}$$

$$\mathbf{H} = -j\omega \mathbf{F} - \frac{j}{\omega\mu\epsilon}\nabla(\nabla \cdot \mathbf{F}) + \frac{1}{\mu}\nabla \times \mathbf{A} \tag{1.26}$$

1.3. INTEGRAL SOLUTIONS FOR THE VECTOR POTENTIALS

An integral solution for the vector potential equation (1.16) for **A** can be constructed by considering an infinitesimal current element at the origin of a spherical coordinate system (Fig. 1.2). For a z-directed current, the potential equation reduces to

$$\nabla^2 A_z + k^2 A_z = -\mu J_z \tag{1.27}$$

For a point source (with source length infinitesimal), A_z must be spherically symmetric. In addition, $J_z = 0$ everywhere except at the origin, and the equation for the potential A_z becomes

$$\frac{1}{r^2}\frac{d}{dr}\left(r^2 \frac{dA_z}{dr}\right) + k^2 A_z = 0 \tag{1.28}$$

This equation has two independent solutions,

$$\frac{1}{r}e^{-jkr} \quad \text{and} \quad \frac{1}{r}e^{+jkr}$$

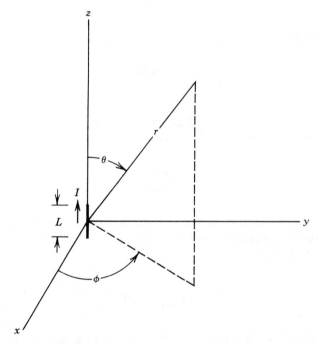

FIGURE 1.2 Current source at origin.

8 AN INTRODUCTION TO ANTENNAS

which represent, respectively, outward- and inward-traveling spherical waves. From physical considerations we reject the inward-traveling wave and choose

$$A_z = \frac{C}{r}e^{-jkr} \tag{1.29}$$

where C is a constant. The constant may be determined by noting that as $k \to 0$, (1.27) reduces to Poisson's equation

$$\nabla^2 A_z = -\mu J_z \tag{1.30}$$

with the well-known solution [5, 6]

$$A_z = \frac{\mu}{4\pi} \iiint \frac{J_z}{r} \, dv' \tag{1.31}$$

where the prime denotes integration around the source point.

If we replace $J_z \, dv'$ by $I \, dz'$ and integrate, the potential is

$$A_z = \frac{\mu IL}{4\pi r} \tag{1.32}$$

and if this is equated to (1.29) with $k = 0$, the constant is

$$C = \frac{\mu IL}{4\pi}$$

Then the solution for the magnetic vector potential of an infinitesimal z-directed current element at the origin is

$$A_z = \frac{\mu IL}{4\pi r} e^{-jkr} \tag{1.33}$$

This solution is readily generalized to the case of the infinitesimal element located at vector distance $\mathbf{r}'$ from the origin and oriented along a line parallel to unit vector $\mathbf{u}$. It is

$$\mathbf{A}(\mathbf{r}) = \mathbf{u}\frac{\mu IL}{4\pi |\mathbf{r} - \mathbf{r}'|} e^{-jk|\mathbf{r} - \mathbf{r}'|} \tag{1.34}$$

Finally, if we have a linear current distribution, a current on a relatively thin wire for example, a solution to (1.16) in the form of an outward-traveling wave is

$$\mathbf{A}(\mathbf{r}) = \frac{\mu}{4\pi} \int \frac{I(\mathbf{r}')}{|\mathbf{r} - \mathbf{r}'|} e^{-jk|\mathbf{r}-\mathbf{r}'|} d\mathbf{L}' \tag{1.35}$$

In terms of the current density $\mathbf{J}$, the appropriate integral is obtained if $I d\mathbf{L}'$ is replaced by $\mathbf{J} dv'$, giving

$$\mathbf{A}(\mathbf{r}) = \frac{\mu}{4\pi} \iiint \frac{\mathbf{J}(\mathbf{r}')}{|\mathbf{r} - \mathbf{r}'|} e^{-jk|\mathbf{r}-\mathbf{r}'|} dv' \tag{1.36}$$

By analogy, an integral form for the electric vector potential $\mathbf{F}$ is

$$\mathbf{F}(\mathbf{r}) = \frac{\epsilon}{4\pi} \iiint \frac{\mathbf{M}(\mathbf{r}')}{|\mathbf{r} - \mathbf{r}'|} e^{-jk|\mathbf{r}-\mathbf{r}'|} dv' \tag{1.37}$$

or if we consider a magnetic current K,

$$\mathbf{F}(\mathbf{r}) = \frac{\epsilon}{4\pi} \int \frac{K(\mathbf{r}')}{|\mathbf{r} - \mathbf{r}'|} e^{-jk|\mathbf{r}-\mathbf{r}'|} d\mathbf{L}' \tag{1.38}$$

1.4. APPROXIMATIONS TO THE POTENTIALS

The evaluation of the vector potentials can be quite difficult to carry out in the general case, and approximations are desirable in the factor $|\mathbf{r} - \mathbf{r}'|$. This can be done by applying the binomial expansion

$$\begin{aligned}|\mathbf{r} - \mathbf{r}'| &= [r^2 - 2\mathbf{r} \cdot \mathbf{r}' + r'^2]^{1/2} \\ &= r - \mathbf{u}_r \cdot \mathbf{r}' + \frac{1}{2r}[r'^2 - (\mathbf{u}_r \cdot \mathbf{r}')^2] + \cdots \end{aligned} \tag{1.39}$$

for $r > r'$, where $\mathbf{u}_r = \mathbf{r}/r$, and terms in r^{-2}, r^{-3}, and so on have been dropped.

Fresnel Zone

At distances from the sources where $r \gg r'$ and $kr \gg 1$ we may approximate $|\mathbf{r} - \mathbf{r}'|$ by r in the amplitude of the potential integrals and by (1.39) in the phase term. Then a typical potential integral simplifies to

$$\mathbf{A}(\mathbf{r}) = \frac{\mu}{4\pi r} e^{-jkr} \iiint \mathbf{J}(\mathbf{r}') \exp\left\{ jk\left[\mathbf{u}_r \cdot \mathbf{r}' + \frac{(\mathbf{u}_r \cdot \mathbf{r}')^2}{2r} - \frac{r'^2}{2r} \right] \right\} dv' \quad (1.40)$$

Fraunhofer Zone (Far Zone)

At still greater distances from the sources, the r^{-1} term of (1.39) can be dropped, and the equation for $\mathbf{A}(\mathbf{r})$, for example, becomes

$$\mathbf{A}(\mathbf{r}) = \frac{\mu}{4\pi r} e^{-jkr} \iiint \mathbf{J}(\mathbf{r}') e^{jk\mathbf{u}_r \cdot \mathbf{r}'} dv' \quad (1.41)$$

where

$$\mathbf{u}_r \cdot \mathbf{r}' = r' \cos \psi$$

with ψ the angle between $\mathbf{r}$ and $\mathbf{r}'$.

The boundaries between the zones are not easily chosen and in fact depend on the distributions $\mathbf{J}$ and $\mathbf{M}$. A commonly used dividing line between the Fresnel and far zones for an antenna with greatest linear dimension L is

$$r = \frac{2L^2}{\lambda} \quad (1.42)$$

With this choice the greatest phase value contributed by the last two terms in (1.39), which were dropped in going from Fresnel to far zone, is

$$\frac{kr'^2}{2r} = \frac{2\pi}{\lambda} L^2 \frac{\lambda}{4L^2} = \frac{\pi}{2}$$

1.5. FAR-ZONE FIELDS

Expressions for the fields can be found from the integral forms of the potentials. From (1.9) and (1.36) we find

$$\mathbf{H}_J = \frac{1}{\mu} \nabla \times \mathbf{A} = \frac{1}{4\pi} \iiint \nabla \times \left[\frac{\mathbf{J}(\mathbf{r}')}{|\mathbf{r} - \mathbf{r}'|} e^{-jk|\mathbf{r} - \mathbf{r}'|} \right] dv'$$

$$= \frac{-1}{4\pi} \iiint \mathbf{J}(\mathbf{r}') \times \nabla \left[\frac{e^{-jk|\mathbf{r} - \mathbf{r}'|}}{|\mathbf{r} - \mathbf{r}'|} \right] dv' \quad (1.43)$$

Then $\mathbf{E}_J$ is found from

$$\mathbf{E}_J = \frac{1}{j\omega\epsilon}\nabla \times \mathbf{H}_J = -\frac{1}{j4\pi\omega\epsilon}\iiint \nabla \times \left[\mathbf{J} \times \nabla\left(\frac{e^{-jk|\mathbf{r}-\mathbf{r}'|}}{|\mathbf{r}-\mathbf{r}'|}\right)\right] dv' \quad (1.44)$$

From the identity

$$\nabla \times (\mathbf{A} \times \mathbf{B}) = \mathbf{A}\nabla \cdot \mathbf{B} - \mathbf{B}\nabla \cdot \mathbf{A} + (\mathbf{B} \cdot \nabla)\mathbf{A} - (\mathbf{A} \cdot \nabla)\mathbf{B}$$

and the fact that $\mathbf{J}$ is a constant vector in the differentiation, $\mathbf{E}_J$ becomes

$$\mathbf{E}_J = \frac{-1}{j4\pi\omega\epsilon}\iiint \left[\mathbf{J}\nabla^2\left(\frac{e^{-jk|\mathbf{r}-\mathbf{r}'|}}{|\mathbf{r}-\mathbf{r}'|}\right) - (\mathbf{J} \cdot \nabla)\nabla\left(\frac{e^{-jk|\mathbf{r}-\mathbf{r}'|}}{|\mathbf{r}-\mathbf{r}'|}\right)\right] dv' \quad (1.45)$$

The function

$$\frac{e^{-jk|\mathbf{r}-\mathbf{r}'|}}{|\mathbf{r}-\mathbf{r}'|}$$

appearing in the integrals is a solution of the scalar Helmholtz equation [2],

$$\nabla^2\left(\frac{e^{-jk|\mathbf{r}-\mathbf{r}'|}}{|\mathbf{r}-\mathbf{r}'|}\right) + k^2\left(\frac{e^{-jk|\mathbf{r}-\mathbf{r}'|}}{|\mathbf{r}-\mathbf{r}'|}\right) = 0 \quad (1.46)$$

and with the use of this, $\mathbf{E}_J$ becomes

$$\mathbf{E}_J = \frac{1}{j4\pi\omega\epsilon}\iiint \left[\mathbf{J}k^2\left(\frac{e^{-jk|\mathbf{r}-\mathbf{r}'|}}{|\mathbf{r}-\mathbf{r}'|}\right) + (\mathbf{J} \cdot \nabla)\nabla\left(\frac{e^{-jk|\mathbf{r}-\mathbf{r}'|}}{|\mathbf{r}-\mathbf{r}'|}\right)\right] dv' \quad (1.47)$$

If this procedure is repeated for the electric vector potential $\mathbf{F}$ with magnetic source $\mathbf{M}$, we obtain

$$\mathbf{E}_M = -\frac{1}{\epsilon}\nabla \times \mathbf{F} = \frac{1}{4\pi}\iiint \mathbf{M}(\mathbf{r}') \times \nabla\left(\frac{e^{-jk|\mathbf{r}-\mathbf{r}'|}}{|\mathbf{r}-\mathbf{r}'|}\right) dv' \quad (1.48)$$

$$\mathbf{H}_M = \frac{-1}{j\omega\mu}\nabla \times \mathbf{E}_M$$

$$= \frac{1}{j4\pi\omega\mu}\iiint \left[\mathbf{M}k^2\left(\frac{e^{-jk|\mathbf{r}-\mathbf{r}'|}}{|\mathbf{r}-\mathbf{r}'|}\right) + (\mathbf{M} \cdot \nabla)\nabla\left(\frac{e^{-jk|\mathbf{r}-\mathbf{r}'|}}{|\mathbf{r}-\mathbf{r}'|}\right)\right] dv' \quad (1.49)$$

Now we use the far-field approximation

$$\frac{e^{-jk|\mathbf{r}-\mathbf{r}'|}}{|\mathbf{r}-\mathbf{r}'|} \approx \frac{e^{-jk(r-\mathbf{u}_r\cdot\mathbf{r}')}}{r} \tag{1.50}$$

If only the terms of order $1/r$ are retained, the gradient of the right side of this equation is

$$\nabla\left(\frac{e^{-jk(r-\mathbf{u}_r\cdot\mathbf{r}')}}{r}\right) \approx \frac{-jk}{r} e^{-jkr} e^{jk\mathbf{u}_r\cdot\mathbf{r}'} \mathbf{u}_r \tag{1.51}$$

If this is used in the equation for $\mathbf{H}_J$, it becomes

$$\mathbf{H}_J = \frac{jk}{4\pi r} e^{-jkr} \iiint \mathbf{J}(\mathbf{r}') \times \mathbf{u}_r e^{jk\mathbf{u}_r\cdot\mathbf{r}'}\, dv' \tag{1.52}$$

which may be written as

$$\mathbf{H}_J = \frac{jk}{4\pi r} e^{-jkr} \iiint (J_\phi \mathbf{u}_\theta - J_\theta \mathbf{u}_\phi) e^{jk\mathbf{u}_r\cdot\mathbf{r}'}\, dv' \tag{1.53}$$

The corresponding value of the electric field for an electric source distribution can be found by substituting (1.50) and (1.51) into (1.47). The result is

$$\mathbf{E}_J = \frac{-jkZ_0}{4\pi r} e^{-jkr} \iint (J_\theta \mathbf{u}_\theta + J_\phi \mathbf{u}_\phi) e^{jk\mathbf{u}_r\cdot\mathbf{r}'}\, dv' \tag{1.54}$$

The fields of a magnetic source distribution are found in the same manner and are

$$\mathbf{E}_M = -\frac{jk}{4\pi r} e^{-jkr} \iint (M_\phi \mathbf{u}_\theta - M_\theta \mathbf{u}_\phi) e^{jk\mathbf{u}_r\cdot\mathbf{r}'}\, dv' \tag{1.55}$$

$$\mathbf{H}_M = -\frac{jk}{4\pi Z_0 r} e^{-jkr} \iint (M_\theta \mathbf{u}_\theta + M_\phi \mathbf{u}_\phi) e^{jk\mathbf{u}_r\cdot\mathbf{r}'}\, dv' \tag{1.56}$$

where

$$Z_0 = \sqrt{\frac{\mu}{\epsilon}} \tag{1.57}$$

is the intrinsic impedance of the medium of interest.

We can summarize the far fields as follows:

ELECTRIC SOURCES

$$E_r = 0 \tag{1.58a}$$

$$E_\theta = -j\omega A_\theta \tag{1.58b}$$

$$E_\phi = -j\omega A_\phi \tag{1.58c}$$

$$H_r = 0 \tag{1.58d}$$

$$H_\theta = \frac{j\omega A_\phi}{Z_0} = -\frac{E_\phi}{Z_0} \tag{1.58e}$$

$$H_\phi = \frac{-j\omega A_\theta}{Z_0} = \frac{E_\theta}{Z_0} \tag{1.58f}$$

MAGNETIC SOURCES

$$H_r = 0 \tag{1.58g}$$

$$H_\theta = -j\omega F_\theta \tag{1.58h}$$

$$H_\phi = -j\omega F_\phi \tag{1.58i}$$

$$E_r = 0 \tag{1.58j}$$

$$E_\theta = -j\omega Z_0 F_\phi = Z_0 H_\phi \tag{1.58k}$$

$$E_\phi = j\omega Z_0 F_\theta = -Z_0 H_\theta \tag{1.58l}$$

These equations show that in the far zone the **E** and **H** fields are orthogonal to each other and to **r**. Such fields are called TEM (transverse electric and magnetic) to **r**. This is true if the sources are electric, magnetic, or a combination of both.

1.6. USE OF THE POTENTIAL INTEGRALS FOR PHYSICAL STRUCTURES

The potential integrals for electric and magnetic sources were developed by considering an infinitesimal current element, with the implication that it was a current existing in a homogeneous medium throughout the region of interest. Antennas, however, have currents flowing in metallic conductors, and we must justify the use of the potential integrals for inhomogeneous media. To do this we use two of the Maxwell equations:

$$\nabla \times \mathbf{H} = (\sigma + j\omega\epsilon)\mathbf{E} + \mathbf{J}^s = \hat{y}\mathbf{E} + \mathbf{J}^s \tag{1.59a}$$

$$-\nabla \times \mathbf{E} = j\omega\mu\mathbf{H} + \mathbf{M}^s = \hat{z}\mathbf{H} + \mathbf{M}^s \tag{1.59b}$$

where the superscript s denotes a source, and the definitions of $\hat{y}$ and $\hat{z}$ are obvious.

The greater part of the region of interest is free space with parameters ϵ_0 and μ_0 and conductivity zero, with a small region (the antenna) having different parameters, ϵ, μ, and σ. We then rewrite these equations as

$$\nabla \times \mathbf{H} = j\omega\epsilon_0\mathbf{E} + (\hat{y} - j\omega\epsilon_0)\mathbf{E} + \mathbf{J}^s = j\omega\epsilon_0\mathbf{E} + \mathbf{J} \quad (1.60a)$$

$$-\nabla \times \mathbf{E} = j\omega\mu_0\mathbf{H} + (\hat{z} - j\omega\mu_0)\mathbf{H} + \mathbf{M}^s = j\omega\mu_0\mathbf{H} + \mathbf{M} \quad (1.60b)$$

where we have defined

$$\mathbf{J} = \mathbf{J}^s + (\hat{y} - j\omega\epsilon_0)\mathbf{E} \quad (1.61a)$$

$$\mathbf{M} = \mathbf{M}^s + (\hat{z} - j\omega\mu_0)\mathbf{H} \quad (1.61b)$$

Formally, (1.60) are the free-space Maxwell equations and may be treated by the potential integrals with sources $\mathbf{J}$ and $\mathbf{M}$ as if the region were homogeneous. The sources are now unknown, but we will see how this problem can be handled for a linear antenna.

Assume that a linear metallic antenna is fed by a current source $\mathbf{J}^s$. Further, for most conductors $\mu = \mu_0$ and $\epsilon = \epsilon_0$ (the approximation for ϵ is unnecessary since $\sigma \gg \omega\epsilon$). Then the densities of (1.61) become

$$\mathbf{J} = \mathbf{J}^s + \sigma\mathbf{E} \quad (1.62a)$$

$$\mathbf{M} = 0 \quad (1.62b)$$

Under these circumstances we can write for the wire antenna

$$\mathbf{A} = \frac{\mu}{4\pi} \iiint \frac{\mathbf{J}^s + \sigma\mathbf{E}}{|\mathbf{r} - \mathbf{r}'|} e^{-jk|\mathbf{r}-\mathbf{r}'|} \, dv' \quad (1.63)$$

This is a valid potential integral applicable to the antenna problem even though we deal with an inhomogeneous region of space. The price that must be paid for obtaining this form is that the integrand is unknown.

Now the current source $\mathbf{J}^s$ is applied to the feed gap of a linear antenna, and $\sigma\mathbf{E}$ is the physical current density in the conducting structure of the antenna. We may then replace the integrand of the potential integral by measured currents on the antenna or by currents obtained by some analytical method and write, for a thin antenna,

$$\mathbf{A}(\mathbf{r}) = \frac{\mu}{4\pi} \int \frac{I(\mathbf{r}')}{|\mathbf{r} - \mathbf{r}'|} e^{-jk|\mathbf{r}-\mathbf{r}'|} \, d\mathbf{L}' \quad (1.64)$$

which agrees with (1.35). The use of the potential integrals to find the potentials produced by currents flowing in metallic antennas is therefore justified [7].

1.7. RADIATION PATTERN

The radiation pattern of an antenna and some related parameters can now be defined. We will use IEEE Standard 145-1983 definitions for these and other quantities unless otherwise indicated [8].

In Section 1.3 we found the magnetic vector potential of an infinitesimal current element at the coordinate origin in order to develop the equations for the vector potentials of more general source distributions. We now let the infinitesimal current element do double duty by using it as a basis for the definition of radiation pattern.

The magnetic vector potential for the current element of Figure 1.2, obtained earlier, is

$$A_z = \frac{\mu IL}{4\pi r} e^{-jkr} \quad (1.33)$$

It is desirable to use spherical coordinate components, and the relationship is

$$A_r = A_z \cos\theta = \frac{\mu IL}{4\pi r} \cos\theta\, e^{-jkr} \quad (1.65a)$$

$$A_\theta = -A_z \sin\theta = -\frac{\mu IL}{4\pi r} \sin\theta\, e^{-jkr} \quad (1.65b)$$

$$A_\phi = 0 \quad (1.65c)$$

To define pattern and gain, only the far-field components obtainable from (1.58) are needed. They are

$$E_r = 0 \quad (1.66a)$$

$$E_\theta = -j\omega A_\theta = \frac{j\omega\mu IL}{4\pi r} \sin\theta\, e^{-jkr} \quad (1.66b)$$

$$E_\phi = -j\omega A_\phi = 0 \quad (1.66c)$$

$$H_r = 0 \quad (1.66d)$$

$$H_\theta = -\frac{E_\phi}{Z_0} = 0 \quad (1.66e)$$

$$H_\phi = \frac{E_\theta}{Z_0} = \frac{j\omega\mu IL}{4\pi Z_0 r} \sin\theta\, e^{-jkr} \quad (1.66f)$$

where

$$k = 2\pi/\lambda$$

16 AN INTRODUCTION TO ANTENNAS

The time-average Poynting vector in the far field becomes

$$\mathscr{P} = \tfrac{1}{2}\operatorname{Re}(\mathbf{E}\times\mathbf{H}^*) = \tfrac{1}{2}\operatorname{Re}(E_\theta H_\phi^*)\mathbf{u}_r = \frac{Z_0|I|^2 L^2}{8\lambda^2 r^2}\sin^2\theta\,\mathbf{u}_r \quad (1.67)$$

The script letter is used for the time-average Poynting vector to distinguish it from a polarization ratio to be introduced later. Note that even though $\mathscr{P}$ is a time average over a time comparable to a period of the harmonic wave, it can vary with time over a longer time period.

We found here the time-average Poynting vector in the far field. Had we used the more general field relationships valid for the fields closer to the sources, we would have found the time-average Poynting vector for the fields that vary as $1/r^2$ and $1/r^3$ to be zero, so that (1.67) is correct everywhere in the field. Of course, energy can be transferred to conducting objects by field terms other than the far fields (e.g., a transformer). The difference is that, in the absence of nearby lossy objects, the energy in the $1/r$ field terms is lost to the antenna system, whereas energy represented by near-field terms is stored.

Radiation Intensity

Radiation intensity in a given direction is the power radiated from the antenna per unit solid angle. A bundle of rays, not all lying in a common plane, and intersecting at a common point, forms a solid angle, measured in steradians (dimensionless). If a sphere of radius r is constructed as in Figure 1.3a, with center at the ray intersection, the rays subtend area A on the sphere surface. The ratio

$$\Omega = \frac{A}{r^2} \quad (1.68)$$

is independent of the sphere radius and defines the solid angle formed by the rays. The infinitesimal element of solid angle can also be written in terms of the elementary area dA on any surface, as in Figure 1.3b. The projection of the surface area element onto a sphere centered at the ray intersection is $\mathbf{u}_r \cdot \mathbf{n}\, dA$, where $\mathbf{n}$ is the unit normal vector to the surface and $\mathbf{u}_r$ is the unit vector in the direction of $\mathbf{r}$, the vector from the ray intersection to the surface element. Then the element of solid angle is

$$d\Omega = \frac{dA}{r^2}\mathbf{u}_r \cdot \mathbf{n} \quad (1.69)$$

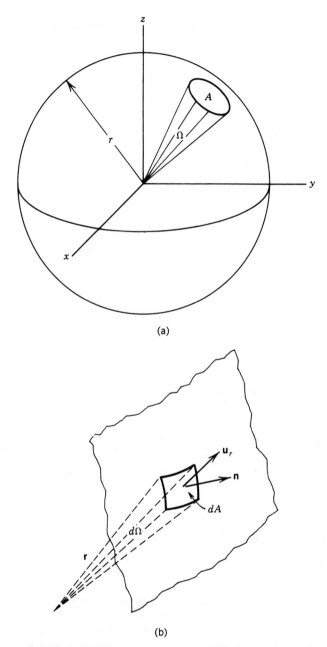

FIGURE 1.3 Illustration of solid angles: (*a*) solid angle and subtended area on a sphere; (*b*) elementary solid angle with general surface.

The radiation intensity U is related to power radiated within solid angle Ω by

$$U = \frac{W}{\Omega} \tag{1.70}$$

if Ω is small enough that U is constant. More generally

$$W = \int U \, d\Omega = \int\int U \sin\theta \, d\theta \, d\phi \tag{1.71}$$

The radiation intensity is readily related to $\mathscr{P}$, the magnitude of the time-average Poynting vector, since

$$W = U\Omega = \mathscr{P}A = \mathscr{P}r^2\Omega$$

Then

$$U(\theta,\phi) = r^2\mathscr{P}(\theta,\phi) \tag{1.72}$$

where the functional notation is used to stress the direction dependence of both U and $\mathscr{P}$.

Radiation Pattern

The *radiation pattern* (antenna pattern) represents the spatial distribution of a quantity that characterizes the electromagnetic field generated by an antenna. In the usual case the radiation pattern is determined in the far-field region and is represented as a function of directional coordinates. Radiation properties include power flux density (magnitude of the Poynting vector, normally called power density here), radiation intensity, phase, polarization, and field strength.

The radiation pattern is measured as a function of direction at a constant radius from the antenna. Absolute or relative measurements may be made, although it is more common to see plots of relative power density or field magnitude. It is usual to present the three-dimensional information as a group of two-dimensional plots, in either polar or rectangular form. Figure 1.4 shows the relative field magnitude and power density in constant-azimuth planes for the z-directed infinitesimal current element or *elementary antenna*.

One two-dimensional plot is sufficient to describe the radiation pattern of the infinitesimal current element since the fields and power density are not

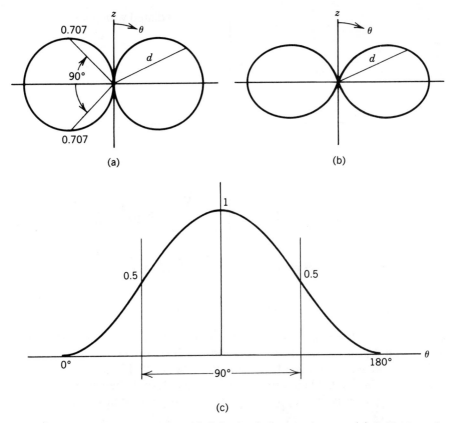

FIGURE 1.4 Radiation patterns of infinitesimal current element: (a) field strength, $d = |\sin \theta|$; (b) power density, $d = \sin^2 \theta$; (c) power density.

functions of azimuth angle ϕ. For more general patterns, plots may be shown for variable polar angle θ in planes of constant azimuth angle ϕ, or vice versa. Two patterns of considerable importance for many antennas are the *principal E-plane* and *H-plane* patterns. The principal E plane is a plane containing the electric field vector and the direction of maximum radiation, and the principal H plane is one containing the magnetic field vector and the direction of maximum radiation [6]. Jointly, these are called the *principal planes*. The definitions apply only to antennas whose radiated wave is linearly polarized in the direction of maximum power density. Any constant-azimuth plane is an E plane for the z-directed current element, so Figure 1.4 is an E-plane pattern. The H plane for the element is the plane $\theta = \pi/2$, or the xy plane. The H-plane pattern for the current element is shown in Figure 1.5.

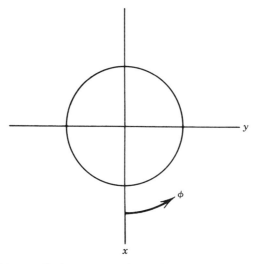

FIGURE 1.5 *H*-plane pattern for *z*-directed current element.

Beamwidth

The *half-power beamwidth*, in a plane containing the direction of the maximum of a beam, is the angle between the two directions in which the radiation intensity is one-half the maximum value of the beam. Half-power beamwidths in the E plane are shown in Figure 1.4 for the current element. In the H plane it is inappropriate to speak of a beamwidth since the

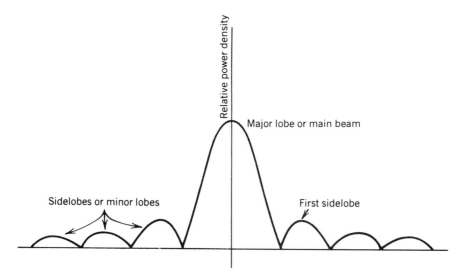

FIGURE 1.6 Radiation lobes of general antenna.

radiation intensity is constant. The current element is an *omnidirectional antenna*, one having a nondirectional antenna pattern in a given plane (azimuth) and a directional pattern in any orthogonal plane.

Radiation Lobe

This is a portion of the radiation pattern bounded by regions of relatively weak radiated intensity [9]. The current element fails as an illustration, and we must go to a more general antenna with the principal plane pattern of Figure 1.6. The names commonly given to the radiation lobes are indicated on the figure.

1.8. GAIN AND DIRECTIVITY

The radiation patterns discussed in the previous section show that the radiation intensity from an antenna is greater in some directions than in others, and this is of benefit to the user in many applications. This characteristic is usefully described by the directivity of the antenna. It is in essence a comparison of the radiation intensity in a specified direction with the intensity that would exist if the antenna radiated the same total power equally in all directions, and can be greater than unity in some directions if it is less than unity in others. The *directivity* is the ratio of the radiation intensity in a given direction to the radiation intensity averaged over all directions. The average radiation intensity is equal to the total power radiated by the antenna divided by 4π (the solid angle subtended by a sphere). If the direction is not specified, the direction of maximum radiation intensity is assumed [8].*

The directivity of the infinitesimal current element of Section 1.7 is easily determined. From (1.67) the radiation intensity is

$$U(\theta, \phi) = \frac{Z_0 |I|^2 L^2}{8\lambda^2} \sin^2 \theta \qquad (1.73)$$

Then the total radiated power is determined by integrating over the surface of a sphere of very large radius, thus

$$W_{\text{rad}} = \int U(\theta, \phi) \, d\Omega = \frac{Z_0 |I|^2 L^2}{8\lambda^2} \int_{\theta=0}^{\pi} \int_{\phi=0}^{2\pi} \sin^3 \theta \, d\theta \, d\phi = \frac{\pi Z_0 |I|^2 L^2}{3\lambda^2} \qquad (1.74)$$

---

*This is a change from previous IEEE Standard definitions [9, 10] that used *directive gain* for this function and *directivity* for the maximum value of the directive gain. The usage in this text will conform to the current Standard.

22 AN INTRODUCTION TO ANTENNAS

The average radiation intensity is

$$U_{av} = \frac{W_{rad}}{4\pi} = \frac{Z_0|I|^2 L^2}{12\lambda^2} \qquad (1.75)$$

and the directivity is

$$D(\theta,\phi) = \frac{U(\theta,\phi)}{U_{av}} = \frac{3}{2}\sin^2\theta \qquad (1.76)$$

Radiation Resistance

The *gain* of an antenna is closely related to its directivity. Before considering antenna gain, however, it is useful to examine power losses in the antenna and also to anticipate a later study of antenna impedance by discussing *radiation resistance*. Power radiated is lost to the generator–transmission line–antenna system, and to the generator the loss is indistinguishable from heat loss in a resistance of appropriate value. We therefore define an equivalent resistance, called the *radiation resistance*, which is the power radiated by the antenna divided by the square of the rms current referred to a specified point.

The radiation resistance of the elementary antenna (infinitesimal current source) whose radiated power is given by (1.74) is obviously

$$R_r = \frac{2\pi Z_0 L^2}{3\lambda^2} = 80\pi^2 \left(\frac{L}{\lambda}\right)^2 \qquad (1.77)$$

with the characteristic impedance of free space 120π. See Section 1.12 for a further discussion of radiation resistance.

The current is the same at all points of the elementary antenna, and it is clear that radiated power is divided by the square of that constant current to obtain the radiation resistance. A widely used antenna that does not have a constant current throughout it is the circular cylindrical center-fed dipole shown in Figure 1.7. If it is made of a wire whose radius is much smaller than both a wavelength and the dipole length, the current, to a good approximation, is sinusoidal [11]:

$$I(z') = I_m \sin\left[k\left(\tfrac{1}{2}L - |z'|\right)\right] \qquad -\tfrac{1}{2}L \le z' \le \tfrac{1}{2}L \qquad (1.78)$$

At the antenna feed point

$$I_{in} = I(0) = I_m \sin\left(\tfrac{1}{2}kL\right) \qquad (1.79)$$

The last phrase of the definition of radiation resistance is now clear. We may

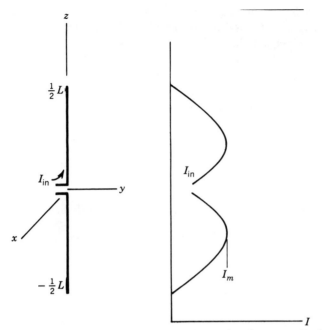

FIGURE 1.7 Center-fed dipole antenna and associated current distribution.

define the radiation resistance of the dipole referred to a current maximum,

$$R_r = \frac{2W_{rad}}{|I_m|^2} \qquad (1.80)$$

or we may define it with reference to the feed point,

$$R_{in} = \frac{2W_{rad}}{|I_{in}|^2} \qquad (1.81)$$

The relationship between the two definitions for the dipole is

$$R_r = R_{in} \sin^2(\tfrac{1}{2}kL) \qquad (1.82)$$

Both definitions are used in the literature.

Antenna Losses

Antennas are constructed of conductors, with finite conductivity, and lossy dielectric materials. It is clear that not all of the power accepted by the antenna from its feed system is radiated; some is lost as heat. The determination of the losses is normally quite tedious and requires a knowledge of tangential magnetic fields (or surface current densities) at conducting surfaces and the electric fields in lossy dielectrics. We consider here one of the simplest cases, the losses in a circular cylindrical antenna made of a wire with conductivity σ and carrying a known current distribution $I(z')$.

For a wire of radius a, we may treat an axial high-frequency current as though it flows with constant density to a depth δ at the wire surface, where δ is the *skin depth*, given by

$$\delta = \frac{1}{\sqrt{\pi f \mu \sigma}} \tag{1.83}$$

Then the high-frequency resistance per unit length of such a wire is*

$$R_{hf} = \frac{1}{2\pi a \delta \sigma} \tag{1.84}$$

Power loss per unit length is

$$\tfrac{1}{2} I^2 R_{hf}$$

and for a finite length antenna,

$$W_{loss} = \tfrac{1}{2} \int_{-L/2}^{L/2} |I(z')|^2 R_{hf} \, dz' \tag{1.85}$$

When applied to the elementary antenna used for most of the illustrations,

$$W_{loss} = \tfrac{1}{2} |I|^2 R_{hf} L \tag{1.86}$$

An equivalent loss resistance for this antenna can be defined as

$$R_{loss} = \frac{2 W_{loss}}{|I|^2} = R_{hf} L \tag{1.87}$$

More generally, the loss resistance for a dipole antenna, referred to the input, is

$$R_{loss} = \frac{1}{|I_{in}|^2} \int_{-L/2}^{L/2} |I(z')|^2 R_{hf} \, dz' \tag{1.88}$$

---

*This is not quite correct, and the equation for a conductor of circular cross section is more complicated. Typically, for an antenna, $\delta \ll a$ and the approximation is excellent.

Radiation Efficiency

In general the transmitter delivers an incident power to the antenna, part of which is reflected because of an impedance mismatch and part of which is *accepted* by the antenna. Of the power accepted, some is radiated and some lost as heat in lossy conductors and dielectrics. *Radiation efficiency* is defined as the ratio of the total power radiated by the antenna to the net power accepted by the antenna from the connected transmitter.

The efficiency may be defined as

$$e = \frac{W_{rad}}{W_{rad} + W_{loss}} = \frac{W_{rad}}{W_{acc}} \quad (1.89)$$

If all of the losses are attributed to a loss resistance, referred to the same point as the radiation resistance, the efficiency is

$$e = \frac{R_r}{R_r + R_{loss}} \quad (1.90)$$

Gain

We are now in a position to define the antenna *gain* (sometimes called *power gain*). It is the ratio of the radiation intensity in a given direction to the radiation intensity that would be obtained if the power accepted by the antenna were radiated isotropically (equally in all directions). Thus

$$G(\theta, \phi) = \frac{U(\theta, \phi)}{(1/4\pi) W_{acc}} \quad (1.91)$$

Gain does not include losses arising from impedance and polarization mismatches. If the direction is not specified, the direction of maximum radiation intensity is implied.

The relation between radiated and accepted power can be used to relate gain and directivity:

$$G(\theta, \phi) = \frac{U(\theta, \phi)}{(1/4\pi) W_{rad}/e} = eD(\theta, \phi) \quad (1.92)$$

1.9. THE DIPOLE ANTENNA: FIELDS

In order to discuss the input impedance of an antenna as we shall do in Section 1.12, it is desirable to obtain the fields of a more complex antenna than the elementary current source. We use the center-fed dipole of Figure

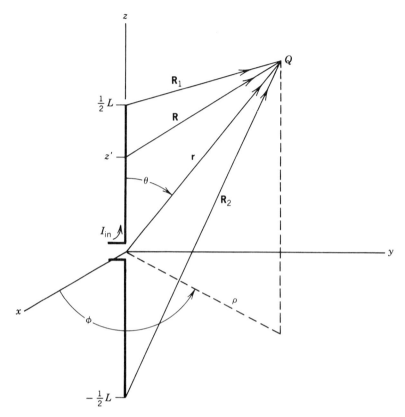

FIGURE 1.8 Geometry for determining dipole fields.

1.8 for this purpose because it is a useful antenna and the equations describing it are readily available. Both the near- and far-field terms are needed, and we will save time by finding the complete fields and specializing as desired.

It was noted earlier that the current distribution on a center-fed dipole is

$$I(z') = I_m \sin\left[k\left(\tfrac{1}{2}L - |z'|\right)\right] \qquad -\tfrac{1}{2}L \le z' \le \tfrac{1}{2}L \quad (1.78)$$

The magnetic vector potential caused by this current is

$$A_z = \frac{\mu I_m}{4\pi} \left\{ \int_{-L/2}^{0} \sin\left[k\left(\frac{1}{2}L + z'\right)\right] \frac{e^{-jkR}}{R} dz' \right.$$
$$\left. + \int_{0}^{L/2} \sin\left[k\left(\frac{1}{2}L - z'\right)\right] \frac{e^{-jkR}}{R} dz' \right\} \quad (1.93)$$

THE DIPOLE ANTENNA: FIELDS   27

If the sine terms of the integrands are replaced by their exponential equivalents, A_z becomes

$$A_z = \frac{\mu I_m}{8\pi j}\left[e^{jkL/2}\int_{-L/2}^{0}\frac{e^{-jk(R-z')}}{R}dz' - e^{-jkL/2}\int_{-L/2}^{0}\frac{e^{-jk(R+z')}}{R}dz'\right.$$
$$\left. + e^{jkL/2}\int_{0}^{L/2}\frac{e^{-jk(R+z')}}{R}dz' - e^{-jkL/2}\int_{0}^{L/2}\frac{e^{-jk(R-z')}}{R}dz'\right]$$

(1.94)

In circular cylindrical coordinates R is

$$R = \sqrt{(z-z')^2 + \rho^2} \tag{1.95}$$

The magnetic field intensity is

$$H_\phi = -\frac{1}{\mu}\frac{\partial A_z}{\partial \rho} \tag{1.96}$$

Differentiating (1.94) according to this equation gives

$$H_\phi = -\frac{I_m}{8\pi j}\left[e^{jkL/2}\int_{-L/2}^{0}\frac{\partial}{\partial\rho}\frac{e^{-jk(R-z')}}{R}dz' - e^{-jkL/2}\int_{-L/2}^{0}\frac{\partial}{\partial\rho}\frac{e^{-jk(R+z')}}{R}dz'\right.$$
$$\left. + e^{jkL/2}\int_{0}^{L/2}\frac{\partial}{\partial\rho}\frac{e^{-jk(R+z')}}{R}dz' - e^{-jkL/2}\int_{0}^{L/2}\frac{\partial}{\partial\rho}\frac{e^{-jk(R-z')}}{R}dz'\right]$$

(1.97)

The derivative in the integrands is

$$\frac{\partial}{\partial\rho}\frac{e^{-jk(R\pm z')}}{R} = -\rho\left(\frac{jk}{R^2} + \frac{1}{R^3}\right)e^{-jk(R\pm z')} \tag{1.98}$$

This term is an exact differential, and it may be verified that

$$\frac{d}{dz'}\frac{e^{-jk(R-z')}}{R(R-z'+z)} = \left(\frac{jk}{R^2} + \frac{1}{R^3}\right)e^{-jk(R-z')} \tag{1.99a}$$

$$\frac{d}{dz'}\frac{e^{-jk(R+z')}}{R(R+z'-z)} = -\left(\frac{jk}{R^2} + \frac{1}{R^3}\right)e^{-jk(R+z')} \tag{1.99b}$$

28 AN INTRODUCTION TO ANTENNAS

With appropriate substitutions, the first integral in (1.97) is

$$-\rho \left[\frac{e^{-jk(R-z')}}{R(R-z'+z)} \right]_{-L/2}^{0}$$

$$= -\rho \left[\frac{e^{-jkr}}{r(r+z)} - \frac{e^{-jk(R_2+L/2)}}{R_2(R_2+L/2+z)} \right]$$

$$= -\rho \left\{ \frac{(r-z)e^{-jkr}}{r(r^2-z^2)} - \frac{(R_2-L/2-z)e^{-jk(R_2+L/2)}}{R_2\left[R_2^2-(L/2+z)^2\right]} \right\}$$

But

$$r^2 = z^2 + \rho^2 \tag{1.100a}$$

$$R_1^2 = (z - L/2)^2 + \rho^2 \tag{1.100b}$$

$$R_2^2 = (z + L/2)^2 + \rho^2 \tag{1.100c}$$

and the integral may be put into the form

$$-\frac{1}{\rho}\left[\left(1 - \frac{z}{r}\right)e^{-jkr} - \left(1 - \frac{L/2+z}{R_2}\right)e^{-jk(R_2+L/2)}\right]$$

The remaining integrals in the equation for H_ϕ are, in order,

$$\frac{1}{\rho}\left[\left(1 + \frac{z}{r}\right)e^{-jkr} - \left(1 + \frac{L/2+z}{R_2}\right)e^{-jk(R_2-L/2)}\right]$$

$$\frac{1}{\rho}\left[\left(1 - \frac{L/2-z}{R_1}\right)e^{-jk(R_1+L/2)} - \left(1 + \frac{z}{r}\right)e^{-jkr}\right]$$

$$-\frac{1}{\rho}\left[\left(1 + \frac{L/2-z}{R_1}\right)e^{-jk(R_1-L/2)} - \left(1 - \frac{z}{r}\right)e^{-jkr}\right]$$

If these four terms are substituted into the equation for H_ϕ, it becomes

$$H_\phi = -\frac{I_m}{4\pi j\rho}\left(e^{-jkR_1} + e^{-jkR_2} - 2\cos\frac{kL}{2}e^{-jkr}\right) \tag{1.101}$$

which is a remarkably simple equation for the field close to the antenna [12].

The electric field components are readily found from the Maxwell equations in circular cylinder coordinates to be

$$E_\rho = -\frac{1}{j\omega\epsilon}\frac{\partial H_\phi}{\partial z}$$

$$= \frac{jZ_0 I_m}{4\pi\rho}\left[\left(z - \frac{L}{2}\right)\frac{e^{-jkR_1}}{R_1} + \left(z + \frac{L}{2}\right)\frac{e^{-jkR_2}}{R_2} - 2z\cos\frac{kL}{2}\frac{e^{-jkr}}{r}\right]$$
(1.102a)

$$E_\phi = 0 \qquad (1.102b)$$

$$E_z = -\frac{1}{j\omega\epsilon\rho}\frac{\partial(\rho H_\phi)}{\partial\rho} = -\frac{jZ_0 I_m}{4\pi}\left[\frac{e^{-jkR_1}}{R_1} + \frac{e^{-jkR_2}}{R_2} - 2\cos\frac{kL}{2}\frac{e^{-jkr}}{r}\right]$$
(1.102c)

The fields are readily specialized to the far zone by substituting in H_ϕ,

$$\rho = r\sin\theta \qquad (1.103a)$$
$$R_1 = r - \tfrac{1}{2}L\cos\theta \qquad (1.103b)$$
$$R_2 = r + \tfrac{1}{2}L\cos\theta \qquad (1.103c)$$

After the substitution, H_ϕ and E_θ in the far zone are

$$H_\phi = \frac{jI_m e^{-jkr}}{2\pi r}\frac{\cos(\tfrac{1}{2}kL\cos\theta) - \cos(\tfrac{1}{2}kL)}{\sin\theta} \qquad (1.104a)$$

$$E_\theta = Z_0 H_\phi = \frac{jZ_0 I_m e^{-jkr}}{2\pi r}\frac{\cos(\tfrac{1}{2}kL\cos\theta) - \cos(\tfrac{1}{2}kL)}{\sin\theta} \qquad (1.104b)$$

The power density is readily formed from these far-zone equations. It is true for this antenna, just as for the infinitesimal current source of Section 1.7, that only the far fields contribute to the time-average Poynting vector. The power density thus formed may be integrated over a sphere of large radius to find the total power radiated and the radiation resistance. The integration is not easily done. Kraus gives the process for the half-wave antenna [11], and Balanis gives the results for an antenna of general length [6]. The resistance will not be given now but deferred to Section 1.12 where the input impedance is developed.

1.10. RECIPROCITY THEOREM

In order to complete the center-fed dipole description by finding its input impedance, we need two theorems, one on reciprocity and one on equivalent sources. The reciprocity theorem will also be useful when we consider the receiving pattern of an antenna. We therefore pause to develop these theorems.

Consider two sets of courses, $\mathbf{J}^1, \mathbf{M}^1$ and $\mathbf{J}^2, \mathbf{M}^2$ in a linear isotropic medium. For these sources, the Maxwell curl equations are

$$\nabla \times \mathbf{H}^1 = \mathbf{J}^1 + j\omega\epsilon\mathbf{E}^1 \tag{1.105a}$$

$$\nabla \times \mathbf{E}^1 = -\mathbf{M}^1 - j\omega\mu\mathbf{H}^1 \tag{1.105b}$$

$$\nabla \times \mathbf{H}^2 = \mathbf{J}^2 + j\omega\epsilon\mathbf{E}^2 \tag{1.105c}$$

$$\nabla \times \mathbf{E}^2 = -\mathbf{M}^2 - j\omega\mu\mathbf{H}^2 \tag{1.105d}$$

where $\mathbf{E}^1, \mathbf{H}^1$ and $\mathbf{E}^2, \mathbf{H}^2$ are the fields produced by sources 1 and 2, respectively. Multiplying (1.105a) by $\mathbf{E}^2$ and (1.105d) by $\mathbf{H}^1$, adding, and using the identity

$$\nabla \cdot (\mathbf{A} \times \mathbf{B}) = \mathbf{B} \cdot (\nabla \times \mathbf{A}) - \mathbf{A} \cdot (\nabla \times \mathbf{B})$$

and repeating, except with the multiplication of (1.105b) by $\mathbf{H}^2$ and (1.105c) by $\mathbf{E}^1$, leads to [5]

$$-\nabla \cdot (\mathbf{E}^1 \times \mathbf{H}^2 - \mathbf{E}^2 \times \mathbf{H}^1) = \mathbf{E}^1 \cdot \mathbf{J}^2 - \mathbf{E}^2 \cdot \mathbf{J}^1 + \mathbf{H}^2 \cdot \mathbf{M}^1 - \mathbf{H}^1 \cdot \mathbf{M}^2 \tag{1.106}$$

Integrating over a volume and using the divergence theorem on the left side yields

$$\oint (\mathbf{E}^1 \times \mathbf{H}^2 - \mathbf{E}^2 \times \mathbf{H}^1) \cdot d\mathbf{A}$$
$$= \iiint (\mathbf{E}^1 \cdot \mathbf{J}^2 - \mathbf{E}^2 \cdot \mathbf{J}^1 + \mathbf{H}^2 \cdot \mathbf{M}^1 - \mathbf{H}^1 \cdot \mathbf{M}^2) \, dv \tag{1.107}$$

This equation represents the *Lorentz reciprocity theorem*. In a source-free region it reduces to

$$\oint (\mathbf{E}^1 \times \mathbf{H}^2 - \mathbf{E}^2 \times \mathbf{H}^1) \cdot d\mathbf{A} = 0 \tag{1.108}$$

1.11. AN EQUIVALENCE THEOREM

The electromagnetic fields produced by a given source distribution are unique, but the reverse is not true. A given field within a region can be produced by more than one source distribution. An electric current above an infinite conducting plane produces the same field above the plane as the

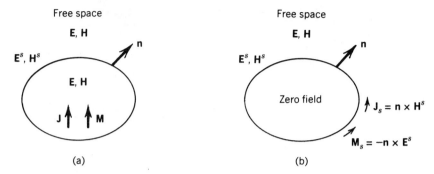

FIGURE 1.9 Fields of a source distribution and equivalent sources.

current and its image acting in free space, for example. Two sources that produce the same fields within a region are *equivalent* in that region [5].

Consider sources **J** and **M** within a region bounded by surface S, as shown in Figure 1.9a, producing fields **E** and **H** internal and external to S. The internal region may contain matter (conceivably nonlinear and nonisotropic), but external to S there is free space without sources. Field values at the surface are $\mathbf{E}^s$ and $\mathbf{H}^s$. Now consider a second case, shown in Figure 1.9b, with the same surface S the boundary between internal and external regions. We require that the fields external to S remain the same as for the first case, but that the internal fields be zero. Further, we assume free space both internal and external to S for the second case. This field configuration will be established if surface currents

$$\mathbf{J}_s = \mathbf{n} \times \mathbf{H}^s \tag{1.109a}$$

$$\mathbf{M}_s = -\mathbf{n} \times \mathbf{E}^s \tag{1.109b}$$

flow on surface S. The surface currents $\mathbf{J}_s$ and $\mathbf{M}_s$ are equivalent, for the fields external to S, to the original current distributions **J** and **M**. The equivalence expressed by (1.109) and its interpretation are commonly called the *Love equivalence principle* [2]. It may be expressed more generally, with nonzero fields internal to S in the equivalent formulation [2, 5], but the form given here is the one most used.

The fields $\mathbf{E}^s$ and $\mathbf{H}^s$ must be found from the original problem before the equivalent surface currents can be determined. In some cases, such as apertures in conducting planes, good approximations can be made to the fields in the aperture. Once the fields over S are known, the potential integrals may be used, since free space exists everywhere in the equivalent problem, to find the fields at all points external to S.

Two useful variations to the equivalence principle are possible. Since the fields internal to S are zero, we can place a perfectly conducting (for electric currents) surface just inside S or fill the entire internal region with a perfect

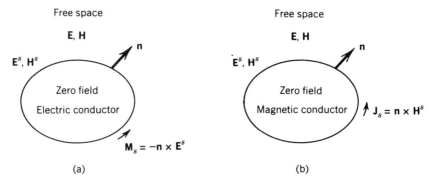

FIGURE 1.10 Equivalent sources in the presence of conductors.

electric conductor. The internal fields remain zero, and the external fields are unchanged. The conducting surface short circuits the electric current $\mathbf{J}_s = \mathbf{n} \times \mathbf{H}^s$ and leaves only the magnetic surface current to radiate, as shown in Figure 1.10a. The region may be filled instead with a perfect magnetic conductor (fictitious) that short circuits the magnetic surface current $\mathbf{M}_s = -\mathbf{n} \times \mathbf{E}^s$, leaving only the electric surface current to radiate, as in Figure 1.10b. If either step is taken, the potential integrals cannot be used in general to determine the fields produced by the equivalent surface current distributions of Figure 1.10 because the currents do not radiate into a homogeneous medium. If the fields of the original problem are insignificant over the greater part of surface S, as in radiation from an aperture in a conducting plane, for example, and if the radii of curvature of S are large enough, where $\mathbf{E}^s$ and $\mathbf{H}^s$ are significant, to use image theory, then one of the latter formulations of the equivalence principle can be useful. We use this concept in another section to determine the fields of an aperture antenna.

1.12. THE DIPOLE ANTENNA: INPUT IMPEDANCE

In this section we develop the input impedance of the linear center-fed antenna using the *induced emf method* commonly associated with Carter [13], although Elliott [1] points out that it was introduced by Brillouin. The treatment here is similar to that of Elliott, with results in the form used by Balanis [6].

The dipole has a circular cross section of radius a and length L, as shown in Figure 1.11. It is fed by an ideal source in an infinitesimally thin gap at the center. Because of skin effect, the current will flow in a thin layer at the conductor surface. This is approximated by assuming the layer to be of infinitesimal thickness with surface current density $J_{sz}^a(z')$. We also assume a surface current density in the feed gap given by the same function. Now, in accordance with concepts developed in Section 1.6, we can remove the

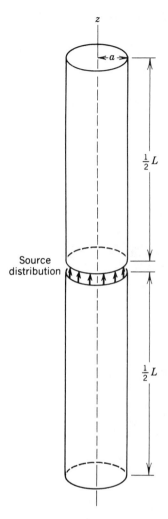

FIGURE 1.11 The dipole antenna of finite radius.

conductors and leave the surface current distribution in free space without altering the external fields. If we surround the current distribution by a cylindrical surface S infinitesimally greater in length and radius than the original antenna, the tangential electric field along the surface, $E_z^a(a, z')$, must be zero except at the feed gap. In the original problem, currents flow on the end caps, $z' = \pm L/2$, for a finite-radius antenna, but if the dipole is thin, with $a \ll \lambda$ and $a \ll L$, we can ignore the contributions of the end caps.

Consider next a line current $I^b(z')$ in free space along the z axis and apply the reciprocity theorem (1.108) for a source-free region,

$$\oint (\mathbf{E}^a \times \mathbf{H}^b - \mathbf{E}^b \times \mathbf{H}^a) \cdot d\mathbf{A} = 0 \qquad (1.110)$$

34   AN INTRODUCTION TO ANTENNAS

where $\mathbf{E}^a, \mathbf{H}^a$ are the fields produced by the surface current J_{sz}^a, and $\mathbf{E}^b, \mathbf{H}^b$ are those produced by I^b. If we note that E_ϕ and H_z are zero whether produced by the generator in the infinitesimal gap or by the line current on the axis, we get on surface S,

$$\int_{-L/2}^{L/2}\int_0^{2\pi}\left[E_z^a(a,z')H_\phi^b(a,z') - E_z^b(a,z')H_\phi^a(a,z')\right]a\,d\phi\,dz' = 0 \tag{1.111}$$

Now E_z^a is zero on S except at the infinitesimal feed gap. Assume further that

$$\int_{\text{gap}} E_z^a(a,z')\,dz' = -1 \tag{1.112}$$

so that E_z^a is a Dirac delta function, and the source voltage is 1 V. In addition, because of symmetry all scalar quantities in the problem are independent of ϕ. Then the integration of (1.111) yields

$$H_\phi^b(a,0) = -\int_{-L/2}^{L/2} E_z^b(a,z')H_\phi^a(a,z')\,dz' \tag{1.113}$$

Now we note that

$$H_\phi^a(a,z') = J_{sz}^a(z') \tag{1.114}$$

and

$$I^a(z') = 2\pi a J_{sz}^a(z') \tag{1.115}$$

With these substitutions, the magnetic field becomes

$$H_\phi^b(a,0) = -\frac{1}{2\pi a}\int_{-L/2}^{L/2} E_z^b(a,z')I^a(z')\,dz' \tag{1.116}$$

We can develop a second expression for the magnetic field $H_\phi^b(a,0)$. Consider the thin disk lying between the arms of the dipole (Fig. 1.12) and

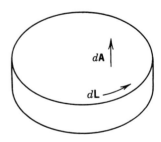

FIGURE 1.12   Disk containing source distribution.

integrate the Maxwell equation

$$\nabla \times \mathbf{H}^b = \mathbf{J}^b + j\omega \mathbf{E}^b \tag{1.117}$$

over one of its flat surfaces. This leads to

$$\int \mathbf{H}^b \cdot d\mathbf{L} = \int\int \mathbf{J}^b \cdot d\mathbf{A} + j\omega\epsilon \int\int \mathbf{E}^b \cdot d\mathbf{A} \tag{1.118}$$

which becomes

$$2\pi a H_\phi^b(a,0) = I^b(0) + j2\pi\omega\epsilon \int_0^a E_z^b(\rho,0)\rho\, d\rho \tag{1.119}$$

where the top surface of the infinitesimally thin disk lies at $z = 0$.

Now we assumed that $a \ll L$ and $a \ll \lambda$. The integral in this equation is therefore negligible compared to $I^b(0)$. This may be readily verified for specified fields. If the integral is neglected, comparison of the resulting equation with (1.116), the previously developed equation for the magnetic field, leads to

$$I^b(0) = -\int_{-L/2}^{L/2} E_z^b(a, z') I^a(z')\, dz' \tag{1.120}$$

Now $I^b(z')$ has not been constrained in any way (in contrast to I^a, which is the unknown current in a physical problem). It may therefore be chosen freely, and we choose it equal to I^a and drop the superscript. The result is

$$I(0) = -\int_{-L/2}^{L/2} E_z(a, z') I(z')\, dz' \tag{1.121}$$

where E_z is established by the axial current I^b and not by the surface current distribution of the physical problem.

The input impedance of the dipole antenna can now be found from

$$Z_{in} = \frac{V}{I(0)} = \frac{VI(0)}{I^2(0)} = -\frac{1}{I^2(0)} \int_{-L/2}^{L/2} E_z(a, z') I(z')\, dz' \tag{1.122}$$

since V was assumed earlier to be 1 V. Bear in mind that $I(z')$ is not known. It may be measured or its form assumed (see also Section 2.2). Once $I(z')$ is obtained, the field $E_z(a, z')$ can be determined in a straightforward manner since $I(z')$ is a filamentary current in free space, for which the potential integral can be used.

36   AN INTRODUCTION TO ANTENNAS

In Section 1.9 we assumed a sinusoidal distribution of current on the z axis

$$I(z') = I_m \sin\left[k\left(\tfrac{1}{2}L - |z'|\right)\right] \qquad -\tfrac{1}{2}L \le z' \le \tfrac{1}{2}L \quad (1.78)$$

and found E_z to be

$$E_z = -\frac{jZ_0 I_m}{4\pi}\left[\frac{e^{-jkR_1}}{R_1} + \frac{e^{-jkR_2}}{R_2} - 2\cos\frac{kL}{2}\frac{e^{-jkr}}{r}\right] \quad (1.102c)$$

Substituting these functions into the integral form for the input impedance allows it to be found for a sinusoidal current distribution. In E_z, we use for surface S

$$R_1 = \sqrt{(L/2 - z')^2 + a^2} \quad (1.123a)$$

$$R_2 = \sqrt{(L/2 + z')^2 + a^2} \quad (1.123b)$$

$$r = \sqrt{z'^2 + a^2} \quad (1.123c)$$

In the integration for the real part of Z_{in}, the approximation $a = 0$ can be made, but not for the imaginary part since it causes the imaginary part to become infinite, except for special antenna lengths.

The results of the integration are [6]

$$R_{in} = \text{Re}(Z_{in}) = \frac{Z_0}{2\pi \sin^2(kL/2)}$$
$$\times \{C + \ln(kL) - Ci(kL) + \tfrac{1}{2}\sin(kL)[Si(2kL) - 2Si(kL)]$$
$$+ \tfrac{1}{2}\cos(kL)[C + \ln(kL/2) + Ci(2kL) - 2Ci(kL)]\} \quad (1.124)$$

$$X_{in} = \text{Im}(Z_{in}) = \frac{Z_0}{4\pi \sin^2(kL/2)}\{2Si(kL) + \cos(kL)[2Si(kL) - Si(2kL)]$$
$$- \sin(kL)[2Ci(kL) - Ci(2kL)$$
$$- Ci(2ka^2/L)]\} \quad (1.125)$$

where

$$C = \text{Euler's constant} \cong 0.5772$$

$$Ci(x) = \text{cosine integral} = -\int_x^\infty \frac{\cos u}{u}\,du \quad (1.126)$$

$$Si(x) = \text{sine integral} = \int_0^x \frac{\sin u}{u}\,du$$

A commonly used antenna is the *half-wave dipole*, with $L = \lambda/2$, for which the impedance, if the dipole is thin, is

$$Z_{in} = 73.1 + j42.5 \, \Omega$$

For a sinusoidal current distribution it is common to reference the impedance (particularly the resistance) not to the input but to the position of maximum current (even if the antenna is so short that the maximum current is not reached at any point on the antenna). This is done by using I_m^2 rather than $I^2(0)$ in (1.122). The relationship between impedance referred to maximum current and input impedance is

$$Z_{in} = \frac{I_m^2}{I^2(0)} Z_m = \frac{Z_m}{\sin^2(kL/2)} \qquad (1.127)$$

The input resistance R_{in} for the dipole obtained by the induced emf method is the radiation resistance of the dipole, as defined in Section 1.8. It can also be obtained by forming the power density from the dipole far fields given by (1.104) and integrating over a sphere to determine the total power radiated. The result is the same as that found by the induced emf method.

Equivalent Circuit for an Antenna

We now recognize that *at one frequency* an antenna is seen by a generator and associated transmission line as an impedance, and the equivalent circuit of Figure 1.13 may be used to determine power accepted by the antenna, power radiated, and so on. Even though we developed the impedance by considering a linear wire antenna in which it is reasonable to think of an ohmic loss resistance in series with a radiation resistance, losses may come from dielectrics or from induced ground currents if the antenna is near a

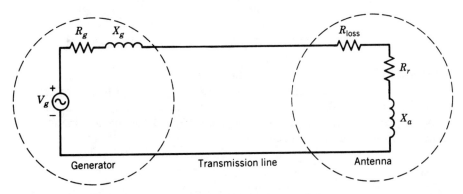

FIGURE 1.13 Equivalent circuit of transmitting antenna.

38 AN INTRODUCTION TO ANTENNAS

lossy ground. In such cases a parallel admittance representation of the antenna (Norton equivalent) may have elements that vary less with frequency than does the Thevenin circuit of Figure 1.13. In general, however, any equivalent circuit representation of an antenna is valid only over a narrow frequency range.

In considering the dipole antenna it is reasonable to consider feeding it with a two-wire transmission line with TEM mode fields. Now the concept of an antenna impedance is clearly dependent on our defining a driving point, or input port, for the antenna. Silver points out that the current distribution in the line must be that characteristic of a transmission line up to the assigned driving point [14]. At high frequencies, interaction between the radiating system and the line may disturb the line currents back over a considerable distance, and there is no definite transition between transmission line currents and antenna currents. In such a case the concept of "antenna impedance" is ambiguous. Some antennas are fed by waveguides that do not propagate the TEM mode. If the waveguide propagates a single mode, as most are designed to do, it is equivalent to a two-wire line, and a mode impedance can be defined. The antenna impedance can be expressed in terms of this mode impedance, but as before the validity of the impedance concept depends on our ability to define an antenna driving point with only a single waveguide mode on one side of this driving point. (A single mode on the other side is not precluded.) We assume here that the antenna impedance is clearly defined and that it does not matter if the feeding transmission system is a two-wire or coaxial line carrying the TEM mode or a waveguide propagating some other single mode. Note that many antennas are so complex that expressions for their impedance have not been developed satisfactorily and measurements are necessary to determine the impedance.

Finally, it should be noted that matching networks are commonly inserted between the antenna and transmission line, and between the generator and transmission line. Transmission lines and matching network principles are outside the scope of this book.

1.13. WAVEGUIDE OPENING INTO INFINITE GROUND PLANE

In this section we examine one of the many possible aperture antennas to see how the equivalence theorem of Section 1.11 may be used to determine the fields. Figure 1.14 shows a rectangular waveguide opening into an infinite ground plane. Assume that the waveguide allows only the dominant TE_{10} mode to propagate, with electric field

$$E_y^i = E_0' \cos \frac{\pi x}{a} e^{-k_g z} \quad (1.128)$$

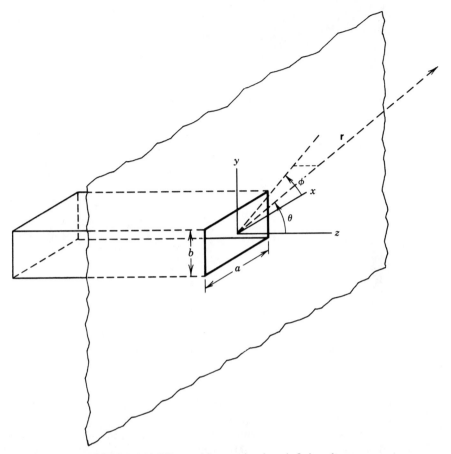

FIGURE 1.14 Waveguide opening into infinite plane.

where k_g is the propagation constant in the guide. At the guide opening a portion of the incident wave is reflected, and we take this to be the TE_{10} mode also,

$$E_y^r = \Gamma E_0' \cos \frac{\pi x}{a} e^{jk_g z} \tag{1.129}$$

At the guide opening, taken as the $z = 0$ plane, the aperture field is the sum of incident and reflected waves,

$$E_y = (1 + \Gamma) E_0' \cos \frac{\pi x}{a} = E_0 \cos \frac{\pi x}{a} \quad \begin{cases} -\frac{1}{2}a \leq x \leq \frac{1}{2}a \\ -\frac{1}{2}b \leq y \leq \frac{1}{2}b \end{cases} \tag{1.130}$$

Along the ground plane the tangential electric field is zero.

40 AN INTRODUCTION TO ANTENNAS

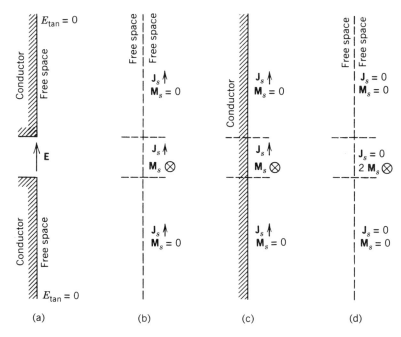

FIGURE 1.15 Equivalent surface currents for waveguide opening into plane.

We now apply the equivalence principle of Section 1.11, as illustrated by Figure 1.15, to find the fields produced.

Figure 1.15a shows the waveguide opening into the ground plane with the tangential component of **E** (and **H**) nonzero in the aperture. In Figure 1.15b the equivalence principle is used to establish a mathematical surface (an infinite plane) with equivalent surface currents $\mathbf{J}_s$ and $\mathbf{M}_s$ found from

$$\mathbf{J}_s = \mathbf{n} \times \mathbf{H}^s \quad (1.109a)$$

$$\mathbf{M}_s = -\mathbf{n} \times \mathbf{E}^s \quad (1.109b)$$

where $\mathbf{n} = \mathbf{u}_z$. Since we do not use $\mathbf{J}_s$, its value is not found. In Figure 1.15c a variation of the equivalence principle discussed in Section 1.11 is used to fill the region to the left of the surface (the "internal" region) with a perfect electric conductor. Finally, image theory is applied since we have an infinite conducting plane. The surface electric current that results from the application is zero since the image of a surface current density vector is an oppositely directed vector lying just inside the conductor, and the two add to zero. In the same way the surface magnetic current density is doubled, as shown in Figure 1.15d.

The magnetic surface current density for use in the potential integral is

$$2\mathbf{M}_s = -2\mathbf{u}_z \times \mathbf{u}_y E_0 \cos\frac{\pi x}{a} = \mathbf{u}_x 2 E_0 \cos\frac{\pi x}{a} \quad \begin{cases} |x| \leq \tfrac{1}{2}a \\ |y| \leq \tfrac{1}{2}b \end{cases} \quad (1.131)$$

and if this is substituted into the electric vector potential specialized to the far zone,

$$\mathbf{F}(\mathbf{r}) = \frac{\epsilon e^{-jkr}}{4\pi r} \iint \mathbf{M}_s(\mathbf{r}') e^{jk\mathbf{u}_r \cdot \mathbf{r}'} \, dA' \quad (1.132)$$

we obtain

$$\mathbf{F}(\mathbf{r}) = \frac{\epsilon e^{-jkr}}{4\pi r} \int_{-a/2}^{a/2} \int_{-b/2}^{b/2} \mathbf{u}_x 2 E_0 \cos\frac{\pi x'}{a} e^{jk\mathbf{u}_r \cdot \mathbf{r}'} \, dx' \, dy' \quad (1.133)$$

which becomes in spherical coordinates

$$F_\theta(\mathbf{r}) = \frac{2\epsilon E_0}{4\pi r} e^{-jkr} \cos\theta \cos\phi \int_{-a/2}^{a/2} \int_{-b/2}^{b/2} \cos\frac{\pi x'}{a}$$
$$\times \exp[jk \sin\theta (x' \cos\phi + y' \sin\phi)] \, dx' \, dy' \quad (1.134\text{a})$$

$$F_\phi(\mathbf{r}) = -\frac{2\epsilon E_0}{4\pi r} e^{-jkr} \sin\phi \int_{-a/2}^{a/2} \int_{-b/2}^{b/2} \cos\frac{\pi x'}{a}$$
$$\times \exp[jk \sin\theta (x' \cos\phi + y' \sin\phi)] \, dx' \, dy' \quad (1.134\text{b})$$

where θ and ϕ are the polar and azimuth angles of Figure 1.14. The integral common to both equations of this set is [1]

$$2\pi ab \frac{\cos(\pi X)}{\pi^2 - 4(\pi X)^2} \frac{\sin(\pi Y)}{\pi Y}$$

where

$$X = \frac{a}{\lambda} \sin\theta \cos\phi$$

$$Y = \frac{b}{\lambda} \sin\theta \sin\phi$$

If these values are substituted into (1.58) the far fields of the waveguide carrying the TE_{10} mode opening into an infinite ground plane are

$$E_\theta = \frac{\omega ab E_0}{cr} e^{-jkr} \sin\phi \frac{\cos[(\pi a/\lambda)\sin\theta\cos\phi]}{\pi^2 - 4[(\pi a/\lambda)\sin\theta\cos\phi]^2}$$

$$\times \frac{\sin[(\pi b/\lambda)\sin\theta\sin\phi]}{(\pi b/\lambda)\sin\theta\sin\phi} \quad (1.135a)$$

$$E_\phi = \frac{\omega ab E_0}{cr} e^{-jkr} \cos\theta\cos\phi \frac{\cos[(\pi a/\lambda)\sin\theta\cos\phi]}{\pi^2 - 4[(\pi a/\lambda)\sin\theta\cos\phi]^2}$$

$$\times \frac{\sin[(\pi b/\lambda)\sin\theta\sin\phi]}{(\pi b/\lambda)\sin\theta\sin\phi} \quad (1.135b)$$

1.14. THE RECEIVING ANTENNA

Impedance

If all sources and matter are of finite extent, the fields far from the source and material objects are related by

$$E_\theta = Z_0 H_\phi \quad (1.136a)$$

$$E_\phi = -Z_0 H_\theta \quad (1.136b)$$

and the left side of (1.107), representing the Lorentz reciprocity theorem, becomes, with integration over an infinitely large sphere,

$$-Z_0 \oint \left(H_\theta^1 H_\theta^2 + H_\phi^1 H_\phi^2 - H_\theta^2 H_\theta^1 - H_\phi^2 H_\phi^1 \right) dA = 0$$

Equation (1.107) then becomes

$$\iiint (\mathbf{E}^1 \cdot \mathbf{J}^2 - \mathbf{H}^1 \cdot \mathbf{M}^2) \, dv = \iiint (\mathbf{E}^2 \cdot \mathbf{J}^1 - \mathbf{H}^2 \cdot \mathbf{M}^1) \, dv \quad (1.137)$$

The integrals in this equation have been called *reaction* integrals [15]. The reaction of field 1 on source 2 is

$$\langle 1, 2 \rangle = \iiint (\mathbf{E}^1 \cdot \mathbf{J}^2 - \mathbf{H}^1 \cdot \mathbf{M}^2) \, dv \quad (1.138)$$

and in this notation (1.137) is

$$\langle 1, 2 \rangle = \langle 2, 1 \rangle \quad (1.139)$$

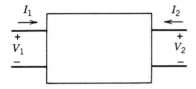

FIGURE 1.16 Two-port network.

showing that the reaction of field 1 on source set 2 is equal to the reaction of field 2 on source set 1.

Consider a current source I_2 with $\mathbf{M}^2 = 0$. Then the reaction $\langle 1,2 \rangle$ becomes

$$\langle 1,2 \rangle = \int\int\int \mathbf{E}^1 \cdot \mathbf{J}^2 \, dv = \int \mathbf{E}^1 \cdot I_2 \, d\mathbf{L} = I_2 \int \mathbf{E}^1 \cdot d\mathbf{L} = -V_1 I_2 \quad (1.140)$$

where V_1 is the voltage across source 2 due to the fields produced by some source 1 (which may be a voltage or current source, or both).

A linear, two-port network with voltages and currents shown in Figure 1.16 may be represented by

$$\begin{bmatrix} V_1 \\ V_2 \end{bmatrix} = \begin{bmatrix} Z_{11} & Z_{12} \\ Z_{21} & Z_{22} \end{bmatrix} \begin{bmatrix} I_1 \\ I_2 \end{bmatrix} \quad (1.141)$$

where the Z matrix is the impedance matrix. If current sources are applied at ports 1 and 2, the *partial voltage* at port 1 due to the current source at port 2 is

$$V_{12} = Z_{12} I_2$$

But

$$V_{12} = -\frac{\langle 2,1 \rangle}{I_1}$$

and combining the two equations gives

$$Z_{12} = -\frac{\langle 2,1 \rangle}{I_1 I_2}$$

In the same way, if a current source is applied to port 1 and the partial

44 AN INTRODUCTION TO ANTENNAS

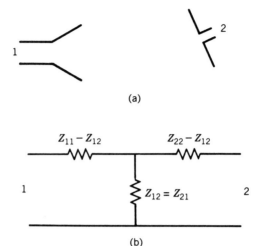

FIGURE 1.17 Polarization-matched antennas in transmit–receive configuration: (a) the antennas; (b) equivalent circuit for two antennas.

voltage at port 2 is considered,

$$Z_{21} = -\frac{\langle 1,2 \rangle}{I_2 I_1}$$

It follows from the equality of the reactions $\langle 1,2 \rangle$ and $\langle 2,1 \rangle$ that

$$Z_{12} = Z_{21} \qquad (1.142)$$

The linear two-port network considered may be the two antennas in a transmit–receive configuration shown in Figure 1.17a with the equivalent circuit of Figure 1.17b. It is important to note that (1.141) and the equivalent circuit of Figure 1.17b hold if the antenna is used either to transmit or receive. If the antennas are widely separated, Z_{12} is small, and an equivalent circuit for the two antennas, with one of them (say 1) transmitting and the other receiving is shown in Figure 1.18 [16]. We assume that the antennas are matched in polarization and defer to Chapter 4 a discussion of polarization matching.

We may consider Z_{11} the input impedance of the transmitting antenna, 1, neglecting the effects of the receiving antenna on the transmitting antenna, an excellent approximation for widely separated antennas. Likewise Z_{22}, the impedance of the receiving antenna in the approximate equivalent circuit, would be the input impedance of antenna 2 if it were transmitting. We can call Z_{11} and Z_{22} the *self-impedances* of the antennas, and the approximate equivalent circuit shows that the self-impedance of an antenna is the same

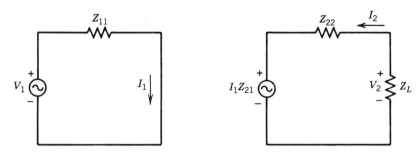

FIGURE 1.18 Approximate equivalent circuit for transmitting and receiving antenna system.

with the antenna transmitting and receiving. An interesting aspect of this equality is that it must hold for lossless antennas and for antennas with losses. If the self-impedance consists of a radiation resistance in series with a loss resistance, the total is the same with the antenna transmitting and receiving. Also, the radiation resistance alone is the same with the antenna transmitting and receiving (this is the lossless case). It follows that the loss resistance and the antenna efficiency are the same with the antenna transmitting and receiving.

Although the loss resistance computation of Section 1.8 was done only for a wire antenna, these concepts can be extended to antennas in general if we consider efficiency rather than loss resistance. In general, then, the self-impedance and efficiency of an antenna are the same when the antenna is receiving as they are when it is transmitting.

Receiving Pattern

It is obvious that an antenna whose radiation pattern is directional in nature will also receive a wave in the same manner. In other words, an antenna has a *receiving pattern* as well as a radiation pattern. We define the receiving pattern here as the spatial distribution of the received power when a polarization-matched plane wave is incident on the antenna. In some cases the received voltage may be measured rather than power. Consider two different positions for antenna 2 of the transmit–receive antenna configuration of Figure 1.19. The movement of antenna 2 from position a to position b is such that the distance from antenna 1 to antenna 2 is the same, and the orientation of antenna 2 with respect to a line drawn between the antennas is the same. In other words, antenna 2 is moved along the surface of a sphere centered at antenna 1. If the sphere radius is large and if the absolute phase of the signal at antenna 2 is not important, the exact location of the sphere center is unimportant. We assume, as we did earlier, that the antennas are polarization matched, but defer a discussion of what that means. We also

46   AN INTRODUCTION TO ANTENNAS

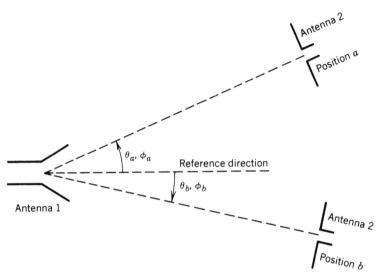

FIGURE 1.19   Measurement of antenna patterns.

assume that the antennas are far enough apart that a wave transmitted by one is for all practical purposes a plane wave at the other.

First let antenna 1 serve as the transmitting antenna and use the equivalent circuit of Figure 1.18. Note that the mutual impedance term Z_{21} in Figure 1.18 is a function of θ and ϕ. The ratio of powers received (power to Z_L) in positions a and b is, from Figure 1.18,

$$\frac{W_{2b}}{W_{2a}} = \frac{|Z_{21}(\theta_b, \phi_b)|^2}{|Z_{21}(\theta_a, \phi_a)|^2} \quad (1.143)$$

If position a is a reference position (it may be the position for maximum received power if we choose), this equation describes the relative radiation pattern of antenna 1 as θ_b and ϕ_b take on arbitrary values.

In addition to the measurement of power to the load of antenna 2, as described above, at each position of antenna 2 we connect the generator to antenna 2 and load to antenna 1 and measure the load power. An equivalent circuit similar to Figure 1.18, with generator $Z_{12}I_2$, leads to a ratio of load powers in positions a and b:

$$\frac{W_{1b}}{W_{1a}} = \frac{|Z_{12}(\theta_b, \phi_b)|^2}{|Z_{12}(\theta_a, \phi_a)|^2} \quad (1.144)$$

This equation is the relative receiving pattern of antenna 1.

It was shown that $Z_{12}(\theta_b, \phi_b) = Z_{21}(\theta_b, \phi_b)$ and $Z_{12}(\theta_a, \phi_a) = Z_{21}(\theta_a, \phi_a)$. We conclude that the relative radiation and receiving patterns of the antenna are equal.

Effective Area

The *effective area* of a receiving antenna in a given direction is the ratio of the available power at the terminals of the antenna to the power density of a polarization-matched plane wave incident on the antenna from that direction. The available power is the power that would be supplied to an impedance-matched load on the antenna terminals. In Figure 1.18 impedance matching means that $Z_L = Z_{22}^*$, where Z_{22} includes radiation resistance and loss resistance of the antenna. The effective area of an antenna is normally a more useful concept than the transmitter current and mutual impedance of Figure 1.18 because it is independent of the transmitter parameters and distance between the antennas. In addition, for aperture antennas it appears to be a natural characteristic. For wire antennas the effective area seems somewhat artificial since it does not correspond to any physical area of the antenna; nonetheless, it is a dimensionally correct and highly useful way to describe even a wire antenna.

We saw that the relative radiation and receiving patterns of an antenna are the same. It follows that the gain G and effective area A_e of the antenna are related by a constant, that is,

$$G(\theta, \phi) = CA_e(\theta, \phi) \tag{1.145}$$

In fact, we find that C is a universal constant for antennas.

Consider two antennas in a transmit–receive configuration, as shown in Figure 1.20. The type of antennas used is arbitrary and they are oriented arbitrarily with respect to their coordinate systems and to each other. Antenna 1 is transmitting and antenna 2 is receiving with a matched load. They are separated from each other by a sufficient distance r to cause the wave from antenna 1 to be effectively a plane wave at antenna 2 and to make the equivalent circuit of Figure 1.18 valid. The antennas need not be lossless.

If antenna 1 accepts power W_{a1} from its generator (and radiates a portion of it) and has gain G_1, the power density at antenna 2 is

$$\mathscr{P} = \frac{W_{a1} G_1(\theta_1, \phi_1)}{4\pi r^2} \tag{1.146}$$

and the power to the impedance-matched load is

$$W_{L2} = \mathscr{P} A_{e2}(\theta_2, \phi_2) \tag{1.147}$$

48 AN INTRODUCTION TO ANTENNAS

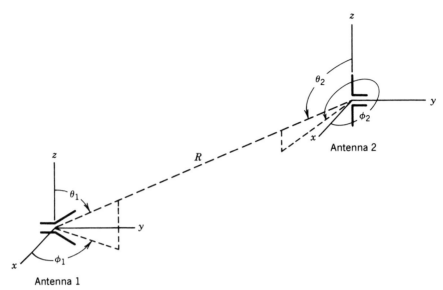

FIGURE 1.20 Transmission between two antennas.

where A_{e2} is the effective area of antenna 2. These equations show that

$$G_1(\theta_1,\phi_1)A_{e2}(\theta_2,\phi_2) = \frac{W_{L2}}{W_{a1}}(4\pi r^2) \qquad (1.148)$$

If we reverse the transmitting and receiving roles of the antennas by connecting a generator to antenna 2 and causing it to accept power W_{a2}, part of which is radiated, the power to a load that is impedance matched to antenna 1 is

$$W_{L1} = \frac{W_{a2}G_2(\theta_2,\phi_2)A_{e1}(\theta_1,\phi_1)}{4\pi r^2}$$

which gives

$$G_2(\theta_2,\phi_2)A_{e1}(\theta_1,\phi_1) = \frac{W_{L1}}{W_{a2}}(4\pi r^2) \qquad (1.149)$$

We make use once more of the valuable equivalent circuit of Figure 1.18. With antenna 1 transmitting and power W_{a1} supplied to the impedance Z_{11}, the ratio of load power (to load $Z_L = Z_{22}^*$) to the power accepted by Z_{11} is

$$\frac{W_{L2}}{W_{a1}} = \frac{|Z_{21}|^2}{4\,\text{Re}(Z_{11})\text{Re}(Z_{22})} \qquad (1.150)$$

If the roles of transmitter and receiver are reversed, the equivalent circuit shows that

$$\frac{W_{L1}}{W_{a2}} = \frac{|Z_{12}|^2}{4\,\mathrm{Re}(Z_{11})\mathrm{Re}(Z_{22})} \tag{1.151}$$

and from the equality of Z_{12} and Z_{21} it follows that

$$\frac{W_{L2}}{W_{a1}} = \frac{W_{L1}}{W_{a2}} \tag{1.152}$$

Using this relation, a comparison of (1.148) and (1.149) results in

$$G_1(\theta_1, \phi_1) A_{e2}(\theta_2, \phi_2) = G_2(\theta_2, \phi_2) A_{e1}(\theta_1, \phi_1)$$

or

$$\frac{A_{e1}(\theta_1, \phi_1)}{G_1(\theta_1, \phi_1)} = \frac{A_{e2}(\theta_2, \phi_2)}{G_2(\theta_2, \phi_2)} \tag{1.153}$$

In this equation the angles are arbitrary and have been retained to show that for one antenna the effective area in a particular direction is being compared to the gain in the same direction. The equation holds for lossy antennas, and will hold also for lossless antennas if $G(\theta, \phi)$ is replaced by the directivity $D(\theta, \phi)$ and $A_e(\theta, \phi)$ is determined for the lossless case. Antenna types were not specified, and it follows that if we can find the ratio A_e/G for one antenna, lossless or lossy, we know it for all.

In Section 1.8 we found the directivity of the infinitesimal z-directed current source of Figure 1.21:

$$D = \tfrac{3}{2} \sin^2 \theta \tag{1.76}$$

and the radiation resistance,

$$R_r = 80\pi^2 \left(\frac{L}{\lambda}\right)^2 \tag{1.77}$$

(Note: In Fig. 1.2, also representing the infinitesimal current element, the center feed point was not shown, but the difference is irrelevant since the current is constant throughout the element for both Figs. 1.2 and 1.21.)

Consider now that the antenna of Figure 1.21 is receiving, with a wave **E** incident on it. The open-circuit voltage induced at the antenna terminals is

$$V_{oc} = EL \sin \theta$$

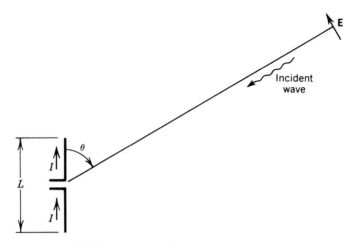

FIGURE 1.21 Infinitesimal current source.

where V_{oc} and E are peak values. Then the power to a matched load is

$$W = \frac{|V_{oc}|^2}{8R_r} = \frac{|E|^2 L^2 \sin^2\theta}{8R_r}$$

The power density at the antenna is

$$\mathcal{P} = \frac{1}{2Z_0}|E|^2$$

and therefore

$$W = \frac{Z_0 \mathcal{P} L^2 \sin^2\theta}{4R_r}$$

This gives an effective area for the lossless infinitesimal antenna of

$$A_e = \frac{W}{\mathcal{P}} = \frac{3\lambda^2 \sin^2\theta}{8\pi}$$

and a ratio of effective area to directivity of $\lambda^2/4\pi$. This ratio was obtained for a lossless example, but we saw earlier that it holds for the lossy case also. As a general rule, therefore, the effective area and gain of an antenna are related by

$$\frac{A_e(\theta,\phi)}{G(\theta,\phi)} = \frac{\lambda^2}{4\pi} \tag{1.154}$$

1.15. TRANSMISSION BETWEEN ANTENNAS

We have defined the necessary terms and developed the equations to determine the power in a receiver load if the power accepted by the transmitting antenna is known. Let the power accepted by antenna 1 in Figure 1.20 be W_{at}, with subscript t denoting the transmitting antenna. If antenna 1 radiated isotropically, the power density at 2 would be

$$\frac{W_{at}}{4\pi r^2}$$

Since it does not radiate isotropically, but has gain G_t, the actual power density at 2 is

$$\frac{W_{at} G_t(\theta_t, \phi_t)}{4\pi r^2}$$

The load power W_r in the load on the receiving antenna then is

$$W_r = \frac{W_{at} G_t(\theta_t, \phi_t) A_{er}(\theta_r, \phi_r)}{4\pi r^2} \qquad (1.155)$$

with subscript r indicating the receiving antenna.

Many variations on this equation are possible, but will not be given here. Instead, some of the more common alterations are:

1. G_t may be replaced by $e_t D_t$, where e_t is the efficiency of the transmitting antenna and D_t is its directivity.

2. W_{at} may be replaced by $(1 - |\Gamma_t|^2) W_{it}$ where W_{it} is the power incident on the (mismatched) transmitting antenna and Γ_t is the reflection coefficient obtained by treating the transmitting antenna as a load on the feeding transmission line.

3. A_{er} may be replaced by the gain G_r of the receiving antenna, according to (1.154).

4. If the receiving antenna is not terminated by a matched load, (1.155) must be multiplied by an impedance match factor, or efficiency, ranging from 0 to 1, to account for the mismatch loss. If the receiving antenna is represented by the series combination of R_a, including both radiation and loss resistances, and X_a, the antenna reactance, and the load impedance is $R_L + jX_L$, it is easy to show that the impedance match factor is

$$M_z = \frac{4 R_a R_L}{(R_a + R_L)^2 + (X_a + X_L)^2} \qquad (1.156)$$

52 AN INTRODUCTION TO ANTENNAS

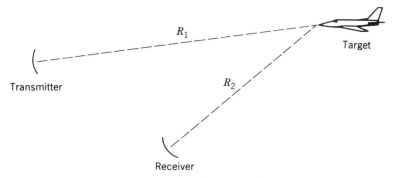

FIGURE 1.22 Bistatic radar and target.

5. If the antennas are not polarization matched, (1.155) must be multiplied by a polarization match factor or polarization efficiency. An equation for the polarization efficiency will be developed in another chapter.

1.16. THE RADAR EQUATION

Figure 1.22 shows a *bistatic* radar and target, with transmitting and receiving antennas separated. If the two antennas are colocated, the radar is a *monostatic* or *backscattering* radar. In many radars, the same antenna is used for transmitting and receiving, and this is a special case of a monostatic radar. In the configuration shown, the power density incident on the target is

$$\mathscr{P}_i = \frac{W_t G_t(\theta_t, \phi_t)}{4\pi r_1^2}$$

where W_t is the power accepted by the transmitting antenna. The transmitted signal may be a pulsed or continuous wave, but its characteristics do not affect this development.

The wave striking the target is reradiated in a directional manner, and a portion of the reradiated, or scattered, power is intercepted by the receiving antenna. The power received depends on the transmitted power, the antenna gains, and the *radar cross section* of the target [8]. The radar cross section may be bistatic or monostatic (backscattering).

It is desirable to define another target cross section, the *scattering cross section* [8] before defining the radar cross section. The scattering cross section of a target is the area of an equivalent target that intercepts a power equal to its area multiplied by the power density of an incident plane wave and reradiates it isotropically to produce at the receiving antenna a power density equal to that produced by the real target. An alternate form of the definition is that the scattering cross section is 4π times the radiation intensity of the scattered wave in a specified direction divided by the power density of the incident plane wave. The following reasoning makes clearer the reasons for this definition in terms of a fictitious equivalent target: An observer at the receiver can determine the power density of the scattered wave at the receiver and (multiplying by r_2^2) the radiation intensity of the scattering wave in the direction of the receiver. The observer does not know how the target scatters the incident wave (without more information than can be obtained by one measurement), and yet to describe the target a commonly accepted assumption is necessary. This assumption is that of isotropic scattering. With this assumption the observer determines that the total scattered power is 4π times the radiation intensity in his direction. It is then reasonable to say that this total power is scattered as a result of the target with an area σ_s intercepting an incident plane wave with power density established at the target by the radar transmitter. Note that the transmitting antenna characteristics enter into the definition of scattering cross section since the power reradiated in a particular direction may depend on the incident wave (e.g., the direction of the electric field). It does not depend on a receiving antenna since the definition is in terms of a power density or radiation intensity, not receiver power.

The radar cross section σ_r of a target is defined like the scattering cross section except that both transmitting and receiving antennas are specified. Then only that part of the scattered power density that can be received by the specified receiving antenna is considered. The definitions of both cross sections should be clearer as the *radar equations* are developed.

The power intercepted by the equivalent target with scattering cross section σ_s is

$$W_{\text{int}} = \sigma_s \mathscr{P}_i = \frac{W_t G_t(\theta_t, \phi_t) \sigma_s}{4\pi r_1^2}$$

If this power is scattered isotropically, as we assumed in defining the cross section, the power density at the receiving antenna is

$$\mathscr{P}_r = \frac{W_{\text{int}}}{4\pi r_2^2} = \frac{W_t G_t(\theta_t, \phi_t) \sigma_s}{(4\pi r_1 r_2)^2}$$

The receiving antenna may not be able to extract the maximum available power from the wave whose power density is given by this equation (the scattered wave may have a vertical electric field, for example, with the receiving antenna capable of receiving only a horizontal field). If we consider only that part of the received power density which is effective in producing a power in the receiver load, it is given by the same equation using the *radar cross section* rather than the scattering cross section, thus

$$\mathscr{P}'_r = \frac{W_t G_t(\theta_t, \phi_t) \sigma_r}{(4\pi r_1 r_2)^2}$$

The power to an impedance-matched load at the receiver is then

$$W_r = \frac{W_t G_t(\theta_t, \phi_t) A_{er}(\theta_r, \phi_r) \sigma_r}{(4\pi r_1 r_2)^2} \tag{1.157}$$

This is a common form of the *radar equation*. In defining the radar cross section σ_r, the polarization characteristics of both transmitting and receiving antennas must be specified, for example, a linear vertical electric field for both transmitted and received waves. The cross section is then valid only for those polarizations. The scattering cross section can be used in this equation if a multiplier is included to account for the polarization mismatch of the receiving antenna. Neither form of the equation is sufficient to describe completely the scattering behavior of a target. The radar equation will be reconsidered in Chapter 6 after the polarization behavior of antennas and targets is analyzed.

If the transmitter and receiver for a radar are at the same site, the radar equation simplifies to

$$W_r = \frac{W_t G_t(\theta, \phi) A_{er}(\theta, \phi) \sigma_r}{(4\pi r^2)^2} \tag{1.158}$$

where σ_r is the monostatic or backscattering radar cross section of the target. In many cases the same antenna is used for transmitting and receiving, and the relation between gain and effective receiving area for the antenna causes the radar equation to be

$$W_r = \frac{W_t G^2(\theta, \phi) \lambda^2 \sigma_r}{(4\pi)^3 r^4} \tag{1.159}$$

The bistatic radar cross section depends on the direction of the transmitter and receiver from the target, not merely on the difference in their directions. For a target as complex as an aircraft a change of transmitter or receiver direction of as little as a degree can change the cross section by many decibels. That is also true for a monostatic radar. The cross section is independent of the distances r_1 and r_2 if they are sufficiently large to cause a wave from either antenna to be plane at the target.

REFERENCES

1. R. S. Elliott, *Antenna Theory and Design*, Prentice-Hall, Englewood Cliffs, NJ, 1981.
2. R. E. Collin and F. J. Zucker, *Antenna Theory*, McGraw-Hill, New York, 1969.
3. A. Sommerfeld, *Electrodynamics*, Academic, New York, 1952.
4. W. K. H. Panofsky and M. Phillips, *Classical Electricity and Magnetism*, 2nd ed., Addison-Wesley, Reading, MA, 1962.
5. R. F. Harrington, *Time-Harmonic Electromagnetic Fields*, McGraw-Hill, New York, 1961.
6. C. A. Balanis, *Antenna Theory*, Harper & Row, New York, 1982.
7. H. Mott and D. N. McQuiddy, "On the Use of the Potential Integral for Determining the Fields of Physical Radiating Structures," *IEEE Transactions on Education*, **E-10**(4), 237–239, December 1967.
8. *IEEE Transactions on Antennas and Propagation*, **AP-31**(6), November 1983.
9. *IEEE Transactions on Antennas and Propagation*, **AP-22**(1), January 1974.
10. *IEEE Transactions on Antennas and Propagation*, **AP-17**(3), May 1969.
11. J. D. Kraus, *Antennas*, 2nd ed., McGraw-Hill, New York, 1988.
12. E. C. Jordan, *Electromagnetic Waves and Radiating Systems*, Prentice-Hall, Englewood Cliffs, NJ, 1950.
13. P. S. Carter, "Circuit Relations in Radiating Systems and Applications in Antenna Problems," *Proc. IRE*, **20**(6), 1004–1041, June 1932.
14. S. Silver, *Microwave Antenna Theory and Design*, Boston Technical Lithographers, Lexington, MA, 1963. (MIT Radiation Laboratory Series, Vol. 12, McGraw-Hill, New York, 1949.)
15. V. H. Rumsey, "The Reaction Concept in Electromagnetic Theory," *Phys. Rev.*, Ser. 2, **94**(6), 1483–1491, June 15, 1954.
16. S. Ramo, J. R. Whinnery, and T. Van Duzer, *Fields and Waves in Communication Electronics*, Wiley, New York, 1984.

PROBLEMS

1.1. Show that the integral of (1.31) is a solution to Poisson's equation (1.30).

1.2. Prove that (1.51) is correct.

1.3 Develop (1.54) by substituting (1.50) and (1.51) into (1.47), keeping only terms that vary as $1/r$.

1.4. Find the general **E** and **H** fields of the infinitesimal electric current element from the magnetic vector potential of (1.33). Show that the time-average Poynting vector formed from these fields is the same as that given by (1.67).

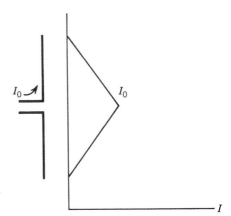

FIGURE P1.5 Short dipole and associated current distribution.

1.5. A practical antenna is the very short center-fed dipole shown in Figure P1.5. If length $L \ll \lambda$, the current distribution on the antenna is, to a good approximation,

$$I = I_0\left(1 - \frac{2}{L}z'\right) \quad 0 \leq z' \leq \frac{L}{2}$$

$$I = I_0\left(1 + \frac{2}{L}z'\right) \quad -\frac{L}{2} \leq z' \leq 0$$

Assume that in (1.35) $|\mathbf{r} - \mathbf{r}'|$ can be approximated by r in both the phase and amplitude of the integrand. Find the vector potential and far fields of the dipole.

1.6. Find the directivity of the short dipole of Problem 1.5.

1.7. Find the radiation resistance of the short dipole of Problem 1.5. The antenna is made of copper with a length of 10 cm and a diameter of 1 mm. The frequency is 300 MHz. $I_0 = 0.5$ A. Find its loss resistance. Determine the radiation efficiency.

1.8. In the equivalent circuit for two antennas in a transmitting–receiving system, the circuit parameters are: $Z_{11} = 73 + j42.5$ Ω; $Z_{22} = 60 + j0$ Ω; $Z_{12} = 10^{-4} + j0$ Ω.

Antenna 1 is fed by a source with internal impedance $73 - j42.5$ Ω and open-circuit voltage 100 V rms. The load impedance on antenna 2 is $60 + j40$ Ω. Find the power to the load on antenna 2. Can the antenna gains be found from the equivalent circuit?

1.9. A radar uses the same antenna for transmitting and receiving. Its gain is 20 dB at the radar frequency of 10 GHz. A target with a radar cross section of 7 m² is 10 km from the radar. The power accepted by the transmitting antenna is 50 kW. Find the power to the radar receiver, assuming an impedance match.

1.10. The transmitting antenna in a communications link accepts a power of 1 kW and has a gain of 40 in the receiver direction. The receiving antenna has a gain of 30 at the system frequency of 10 GHz. The antenna separation is 10 km. The antennas are polarization matched.

If the receiving antenna has impedance $25 + j40$ Ω and the load impedance is $40 - j40$ Ω, find the power to the load.

1.11. A paraboloidal reflector antenna is formed by rotating a parabola around an axis to form a bowl-shaped reflector. If the antenna diameter is 40 cm and it is used for a frequency of 8 Ghz, find the minimum distance from the antenna to a point considered in the far zone.

1.12. Sketch the radiated electric field intensity patterns of the waveguide opening into a plane for the principal E plane, $\phi = \pi/2$, and the principal H plane, $\theta = \pi/2$. Waveguide dimensions are $a/\lambda = 0.762$ and $b/\lambda = 0.339$.

1.13. Sketch in polar form the electric field intensity pattern in a constant azimuth plane of a center-fed dipole antenna whose length is $\lambda/2$. Note that the magnetic field intensity pattern is the same as the electric field intensity pattern and sketch the pattern of the radiated power density.

1.14. Use the sinusoidal current distribution of (1.78) to find the loss resistance of a center-fed halfwave dipole antenna made of copper and operating at 200 MHz. The antenna diameter is 0.5 cm. Find the radiation efficiency of the antenna.

1.15. The radiation intensity of an antenna is

$$U(\theta, \phi) = U_0 \sin^2 \theta \sin^2 \phi \quad \begin{cases} 0 \leq \theta \leq \pi \\ 0 \leq \phi \leq 2\pi \end{cases}$$

Find the average radiation intensity of the antenna. Find the directivity $D(\theta, \phi)$. If the radiation efficiency is 0.98 and the design frequency is 400 MHz, find the effective receiving area.

1.16. The gain of an antenna is measured by comparing its received power to that of a standard antenna with a gain of 12 dB. Both antennas are impedance and polarization matched. The test antenna receives a power of 20 μW from a third antenna used as a transmitter. When the test antenna is replaced by the standard antenna, the received power is 56.4 μW. Find the gain of the test antenna.

CHAPTER TWO

Arrays, Broadband Antennas, and Noise

2.1. INTRODUCTION

In this chapter more advanced antenna topics are considered. These include the moment method, the geometrical theory of diffraction, arrays, adaptive arrays, broadband antennas, and antenna environmental noise.

2.2. MOMENT METHOD AND GEOMETRICAL THEORY OF DIFFRACTION

The emphasis in the first chapter was on introducing concepts and developing equations for the radiated fields of some antennas for illustrative purposes. No attempt was made to describe all of the methods used to obtain field equations. Nevertheless, a brief examination of two methods of analysis that have become important in recent years is desirable. The *moment method* is most suitable for structures that are small in terms of a wavelength because of computation time required. The *geometrical theory of diffraction* is an extension of geometrical optics and is therefore more suitable for large structures.

The Moment Method

In Section 1.9 the magnetic vector potential

$$\mathbf{A}(\mathbf{r}) = \frac{\mu}{4\pi} \int \frac{I(\mathbf{r}')}{|\mathbf{r} - \mathbf{r}'|} e^{-jk|\mathbf{r} - \mathbf{r}'|} d\mathbf{L}' \qquad (1.35)$$

was used to find the fields of a center-fed dipole. The process revealed a basic problem in antenna theory; the current distribution $I(\mathbf{r}')$ on the dipole was not known, so based on experience we assumed that it was sinusoidal. In a more general case, the *a priori* knowledge is not available, and the current function must be found.

If $\mathbf{A}(\mathbf{r})$ in (1.35) is known at N points and if $I(\mathbf{r}')$ is expanded in a series of N terms, a system of N linear equations results that can be solved for the expansion coefficients of I. Thus an approximate form for I can be found [1, 2]. In practice, it is more convenient to use an integral for a field intensity rather than for the vector potential. A representative equation is *Pocklington's equation*,

$$-j\omega\epsilon_0 E_z^i\big|_{\rho=a} = \int_{-d/2}^{d/2} I(z') \left[\left(\frac{\partial^2}{\partial z^2} + k^2 \right) \frac{e^{-jkR}}{4\pi R} \right] dz' \qquad (2.1)$$

relating the current on a thin conducting wire of length d and radius a on

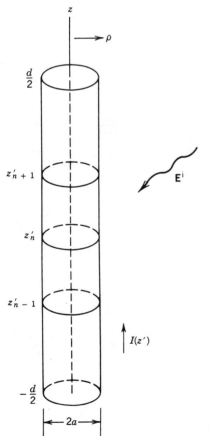

FIGURE 2.1 Cylindrical wire antenna.

the z axis to an incident electric field $\mathbf{E}^i$ (Fig. 2.1). The incident electric field may exist only at a small feed gap in the wire, in which case it becomes a radiating antenna.

Pocklington's equation is a special case of the general form

$$L[I(z')] = E(\rho, z) \tag{2.2}$$

where L is a linear integral operator. The linearity of L makes the inversion of this equation to find $I(z')$ possible. The process begins by expanding

$$I(z') = \sum_1^N c_n I_n(z') \tag{2.3}$$

where the $I_n(z')$ are known functions called *basis functions*. An example of basis functions for the thin-wire antenna is the piecewise constant set

$$I_n(z') = \begin{cases} 1 & z'_{n-1} < z' < z'_n \\ 0 & \text{elsewhere} \end{cases}$$

which leads to a staircase representation of the unknown current. After the expansion, the general equation becomes

$$\sum_1^N c_n L[I_n(z')] = E(\rho, z) \tag{2.4}$$

Since the $I_n(z')$ are known functions, $L[I_n(z')]$ can be evaluated either in closed form or numerically.

Now suppose that the right side of this equation is known at N points,

$$z = z_m \quad 1 \leq m \leq N$$

If it is noted that $L[I_n(z')]$ is dependent on z_m (e.g., because of the distance R in Pocklington's equation), we may write

$$L_{mn} = L[I_n(z')] \tag{2.5}$$

and the equation of interest becomes

$$\sum_1^N c_n L_{mn} = E(\rho, z_m) \tag{2.6}$$

or in matrix form

$$\mathbf{Lc} = \mathbf{E} \tag{2.7}$$

with $\mathbf{L}$ an $N \times N$ matrix and $\mathbf{c}$ and $\mathbf{E}$ N-element vectors. This form yields a set of N independent equations, if the z_m are chosen properly, that can be solved for the c_m.

Care must be taken in choosing the z_m in order to avoid singularities in the evaluation of the L_{mn}. For the wire antenna, for example, letting the current $I(z')$ flow in an equivalent filament along the z axis at $\rho = 0$ and choosing the field matching points in $E(\rho, z)$ at $\rho = a$ insures that R in the Pocklington equation is not zero.

The pulse functions for the expansion of $I(z')$ are not the only basis functions that can be used. Other functions, such as low-order spline functions that are nonzero over only part of the structure, can be used, and also $I(z')$ can be expanded with basis functions nonzero over the entire structure, such as continuous sinusoidal functions. Other functions may give a better representation for the current, but may require more computational effort to evaluate L_{mn}.

The procedure outlined here is a specialization of the moment method known as *point matching*. With its use the boundary conditions (e.g., the vanishing of the tangential electric field at the wire antenna) are met at N discrete points, but not between the points. In a more general approach, weighting functions are chosen that reduce the difference between computed and actual fields between the discrete matching points. The point matching technique is equivalent to using weighting functions that are Dirac delta functions.

Geometrical Theory of Diffraction

Geometrical optics is a widely used method for determining the radiated fields of certain antenna types, such as the parabolic reflector discussed in Chapter 5 of this text. For vanishingly small wavelengths, solutions to the Maxwell equations imply that electromagnetic energy is transported along a family of rays that can be traced through dielectric media and after reflections at conducting boundaries [1, 3]. See Section 5.10 for an example of the process.

In regions far from sources, the fields may be written as

$$\mathbf{E}(\mathbf{r}) = \mathbf{E}_0(\mathbf{r})e^{-jk_0\Psi(\mathbf{r})} \tag{2.8a}$$

$$\mathbf{H}(\mathbf{r}) = \mathbf{H}_0(\mathbf{r})e^{-jk_0\Psi(\mathbf{r})} \tag{2.8b}$$

where k_0 is the free-space propagation constant. If these fields are substituted into the Maxwell equations and the frequency is allowed to approach infinity, the following basic equations of geometric optics can be developed:

$$|\nabla\Psi|^2 = n^2(\mathbf{r}) \tag{2.9}$$

$$n\frac{d\mathbf{r}}{ds} = \nabla\Psi \tag{2.10}$$

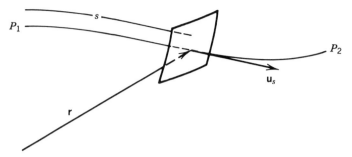

FIGURE 2.2 Geometric optics ray path.

where

$$n = \sqrt{\frac{\epsilon}{\epsilon_0}} \tag{2.11}$$

is the local index of refraction of the medium in which the wave travels, **r** is the position of a point with respect to an arbitrary origin, and s is the distance along the ray path (generally curved) from some arbitrary point, as in Figure 2.2.

The first equation relates the equiphase surfaces, or wavefronts, to the local index of refraction. The function Ψ is sometimes called the *eikonal*, and the equation is the *eikonal equation*. The second is the equation of a ray path and shows the local direction of energy transport, or Poynting vector direction. It is clear from the second equation that a ray path is perpendicular to an equiphase surface. It may also be shown from the second equation that in a homogeneous medium, with n constant, the rays are straight lines. Finally, the development leading to (2.9) and (2.10) shows $\mathbf{E}_0$ and $\mathbf{H}_0$ to be perpendicular to each other and to the ray direction at each point.

If the index of refraction has a discontinuity, the foregoing equations must be supplemented by *Snell's laws*. These are developed in Chapter 6 for any frequency and for a *plane* interface between two media having different refractive indices. However, if the radii of curvature of the interface in Figure 2.3 are large compared to a wavelength, Snell's laws are valid for nonplanar interfaces [4]. At the discontinuity, an incident wave gives rise to a reflected wave whose ray path lies in the *plane of incidence* defined by the incident ray and a local normal **n**, and

$$\theta_r = \theta_i \tag{2.12}$$

A transmitted or refracted wave in the second medium lies in the same

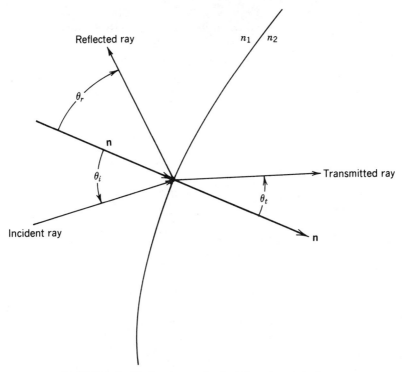

FIGURE 2.3 Reflection and refraction at an interface.

plane, with an angle in Figure 2.3 given by

$$n_2 \sin \theta_t = n_1 \sin \theta_i \tag{2.13}$$

If the second medium is perfectly conducting, there is no transmitted wave, but the angle of the reflected wave direction is still the same as the incident ray angle.

If $\mathbf{u}_s$ is the unit vector in the ray direction, as in Figure 2.2, the rate of change of Ψ on the ray path is

$$\frac{d\Psi}{ds} = \nabla\Psi \cdot \mathbf{u}_s = \nabla\Psi \cdot \frac{\nabla\Psi}{|\nabla\Psi|} = |\nabla\Psi| = n \tag{2.14}$$

It follows that the phase difference between the points P_1 and P_2 of Figure 2.2 is

$$\Psi_{12} = \Psi(P_2) - \Psi(P_1) = \int_{P_1}^{P_2} n\, ds \tag{2.15}$$

64 ARRAYS, BROADBAND ANTENNAS, AND NOISE

The integral is known as the *optical path length* between the points P_1 and P_2. The optical path length between any two equiphase surfaces is independent of the ray chosen to move from one surface to the other. *Fermat's principle* asserts that the optical path length of an actual ray between P_1 and P_2 is shorter than the optical path length of any other curve joining the points. In a homogeneous medium, the optical path length for path length s

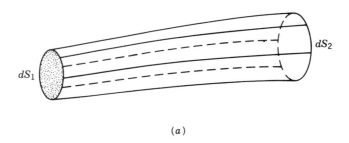

(a)

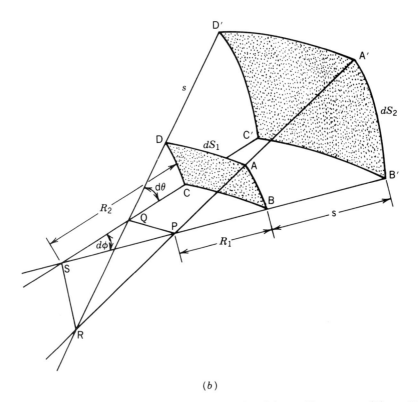

(b)

FIGURE 2.4 The intensity law of geometric optics: (*a*) curvilinear rays; (*b*) rectilinear rays.

is ns and the phase change along path s is

$$k_0 \Psi = k_0 ns = ks \tag{2.16}$$

As a final topic in geometric optics, consider the *intensity law*. Figure 2.4a is a tube formed by all the rays passing through surface element dS_1 on equiphase surface Ψ_1 and element dS_2 on equiphase surface Ψ_2. Since the energy is transported in the direction of these geometric rays, all the power through dS_1 also passes through dS_2, or

$$\mathscr{P}_1 \, dS_1 = \mathscr{P}_2 \, dS_2$$

where $\mathscr{P}_1$ and $\mathscr{P}_2$ are the Poynting vector magnitudes.

In a homogeneous medium the rays are straight lines as shown in Figure 2.4b. All distances along a ray between the equiphase surfaces are equal and are denoted by s. If rays are traced from corner A′ through A and from B′ through B, they intersect at P with included angle $d\theta$. Likewise the rays CC′ and DD′ intersect at Q with the same included angle. All rays through the surfaces intersect the line PQ, which is called a *caustic*. Similarly, the rays AA′ and DD′ intersect at R with included angle $d\phi$, and rays BB′ and CC′ intersect at S. Line RS is also a caustic. The distances to these points of intersection are the principal radii of curvature, $R_1, R_2, R_1 + s$, and $R_2 + s$, of the wavefronts. From the figure,

$$dS_1 = R_1 \, d\theta \, R_2 \, d\phi$$
$$dS_2 = (R_1 + s) \, d\theta (R_2 + s) \, d\phi$$

Then the intensity law of geometric optics is

$$\frac{\mathscr{P}_2}{\mathscr{P}_1} = \frac{dS_1}{dS_2} = \frac{R_1 R_2}{(R_1 + s)(R_2 + s)} \tag{2.17}$$

If a plane wave is incident on a two-dimensional conducting wedge, as in Figure 2.5, geometric optics predicts field discontinuities at the shadow boundaries for the incident and reflected waves. Above the boundary surface for the reflected rays, the total field is the sum of incident and reflected waves; between the boundary surfaces shown, only the incident wave exists; and below the incident ray shadow boundary the field is zero. Physically, the discontinuities do not exist, so geometric optics fails near the boundaries shown in Figure 2.5.

For a relatively simple problem such as this two-dimensional wedge, a solution to a boundary-value problem can be found, often in a series form that can be evaluated to any desired accuracy. A geometry susceptible to an exact solution is commonly called a *canonical* problem. The difference between the exact field and that predicted by geometric optics is caused by *diffraction*, the scattering of the incident finite-frequency wave from conduc-

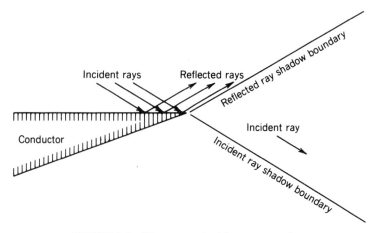

FIGURE 2.5 Plane wave incident on a wedge.

tor surfaces near, or at, discontinuities in the surface normal vector of the conducting body such as edges or points. If the exact solution is compared to the geometric optics solution, the diffracted field can be obtained. Then a coefficient, called the *diffraction coefficient*, can be obtained, which, when used as a multiplier of the incident field, gives (approximately) the diffracted field. The diffraction coefficient depends on the problem geometry and the incident field type. A typical relationship between incident and diffracted fields is that for the two-dimensional wedge of Figure 2.6,

$$\mathbf{E}^d = \mathbf{E}^i(Q) \cdot \overline{\mathbf{D}} A(s', s) e^{-jks} \tag{2.18}$$

where s' is the distance from the source point to the diffraction point Q, s is

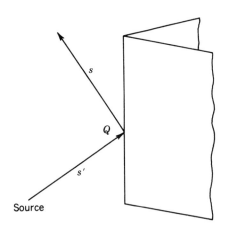

FIGURE 2.6 Diffraction by a wedge.

the distance from Q to the field point, and

$$A(s', s) = \begin{cases} \dfrac{1}{\sqrt{s}} \\ \sqrt{\dfrac{s'}{s(s+s')}} \end{cases} \qquad (2.19)$$

for incident plane and spherical waves respectively [1]. For a curved edge, A is more complicated and involves an intensity relationship similar to (2.17).

In the equation for the diffracted field, $\overline{\mathbf{D}}$ is a *dyadic function* or *dyad* [5]. A dyadic function is formed by two vector functions, thus,

$$\overline{\mathbf{D}} = \mathbf{AB} \qquad (2.20)$$

where $\mathbf{A}$ and $\mathbf{B}$ are field vectors. The dyadic function is used in a product form in conjunction with a vector. For example, the *anterior scalar product* is defined by

$$\mathbf{C} \cdot \overline{\mathbf{D}} = (\mathbf{C} \cdot \mathbf{A})\mathbf{B} \qquad (2.21)$$

which, despite its name, is a vector. The operation is not commutative, that is

$$\mathbf{C} \cdot \overline{\mathbf{D}} \neq \overline{\mathbf{D}} \cdot \mathbf{C} \qquad (2.22)$$

Vector products, such as

$$\mathbf{C} \times \overline{\mathbf{D}} = (\mathbf{C} \times \mathbf{A})\mathbf{B} \qquad (2.23)$$

which are themselves dyadic functions, can also be formed. In the equation for the diffracted field, the scalar product of incident field and diffraction coefficient yields a vector diffracted field.

The extension of geometric optics in this manner to include the diffracted field is called the *geometrical theory of diffraction*.

Exact boundary-value solutions do not exist for complex scattering-body shapes. To determine the scattering from such a body, it is decomposed into an assembly of canonical shapes whose exact solutions are known and whose diffraction coefficients can therefore be found. Diffraction is essentially a local phenomenon, so the diffraction coefficients can be developed and used independently for each canonical subscatterer. The total scattered field is then the sum of the reflected and diffracted fields from the canonical scatterers.

68 ARRAYS, BROADBAND ANTENNAS, AND NOISE

2.3. ANTENNA ARRAYS

An antenna configuration of identical radiating elements at specified positions, fed by currents whose amplitudes and phases can be varied, is widely used in radars. With an *antenna array* the main lobe of the radiation pattern can be moved at will, or scanned, by electronic phase shifters for the feed currents. This beam motion is more rapid and flexible than a mechanical scan motion. An array element is smaller than a single element having the same gain as the array, and in many cases the overall array has a more convenient size and shape than an equivalent single antenna requiring mechanical movement. In this section we consider linear and planar arrays.

Linear Arrays

Consider a *linear array* of N elements in the xy plane, as shown in Figure 2.7. The spacing between individual antenna elements is d, and each antenna has an electric field radiation pattern, or *element pattern*, $\mathbf{f}(\theta, \phi)$, so that if it were located at the origin its radiated field would be

$$\mathbf{E}(\mathbf{r}) = \frac{I\mathbf{f}(\theta, \phi)}{\sqrt{4\pi}\, r} e^{-jkr} \qquad (2.24)$$

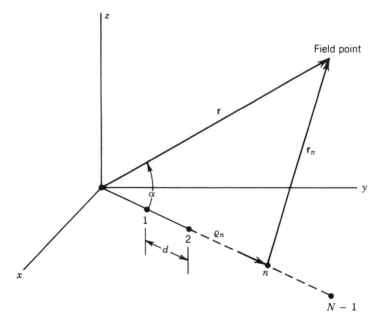

FIGURE 2.7 Linear array.

The field produced by the array is the phasor sum

$$\mathbf{E}(\mathbf{r}) = \frac{1}{\sqrt{4\pi}} \sum_{0}^{N-1} \frac{I_n \mathbf{f}(\theta_n, \phi_n)}{r_n} e^{-jkr_n} \qquad (2.25)$$

If the field point is far from the array, the distance term r_n in the denominator has virtually no effect on the amplitude and can be removed from the summation. Also, the angles θ_n and ϕ_n differ negligibly from one element to another for large field point distances. Then the overall field is

$$\mathbf{E}(\mathbf{r}) = \frac{\mathbf{f}(\theta, \phi)}{\sqrt{4\pi}\, r} \sum_{0}^{N-1} I_n e^{-jkr_n} \qquad (2.26)$$

It may be written as

$$\mathbf{E}(\mathbf{r}) = \mathbf{f}(\theta, \phi) F(\theta, \phi) \qquad (2.27)$$

where

$$F(\theta, \phi) = \frac{1}{\sqrt{4\pi}\, r} \sum_{0}^{N-1} I_n e^{-jkr_n} \qquad (2.28)$$

The function $F(\theta, \phi)$ is called the *array factor* or *array pattern* and is the field pattern radiated by an array of isotropic sources (independent of θ and ϕ) where the field of each source is a scalar and has no directional characteristics. This is a significant result; *the polarization characteristics of an array of identical radiators are determined by the elements and not by the array.*

In the phase term, the differences in r_n from one element to another are significant. Each distance can be determined from

$$\mathbf{r}_n = \mathbf{r} - \boldsymbol{\rho}'_n = \mathbf{u}_x(x - x'_n) + \mathbf{u}_y(y - y'_n) + \mathbf{u}_z z$$

where

$$\boldsymbol{\rho}'_n = \mathbf{u}_x x'_n + \mathbf{u}_y y'_n = nd\mathbf{u}_\rho$$

is the distance from the origin to the nth element. Then

$$r_n = \left[r^2 - 2xx'_n - 2yy'_n + x'^2_n + y'^2_n \right]^{1/2}$$

If the last two terms are dropped as negligible and an approximation is used for the square root, a distance term becomes

$$r_n = r - \frac{\mathbf{r} \cdot \boldsymbol{\rho}'_n}{r} = r - \rho'_n \cos\alpha = r - nd\cos\alpha$$

70 ARRAYS, BROADBAND ANTENNAS, AND NOISE

where α is the angle between the line of the array and the line of sight from array to field point. Using this approximation in the field equation causes it to become

$$\mathbf{E}(\mathbf{r}) = \frac{1}{\sqrt{4\pi}\,r}\mathbf{f}(\theta,\phi)e^{-jkr}\sum_{0}^{N-1} I_n e^{jknd\cos\alpha} \tag{2.29}$$

It is common to operate arrays with a constant phase difference in the feed currents between adjacent elements. If this phase difference is δ and the element at the origin is the reference,

$$\mathbf{E}(\mathbf{r}) = \frac{1}{\sqrt{4\pi}\,r}\mathbf{f}(\theta,\phi)e^{-jkr}\sum_{0}^{N-1} |I_n| e^{jn\delta} e^{jknd\cos\alpha} \tag{2.30}$$

If the feed currents all have the same phase, the effect of the array (separate from the element pattern) is to produce a beam maximum in a plane perpendicular to the line of the array (and other maxima for large values of d). If $\delta \neq 0$ the beam maximum is at an angle away from broadside. It is readily seen to occur where

$$kd\cos\alpha + \delta = 0$$

By varying δ the beam maximum may be scanned to any angle. Note that the pattern is rotationally symmetric about the line of the array. The directional characteristics in a plane transverse to the line of the array must be provided by the element pattern.

If all feed currents have the same amplitude, the resulting array pattern may have unacceptably high sidelobes for radar. *Tapering* the current distribution over the line of the array, using larger current amplitudes near the array center than near the ends, can reduce the sidelobes significantly, although it broadens the main beam. The element pattern may also be used to reduce the sidelobe levels.

Planar Arrays

Figure 2.8 shows the elements of a *planar array* on a rectangular grid in the xy plane. The element separations are d_x and d_y in the x and y directions. If an array is constructed in a regular, but nonrectangular, pattern it is convenient to treat it as being on a rectangular grid with some element currents zero.

The field produced by the array is

$$\mathbf{E}(\mathbf{r}) = \frac{1}{\sqrt{4\pi}}\sum_{0}^{M-1}\sum_{0}^{N-1}\frac{I_{mn}\mathbf{f}(\theta_{mn},\phi_{mn})}{r_{mn}}e^{-jkr_{mn}} \tag{2.31}$$

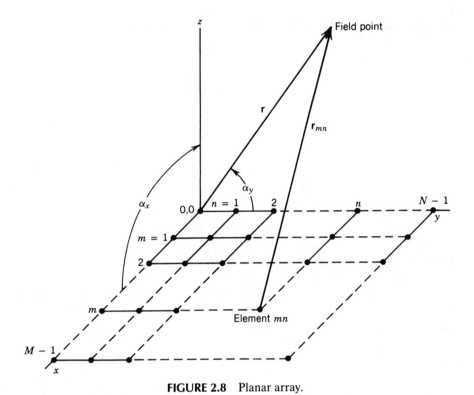

FIGURE 2.8 Planar array.

where I_{mn} is the feed current in the mn element and $\mathbf{f}$ is the element pattern. Proceeding as with the linear array allows this to be written as

$$\mathbf{E}(\mathbf{r}) = \frac{\mathbf{f}(\theta,\phi)}{\sqrt{4\pi}\,r} \sum_{0}^{M-1} \sum_{0}^{N-1} I_{mn} e^{-jkr_{mn}} \qquad (2.32)$$

and, like that of the linear array, the field is a product of the element pattern, which contains all the information about the direction of the electric field vector (the polarization), and an array pattern,

$$F(\theta,\phi) = \frac{1}{\sqrt{4\pi}\,r} \sum_{0}^{M-1} \sum_{0}^{N-1} I_{mn} e^{-jkr_{mn}} \qquad (2.33)$$

which is the pattern of an array of isotropic sources. It contains no polarization information.

In the usual manner of operation, the current feed phase of a row of elements is advanced by δ_x over the phase of the preceding row, and the phase of each column is advanced by δ_y. Then the current in element mn is

$$I_{mn} = |I_{mn}|e^{j(m\delta_x + n\delta_y)} \tag{2.34}$$

The distance r_{mn} in the phase term can be approximated by

$$\begin{aligned}
r_{mn} &= \left[(x - md_x)^2 + (y - nd_y)^2 + z^2\right]^{1/2} \\
&= \left[r^2 - 2mxd_x - 2nyd_y + m^2d_x^2 + n^2d_y^2\right]^{1/2} \\
&\approx r - \frac{mxd_x + nyd_y}{r} \\
&\approx r - md_x \cos\alpha_x - nd_y \cos\alpha_y
\end{aligned}$$

where α_x and α_y are angles measured from the x and y axes, respectively, to the line from the array to the field point.

The cosines in the equation are projections of a unit length along **r** onto the axes, so

$$\begin{aligned}
\cos\alpha_x &= \sin\theta\cos\phi \\
\cos\alpha_y &= \sin\theta\sin\phi
\end{aligned} \tag{2.35}$$

where θ and ϕ are the polar and azimuth angles measured to the field point. If the relevant information is combined, the field of a planar array is

$$\mathbf{E}(\mathbf{r}) = \frac{1}{\sqrt{4\pi}\, r}\mathbf{f}(\theta,\phi)e^{-jkr}\sum_{0}^{M-1}\sum_{0}^{N-1}\{|I_{mn}|\exp[j(m\delta_x + n\delta_y)]$$

$$\times \exp[jk(md_x\sin\theta\cos\phi + nd_y\sin\theta\sin\phi)]\} \tag{2.36}$$

The array factor is maximum if

$$(kmd_x\sin\theta\cos\phi + m\delta_x) + (knd_y\sin\theta\sin\phi + n\delta_y) = 0$$

This can be satisfied by separately requiring

$$\begin{aligned}
kd_x\sin\theta\cos\phi + \delta_x &= 0 \\
kd_y\sin\theta\sin\phi + \delta_y &= 0
\end{aligned}$$

If the effect of the element pattern is neglected, a beam maximum occurs at angles satisfying these equations. It can be seen that varying the current feed phases allows the main beam of the array to be scanned in both θ and ϕ.

SCANNING WITH PLANAR ARRAYS

Note that in contrast to the linear array, whose array factor is rotationally symmetric about the line of the array, the planar array is capable of producing a pencil beam with small beamwidth, even if the directional characteristics of each array element are not considered.

2.4. SCANNING WITH PLANAR ARRAYS

We saw earlier that the array pattern of a planar array is maximum in a direction for which

$$kd_x \sin\theta \cos\phi + \delta_x = 0$$
$$kd_y \sin\theta \sin\phi + \delta_y = 0$$

where δ_x is the phase progression between adjacent array elements along the x direction and δ_y is the phase progression in the y direction. It is clear that if the feed phases of the array elements are changed the beam maximum can be scanned, or moved to an arbitrary direction.

Now replace the polar and azimuth angles by the angles α_x and α_y measured from the axes, according to (2.35). Note also that the array factor is maximum for phase multiples of 2π. Then, more generally, beam maxima occur at

$$kd_x \cos\alpha_x + \delta_x = 2p\pi \quad p = 0, \pm 1, \pm 2, \ldots$$
$$kd_y \cos\alpha_y + \delta_y = 2q\pi \quad q = 0, \pm 1, \pm 2, \ldots \tag{2.37}$$

If these equations are solved for the direction cosines there results

$$\cos\alpha_x = \frac{2p\pi}{kd_x} - \frac{\delta_x}{kd_x} = \frac{\lambda}{d_x}p - \frac{\delta_x}{kd_x}$$
$$\cos\alpha_y = \frac{\lambda}{d_y}q - \frac{\delta_y}{kd_y} \tag{2.38}$$

It may thus be seen that in the array pattern there are an infinite number of beam maxima spaced λ/d_x in $\cos\alpha_x$ and λ/d_y in $\cos\alpha_y$. The locations of some of these on a plane with $\cos\alpha_x$ and $\cos\alpha_y$ as orthogonal coordinates are shown in Figure 2.9 for array spacing $d_x = d_y = \lambda$.

Also shown in Figure 2.9 is a circular contour for which

$$\cos^2\alpha_x + \cos^2\alpha_y = 1$$

The region inside this circle corresponds to $\cos\alpha_x \leq 1$ and $\cos\alpha_y \leq 1$, which is seen from Figure 2.8 to represent the *visible region* of space; that is, a point

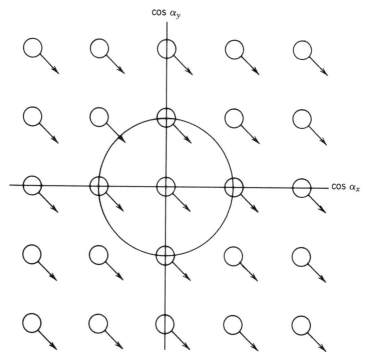

FIGURE 2.9 Grating lobes for the planar array, $d_x = d_y = \lambda$.

in the region corresponds to a pair of real values of the polar and azimuth angles θ and ϕ. A point outside the unit circle does not have a corresponding pair of angles θ and ϕ in real space and this region is called the *invisible region* or *invisible space*.

The locations of the beam maxima in Figure 2.9 are given by circles rather than points. A contour of constant radiation intensity near the beam maximum is a circle in the $\cos \alpha_x - \cos \alpha_y$ plane for a square array [6], and these contours are suggested by Figure 2.9. The figure is drawn for all array elements fed in phase, or $\delta_x = \delta_y = 0$, giving a main beam broadside to the array.

The beam maxima, other than the main beam, of the same amplitude as the main beam (not minor lobes) are called *grating lobes*. Note that in Figure 2.9 four grating lobes appear at the edge of the visible region.

Consider that the main beam is scanned to a new position, shown by the tip of the arrow from the origin of Figure 2.9, by selecting nonzero values for δ_x and δ_y. Equation (2.38) shows that the grating lobes move by exactly the same amount as the main beam. Grating lobes originally not in visible space can move there when the main beam is scanned (and grating lobes in visible space can move out and disappear). Figure 2.10 shows the main beam and

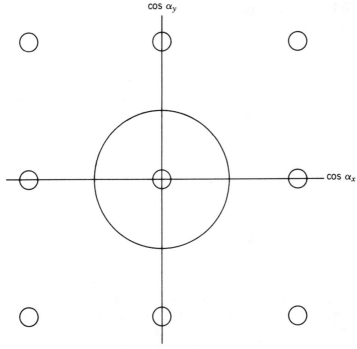

FIGURE 2.10 Grating lobes for the planar array, $d_x = d_y = \lambda/2$.

grating lobes for an array spacing $d_x = d_y = \lambda/2$. Note that a grating lobe does not appear at the edge of the visible region and will not appear with scanning until the main beam is scanned (in one of four directions) to the edge of visible space.

It is shown next that the equal-intensity contour of the beam from an array broadens as the main beam is scanned away from broadside [6]. This beam-shape change can be significant as the beam is scanned to the edge of the visible region.

Beam Cross Section

In the equations for locating the beam maxima of the planar array,

$$kd_x \cos \alpha_x + \delta_x = 2p\pi$$
$$kd_y \cos \alpha_y + \delta_y = 2q\pi \tag{2.37}$$

let

$$\delta_x = -kd_x \cos \alpha_{xs}$$
$$\delta_y = -kd_y \cos \alpha_{ys} \tag{2.39}$$

Then the beam maxima occur at

$$kd_x(\cos \alpha_x - \cos \alpha_{xs}) = 2p\pi$$
$$kd_y(\cos \alpha_y - \cos \alpha_{ys}) = 2q\pi \quad (2.40)$$

and it may be seen that α_{xs} and α_{ys} represent the position of the main beam.

In terms of these angles, the antenna array factor may be extracted from (2.36) and written as

$$F(\theta, \phi) = \frac{e^{-jkr}}{\sqrt{4\pi}\, r} \sum_0^{M-1} \sum_0^{N-1} |I_{mn}| e^{jm\Psi_x} e^{jn\Psi_y} \quad (2.41)$$

where

$$\Psi_x = kd_x(\cos \alpha_x - \cos \alpha_{xs})$$
$$\Psi_y = kd_y(\cos \alpha_y - \cos \alpha_{ys}) \quad (2.42)$$

Consider now the special case of an array having all elements fed by the same current amplitude. If this is 1, and if the constant phase and amplitude of the array factor are neglected, it becomes

$$F(\theta, \phi) = \sum_0^{M-1} e^{jm\Psi_x} \sum_0^{N-1} e^{jn\Psi_y} \quad (2.43)$$

It is not difficult to show that, apart from a phase factor,

$$\sum_0^{N-1} e^{jn\Psi} = \frac{\sin\left(N\dfrac{\Psi}{2}\right)}{\sin\left(\dfrac{\Psi}{2}\right)}$$

If this is used and if the array factor is normalized to have a maximum value of 1, it becomes

$$F(\theta, \phi) = \frac{\sin\left(M\dfrac{\Psi_x}{2}\right) \sin\left(N\dfrac{\Psi_y}{2}\right)}{M \sin\left(\dfrac{\Psi_x}{2}\right) N \sin\left(\dfrac{\Psi_y}{2}\right)} \quad (2.44)$$

We are interested in the shape of the beam cross section near a maximum, for which Ψ_x and Ψ_y are small. If the sine terms in the numerator and denominator of F are approximated by two and one terms, respectively, of a

Taylor's series and F is used as

$$F = 1 - \epsilon \qquad (2.45)$$

with ϵ a small constant, the result,

$$\left(\frac{M\Psi_x}{2}\right)^2 + \left(\frac{N\Psi_y}{2}\right)^2 = 6\epsilon \qquad (2.46)$$

is a circle with radius proportional to the square root of ϵ. Thus a contour of constant array factor, which is the beam cross section shape in the absence of a shaping factor in the element pattern, is a circle on a plane with coordinates $M\Psi_x/2$ and $N\Psi_y/2$. If the coordinate axes are taken as Ψ_x and Ψ_y, the beam cross-section shape, for contours near the beam maximum, is elliptical. The major axis is transverse to the smaller array dimension. Away from the beam maximum, it may be shown that the contours become somewhat rectangular in form.

Now consider the special case $M = N$ and $d_x = d_y$. For this, (2.42) and (2.46) combine to give

$$(\cos\alpha_x - \cos\alpha_{xs})^2 + (\cos\alpha_y - \cos\alpha_{ys})^2 = R^2$$

which is a circle on the $\cos\alpha_x - \cos\alpha_y$ plane. We are restricted to the region near the beam maximum, located at $(\cos\alpha_{xs}, \cos\alpha_{ys})$, and the circle radius R is small.

Define a complex variable

$$T = \cos\alpha_x + j\cos\alpha_y \qquad (2.47)$$

to locate a point on the $\cos\alpha_x - \cos\alpha_y$ plane (or T plane). From (2.35) this becomes

$$T = \sin\theta\cos\phi + j\sin\theta\sin\phi = \sin\theta e^{j\phi} \qquad (2.48)$$

where θ and ϕ are the polar and azimuth angles in physical space of the field point. It is readily seen from Figure 2.11 that T is the projection of a point on a unit-radius sphere onto the $\cos\alpha_x - \cos\alpha_y$ plane.

A beam contour, which is a circle on the $\cos\alpha_x - \cos\alpha_y$ plane near the beam maximum, can be projected upward to a unit sphere by the transformation (2.48) and on the unit sphere is the beam contour in physical space. Figure 2.12 shows two such projections, one for the broadside case, $\alpha_{xs} = \alpha_{ys} = \pi/2$, and one for the main beam scanned away from broadside. At

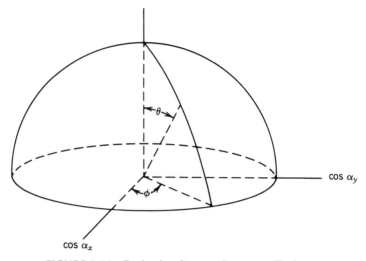

FIGURE 2.11 Projection from real space to T plane.

broadside the circular beam contour on the T plane is also circular in physical space. When the beam is scanned away from broadside, however, Figure 2.12 shows that it is broadened in the scanning direction. It may be shown that the beamwidth in the scan direction is

$$\theta_{3\text{ dB}} = \frac{\theta_{3\text{ dB (broadside)}}}{\cos \theta}$$

where θ is the scan angle away from broadside.

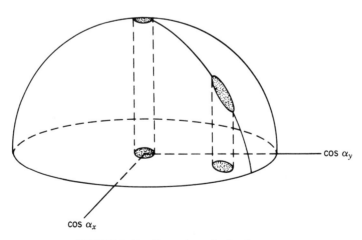

FIGURE 2.12 Beam broadening by scan.

2.5. ADAPTIVE ARRAYS

Adaptive antenna arrays are able, by feedback control processes, to alter their pattern while receiving a signal. This pattern change can be used to maximize the power received from a signal source whose direction may initially be unknown. If an interfering source, such as a jammer, and environmental noise are present, the adaptive array can maximize the signal to interference-plus-noise ratio. Another advantage is that the elements need not be disposed in a plane but can be placed conformally on, for example, an aircraft surface. In this section we consider briefly an early type of adaptive array, the phase-lock loop, because of its conceptual simplicity, and the LMS array in its continuous form. Other algorithms are used but are not discussed here.

The Phase-Lock Loop

Figure 2.13 shows a phase-lock-loop array for receiving. The phase-lock loop with each array element shifts the phase of each IF signal $s_i(t)$ until all the s_i are in phase, making the summer output a maximum. The reference signal is at the intermediate frequency of the receiver and the $s_i(t)$ are modulated IF signals. The phase of each s_i is compared to that of the reference in the phase detector. If they are not in phase, the phase detector alters the

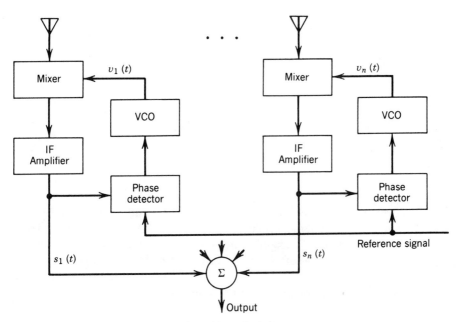

FIGURE 2.13 Phase-lock-loop array.

frequency of the voltage-controlled oscillator (VCO) that is the local oscillator for an antenna channel. This small change in the frequency of the VCO output, v_i, which may be looked at as a phase change, acts in concert with the antenna signal in the mixer to change the phase of $s_i(t)$. Thus all the $s_i(t)$ are phase aligned with the reference signal.

If only one wave is incident on the array, coming from a particular direction, this phase alignment by the control loops has the effect of pointing the array main beam in the source direction. The array thus automatically forms a pattern with the main beam aligned with the source, even if the source is in motion. A disadvantage of this system is that it can track only one signal at a time. A strong interfering signal from other than the source direction can capture the array main beam and prevent reception of the desired signal. For this reason, phase-lock loops are not widely used for beam pointing in adaptive arrays.

The LMS Array

Figure 2.14 shows an adaptive array that can be used to implement the continuous least-mean-square (LMS) algorithm. The 3-dB hybrid splits the signal from an antenna element so that one output is in phase and the second is retarded in phase by 90°, so that, for example,

$$s_{r1} = \frac{s_1}{\sqrt{2}} \quad s_{i1} = -j\frac{s_1}{\sqrt{2}}$$

After each of these signals is multiplied by the real weights w_{r1} and w_{i1}, thus,

$$v_{r1} = w_{r1} s_{r1} \quad v_{i1} = w_{i1} s_{i1}$$

the resulting signals are summed with those of the other elements to give output V. That part of V due to antenna 1 is then

$$V_1 = v_{r1} + v_{i1} = \frac{s_1}{\sqrt{2}}(w_{r1} - jw_{i1})$$

If both weights were equal to 1, then V_1 would be s_1 delayed in phase by 45°. It is seen that for general weight values, however, the output signal due to each antenna element is changed in amplitude and phase by the weights. This allows an antenna pattern to be formed to meet desired criteria.

The weights are determined by comparing the array output signal $V(t)$ to a reference $R(t)$ and obtaining the error signal

$$\epsilon(t) = R(t) - V(t) \tag{2.49}$$

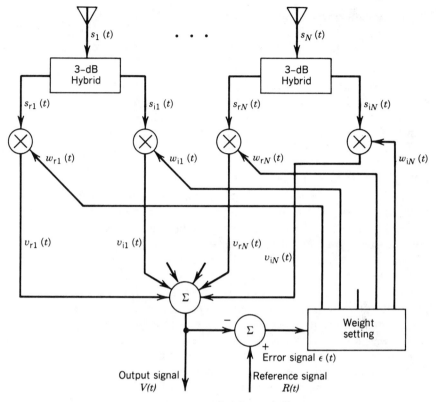

FIGURE 2.14 The LMS array.

In the LMS adaptive array, the weights are chosen to minimize $E[\epsilon^2(t)]$, the expected value of the square of the error signal.

If the random error signal is ergodic, the expectation can be replaced by a time average, so that

$$E[\epsilon^2(t)] = \langle \epsilon^2(t) \rangle = \lim_{T \to \infty} \frac{1}{T} \int_{-T/2}^{T/2} \epsilon^2(t)\, dt \qquad (2.50)$$

If noise $n(t)$ and an interfering signal $V_i(t)$ are also present in the array output, so that

$$V(t) = V_d(t) + n(t) + V_i(t) \qquad (2.51)$$

with V_d the desired signal, then the error is

$$\epsilon(t) = R(t) - V_d(t) - n(t) - V_i(t) \qquad (2.52)$$

and the mean-square error is

$$E[\epsilon^2(t)] = E\{[R(t) - V_d(t)]^2\} + E[n^2(t)] + E[V_i^2(t)] \quad (2.53)$$

if the noise, interfering signal, and difference between reference and desired signal are zero-mean and uncorrelated.

The array weights affect $E[V_i^2(t)]$ by forming an antenna pattern with some effective area in the directions of interfering sources. If this effective area is made small by the weight choices, the mean-square error is reduced. In the same way $E\{[R(t) - V_d(t)]^2\}$ can be made small by the weight selection if $R(t)$ and V_d are correlated.

The LMS array can make $E[V_i^2(t)]$ small by forming a pattern with nulls in the directions of interfering (and noise) sources and at the same time can maximize the desired signal power by establishing the main beam in the desired signal direction. This causes $V_d(t)$ to approach $R(t)$ more closely in amplitude and phase, and $E\{[R(t) - V_d(t)]^2\}$ will be small.

We see then that if $R(t)$ is chosen properly, the mean-square error will be small only if the power from interfering sources and noise is near or at a minimum and the power from the desired signal is near or at a maximum. Two comments are in order. A deliberately-introduced interfering (jamming) signal may not be uncorrelated with the desired signal; then the adaptive process will be more difficult. The reference signal should be identical to the desired signal for ideal performance, but that is obviously impossible. It can normally be chosen to have a high degree of correlation with the desired signal, however. For example, if the desired signal is a modulated carrier, the reference can be an unmodulated signal of the same frequency.

Optimum Weights

The error signal of the array of Figure 2.14 is

$$\epsilon(t) = R(t) - V(t) = R(t) - \sum_{\substack{j=1 \\ p=r,i}} w_{pj} s_{pj}(t) \quad (2.54)$$

where p successively represents both r and i, and $V(t)$ is the sum of desired signal, noise, and interfering signal. The mean-square error is

$$E[\epsilon^2(t)] = E[R^2(t)] - \sum_{\substack{j=1 \\ p=r,i}}^{N} w_{pj} E[R(t) s_{pj}(t)]$$

$$+ \sum_{\substack{j=1 \\ p=r,i}}^{n} \sum_{\substack{k=1 \\ q=r,i}}^{N} w_{pj} w_{qk} E[s_{pj}(t) s_{qk}(t)] \quad (2.55)$$

This equation can be written in a more convenient form if we define the real $2N$-dimensional vectors (column matrices)

$$\mathbf{W}(t) = \begin{bmatrix} w_{r1}(t) & w_{i1}(t) & w_{r2}(t) & w_{i2}(t) & \cdots \end{bmatrix}^T \quad (2.56)$$

and

$$\mathbf{S}(t) = \begin{bmatrix} s_{r1}(t) & s_{i1}(t) & s_{r2}(t) & s_{i2}(t) & \cdots \end{bmatrix}^T \quad (2.57)$$

We also define the vector

$$\mathbf{Q} = E[\mathbf{S}(t)R(t)] \quad (2.58)$$

The first two vectors are an array weight vector and an array signal vector. The vector $\mathbf{Q}$ has elements that are correlations of the reference signal and the individual array signals following the 3-dB hybrids and before weighting. It may be called a correlation vector.

With these definitions, the mean-square error can be written

$$E[\epsilon^2(t)] = E[R^2(t)] - 2\mathbf{W}^T\mathbf{Q} + \mathbf{W}^T\mathbf{\Phi}\mathbf{W} \quad (2.59)$$

where

$$\mathbf{\Phi} = E[\mathbf{S}(t)\mathbf{S}^T(t)]$$

$$= E\begin{bmatrix} s_{r1}(t)s_{r1}(t) & s_{r1}(t)s_{i1}(t) & s_{r1}(t)s_{r2}(t) & s_{r1}(t)s_{i2}(t) & \cdots \\ s_{i1}(t)s_{r1}(t) & s_{i1}(t)s_{i1}(t) & s_{i1}(t)s_{r2}(t) & s_{i1}(t)s_{i2}(t) & \cdots \\ s_{r2}(t)s_{r1}(t) & s_{r2}(t)s_{i1}(t) & s_{r2}(t)s_{r2}(t) & s_{r2}(t)s_{i2}(t) & \cdots \\ \vdots & \vdots & \vdots & \vdots & \vdots \end{bmatrix}$$

$$(2.60)$$

In a real, three-dimensional space, the gradient of a scalar function of position is a vector,

$$\nabla F = \mathbf{u}_x \frac{\partial F}{\partial x} + \mathbf{u}_y \frac{\partial F}{\partial y} + \mathbf{u}_z \frac{\partial F}{\partial z}$$

which may be written in matrix form as

$$\nabla F = \begin{bmatrix} \frac{\partial F}{\partial x} & \frac{\partial F}{\partial y} & \frac{\partial F}{\partial z} \end{bmatrix}^T$$

This concept can be extended to a function of any number of variables, and the variables need not be considered space coordinates.

84   ARRAYS, BROADBAND ANTENNAS, AND NOISE

The mean-square error is a scalar function of the $2N$ weight variables, and its gradient can be obtained. Then $E[\epsilon^2(t)]$ is an extremum if all the elements of the gradient are zero, thus,

$$\nabla E[\epsilon^2(t)] = \begin{bmatrix} \dfrac{\partial E[\epsilon^2(t)]}{\partial w_{r1}} \\ \dfrac{\partial E[\epsilon^2(t)]}{\partial w_{i1}} \\ \dfrac{\partial E[\epsilon^2(t)]}{\partial w_{r2}} \\ \vdots \end{bmatrix} = \mathbf{0} \qquad (2.61)$$

The extremum is a minimum since (2.55) shows clearly that the mean-square error can be made arbitrarily large by increasing w_{pj} and w_{qk}. It may be shown that the function has no saddlepoints or relative minima, but only one well-defined minimum [7]. Then the optimal weight setting is at this minimum, determined by (2.61).

From (2.59) the gradient of the mean-square error is

$$\nabla E[\epsilon^2(t)] = -2\nabla(\mathbf{W}^T \mathbf{Q}) + \nabla(\mathbf{W}^T \mathbf{\Phi} \mathbf{W}) \qquad (2.62)$$

But

$$\nabla(\mathbf{W}^T \mathbf{Q}) = \nabla \left(\begin{bmatrix} w_{r1} & w_{i1} & w_{r2} & \cdots \end{bmatrix} \begin{bmatrix} Q_{r1} \\ Q_{i1} \\ Q_{r2} \\ \vdots \end{bmatrix} \right) = \begin{bmatrix} Q_{r1} \\ Q_{i1} \\ Q_{r2} \\ \vdots \end{bmatrix} = \mathbf{Q}$$

Consider

$$F = \mathbf{X}^T \mathbf{Y} \mathbf{X} = \sum_{j=1}^{N} \sum_{k=1}^{N} x_j x_k y_{jk}$$
$$= x_1 x_1 y_{11} + x_1 x_2 y_{12} + \cdots + x_1 x_N y_{1N}$$
$$+ x_2 x_1 y_{21} + x_2 x_2 y_{22} + \cdots + x_2 x_N y_{2N}$$
$$\vdots$$
$$+ x_N x_1 y_{N1} + x_N x_2 y_{N2} + \cdots + x_N x_N y_{NN}$$

Then

$$\frac{\partial F}{\partial x_1} = 2 x_1 y_{11} + x_2 (y_{12} + y_{21}) + \cdots + x_N (y_{1N} + y_{N1})$$

and if **Y** is symmetric,

$$\frac{\partial F}{\partial x_1} = 2\sum_{1}^{N} y_{1j} x_j$$

Other derivatives are found similarly. Then

$$\nabla F = \left[\frac{\partial F}{\partial x_1} \quad \frac{\partial F}{\partial x_2} \quad \cdots \right]^T = 2\left[\sum_{1}^{N} y_{1j} x_j \quad \sum_{1}^{N} y_{2j} x_j \quad \cdots \right]^T = 2\mathbf{YX}$$

From this development, it is seen that

$$\nabla E[\epsilon^2(t)] = -2\mathbf{Q} + 2\mathbf{\Phi W} \tag{2.63}$$

and if this is set to zero, the optimum weights that minimize the mean-square error in the LMS adaptive array are found to be

$$\mathbf{W}_{\text{opt}} = \mathbf{\Phi}^{-1}\mathbf{Q} \tag{2.64}$$

if $\mathbf{\Phi}$ is nonsingular so that it has an inverse. The presence of thermal noise insures that it is nonsingular [7].

Setting the Weights

Initially the weight settings are not optimum and the mean-square error is not minimum. To insure that the error moves toward a minimum, we require that it decrease with time, or

$$\frac{d}{dt}\{E[\epsilon^2(t)]\} = \sum_{\substack{j=1 \\ p=r,i}}^{N} \frac{\partial E[\epsilon^2(t)]}{\partial w_{pj}} \frac{dw_{pj}}{dt} \tag{2.65}$$

To insure that the mean-square-error derivative is always negative, we require that

$$\frac{dw_{pj}}{dt} = -k\frac{\partial E[\epsilon^2(t)]}{\partial w_{pj}} \tag{2.66}$$

with k a positive real constant. This corresponds, for the full array, to

$$\frac{d\mathbf{W}}{dt} = -k\nabla\{E[\epsilon^2(t)]\}$$

This requirement leads to

$$\frac{d}{dt}\{E[\epsilon^2(t)]\} = -k\sum_{\substack{j=1\\p=r,i}}^{N}\left\{\frac{\partial E[\epsilon^2(t)]}{\partial w_{pj}}\right\}^2$$

which is always negative.

Equation (2.66) for the time derivative of the weights can be simplified. From the mean-square error, (2.55),

$$\frac{\partial E[\epsilon^2(t)]}{\partial w_{pj}} = -2E[R(t)s_{pj}(t)] + \frac{\partial}{\partial w_{pj}}\sum_{\substack{j=1\\p=r,i}}^{N}\sum_{\substack{k=1\\q=r,i}}^{N}w_{pj}w_{qk}E[s_{pj}(t)s_{qk}(t)]$$

$$= -2E[R(t)s_{pj}(t)] + 2\sum_{\substack{k=1\\q=r,i}}^{N}w_{qk}E[s_{pj}(t)s_{qk}(t)]$$

$$= -2E\left\{s_{pj}(t)\left[R(t) - \sum_{\substack{k=1\\q=r,i}}^{N}w_{qk}s_{qk}(t)\right]\right\}$$

The bracketed term is the error $\epsilon(t)$, so

$$\frac{\partial E[\epsilon^2(t)]}{\partial w_{pj}} = -2E[s_{pj}(t)\epsilon(t)]$$

If this is used in (2.66), the time derivative of the weights becomes

$$\frac{dw_{pj}}{dt} = 2kE[s_{pj}(t)\epsilon(t)] \tag{2.67}$$

or in vector form

$$\frac{d\mathbf{W}}{dt} = 2kE[\epsilon(t)\mathbf{S}(t)] \tag{2.68}$$

In practice, because of the long time required to obtain the expected values in (2.68) (by a time average), it is common to estimate $E[s_{pj}\epsilon(t)]$ by $s_{pj}\epsilon(t)$, leading to a control equation

$$\frac{dw_{pj}}{dt} = 2ks_{pj}\epsilon(t) \tag{2.69}$$

This equation is commonly called the Widrow LMS algorithm [8].

In general the array signals s_{pj} are stochastic and so is $\epsilon(t)$. Then the weights as given by (2.69) will vary randomly about their mean values. If k in (2.69) is small, the weight fluctuations can be kept within acceptable limits.

Figure 2.15 shows an implementation of the LMS algorithm for one antenna element of the array. Each feedback loop forms the product $2ks_{pj}(t)\epsilon(t)$ and integrates it to give the array weights

$$w_{pj}(t) = 2k \int s_{pj}(t)\epsilon(t)\, dt \tag{2.70}$$

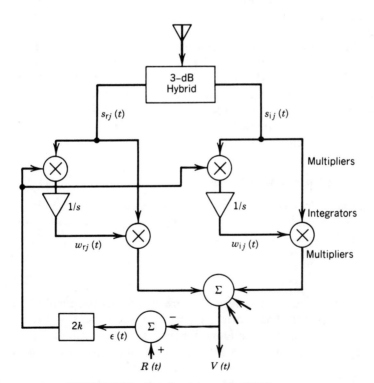

FIGURE 2.15 Feedback loops for LMS array.

88 ARRAYS, BROADBAND ANTENNAS, AND NOISE

Other Adaptive Arrays

The LMS array is not the only adaptive array in use. The Applebaum array adapts to maximize the signal-to-interference-plus-noise ratio at the array output, and the Shor array (not widely used) also acts to maximize the signal-to-noise ratio [7]. The LMS array was considered here in continuous form, but a discrete LMS array form is widely used. Finally, the adaptive array was treated here with the goal of maximizing a desired signal from a particular direction and minimizing an undesired signal from another direction by setting weights to create an antenna pattern. Polarization was not considered, but it would be appropriate to use polarized elements (in conjunction with pattern forming) to selectively prefer one polarization of an incoming wave and reject another. This idea is considered in more detail in Chapter 5.

2.6. BROADBAND AND FREQUENCY-INDEPENDENT ANTENNAS

Antennas that can transmit and receive over a wide frequency range without unacceptable changes in pattern or impedance are desirable for some applications. For example, commercial VHF television covers the range 54–216 MHz and it is desirable to use one receiving antenna for all the VHF channels. Many antennas (the linear dipole we considered previously for instance) cannot cover this relatively wide band, and wideband antennas are necessary. In this section we shall consider three categories of broadband antennas: a class that is theoretically independent of frequency in both pattern and impedance, a class that has the same pattern and impedance for any two frequencies having a specified ratio (log periodic antennas), and a group that is not theoretically frequency independent but nevertheless exhibits broadband characteristics.

Frequency-Independent Antennas

Consider an antenna, made of a perfect conductor surrounded by an infinite lossless, homogeneous, isotropic medium, whose surface can be described in spherical coordinates by

$$r = F(\theta, \phi) \tag{2.71}$$

Assume that the antenna terminals are spaced infinitesimally apart on the z axis with the coordinate origin midway between them. If the antenna is scaled to create a new antenna for use at a wavelength K times the wavelength of the first antenna, it must be K times as large, with surface

$$r' = Kr = KF(\theta, \phi) \tag{2.72}$$

Now suppose that the new antenna is identical to the old, although it need not be superposed on the old. If it is not superposed, the failure cannot be because the new antenna is translated, since the same coordinate origin is used for both. The failure cannot be because the new antenna is rotated in polar angle θ from the old, either, because both antennas have terminals on the same axis, from which θ is measured. Thus if they are not superposed, it is because the new antenna is rotated in azimuth angle ϕ from the old, and they can be superposed by a compensatory azimuth angle rotation.

It is clear that the two antennas can only be identical if the feed gap of each is infinitesimal and if the antennas are infinitely large.

Since the two antennas can be superposed by a rotation in azimuth angle, say α, then

$$KF(\theta, \phi) = F(\theta, \phi + \alpha) \tag{2.73}$$

Now the angle of rotation α depends on the value of K, but neither depends on θ or ϕ. Then K can be considered a function of α and its derivative found. If (2.73) is differentiated with respect to α and then with respect to ϕ, the values found are

$$\frac{dK}{d\alpha} F(\theta, \phi) = \frac{\partial}{\partial \alpha} F(\theta, \phi + \alpha) = \frac{\partial}{\partial(\phi + \alpha)} F(\theta, \phi + \alpha)$$

and

$$K \frac{\partial}{\partial \phi} F(\theta, \phi) = \frac{\partial}{\partial \phi} F(\theta, \phi + \alpha) = \frac{\partial}{\partial(\phi + \alpha)} F(\theta, \phi + \alpha)$$

The two terms on the left are therefore equal, or

$$\frac{dK}{d\alpha} F(\theta, \phi) = K \frac{\partial}{\partial \phi} F(\theta, \phi)$$

which may be written, using (2.71), as

$$\frac{\partial r}{\partial \phi} - ar = 0 \tag{2.74}$$

where

$$a = \frac{1}{K} \frac{dK}{d\alpha} \tag{2.75}$$

is independent of θ and ϕ.

The solution to this equation is

$$r = f(\theta)e^{a\phi} \tag{2.76}$$

We began the discussion with the requirement that the new antenna have the same impedance and radiation pattern at frequency f/K that the old antenna has at frequency f. Equation (2.76) is the requirement that the new antenna be identical to the old. Therefore, it describes an antenna that is independent of frequency if the antenna size is infinite [3, 9]. Since, in practice, an antenna cannot be infinitely large nor have an infinitesimal feed gap, it cannot be truly frequency-independent, but it can be almost so over a large frequency range.

Two special cases of (2.76) are interesting. For the first let

$$\frac{df}{d\theta} = A\delta\left(\frac{\pi}{2} - \theta\right)$$

where A is a positive constant and δ is the Dirac delta function. Then

$$f(\theta) = A \quad \theta = \pi/2$$
$$f(\theta) = 0 \quad \theta \neq \pi/2$$

This is a plane surface and cylindrical coordinates can be used, giving a logarithmic or *equiangular* spiral form

$$\rho = \rho_0 e^{a(\phi - \phi_0)} \tag{2.77}$$

where

$$\rho_0 e^{-a\phi_0}$$

is used to replace the constant A, and ϕ_0 is regarded as a parameter, with ρ_0 fixed.

Let ϕ_0 take on successively the four values 0, δ, π, and $\pi + \delta$, with δ normally less than or equal to $\pi/2$. The four spirals generated by the equation for the antenna surface are shown in Figure 2.16 (for $\delta = \pi/4$). If the surfaces between the lines for $\phi_0 = 0$ and δ and between the π and $\pi + \delta$ contours are made conducting and a small feed gap is created at the origin as indicated by Figure 2.16, the resulting planar antenna will be relatively independent of frequency over a wide range. In practice, the spiral arms must be truncated (gradually as shown) and this sets a low-frequency limit to the antenna operation. The finite size of the feed gap sets a high-frequency limit.

Radiation from the antenna is bidirectional and normal to the antenna plane. The gain is relatively low, a few decibels, and near the peak of the

FIGURE 2.16 Logarithmic spiral antenna.

main beam the radiated field is circularly polarized in the useful frequency range.

A spiral slot antenna can also be constructed by using a metallic conductor for the entire plane of the array and cutting slots to correspond to the metallic areas of Figure 2.16. The feed is a balanced structure at the narrow ends of the slots.

ARRAYS, BROADBAND ANTENNAS, AND NOISE

Another interesting antenna results if in (2.76) we let

$$\frac{df}{d\theta} = A\delta(\theta_0 - \theta)$$

so that

$$r = Ae^{a\phi} = r_0 e^{a(\phi-\phi_0)} \quad \theta = \theta_0$$
$$= 0 \quad \theta \neq \theta_0$$

An antenna described by this equation can be formed of spiral conducting arms, or spiral slots, like the planar antenna of Figure 2.16, but wrapped on a cone of half angle θ_0. Radiation from this conical spiral antenna is directed primarily in the direction of the cone apex. It is circularly polarized in the useful frequency range.

Log-Periodic Antennas [3, 10]

Figure 2.17 shows a planar antenna that is approximately frequency independent over a wide frequency range. The structure teeth have successive radii

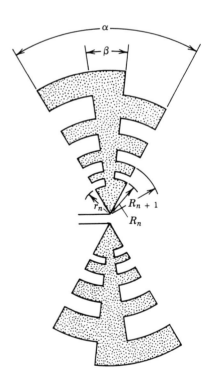

FIGURE 2.17 Log-periodic antenna.

BROADBAND AND FREQUENCY-INDEPENDENT ANTENNAS

in the ratio

$$\frac{r_{n+1}}{r_n} = \frac{R_{n+1}}{R_n} = \tau \tag{2.78}$$

The first radial tooth from the origin shown is labeled r_n because an infinite number of smaller, closer teeth are omitted for construction reasons.

Suppose that an infinite structure with infinitesimal feed gap like Figure 2.17 exists with particular values of impedance and pattern for frequency f. Now consider a new antenna scaled to have the same impedance and pattern at the frequency f/K. The new antenna will be larger by factor K. Then the tooth radii of the new antenna are related to the old by

$$\begin{aligned} R'_n &= KR_n \\ r'_n &= Kr_n \end{aligned} \tag{2.79}$$

Now suppose that the new antenna is identical to the old. This can occur only for infinite structures and if

$$\begin{aligned} R'_n &= R_{n+m} \\ r'_n &= r_{n+m} \end{aligned} \tag{2.80}$$

where m is some integer.

From these two equation sets it follows that

$$\begin{aligned} KR_n &= R_{n+m} \\ Kr_n &= r_{n+m} \end{aligned}$$

But from (2.78),

$$R_{n+m} = R_n \tau^m \qquad r_{n+m} = r_n \tau^m \tag{2.81}$$

and it follows that

$$K = \tau^m \tag{2.82}$$

Now K is the ratio of two frequencies for which the antenna has the same impedance and radiation pattern. Therefore the antenna properties are the same for frequencies f_0 and f_m if they are related by

$$\frac{f_m}{f_0} = \tau^m \tag{2.83}$$

In particular, let $m = 1$, so that $f_1/f_0 = \tau$. If the antenna is approximately frequency independent in the frequency range f_0 to f_1, it will then be frequency independent over all frequencies (until the limits given below come into play). Frequency independence between f_0 and f_1 is more easily obtained for closely spaced frequencies. It is then desirable to make τ small (but greater than 1) subject to the required bandwidth and maximum antenna size.

If (2.83) is rewritten as

$$\frac{f_m}{f_0} = \tau^m = (e^a)^m$$

or

$$\ln f_m - \ln f_0 = ma$$

it is seen that the logarithm of the frequency is periodic with period a. The antenna is therefore called *log-periodic*.

In practice the antenna must be truncated, and its maximum size sets the low-frequency limit of operation. The finite size of the feed gap and the lack of structural detail near the origin determine the high-frequency limit.

The log-periodic antenna may also be realized in a slot form. In Figure 2.17 the metal area is replaced by slots cut in a conducting ground plane. The slot antenna may be backed by a cavity to make the radiation pattern unidirectional, at some cost in bandwidth. It has also been determined that the currents in Figure 2.17 are concentrated near the metal edges so that a wire antenna lighter in weight and less subject to wind loading can be used. Figure 2.18 shows such an antenna with straight wires instead of curved since no significant loss in performance occurs with straight wires.

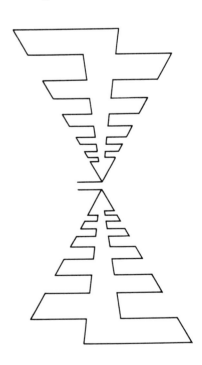

FIGURE 2.18 Wire antenna.

Radiation from the log-periodic antenna is bidirectional and transverse to the antenna plane, and the gain is a few decibels over a frequency range of 10–1 or greater. Polarization is linear. The radiation can be made unidirectional and the gain increased by folding the two antenna halves at the feed point so that they appear to lie on the surface of an imaginary wedge. Radiation is toward the antenna apex.

Broadband Antennas

The useful bandwidth of an antenna is determined both by its input impedance and its radiation pattern (primarily by its maximum gain). Both must remain within acceptable limits as frequency changes. To illustrate this, consider the thin dipole of Section 1.9. At a frequency f for which the dipole length is λ, the free-space wavelength, the maximum gain is in a direction perpendicular to the line of the dipole, at $\theta = \pi/2$. If the frequency is increased, the gain at $\theta = \pi/2$ decreases while increasing in other directions. At a frequency $2f$ where the dipole length is 2λ, the gain is zero at $\theta = \pi/2$ and the antenna cannot be used in the same manner as at frequency f. As the frequency changes from f to $2f$ the input impedance also changes significantly, and the mismatch between transmission line and antenna may affect radiated power.

The ratio of power accepted by an antenna to the incident power to the antenna from its feed line is $1 - |K|^2$, where K is the voltage reflection coefficient at the antenna (see Section 3.16). The reflection coefficient is related to S, the voltage standing wave ratio (VSWR), on the line by

$$S = \frac{1 + |K|}{1 - |K|}$$

Since both power accepted and gain in a specified direction affect the radiation intensity in that direction, a reasonable criterion for the broadband behavior of an antenna is that the factor

$$(1 - |K|^2)G = \frac{4S}{(S + 1)^2}G$$

not change unacceptably for the desired direction over the frequency range of interest.

The Thick Dipole A half-wave dipole of length L has a pattern similar to Figure 1.4 for L less than a free-space wavelength. As frequency increases the pattern becomes sharper, but additional lobes are not formed if $L < \lambda$. For $L = \lambda/4$ the 3-dB beamwidth is about 87° and for $L = \lambda$ it is about 48°. This pattern change is acceptable for many applications, and if gain were the only criterion, the dipole could be used over a frequency range of 4–1 or greater (for frequencies lower than that for which $L = \lambda$). The impedance change with frequency over this range may make the dipole unsatisfactory, however.

The input impedance of a thin cylindrical dipole antenna changes greatly with frequency, but the change is much less for a thicker antenna. For a ratio of dipole length to cylinder diameter of 472, the measured impedance magnitude ranges from less than 100 Ω to more than 1600 Ω for dipole lengths in the range $\lambda/4$ to λ. If the length-to-diameter ratio is 20, the minimum impedance magnitude does not change greatly, but the maximum impedance is less than 500 Ω [11]. Thus the thicker antenna is more suitable for broadband applications. It has been said that a dipole with a length–diameter ratio of 2000 has an "acceptable" bandwidth of about 3%, while one with a ratio of 260 has a bandwidth of about 30% [12]. Clearly the one with ratio 20 would have an even greater bandwidth.

Bow-Tie and Biconical Antennas Figure 2.19 shows a center-fed "bow-tie" antenna fabricated of thin conducting sheets. If the antenna shown were rotated about an axis passing through the apices of the two triangles composing the bow-tie, a biconical antenna would be formed. The flat sheets of the bow-tie can be replaced by triangular wire outlines of the sheet edges to form a lighter-weight antenna less susceptible to wind loading. If this wire antenna is rotated about an axis through the wire triangles to discrete azimuth angles, a wire simulation of the biconical antenna is formed, as in Figure 2.20.

Measurements of a bow-tie antenna over almost a 2–1 frequency range (480–900 MHz) have shown a VSWR less than 2 and a gain variation of

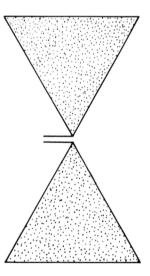

FIGURE 2.19 Bow-tie antenna.

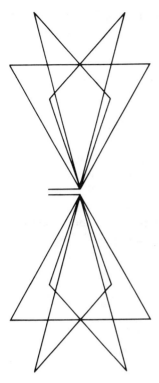

FIGURE 2.20 Wire approximation to biconical antenna.

about 2 dB [13]. Gain measurements of a biconical antenna have been reported to vary only about 1 dB over a frequency range of 3.5–1 [13].

The Ridged Horn The useful bandwidth of a pyramidal horn antenna (see Section 5.9) can be increased by adding conducting center ridges, as shown by the front view and side cross section in Figure 2.21. The ridge on the rectangular waveguide horn feed lowers the cutoff frequency of the dominant

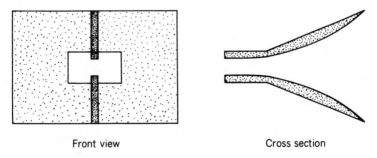

Front view　　　　　　　　　　　Cross section

FIGURE 2.21 Ridged pyramidal horn.

waveguide mode [14]. Measurements on a ridged horn have shown a VSWR of less than 2 over a frequency range of 2–18 GHz. The horn gain was measured as 6 dB at 2 GHz and 19 dB at 18 GHz [15]. The gain increase with frequency is unavoidable, but it is not accompanied by changes in the pattern lobe structure, so the horn is still useful over the full frequency range.

2.7. ANTENNA NOISE

A limiting factor in maximum radar range or the maximum distance between antennas in a communication system is the noise generated in a receiver or the electromagnetic noise incident on a receiving antenna. We address the latter noise problem in this section. It is desirable to introduce the concepts of brightness and brightness temperature to describe certain noise sources and system noise figure and system noise temperature to describe additional noise generated before and in the antenna. This is done in the following subsections.

Brightness

The power received from a radiating body extended in space is usefully described by its *brightness*, a term from radio astronomy. The power flux density per Hertz reaching the receiver of Figure 2.22 from the radiating source region subtended by solid angle $d\Omega$ is the brightness b of that portion of the radiator. Then the total power flux density (Poynting vector magnitude) at the receiver is

$$\mathscr{P} = \int_\Omega \int_f b(\theta,\phi,f)\, d\Omega\, df = \int_\theta \int_\phi \int_f b \sin\theta\, d\theta\, d\phi\, df \qquad (2.84)$$

The units of b are watts per square meter Hertz steradian. An impedance-matched receiving antenna with effective area $A_e(\theta,\phi,f)$ will receive from

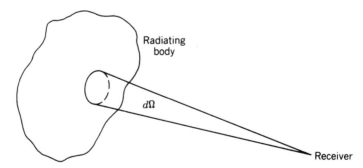

FIGURE 2.22 Power from extended radiating body.

the radiating body a total power

$$W = \int_\Omega \int_f b(\theta,\phi,f) A_e(\theta,\phi,f) \, d\Omega \, df \qquad (2.85)$$

if the antenna is matched in polarization to the received wave. If the incident wave is unpolarized, as it is for most noiselike power, the antenna receives half the power available in the wave regardless of the polarization properties of the antenna, as shown in Chapter 7. Then the proper equation for received power from an unpolarized wave is

$$W = \tfrac{1}{2} \int_\Omega \int_f b(\theta,\phi,f) A_e(\theta,\phi,f) \, d\Omega \, df \qquad (2.86)$$

In many cases both b and A_e are independent of frequency in the bandwidth of interest. Then the received power in bandwidth B simplifies to

$$W = \frac{B}{2} \int_\Omega b(\theta,\phi,f_0) A_e(\theta,\phi,f_0) \, d\Omega \qquad (2.87)$$

where f_0 is the center frequency of the receiver band.

Two special cases of these equations are of interest. If the solid angle subtended by the radiating region is greater than that of the antenna beam, then the integration is carried out only over the beam solid angle. Typically this might be noise from the entire sky with a high-gain radar antenna.

In another case of considerable interest, the radiating body is so small that the effective area of the antenna is constant over the subtended solid angle Ω_s. Then the received power is

$$W = \tfrac{1}{2} \int_f A_e(\theta_0,\phi_0,f) \int_{\Omega_s} b(\theta,\phi,f) \, d\Omega \, df \qquad (2.88)$$

An average brightness can then be defined as

$$b_0(\theta_0,\phi_0,f) = \frac{1}{\Omega_s} \int_{\Omega_s} b(\theta,\phi,f) \, d\Omega \qquad (2.89)$$

and the received power is

$$W = \tfrac{1}{2} \int_f A_e(\theta_0,\phi_0,f) b_0(\theta_0,\phi_0,f) \Omega_s \, df \qquad (2.90)$$

Finally, if A_e and b_0 are independent of frequency in bandwidth B, centered at f_0,

$$W = \frac{B}{2} A_e(\theta_0,\phi_0,f_0) b_0(\theta_0,\phi_0,f_0) \Omega_s \qquad (2.91)$$

This situation is typified by radiation from a star or the sun received by an antenna whose beam angle is much greater than the solid angle subtended by the sun.

The product

$$\mathscr{S} = b_0 \Omega_s = \iint b \, d\Omega$$

has units of watts per square meter Hertz and is often called the flux density. Since in this book flux density is used for quantities with units of watts per square meter, then $\mathscr{S}$ is called the *spectral flux density*. Measurements of power reaching an antenna from extended regions of the sky are normally expressed in terms of brightness (or *brightness temperature*; see the subsection on antenna noise temperature) and from discrete objects in terms of spectral flux density [16].

Thermal Power Received and Power Produced by a Resistor

All bodies at temperature T radiate electromagnetic energy. They also absorb or reflect any incident electromagnetic waves. It has been shown that good absorbers are also good radiators. An object that absorbs all incident energy and reflects none is commonly called a *blackbody*, even at nonvisible-light frequencies. Such a body is also a perfect radiator.

The brightness of the radiated energy from a blackbody is given by *Planck's radiation law* [17],

$$b = \frac{2hf^3}{c^2} \frac{1}{e^{hf/kT} - 1} \tag{2.92}$$

where h is Planck's constant ($= 6.63 \times 10^{-34}$ Js), k is Boltzmann's constant ($= 1.38 \times 10^{-23}$ J/°K), f is the frequency, and T is the temperature.

At radio frequencies $hf \ll kT$ and the exponential can be approximated by two terms of a Taylor's series so that the radiation brightness becomes

$$b = \frac{2kTf^2}{c^2} = \frac{2kT}{\lambda^2} \tag{2.93}$$

which is the *Rayleigh–Jeans radiation law*.

Now consider a lossless antenna with effective area $A_e(\theta, \phi, f)$ and impedance

$$Z_{ant} = R + jX$$

surrounded by a radiating blackbody at temperature T, as in Figure 2.23. The

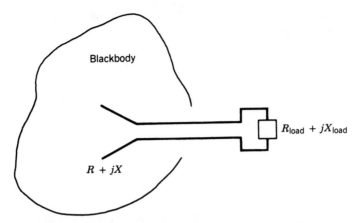

FIGURE 2.23 Antenna surrounded by blackbody radiator.

antenna is connected by a lossless line to an impedance-matched load,

$$Z_{\text{load}} = R_{\text{load}} + jX_{\text{load}} = R - jX$$

at the same temperature T. The blackbody radiation is unpolarized, and the antenna is completely polarized, so that half the available power is received. Under these conditions the power delivered to the matched load impedance from the blackbody is, from (2.86),

$$W = \tfrac{1}{2} \int_\Omega \int \int_f \frac{2kT}{\lambda^2} A_e(\theta, \phi, f)\, d\Omega\, df \qquad (2.94)$$

If A_e is independent of f over the narrow frequency band B considered here, the load power is

$$W = \frac{kT}{c^2} \int_\Omega \int A_e(\theta, \phi, f_0)\, d\Omega \int_{f_0 - B/2}^{f_0 + B/2} f^2\, df = \frac{kTBf_0^2}{c^2} \int_\Omega \int A_e(\theta, \phi, f_0)\, d\Omega$$

Now the effective area of the antenna can be replaced by its gain G according to

$$\frac{A_e}{G} = \frac{\lambda_0^2}{4\pi}$$

and the received power is

$$W = kTB \int_\Omega \int \frac{G(\theta, \phi, f_0)}{4\pi}\, d\Omega$$

For a lossless antenna the gain G is related to the radiation intensity in a specific direction by

$$G(\theta, \phi, f_0) = \frac{U(\theta, \phi, f_0)}{U_{av}}$$

and if this is used in the equation for received power, it becomes

$$W = kTB \int\int_\Omega \frac{U(\theta, \phi, f_0)}{4\pi U_{av}} d\Omega$$

Using the definition of average radiation intensity, this becomes

$$W = kTB \qquad (2.95)$$

The power given by this equation is received from the blackbody radiator through the antenna. If the system is in thermal equilibrium, the resistor R must radiate the same amount of power to the blackbody. The random motion of electrons in the resistor caused by their thermal energy produces a voltage across the resistor terminals. With voltage V across the resistor of value R, the power that can be transferred to the antenna, with radiation resistance R, is $V^2/4R$, and this is the power W that is radiated by the load resistance R, so that

$$W = kTB = \frac{V^2}{4R}$$

Then the RMS voltage across the resistor terminals is

$$V = 2\sqrt{kTBR} \qquad (2.96)$$

These equations for resistor voltage and available power from a resistor were first developed by Nyquist using a different starting point [18].

Equation (2.95) shows that the received power is independent of the frequency of the radiation and depends only on the bandwidth and blackbody temperature. This power is commonly referred to as *white noise*—"noise" because of the random phase relations of its components and "white" because all spectral components are present with the same spectral density (power per Hertz). Note that this frequency independence is characteristic of the received power, not of the brightness or spectral flux density of the incident radiation. Since the load resistor generates a power also given by (2.95), it too is a white noise generator.

Antenna Noise Temperature

The electromagnetic noise reaching an antenna from a source may not be closely related to the physical temperature of the source. An example is man-made noise from sources that are not thermal. Nevertheless, it is useful to assign an equivalent temperature to such sources. This is readily done in the radio-frequency region by using the Rayleigh–Jeans equation. Then the *brightness temperature* of an object is defined by

$$T_b = \frac{b\lambda^2}{2k} \tag{2.97}$$

The brightness temperature may be much greater than the actual temperature of the radiating object (a radar jammer is an example) or less (for an object that radiates thermally with lower efficiency than a blackbody). The brightness temperature of the sky may range from a few degrees for an antenna pointed parallel to the galactic axis to a much larger value for an antenna pointed in a narrow angular range around the direction of the galactic center. The brightness temperature of the sun depends on the frequency of observation and can be quite high, as an example in this section will show.

In a similar manner, an equivalent or effective temperature may be assigned to an antenna. In Figure 2.23, remove the blackbody surrounding the antenna and let the antenna be exposed to an assemblage of radiating objects with brightness $b(\theta, \phi, f)$. The antenna and load remain impedance matched, but the load resistance may be at an arbitrary temperature, since thermal equilibrium is not required (the load resistor temperature may increase, for example, as power is received from the external source). The power in small bandwidth B is given by (2.87). It is also given by (2.95) with T replaced by an equivalent *antenna temperature* T_a. Equating the two powers results in

$$T_a = \frac{W}{kB} = \frac{1}{2k} \int_\Omega \int b(\theta, \phi, f_0) A_e(\theta, \phi, f_0) \, d\Omega \tag{2.98}$$

If the brightness temperature is used, the antenna temperature becomes

$$T_a = \int_\Omega \int \frac{T_b(\theta, \phi, f_0)}{\lambda^2} A_e(\theta, \phi, f_0) \, d\Omega \tag{2.99}$$

or in terms of the gain of the antenna,

$$T_a = \frac{1}{4\pi} \int_\Omega \int T_b(\theta, \phi, f_0) G(\theta, \phi, f_0) \, d\Omega \tag{2.100}$$

Note that the distance to a radiating body does not appear in the equations involving brightness or brightness temperature. Brightness is spectral power density per steradian, and if the distance to a body increases by factor a, the area of the body subtended by infinitesimal solid angle $d\Omega$ increases by factor a^2. This increase compensates for the $1/a^2$ falloff expected in received power from a point source. If there is attenuation in the intervening region, this relationship does not hold.

The primary contributors to antenna noise temperature, apart from manmade noise, are the sun, the moon, discrete cosmic noise from radio stars, extended cosmic noise from ionized interstellar gas, the warm earth as seen by antenna sidelobes, a lossy atmosphere, and a lossy radome covering the antenna. The sun is by far the largest contributor for a high-gain antenna pointed toward it.

Example Determine the antenna temperature and the brightness temperature of the sun if a 40-dB gain X-band antenna is pointed directly at the sun.

Solution The spectral flux density from the "quiet" sun at 10 GHz is about 3×10^{-20} W/m² Hz [16] and can increase by orders of magnitude during solar storms. We use 10^{-19} as a reasonable value.

The sun's disk subtends approximately 1/2°, giving a subtended solid angle of 6.0×10^{-5} steradians. Then the sun's brightness is

$$b = \frac{10^{-19}}{6.0 \times 10^{-5}} = 1.67 \times 10^{-15}$$

and its brightness temperature (at 10 GHz) is

$$T_b = \frac{b\lambda^2}{2k} = \frac{(1.67 \times 10^{-15})(0.03)^2}{2(1.38 \times 10^{-23})} = 54{,}400 \text{ K}$$

The effective area of an antenna having gain 40 dB is

$$A_e = \frac{(0.03)^2}{4\pi}(10^4) = 0.716$$

This gives a received spectral power density

$$\frac{W}{B} = \frac{1}{2}A_e \mathscr{S} = 0.358 \times 10^{-19}$$

and an antenna temperature

$$T_a = \frac{W}{kB} = \frac{0.358 \times 10^{-19}}{1.38 \times 10^{-23}} = 2595 \text{ K}$$

Note that the antenna temperature is proportional to the antenna effective area, and so is the received signal power. If the antenna beamwidth is greater than the solid angle subtended by the noisy radiator, the signal-to-noise ratio is independent of antenna gain.

It is desirable to postpone further discussion of antenna temperature until the noise introduced by losses in the atmosphere and radome and in the line connecting antenna to receiver are considered. This is best done by consideration of the *noise figure* of a two-port network.

Noise Figure

The noise figure of a linear two-port system, such as an amplifier, a lossy line, or even a wave propagation path that attenuates the wave, is defined as

$$F = \frac{S_{in}/N_{in}}{S_{out}/N_{out}} \qquad (2.101)$$

where

S_{in} = available signal power at input
N_{in} = available noise power at input
S_{out} = available signal power at output
N_{out} = available noise power at output

and S/N is the signal-to-noise power ratio. The two-port network is assumed to be matched to the source impedance at the input and to the load impedance at the output.

The input noise power is

$$N_{in} = kTB$$

where B is the bandwidth. If the network gain is G (which may be greater than or less than 1), both input signal power and input noise power are multiplied by G at the output. In addition, the network adds noise ΔN. In early radars, RF amplifiers produced a considerable amount of noise, and it was common to use a mixer following the antenna. Modern amplifiers using traveling-wave tubes, parametric amplifiers, or masers create a relatively small amount of noise and are desirable as the first element after the antenna.

FIGURE 2.24 Two networks in cascade.

With this input noise power and the additional noise power generated in the network, the noise figure becomes

$$F = \frac{S_{in}}{S_{out}} \frac{N_{out}}{N_{in}} = \frac{1}{G} \frac{GN_{in} + \Delta N}{N_{in}} = \frac{kTBG + \Delta N}{kTBG}$$

In order to standardize the definition of noise figure, T is specified as $T = T_0 = 290$ K in all measurements of network noise figure. Then the noise figure of a linear two-port network is

$$F = 1 + \frac{\Delta N}{kT_0 BG} \qquad (2.102)$$

Consider two impedance-matched networks in cascade, as in Figure 2.24. Both have the same bandwidth B, but they have different gains and noise figures. It is clear that

$$S_{out} = G_1 G_2 S_{in}$$

Also, the noise at the input of network 2 is $F_1 G_1 k T_0 B$. This noise is multiplied by G_2 and to it is added, from (2.102),

$$\Delta N_2 = (F_2 - 1) k T_0 B G_2$$

Then the overall noise figure of the cascaded networks is

$$F_0 = \frac{S_{in}}{S_{out}} \frac{N_{out}}{N_{in}} = \frac{1}{G_1 G_2} \frac{F_1 G_1 G_2 k T_0 B + (F_2 - 1) k T_0 B G_2}{k T_0 B}$$

$$= F_1 + \frac{F_2 - 1}{G_1} \qquad (2.103)$$

This procedure can be extended to N cascaded, impedance-matched networks to give the overall noise figure

$$F_0 = F_1 + \frac{F_2 - 1}{G_1} + \frac{F_3 - 1}{G_1 G_2} + \cdots + \frac{F_N - 1}{G_1 G_2 \ldots G_{N-1}} \qquad (2.104)$$

Effective Noise Temperature of a Two-Port Network

The input noise to a two-port network to be considered in the definition of network noise figure is

$$N_{in} = kT_0 B$$

and the noise power added by the network is ΔN. This added noise power can be regarded as being produced by an input resistance at temperature T_e connected to a noise-free equivalent network with the same gain. Then from

$$\Delta N = kT_e BG$$

and the definition (2.102) the noise figure is

$$F = 1 + \frac{kT_e BG}{kT_0 BG} = 1 + \frac{T_e}{T_0} \qquad (2.105)$$

Temperature T_e is called the *effective noise temperature* of the network. For cascaded networks with effective temperatures $T_1, T_2, \ldots, T_N$, the overall effective temperature can be shown, in the same manner as the overall noise figure, to be

$$T_e = T_1 + \frac{T_2}{G_1} + \cdots + \frac{T_N}{G_1 G_2 \ldots G_{N-1}} \qquad (2.106)$$

Noise Figure and Temperature of Lossy Medium

In the discussion of noise figure and noise temperature of a two-port network, no consideration was given to the amount of noise produced by the network. For one important problem, a lossy network or lossy space region through which the wave travels, this noise can be determined from the attenuation of the network or space region.

Consider once more an antenna surrounded by a blackbody. The region between the blackbody and antenna is filled with an absorbing material that attenuates a wave traveling through it. The blackbody, antenna, and absorbing material are in thermal equilibrium at temperature T.

The noise power available to a polarized antenna from the blackbody is kTB and after it passes through the lossy region the noise power at the antenna is

$$kTBG = \frac{kTB}{L}$$

where G is the gain (less than 1) and L, the inverse of G, is the loss. The power absorbed by the region is $kTB - kTBG$. This power is then radiated by the lossy region, and ΔN in the equation for noise figure is

$$\Delta N = kTB(1 - G)$$

Therefore the noise figure of the lossy region is

$$F = 1 + \frac{\Delta N}{kT_0 BG} = 1 + \frac{T}{T_0}\frac{1 - G}{G} \tag{2.107}$$

where again it is noted that T is the actual temperature of the system. If this expression is equated to (2.105), relating effective temperature T_e to noise figure,

$$F = 1 + \frac{T_e}{T_0} = 1 + \frac{T}{T_0}\frac{1 - G}{G}$$

it is seen that

$$T_e = T\left(\frac{1}{G} - 1\right) = T(L - 1) \tag{2.108}$$

This equation shows that the effective temperature of a lossy region may be greater than its real temperature.

If the lossy region is at temperature T_0, the noise figure becomes

$$F = 1 + \frac{1 - G}{G} = \frac{1}{G} = L \tag{2.109}$$

but (2.107) is the more general form.

The region under consideration may be the atmosphere. Then the loss mechanism is the absorption of RF energy by water vapor or oxygen. Another lossy region between radiator and antenna is a lossy radome. The loss may also be due to an RF device, such as a mixer or lossy transmission line, after the wave is received by the antenna, or the lossy antenna itself. In all these cases the effective temperature is given by (2.108). If the two-port network is a transmission line following the antenna, matched both to the antenna and load by lossless matching networks, the gain G for use in (2.108) is given by

$$G = e^{-2\alpha d}$$

where d is the line length and

$$\alpha + j\beta = [(R + j\omega L)(G + j\omega C)]^{1/2}$$

with R, L, G, and C the line parameters.

Example of Effective Antenna Temperature

We are now in a position to put the previous concepts together and approximate the antenna temperature for a realistic example. Consider the 40-dB gain antenna at 10 GHz used in the example of antenna temperature caused by the sun. It is pointed so that noise from the galactic center is not a factor, and it is not directed toward the sun, moon, or radio star.

It is necessary first to find the proportion of the total power that would be received in the main beam of the antenna if power were radiated toward the antenna with equal intensity from all directions. This is called the *beam efficiency* of the antenna and is also equal to the main-beam power radiated by the antenna divided by the total radiated power, or

$$e_b = \frac{\int_{\text{main beam}} \int G(\theta, \phi) \, d\Omega}{\int\int_{4\pi} G(\theta, \phi) \, d\Omega} \qquad (2.110)$$

Beam efficiencies for rectangular tapered-feed apertures have been obtained [19]. A tapered-feed aperture has a greater feed voltage or current amplitude at the aperture center than at the edges. Tapering the feed reduces sidelobe amplitudes and broadens the main beam compared to a uniform feed distribution. For a square aperture lying in the xy plane with center at the origin, the feed amplitude for a commonly used taper is

$$V(x, y) = \left\{ K_1 + K_2 \left[1 - \left(\frac{2x}{D}\right)^2 \right] \right\} \left\{ K_1 + K_2 \left[1 - \left(\frac{2y}{D}\right)^2 \right] \right\} \qquad (2.111)$$

where D is the length of one side of the aperture. For $K_2 = 0$ (no taper) the sidelobes are unacceptably high for radar applications, and for $K_1 = 0$ (full taper with $V = 0$ at the aperture edge) the main beam is in general too broad. A reasonable choice is $K_1 = K_2$, so that

$$V(x, y) = 4\left[1 - 2\left(\frac{x}{D}\right)^2 \right]\left[1 - 2\left(\frac{y}{D}\right)^2 \right]$$

The beam efficiency is critically dependent on the rms phase front displacement from planar over the aperture. For the ideal condition of zero displacement, the beam efficiency was found by Nash to be 0.93, and it dropped to 0.87 for a relatively small displacement [19]. We use the second value in this example.

We assume then that 0.87 of the antenna radiated (or received) power is in the main beam and 0.13 in the sidelobes, with half of the sidelobes pointed toward the sky and half toward the ground.

Cosmic Noise At 10 GHz, cosmic noise is negligible for most purposes. Nevertheless, to show how it is treated, we assume a brightness temperature of 4 K. Further, assume that atmospheric losses are 0.2 dB → 0.955, dissipative losses in the radome are 0.5 dB → 0.891, and dissipative losses in the antenna are 0.1 dB → 0.977. If there were no atmospheric, radome, or antenna losses, the equivalent antenna temperature, with cosmic noise received by a portion of the beam comprising $0.87 + 0.13/2 = 0.935$ of the total, would be

$$T_a = \frac{1}{4\pi}\int\int_\Omega T_b G\, d\Omega = \frac{T_b}{4\pi}\int\int_\Omega G\, d\Omega = 0.935 T_b$$

Since the cosmic noise is attenuated by the atmosphere, radome, and antenna, its contribution to the antenna temperature is

$$T_a = (0.935)(0.955)(0.891)(0.977) T_b = 3.1 \text{ K}$$

Atmospheric Loss The equation for effective temperature of a lossy medium

$$T_e = T\left(\frac{1}{G} - 1\right)$$

where T is the ambient temperature, is difficult to apply to the atmosphere. At high altitudes, the temperature is quite high, but the molecular density and the loss per unit length are small. It is therefore customary to use measurements of the atmospheric brightness temperature, and at 10-GHz temperatures measured by a completely polarized antenna range from about 6 K in the direction of the zenith to about 140 K in the direction of the horizon [20]. We use 20 K as the brightness temperature. Note that this is the effective temperature, and the atmospheric loss $L (= 1/G)$ is not used. Then the atmospheric loss contribution to the antenna temperature, following the procedure used for the cosmic noise, including radome and antenna attenuation but not atmospheric attenuation, is

$$T_a = (0.935)(0.891)(0.977)(20) = 16.3 \text{ K}$$

Radome Loss Let the radome temperature be 290 K. For a radome loss of 0.5 dB, its effective brightness temperature is

$$T_e = 290\left(\frac{1}{0.891} - 1\right) = 35.5 \text{ K}$$

The corresponding antenna temperature, for a completely-polarized antenna, is

$$T_a = \frac{1}{4\pi}\int\int_\Omega T_e G\, d\Omega$$

Since the radome is seen by the main beam and all sidelobes, and its noise is attenuated by the antenna because it is not perfectly efficient,

$$T_a = (0.977)(35.5) = 34.7 \text{ K}$$

Noise from Earth The earth is a good absorber of microwave radiation and therefore a good radiator at microwave frequencies. Note that it is seen by half the sidelobes in this model, and if the ground temperature is 290 K, the effective antenna temperature as seen by a completely polarized antenna, taking radome and antenna losses into account, is

$$T_a = (0.065)(0.891)(0.977)(290) = 16.4 \text{ K}$$

Antenna Losses The antenna itself has losses, and it is reasonable also to assign RF losses between the antenna and first circuit element (amplifier or mixer) to the antenna. If these losses are 0.1 dB and the ambient temperature is 290 K the effective temperature due to them is

$$T_a = T_e = 290 \left(\frac{1}{0.977} - 1 \right) = 6.8 \text{ K}$$

The total antenna temperature caused by all these sources is then

$$T_a = 3.1 + 16.3 + 34.7 + 16.4 + 6.8 = 77.3 \text{ K}$$

Of these five temperature components the last three are most easily controlled by the designer. To minimize noise, the radome, antenna, and RF circuitry should be made as lossless as possible. Earth noise is significant and can be minimized by designing for low sidelobes and, if appropriate, by using conducting screens to cover the earth region seen by the sidelobes.

By way of comparison to this effective antenna temperature, reasonable values of noise figure and noise temperature of three receiver front ends at 10 GHz are:

	F dB	T_e (K)
Maser	0.8	58
Parametric amplifier	4.5	528
Crystal mixer	7.5	1340

The example indicates that the antenna effective temperature is relatively unimportant if the receiver front end is a mixer (except for radiation from the sun), but with a maser amplifier, a strong effort should be made to minimize input noise to the antenna.

The antenna temperature and its effect on the receiver system depend strongly on frequency. At a lower frequency, cosmic noise and atmospheric absorption noise will be greater than for this 10-GHz example, and the effective temperature of a parametric amplifier may be much lower than that found here. Thus the design criteria may be quite different at 1 GHz and 10 GHz.

REFERENCES

1. C. A. Balanis, *Advanced Engineering Electromagnetics*, Wiley, New York, 1989.
2. R. F. Harrington, *Field Computation by Moment Methods*, Macmillan, New York, 1968.
3. R. S. Elliott, *Antenna Theory and Design*, Prentice-Hall, Englewood Cliffs, NJ, 1981.
4. M. Born and E. Wolf, *Principles of Optics*, 3rd ed., Pergamon Press, New York, 1965.
5. C. T. Tai, *Dyadic Green's Functions in Electromagnetic Theory*, Intext, Scranton, PA, 1971.
6. W. H. von Aulock, "Properties of Phased Arrays," *Proc. IRE*, **48**(10), 1715–1727, October 1960.
7. R. T. Compton, Jr., *Adaptive Antennas*, Prentice-Hall, Englewood Cliffs, NJ, 1988.
8. B. Widrow, P. E. Mantey, L. J. Griffiths, and B. B. Goode, "Adaptive Antenna Systems," *Proc. IEEE*, **55**(12), 2143–2159, December 1967.
9. V. H. Rumsey, "Frequency Independent Antennas," *IRE National Convention Record, Part I*, 114–118, March 1957.
10. R. H. DuHamel and D. E. Isbell, "Broadband Logarithmically Periodic Antenna Structures," *IRE National Convention Record*, 119–128, March 1957.
11. J. D. Kraus, *Antennas*, 2nd ed., McGraw-Hill, New York, 1988.
12. C. A. Balanis, *Antenna Theory*, Harper & Row, New York, 1982.
13. G. H. Brown and O. M. Woodward, "Experimentally Determined Radiation Characteristics of Conical and Triangular Antennas," *RCA Review*, **13**, 425–452, December 1952.
14. S. B. Cohn, "Properties of the Ridge Wave Guide," *Proc. IRE*, **35**(8), 783–789, August 1947.
15. A. A. Polkinghorne and J. H. Weitenhagen, *Radio Frequency Simulation System—System Design Handbook*, Document No. D243-10004-1, Boeing Aerospace Company, Huntsville, AL, 1975.
16. M. I. Skolnik, *Introduction to Radar Systems*, McGraw-Hill, New York, 1962.
17. J. D. Kraus, *Radio Astronomy*, McGraw-Hill, New York, 1966.
18. H. Nyquist, "Thermal Agitation of Electric Charge in Conductors," *Phys. Rev.*, **32**(1), 110–113, July 1928.
19. R. T. Nash, "Beam Efficiency Limitations of Large Antennas," *IEEE Transactions on Antennas and Propagation*, **AP-12**(7), 918–923, December 1964.
20. J. C. Greene and M. T. Lebenbaum, *Microwave J.*, **2**, 13–14, October 1959.

PROBLEMS

2.1. A five-element linear array of isotropic elements has a spacing of $\lambda/2$ between elements. Feed currents to all elements have the same amplitude, and the phase advance of the currents from one element to the next is 30°. Find angle α, measured from the line of the array to the main beam maximum.

2.2. For Problem 2.1, find the angles at which the array pattern is zero.

2.3. Prove that

$$\sum_{0}^{N-1} e^{jn\Psi} = \frac{\sin\left(N\frac{\Psi}{2}\right)}{\sin\left(\frac{\Psi}{2}\right)}$$

2.4. Find the angles of all pattern nulls for an N-element linear array of isotropic sources with all elements fed by equal-amplitude current sources with constant phase difference δ. Let

$$\Psi = kd \cos \alpha + \delta$$

and use the result of Problem 2.3.

2.5. For the N-element linear array of Problem 2.4 find the approximate angular locations of all sidelobe maxima.

2.6. Plot the field pattern of a linear array of four isotropic point sources spaced a half wavelength apart and fed by equal in-phase currents. Plot the field pattern of the array if the two inside elements are fed by currents greater by a factor of 3 than the outside elements; that is, the relative currents in the elements are 1 3 3 1. This array is called a *binomial array*. An array of any number of elements fed by currents proportional to the binomial coefficients has no sidelobes.

2.7. A four-element ordinary end-fire array has a spacing of $\lambda/2$ and phase angle $\delta = -\pi$. Find angle α, measured from the line of the array, to the first side-lobe maximum. Find the value of the side-lobe maximum, relative to the main beam maximum.

2.8. A two-element broadside array with $d = \lambda/2$ and $\delta = 0$ has short dipole elements oriented perpendicular to the line of the array. Sketch the field pattern in a plane containing the dipoles. Sketch the pattern in a plane containing the line of the array and transverse to the dipoles.

2.9. A four-element broadside array with half-wave spacing has equal feed currents with the same phase for all elements. Find the half power beamwidth.

2.10. A planar 6×6 array lying in the xy plane has element spacing $d_x = d_y = \lambda/2$. Feed current amplitudes are all equal. It is desired to point the main beam at polar angle $\theta = 15°$ and azimuth angle $\phi = 45°$. Find the feed phases δ_x and δ_y.

2.11. Why is it necessary to use two Taylor's series terms to approximate the numerator terms of (2.44) while only one term suffices for the denominator?

2.12. In (2.44) let $M = 8$, $N = 6$, $d_x = d_y = \lambda/2$. Find the beam contour on the $\Psi_x - \Psi_y$ plane for $F = 0.3$.

2.13. The spectral flux density from the sun, in a noisy state, is 10^{-18} W/m² Hz at a frequency of 1 GHz. Find the equivalent antenna temperature of a 30-dB gain antenna pointed at the sun. Find the total power received in a 1-MHz bandwidth.

2.14. At a frequency of 20 GHz, atmospheric attenuation due to water vapor is 0.1 dB/km. Find the antenna temperature caused by this loss for a radar that observes a target at a range of 4 km.

2.15. For the text example of a 40-dB gain antenna at 10 GHz operating with an effective antenna temperature of 77.3 K, assume a noise bandwidth of 1 MHz. Find the power density of a wave incident on this receiving antenna from the direction of maximum gain that gives a signal-to-noise ratio of 15 dB. Note that combined losses of radome and antenna that affect the signal are 0.6 dB.

2.16. A four-element linear array has a tapered feed with feed currents

$$I_n = 1 - 2\left(\frac{\rho_n}{L}\right)^2$$

where L is the array length and ρ_n is the distance from array center to element n. Sketch the pattern in a plane containing the array if all elements are fed in phase. Compare the half-power beamwidth and sidelobe maximum to those of the uniformly fed array of Problem 2.9.

CHAPTER THREE

Representations of Wave Polarization

3.1. INTRODUCTION

It is necessary to have a good foundation in polarization theory to understand all of the principles of antennas and scattering by a target. This chapter is intended to provide such a foundation. Chapters 4 and 5 then apply polarization concepts to antenna matching and consider the polarization characteristics of some useful antennas.

The electric vector of a harmonic plane wave traces an ellipse in the transverse plane with time. In this chapter we develop the equation of the ellipse for a general, nonplane wave and consider the ellipse and the behavior of the field vectors in detail for a plane wave. The parameters commonly used to describe wave polarization, namely the linear and circular polarization ratios, the ellipse axial ratio, tilt angle, and rotation sense, and the Stokes parameters, are introduced and related to each other. A polarization chart based on the familiar Smith chart of transmission line theory is discussed. The Poincaré sphere is introduced, and mapping of the sphere onto a complex plane is described. Conformal and equal-area mapping of the Poincaré sphere are discussed, and equations are given for Aitoff–Hammer, Mollweide, and other useful maps.

3.2. THE GENERAL HARMONIC WAVE

In this section we show that a general (nonplanar) harmonic wave is elliptically polarized and find the equation of the polarization ellipse [1].

A nonpolar single-frequency wave with components

$$\mathscr{E}_i(\mathbf{r}, t) = a_i(\mathbf{r})\cos[\omega t - g_i(\mathbf{r})] \qquad i = 1, 2, 3 \qquad (3.1)$$

where a_i and g_i are real, may be written as

$$\mathscr{E}(\mathbf{r}, t) = \sum_1^3 \mathbf{u}_i a_i(\mathbf{r}) \cos[\omega t - g_i(\mathbf{r})]$$

$$= \sum_1^3 \mathbf{u}_i a_i(\mathbf{r}) \operatorname{Re}[e^{j[\omega t - g_i(\mathbf{r})]}] \tag{3.2}$$

and if

$$E_i(\mathbf{r}) = a_i(\mathbf{r}) e^{-j g_i(\mathbf{r})} \tag{3.3}$$

is the complex, time-invariant term associated with each real, time-varying electric field component, then

$$\mathscr{E}(\mathbf{r}, t) = \operatorname{Re}\left[\sum_1^3 \mathbf{u}_i E_i(\mathbf{r}) e^{j\omega t}\right] \tag{3.4}$$

where the $\mathbf{u}_i$ are real orthogonal unit vectors. For a plane wave the phase term is given by

$$g_i(\mathbf{r}) = \mathbf{k} \cdot \mathbf{r} - \delta_i \tag{3.5}$$

but at this point we are not restricted to plane waves.

If a complex vector representing the field is defined as

$$\mathbf{E}(\mathbf{r}) = \mathbf{E}'(\mathbf{r}) + j\mathbf{E}''(\mathbf{r}) = \sum_1^3 \mathbf{u}_i E_i(\mathbf{r}) \tag{3.6}$$

the harmonic vector field may be written as

$$\mathscr{E}(\mathbf{r}, t) = \operatorname{Re}[\mathbf{E}(\mathbf{r}) e^{j\omega t}] \tag{3.7}$$

Let us assume that $\mathbf{E}$ can be transformed to a new set of axes defined by the orthogonal real vectors $\mathbf{m}$ and $\mathbf{n}$, using the relation

$$\mathbf{E} = \mathbf{E}' + j\mathbf{E}'' = (\mathbf{m} + j\mathbf{n}) e^{j\theta} \tag{3.8}$$

Equating real and imaginary parts of this equation yields

$$\begin{aligned} \mathbf{E}' &= \mathbf{m} \cos \theta - \mathbf{n} \sin \theta \\ \mathbf{E}'' &= \mathbf{m} \sin \theta + \mathbf{n} \cos \theta \end{aligned} \tag{3.9}$$

and solving for $\mathbf{m}$ and $\mathbf{n}$ leads to

$$\begin{aligned} \mathbf{m} &= \mathbf{E}' \cos \theta + \mathbf{E}'' \sin \theta \\ \mathbf{n} &= -\mathbf{E}' \sin \theta + \mathbf{E}'' \cos \theta \end{aligned} \tag{3.10}$$

If we assume, without loss of generality, that $|\mathbf{m}| \geq |\mathbf{n}|$, and require the orthogonality condition $\mathbf{m} \cdot \mathbf{n} = 0$, these equations can be solved to give

$$\tan 2\theta = \frac{2\mathbf{E}' \cdot \mathbf{E}''}{|\mathbf{E}'|^2 - |\mathbf{E}''|^2} \tag{3.11}$$

Since $\tan(2\theta)$ as given by this equation is real, the assumed transformation may be carried out.

Next (3.8) is substituted into (3.7) to find the real field components. This gives

$$\mathscr{E} = \text{Re}[\mathbf{E} e^{j\omega t}] = \text{Re}\left[(\mathbf{m} + j\mathbf{n}) e^{j\theta} e^{j\omega t}\right] \tag{3.12}$$

and since $\mathbf{m}$ and $\mathbf{n}$ are real,

$$\mathscr{E} = \mathbf{m}\cos(\omega t + \theta) - \mathbf{n}\sin(\omega t + \theta) \tag{3.13}$$

If at each field point we set up a local coordinate system with two of the axes directed along $\mathbf{m}$ and $\mathbf{n}$, the field components are

$$\mathscr{E}_m = m\cos(\omega t + \theta) \tag{3.14a}$$

$$\mathscr{E}_n = -n\sin(\omega t + \theta) \tag{3.14b}$$

$$\mathscr{E}_3 = 0 \tag{3.14c}$$

where $m = |\mathbf{m}|$ and $n = |\mathbf{n}|$. Subscript 3 refers to the third of the three coordinates. Again it is noted that these are local coordinates. For a plane wave they are the same at all points, but not for a wave in general.

The equations for the field components can be combined to give

$$\frac{\mathscr{E}_m^2}{m^2} + \frac{\mathscr{E}_n^2}{n^2} = \cos^2(\omega t + \theta) + \sin^2(\omega t + \theta) = 1 \tag{3.15}$$

This is the equation of an ellipse, in the plane defined by $\mathbf{m}$ and $\mathbf{n}$, with semimajor and semiminor axes m and n. The field intensity ellipse is shown in Figure 3.1. The field vector terminates on the ellipse, since its components $\mathscr{E}_m$ and $\mathscr{E}_n$ are not independent but obey the ellipse equation (3.15). The direction of $\mathscr{E}$ changes with time as its tip moves around the ellipse with a direction and velocity to be determined in another section.

We see then that any harmonic wave, planar or nonplanar, is elliptically polarized. The plane of the ellipse and its shape and orientation in that plane are functions of the coordinates of the field point, but not of time.

This development has been concerned with the time-varying electric field, but it is clear that the magnetic field is also elliptically polarized (see Problem 3.7).

118 REPRESENTATIONS OF WAVE POLARIZATION

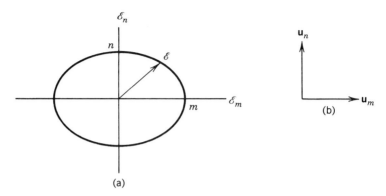

FIGURE 3.1 The polarization ellipse: (*a*) field intensity coordinates; (*b*) space coordinates.

3.3. POLARIZATION ELLIPSE FOR PLANE WAVES

A plane wave traveling in the *z* direction

$$\mathbf{E} = (E_x \mathbf{u}_x + E_y \mathbf{u}_y) e^{-jkz} \tag{3.16}$$

results if the phase term (3.5) is used and if the vector propagation constant is

$$\mathbf{k} = \mathbf{u}_z k \tag{3.17}$$

Both field components are complex and may be written as

$$E_x = |E_x| e^{j\phi_x} \qquad E_y = |E_y| e^{j\phi_y} \tag{3.18}$$

so that

$$\mathbf{E} = (\mathbf{u}_x |E_x| e^{j\phi_x} + \mathbf{u}_y |E_y| e^{j\phi_y}) e^{-jkz} \tag{3.19}$$

and the time-varying field is

$$\mathscr{E} = \mathrm{Re}(\mathbf{E} e^{j\omega t}) = \mathbf{u}_x |E_x| \cos(\omega t - kz + \phi_x) + \mathbf{u}_y |E_y| \cos(\omega t - kz + \phi_y) \tag{3.20}$$

If

$$\beta = \omega t - kz \tag{3.21}$$

POLARIZATION ELLIPSE FOR PLANE WAVES

the time-varying field components become

$$\frac{\mathscr{E}_x}{|E_x|} = \cos\beta \cos\phi_x - \sin\beta \sin\phi_x \tag{3.22a}$$

$$\frac{\mathscr{E}_y}{|E_y|} = \cos\beta \cos\phi_y - \sin\beta \sin\phi_y \tag{3.22b}$$

Multiplying and subtracting as indicated leads to

$$\frac{\mathscr{E}_x}{|E_x|} \sin\phi_y - \frac{\mathscr{E}_y}{|E_y|} \sin\phi_x = \cos\beta \sin(\phi_y - \phi_x) \tag{3.23a}$$

$$\frac{\mathscr{E}_x}{|E_x|} \cos\phi_y - \frac{\mathscr{E}_y}{|E_y|} \cos\phi_x = \sin\beta \sin(\phi_y - \phi_x) \tag{3.23b}$$

Squaring and adding these terms gives

$$\frac{\mathscr{E}_x^2}{|E_x|^2} - 2\frac{\mathscr{E}_x}{|E_x|}\frac{\mathscr{E}_y}{|E_y|}\cos(\phi_y - \phi_x) + \frac{\mathscr{E}_y^2}{|E_y|^2} = \sin^2(\phi_y - \phi_x) \tag{3.24}$$

This is the equation of a conic, and we have already seen in a more general case that it represents an ellipse. Let

$$\phi = \phi_y - \phi_x \tag{3.25}$$

and the ellipse equation becomes

$$\frac{\mathscr{E}_x^2}{|E_x|^2} - 2\frac{\mathscr{E}_x}{|E_x|}\frac{\mathscr{E}_y}{|E_y|}\cos\phi + \frac{\mathscr{E}_y^2}{|E_y|^2} = \sin^2\phi \tag{3.26}$$

From (3.22), which can be rewritten as

$$\frac{\mathscr{E}_x}{|E_x|} = \cos(\beta + \phi_x) \tag{3.27a}$$

$$\frac{\mathscr{E}_y}{|E_y|} = \cos(\beta + \phi_y) \tag{3.27b}$$

it is clear that the greatest values of $\mathscr{E}_x$ and $\mathscr{E}_y$ are, respectively, $|E_x|$ and $|E_y|$. Then the ellipse of the time-varying field components can be inscribed in a rectangle with sides parallel to the x and y axes and dimensions $2|E_x|$ and $2|E_y|$ as shown in Figure 3.2.

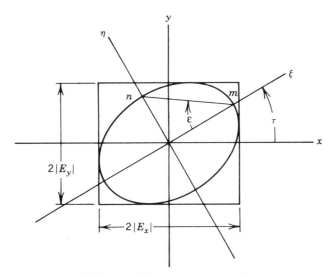

FIGURE 3.2 Tilted polarization ellipse.

From the last equation, $\mathscr{E}_x$ is maximum if

$$\beta + \phi_x = 0$$

$$\beta + \phi_y = \beta + \phi_x + (\phi_y - \phi_x) = \phi$$

and $\mathscr{E}_y$ is maximum if

$$\beta + \phi_y = 0 \qquad \beta + \phi_x = -\phi$$

The ellipse of Figure 3.2 intersects the sides of the rectangle at $\pm|E_x|$, $\pm|E_y|\cos\phi$ and $\pm|E_x|\cos\phi, \pm|E_y|$.

The angle τ of Figure 3.2, measured from the x axis, is called the *tilt angle* of the polarization ellipse. It may be defined over any range of π radians, but perhaps the most common range is

$$-\frac{\pi}{2} \leq \tau \leq \frac{\pi}{2} \tag{3.28}$$

Finding τ is relatively complicated. From Figure 3.3,

$$\xi = x \cos\tau + y \sin\tau \tag{3.29a}$$

$$\eta = y \cos\tau - x \sin\tau \tag{3.29b}$$

POLARIZATION ELLIPSE FOR PLANE WAVES

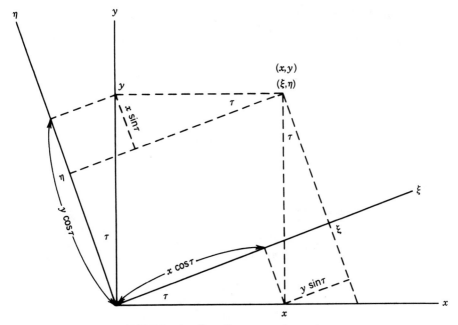

FIGURE 3.3 Coordinate transformations.

and the field components transform as

$$\mathscr{E}_\xi = \mathscr{E}_x \cos \tau + \mathscr{E}_y \sin \tau \tag{3.30a}$$

$$\mathscr{E}_\eta = -\mathscr{E}_x \sin \tau + \mathscr{E}_y \cos \tau \tag{3.30b}$$

Now the field components are also given by

$$\mathscr{E}_\xi = m \cos(\beta + \phi_0) \tag{3.31a}$$

$$\mathscr{E}_\eta = \pm n \sin(\beta + \phi_0) \tag{3.31b}$$

where m and n are the positive semiaxes of the polarization ellipse and ϕ_0 is some phase angle. That this is a correct representation is easily seen by noting that it satisfies

$$\frac{\mathscr{E}_\xi^2}{m^2} + \frac{\mathscr{E}_\eta^2}{n^2} = 1$$

In (3.31) $\mathscr{E}_\eta$ carries the plus or minus sign since the rotation sense of $\mathscr{E}$ has not yet been determined. If we set $\beta + \phi_0 = 0$ in (3.31) and then allow $\beta (= \omega t - kz)$ to increase infinitesimally, we see that the plus sign corresponds to counterclockwise rotation of $\mathscr{E}$ (as we look at Fig. 3.2) as β (or time) increases, and the minus sign to clockwise rotation.

122 REPRESENTATIONS OF WAVE POLARIZATION

Equating (3.30) and (3.31):

$$\mathscr{E}_\xi = m\cos(\beta + \phi_0) = \mathscr{E}_x \cos\tau + \mathscr{E}_y \sin\tau \quad (3.32a)$$
$$\mathscr{E}_\eta = \pm n\sin(\beta + \phi_0) = -\mathscr{E}_x \sin\tau + \mathscr{E}_y \cos\tau \quad (3.32b)$$

Expanding the left side of these two equations and using on the right the wave components given by (3.22) leads to

$$\begin{aligned}m(\cos\beta\cos\phi_0 &- \sin\beta\sin\phi_0)\\ = |E_x|(\cos\beta\cos\phi_x &- \sin\beta\sin\phi_x)\cos\tau\\ + |E_y|(\cos\beta\cos\phi_y &- \sin\beta\sin\phi_y)\sin\tau\end{aligned} \quad (3.33a)$$

$$\begin{aligned}\pm n(\sin\beta\cos\phi_0 &+ \cos\beta\sin\phi_0)\\ = -|E_x|(\cos\beta\cos\phi_x &- \sin\beta\sin\phi_x)\sin\tau\\ + |E_y|(\cos\beta\cos\phi_y &- \sin\beta\sin\phi_y)\cos\tau\end{aligned} \quad (3.33b)$$

Equating the coefficients of $\cos\beta$ and $\sin\beta$ in these equations leads to

$$m\cos\phi_0 = |E_x|\cos\phi_x \cos\tau + |E_y|\cos\phi_y \sin\tau \quad (3.34a)$$
$$m\sin\phi_0 = |E_x|\sin\phi_x \cos\tau + |E_y|\sin\phi_y \sin\tau \quad (3.34b)$$
$$\pm n\cos\phi_0 = |E_x|\sin\phi_x \sin\tau - |E_y|\sin\phi_y \cos\tau \quad (3.34c)$$
$$\pm n\sin\phi_0 = -|E_x|\cos\phi_x \sin\tau + |E_y|\cos\phi_y \cos\tau \quad (3.34d)$$

Squaring and adding these equations results in

$$m^2 + n^2 = |E_x|^2 + |E_y|^2 \quad (3.35)$$

Next we multiply the first and third equations of (3.34) and also the second and fourth, and add the products, obtaining

$$\pm mn = -|E_x||E_y|\sin\phi \quad (3.36)$$

Dividing the third equation by the first and the fourth by the second gives

$$\begin{aligned}\pm\frac{n}{m} &= \frac{|E_x|\sin\phi_x \sin\tau - |E_y|\sin\phi_y \cos\tau}{|E_x|\cos\phi_x \cos\tau + |E_y|\cos\phi_y \sin\tau}\\ &= \frac{-|E_x|\cos\phi_x \sin\tau + |E_y|\cos\phi_y \cos\tau}{|E_x|\sin\phi_x \cos\tau + |E_y|\sin\phi_y \sin\tau}\end{aligned} \quad (3.37)$$

Cross multiplying and collecting terms in the last equation gives

$$\left(|E_x|^2 - |E_y|^2\right)\sin 2\tau = 2|E_x||E_y|\cos 2\tau \cos\phi \quad (3.38)$$

If the auxiliary angle α is defined by

$$\tan \alpha = \frac{|E_y|}{|E_x|} \qquad 0 \leq \alpha \leq \frac{\pi}{2} \qquad (3.39)$$

then tilt angle τ may be found from

$$\tan 2\tau = \tan 2\alpha \cos \phi \qquad (3.40)$$

If $\tan \alpha = 1$, the tilt angle cannot be found in this way. It is best found from a circular polarization ratio to be defined in a later section.

To find the axial ratio of the ellipse and the rotation sense of the $\mathscr{E}$ vector, we define another auxiliary angle by

$$\tan \epsilon = \mp \frac{n}{m} \qquad -\frac{\pi}{4} \leq \epsilon \leq \frac{\pi}{4} \qquad (3.41)$$

If this equation is combined with (3.35) and (3.36), the result, which allows ϵ to be found from the field components, is

$$\sin 2\epsilon = \frac{2|Ex||E_y|}{|E_x|^2 + |E_y|^2} \sin \phi \qquad (3.42)$$

After ϵ is obtained from the field component magnitudes and phase difference the *axial ratio* m/n of the ellipse is readily found from (3.41).

The angle ϵ is the *ellipticity angle* of the polarization ellipse. It is shown in Figure 3.2 as a positive angle.*

Tilt and ellipticity angles are not sufficient to describe the wave; rotation sense is also necessary and will be determined next. The time-varying angle of $\mathscr{E}$, measured from the x axis toward the y axis is

$$\psi = \tan^{-1} \frac{\mathscr{E}_y}{\mathscr{E}_x} = \tan^{-1} \frac{|E_y|\cos(\beta + \phi_y)}{|E_x|\cos(\beta + \phi_x)} \qquad (3.43)$$

where $\beta = \omega t - kz$. Then

$$\frac{\partial \psi}{\partial \beta} = \frac{(|E_y|/|E_x|)[-\cos(\beta + \phi_x)\sin(\beta + \phi_y) + \sin(\beta + \phi_x)\cos(\beta + \phi_y)]}{[1 + |E_y|^2 \cos^2(\beta + \phi_y)/|E_x|^2 \cos^2(\beta + \phi_x)]\cos^2(\beta + \phi_x)} \qquad (3.44)$$

---

*The symbols for tilt angle (τ) and ellipticity angle (ϵ) were chosen so that their initial letters when spelled out would serve as a memory aid. The phase angle (ϕ) was selected for the same reason.

and at some particular value of β, say $\beta = 0$,

$$\frac{\partial \psi}{\partial \beta} = -\frac{|E_x||E_y|\sin\phi}{|E_x|^2 \cos^2\phi_x + |E_y|^2 \cos^2\phi_y} \tag{3.45}$$

From this form, it is seen that

$$\frac{\partial \psi}{\partial \beta} < 0 \quad 0 < \phi < \pi$$

$$> 0 \quad \pi < \phi < 2\pi \tag{3.46}$$

If we look in the direction of wave propagation, in this case the $+z$ direction, $\partial\psi/\partial\beta > 0$ corresponds to clockwise rotation of the $\mathscr{E}$ vector as β (or time) increases. By definition, we call this *right-handed* rotation of the vector. Conversely, a negative value of the derivative corresponds to counterclockwise or *left-handed* rotation. From (3.42) and (3.46) it is seen that the ellipticity angle ranges corresponding to right- and left-handed rotations are

$$\begin{aligned} \epsilon &< 0 \quad \text{right-handed rotation} \\ \epsilon &> 0 \quad \text{left-handed rotation} \end{aligned} \tag{3.47}$$

The ellipticity angle can also be obtained from the auxiliary angle used to find the tilt angle. From (3.39)

$$\sin 2\alpha = \frac{2|E_x||E_y|}{|E_x|^2 + |E_y|^2} \tag{3.48}$$

and, if this is used in (3.42), we get a simple equation,

$$\sin 2\epsilon = \sin 2\alpha \sin\phi \tag{3.49}$$

To summarize, from a knowledge of the field component amplitudes and their phase difference, the auxiliary angle α is found from (3.39), the tilt angle from (3.40), and the ellipticity angle from (3.49). The axial ratio is then obtained from the inverse of (3.41), and a decision about the rotation sense is made from (3.47).

If (3.41) is substituted into (3.31), the field components may be written as

$$\mathscr{E}_\xi = m\cos(\beta + \phi_0) \tag{3.50a}$$

$$\mathscr{E}_\eta = -m\tan\epsilon\cos(\beta + \phi_0 - \pi/2) \tag{3.50b}$$

POWER DENSITY

In phasor form, suppressing time and distance variation,

$$E_\xi = m e^{j\phi_0} \tag{3.51a}$$

$$E_\eta = jm \tan \epsilon \, e^{j\phi_0} \tag{3.51b}$$

These field components may be written in a vector form as

$$\mathbf{E}(\xi, \eta) = m \begin{bmatrix} 1 \\ j \tan \epsilon \end{bmatrix} e^{j\phi_0} = \frac{m}{\cos \epsilon} \begin{bmatrix} \cos \epsilon \\ j \sin \epsilon \end{bmatrix} e^{j\phi_0} \tag{3.52}$$

Although (3.30) relates time-varying fields in the two coordinate systems, it is clearly valid for phasor components, so the electric field intensity in xyz coordinates can be obtained from the components in $\xi\eta\zeta$ coordinates by

$$\mathbf{E}(x, y) = \begin{bmatrix} \cos \tau & -\sin \tau \\ \sin \tau & \cos \tau \end{bmatrix} \mathbf{E}(\xi, \eta) \tag{3.53}$$

Combining the last two equations allows the electric field of a harmonic plane wave to be written in terms of the tilt and ellipticity angles of its polarization ellipse by

$$\mathbf{E}(x, y) = \frac{m}{\cos \epsilon} \begin{bmatrix} \cos \tau & -\sin \tau \\ \sin \tau & \cos \tau \end{bmatrix} \begin{bmatrix} \cos \epsilon \\ j \sin \epsilon \end{bmatrix} e^{j\phi_0} \tag{3.54}$$

The common phase term is normally of little significance and may be omitted.

3.4. LINEAR AND CIRCULAR POLARIZATION

In the special cases of $|E_x| = 0$, $|E_y| = 0$, or $\phi = 0$, the polarization ellipse degenerates to a straight line and the wave is *linearly polarized*. The axial ratio is infinite, the tilt angle can be found by the standard equations, and rotation sense is meaningless.

If $|E_x| = |E_y|$ and $\phi = \pm\pi/2$, the axial ratio found from the equations given here becomes equal to 1, the polarization ellipse degenerates to a circle, and the wave is *circularly polarized*, right circular if $\phi = -\pi/2$ and left circular if $\phi = \pi/2$.

3.5. POWER DENSITY

From

$$\mathbf{E} = (E_x \mathbf{u}_x + E_y \mathbf{u}_y) e^{-jkz} \tag{3.16}$$

and the Maxwell equations we can find the magnetic field

$$\mathbf{H} = \frac{1}{Z_0}(-E_y \mathbf{u}_x + E_x \mathbf{u}_y)e^{-jkz} \tag{3.55}$$

where Z_0 is the characteristic impedance of the medium, defined by

$$Z_0 = \sqrt{\frac{\mu}{\epsilon}}$$

The complex Poynting vector is then

$$\mathscr{P}_c = \frac{1}{2}\mathbf{E} \times \mathbf{H}^* = \frac{|E_x|^2 + |E_y|^2}{2Z_0^*}\mathbf{u}_z \tag{3.56}$$

with the time-average Poynting vector given by

$$\mathscr{P} = \text{Re}[\mathscr{P}_c] \tag{3.57}$$

3.6. ROTATION RATE OF THE FIELD VECTOR

In the $z = 0$ plane, the time-varying electric field reduces to

$$\mathscr{E} = \mathbf{u}_x|E_x|\cos(\omega t + \phi_x) + \mathbf{u}_y|E_y|\cos(\omega t + \phi_y) \tag{3.58}$$

The angle ψ between $\mathscr{E}$ and the positive x axis is given as a function of time by

$$\psi = \tan^{-1}\left[\frac{|E_y|\cos(\omega t + \phi_y)}{|E_x|\cos(\omega t + \phi_x)}\right] \tag{3.59}$$

and the rate of increase of the angle with time is

$$\frac{\partial \psi}{\partial t} = \frac{-\omega|E_x||E_y|\sin\phi}{|E_x|^2 \cos^2(\omega t + \phi_x) + |E_y|^2 \cos^2(\omega t + \phi_y)} \tag{3.60}$$

where

$$\phi = \phi_y - \phi_x \tag{3.25}$$

In general the rotation rate of the field vector is not constant. For the special

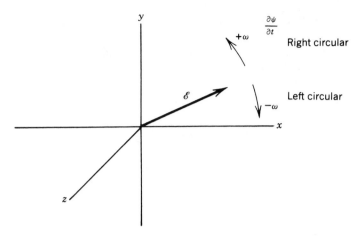

FIGURE 3.4 Rotation relationships for the polarization ellipse.

case of circular polarization,

$$|E_x| = |E_y| \qquad \phi = \pm \frac{\pi}{2}$$

where the upper sign corresponds to left-circular rotation and the lower to right-circular, the rotation rate equation reduces to

$$\frac{\partial \psi}{\partial t} = \mp \omega \tag{3.61}$$

Figure 3.4 indicates that $-\omega$ is consistent with left-circular polarization.

Equation (3.60) can be simplified by noting that its denominator is the square of the magnitude of the time-varying field. Then

$$\frac{\partial \psi}{\partial t} = \frac{-\omega |E_x||E_y|\sin\phi}{|\mathscr{E}|^2} \tag{3.62}$$

On the major axis of the polarization ellipse, $|\mathscr{E}|$ is a maximum, given by m. Thus the rotation rate is a minimum, given by

$$\left.\frac{\partial \psi}{\partial t}\right|_{\min} = \frac{-\omega |E_x||E_y|\sin\phi}{m^2} \tag{3.63}$$

The rotation rate is maximum on the minor axis and is

$$\left.\frac{\partial \psi}{\partial t}\right|_{\max} = \frac{-\omega |E_x||E_y|\sin\phi}{n^2} \tag{3.64}$$

3.7. AREA SWEEP RATE

The area swept by the $\mathscr{E}$ vector in a time dt as it moves through an angle $d\psi$ is shown by Figure 3.5 to be

$$dA = \tfrac{1}{2}|\mathscr{E}|^2\, d\psi$$

Then the rate of increase of swept area is

$$\frac{\partial A}{\partial t} = \frac{1}{2}|\mathscr{E}|^2 \frac{\partial \psi}{\partial t} \tag{3.65}$$

Using (3.62) makes this

$$\frac{\partial A}{\partial t} = -\frac{1}{2}\omega |E_x||E_y|\sin\phi \tag{3.66}$$

A negative value for the rate of area sweep is quite valid and indicates only that for right-handed rotation the rate of area sweep is positive and for left-handed rotation it is negative.

The equation shows that the rate of area sweep is not a function of time or of position of the tip of the electric field vector on the polarization ellipse. This may be considered a kind of Kepler's second law for electromagnetics. The laws are not precisely the same for electromagnetics and planetary motion, however, since the Kepler laws state that the planets move around the sun in ellipses with the sun at one focus, and the radius vector *from the sun* to a planet sweeps out equal areas in equal time intervals [2]. The electric field vector drawn from the ellipse origin, not a focus, sweeps out equal areas in equal intervals of time.

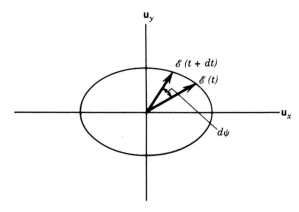

FIGURE 3.5 Area sweep of the $\mathscr{E}$ vector.

Note also that the $\mathscr{E}$ vector completes one rotation in the time

$$T = \frac{2\pi}{\omega} \tag{3.67}$$

3.8. ROTATION OF $\mathscr{E}$ WITH DISTANCE

At time $t = 0$, the position angle of $\mathscr{E}$ as a function of distance z is

$$\psi = \tan^{-1} \frac{|E_y|\cos(-kz + \phi_y)}{|E_x|\cos(-kz + \phi_x)} \tag{3.68}$$

This equation is similar to (3.59) with ω replaced by $-k$ and t by z. The rotation rate with distance can therefore be found by making the same replacements in (3.60), thus

$$\frac{\partial \psi}{\partial z} = \frac{k|E_x||E_y|\sin \phi}{|E_x|^2 \cos^2(-kz + \phi_x) + |E_y|^2 \cos^2(-kz + \phi_y)} \tag{3.69}$$

and since the denominator is obviously the square of the field magnitude at $t = 0$,

$$\frac{\partial \psi}{\partial z} = \frac{k|E_x||E_y|\sin \phi}{|\mathscr{E}|^2} \tag{3.70}$$

This indicates that the rotation rate with z is minimum at the major axis of the polarization ellipse and maximum at the minor axis, just as it was with the time rotation rate.

If the rotation of $\mathscr{E}$ with increasing time in a fixed plane is clockwise, the fact that (3.62) and (3.70) have different signs shows that the rotation with increasing distance at a fixed time is counterclockwise. A right-handed circular wave in space looks like a *left-handed* screw. With increasing time the screw rotates in a clockwise direction as we look in the direction of wave motion. This is shown in Figure 3.6.

It may be seen from (3.68) that the distance between two points of the wave having parallel field vectors at constant time is

$$\Delta z = \frac{2\pi}{k} = \lambda \tag{3.71}$$

Thus the wave appears to rotate once in space in a distance of one wavelength.

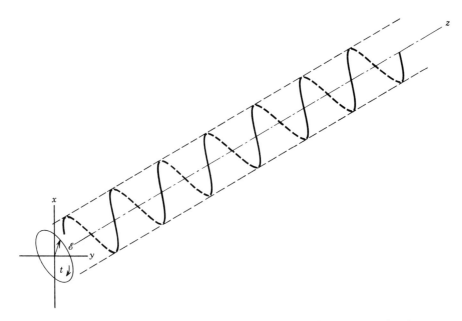

FIGURE 3.6 Rotation of $\mathscr{E}$ with time and distance for a right-handed circular wave.

3.9. THE POLARIZATION RATIO

A description of the elliptically polarized wave in terms of tilt angle, axial ratio, and rotation sense leads to a good physical understanding of the wave, but it is not convenient mathematically. In this and the following sections the wave is characterized by more tractable mathematical terms.

The time-invariant electric field may be written as

$$\mathbf{E} = E_x \mathbf{u}_x + E_y \mathbf{u}_y = E_x \left(\mathbf{u}_x + \frac{E_y}{E_x} \mathbf{u}_y \right) \quad (3.72)$$

where the common phase term $\exp(-jkz)$ of (3.16) has been dropped. Now the value of E_x does not affect the wave polarization in any way, since that is determined only by the relative magnitudes and phase difference of the wave components. Except in questions concerned with power, therefore, it will be neglected. We define a *complex polarization ratio* (or linear polarization ratio, or simply polarization ratio) P, which alone carries all the polarization information, by

$$P = \frac{E_y}{E_x} = \frac{|E_y|}{|E_x|} e^{j\phi} \quad (3.73)$$

Some special values of the polarization ratio are:

Wave	Characteristics	P
Linear vertical	$E_x = 0$	∞
Linear horizontal	$E_y = 0$	0
Right circular	$|E_x| = |E_y|, \phi = -\pi/2$	$-j$
Left circular	$|E_x| = |E_y|, \phi = +\pi/2$	$+j$

3.10. CIRCULAR WAVE COMPONENTS

Consider the complex vectors

$$\mathbf{u}_L = \frac{1}{\sqrt{2}}(\mathbf{u}_x + j\mathbf{u}_y) \qquad (3.74a)$$

$$\mathbf{u}_R = \frac{1}{\sqrt{2}}(\mathbf{u}_x - j\mathbf{u}_y) \qquad (3.74b)$$

They may be written in the form of the field components, thus

$$\mathbf{u}_L = \frac{1}{\sqrt{2}}\mathbf{u}_x + \frac{1}{\sqrt{2}}e^{j\pi/2}\mathbf{u}_y \qquad (3.75a)$$

$$\mathbf{u}_R = \frac{1}{\sqrt{2}}\mathbf{u}_x + \frac{1}{\sqrt{2}}e^{-j\pi/2}\mathbf{u}_y \qquad (3.75b)$$

It is clear that if we think of the vectors as representing fields propagating in the z direction, then $\mathbf{u}_L$ is a left circular wave ($|E_x| = |E_y|, \phi = \pi/2$) and $\mathbf{u}_R$ is a right circular wave. To state this more explicitly, the real time-varying field components associated with the complex time-invariant fields represented by (3.75) are

$$\text{Re}\left[\frac{1}{\sqrt{2}}(\mathbf{u}_L e^{j\omega t} e^{-jkz})\right]$$
$$= \frac{1}{\sqrt{2}}\mathbf{u}_x \cos(\omega t - kz) + \frac{1}{\sqrt{2}}\mathbf{u}_y \cos(\omega t - kz + \pi/2)$$

$$\text{Re}\left[\frac{1}{\sqrt{2}}(\mathbf{u}_R e^{j\omega t} e^{-jkz})\right] \qquad (3.76)$$
$$= \frac{1}{\sqrt{2}}\mathbf{u}_x \cos(\omega t - kz) + \frac{1}{\sqrt{2}}\mathbf{u}_y \cos(\omega t - kz - \pi/2)$$

These may be recognized as real time-varying vectors of constant amplitude rotating, in order, in a left- and right-handed sense.

The electric field may be expanded in terms of the complex vectors $\mathbf{u}_L$ and $\mathbf{u}_R$, thus

$$\mathbf{E} = E_L \mathbf{u}_L + E_R \mathbf{u}_R = |E_L| e^{j\theta_L} \mathbf{u}_L + |E_R| e^{j\theta_R} \mathbf{u}_R$$
$$= e^{j\theta_L} \left(|E_L| \mathbf{u}_L + |E_R| e^{j\theta} \mathbf{u}_R \right) \tag{3.77}$$

where

$$\theta = \theta_R - \theta_L \tag{3.78}$$

We have been interpreting the general elliptical wave as being the sum of two linearly polarized waves whose fields are perpendicular to each other and to the direction of wave travel. We see now that with equal validity we may think of an elliptical wave as the sum of a left- and a right-circular wave. If $|E_L| > |E_R|$, the wave should be left-handed, and examination of (3.77) using the criteria developed previously shows this to be so.

The circular wave components may be found in terms of linear components by equating the field expansions

$$\mathbf{E} = E_x \mathbf{u}_x + E_y \mathbf{u}_y = E_L \mathbf{u}_L + E_R \mathbf{u}_R \tag{3.79}$$

If (3.74) is substituted into this equation and coefficients of like unit vectors are compared, there results

$$E_x = \frac{1}{\sqrt{2}} (E_L + E_R) \tag{3.80a}$$

$$E_y = \frac{j}{\sqrt{2}} (E_L - E_R) \tag{3.80b}$$

which, when inverted, gives

$$E_L = \frac{1}{\sqrt{2}} (E_x - jE_y) \tag{3.81a}$$

$$E_R = \frac{1}{\sqrt{2}} (E_x + jE_y) \tag{3.81b}$$

We defined a polarization ratio, or linear polarization ratio, as the complex ratio of the linear components of a wave. It is consistent to define a ratio of the circular components of the wave, thus

$$q = \frac{E_L}{E_R} = \frac{|E_L|}{|E_R|} e^{-j\theta} \tag{3.82}$$

The parameter q is the inverse of the IEEE Standard definition of the *circular polarization ratio* of the plane wave [3]. As used here, it agrees with Rumsey [4] and facilitates the development of a complex plane presentation of polarization states to be presented later. It is not used extensively later, but when used will be called the inverse circular polarization ratio, to be consistent with IEEE notation. We shall see later that all of the characteristics of the polarization ellipse can be determined from a knowledge of P or q.

3.11. CHANGE OF POLARIZATION BASE: ORTHOGONAL ELLIPTICAL COMPONENTS OF A WAVE

We saw in previous sections that the electric field can be written as the sum of rectangular or circular components. It can also be considered the sum of two orthogonally polarized elliptical waves. Let

$$\mathbf{E} = \mathbf{E}_1 + \mathbf{E}_2 = E_1 \mathbf{u}_1 + E_2 \mathbf{u}_2 \qquad (3.83)$$

where $\mathbf{u}_1$ and $\mathbf{u}_2$ are orthogonal unit vectors obeying

$$\begin{aligned} \mathbf{u}_1 \cdot \mathbf{u}_1^* &= \mathbf{u}_2 \cdot \mathbf{u}_2^* = 1 \\ \mathbf{u}_1 \cdot \mathbf{u}_2^* &= \mathbf{u}_1^* \cdot \mathbf{u}_2 = 0 \end{aligned} \qquad (3.84)$$

If one of the wave components, say $\mathbf{E}_1$, is to have specified polarization characteristics, we may write $\mathbf{u}_1$ in terms of a linear polarization ratio in the xy system, thus

$$\mathbf{u}_1 = u_{1x}\mathbf{u}_x + u_{1y}\mathbf{u}_y = u_{1x}(\mathbf{u}_x + P_1\mathbf{u}_y) \qquad (3.85)$$

The phase of $\mathbf{E}_1$ may be included in the multiplier E_1 and u_{1x} taken as real. Doing so and noting that $\mathbf{u}_1$ has unit length leads to

$$\mathbf{u}_1 = \frac{\mathbf{u}_x + P_1\mathbf{u}_y}{\left(1 + |P_1|^2\right)^{1/2}} \qquad (3.86)$$

In the same way,

$$\mathbf{u}_2 = \frac{\mathbf{u}_x + P_2\mathbf{u}_y}{\left(1 + |P_2|^2\right)^{1/2}} \qquad (3.87)$$

where P_2 is the linear polarization ratio (in xy coordinates) of the second elliptical component, $\mathbf{E}_2$, of the wave.

134 REPRESENTATIONS OF WAVE POLARIZATION

The orthogonality condition

$$\mathbf{u}_1^* \cdot \mathbf{u}_2 = 0$$

leads quickly to

$$1 + P_1^* P_2 = 0$$

or

$$P_2 = -\frac{1}{P_1^*} \tag{3.88}$$

Then base vector $\mathbf{u}_2$ becomes

$$\mathbf{u}_2 = \frac{|P_1|}{\left(1 + |P_1|^2\right)^{1/2}} \left(\mathbf{u}_x - \frac{1}{P_1^*} \mathbf{u}_y \right) \tag{3.89}$$

We have thus seen that a plane harmonic wave can be expressed as the sum of two orthogonal elliptical waves. If one of the waves has a specified polarization state, the unit vectors of the new polarization base vector set are given by (3.86) and (3.89). Chapter 4 shows that if a receiving antenna has a polarization that causes it to receive one of the wave components, say $\mathbf{E}_1$, without loss, then it cannot receive component $\mathbf{E}_2$ at all.

It may readily be seen that the circular component unit vectors $\mathbf{u}_L$ and $\mathbf{u}_R$ of the previous section are specializations, with $P_1 = j$, of the more general orthogonal vectors.

We write the field in two polarization bases

$$\mathbf{E} = E_1 \mathbf{u}_1 + E_2 \mathbf{u}_2 = E_x \mathbf{u}_x + E_y \mathbf{u}_y \tag{3.90}$$

and note that

$$E_1 = \mathbf{E} \cdot \mathbf{u}_1^* \tag{3.91a}$$

$$E_2 = \mathbf{E} \cdot \mathbf{u}_2^* \tag{3.91b}$$

Then we may write

$$\begin{bmatrix} E_1 \\ E_2 \end{bmatrix} = \begin{bmatrix} \mathbf{u}_1^* \cdot \mathbf{u}_x & \mathbf{u}_1^* \cdot \mathbf{u}_y \\ \mathbf{u}_2^* \cdot \mathbf{u}_x & \mathbf{u}_2^* \cdot \mathbf{u}_y \end{bmatrix} \begin{bmatrix} E_x \\ E_y \end{bmatrix} \tag{3.92}$$

or

$$\mathbf{E}_{\mathbf{u}_1, \mathbf{u}_2} = \mathbf{R} \mathbf{E}_{\mathbf{u}_x, \mathbf{u}_y} \tag{3.93}$$

If $\mathbf{u}_1$ and $\mathbf{u}_2$ are expressed in terms of the linear polarization ratio, the matrix

R that transforms E from the $\mathbf{u}_x, \mathbf{u}_y$ polarization base to the $\mathbf{u}_1, \mathbf{u}_2$ base becomes

$$\mathbf{R} = \frac{1}{\left(1 + |P_1|^2\right)^{1/2}} \begin{bmatrix} 1 & P^* \\ |P| & -|P|/P \end{bmatrix} \quad (3.94)$$

It is readily seen that

$$\mathbf{R}\mathbf{R}^{T*} = \begin{bmatrix} 1 & 0 \\ 0 & 1 \end{bmatrix} = \mathbf{I} \quad (3.95)$$

so **R** is a unitary matrix.

If the wave is expressed in the $\mathbf{u}_1, \mathbf{u}_2$ base with the linear polarization ratio of $\mathbf{u}_1$ known in xy coordinates, it may be transformed to the $\mathbf{u}_x, \mathbf{u}_y$ base by

$$\mathbf{E}_{\mathbf{u}_x, \mathbf{u}_y} = \mathbf{R}^{-1} \mathbf{E}_{\mathbf{u}_1, \mathbf{u}_2} = \mathbf{R}^{T*} \mathbf{E}_{\mathbf{u}_1, \mathbf{u}_2} \quad (3.96)$$

It may then be transformed to any other base vector set $\mathbf{v}_1, \mathbf{v}_2$ by multiplication with

$$\mathbf{V} = \begin{bmatrix} \mathbf{v}_1^* \cdot \mathbf{u}_x & \mathbf{v}_1^* \cdot \mathbf{u}_y \\ \mathbf{v}_2^* \cdot \mathbf{u}_x & \mathbf{v}_2^* \cdot \mathbf{u}_y \end{bmatrix} \quad (3.97)$$

where **V** is unitary. Then

$$\mathbf{E}_{\mathbf{v}_1, \mathbf{v}_2} = \mathbf{V}\mathbf{E}_{\mathbf{u}_x, \mathbf{u}_y} = \mathbf{V}\mathbf{R}^{-1}\mathbf{E}_{\mathbf{u}_1, \mathbf{u}_2} \quad (3.98)$$

Now the product of two unitary matrices is itself unitary. It follows then that a plane harmonic wave expressed in one polarization base can be transformed to another base by multiplication with a unitary 2×2 matrix.

Coordinate Rotation

Of special interest is the change in base caused by a coordinate rotation. Then $\mathbf{u}_1$ is a real unit vector rotated by angle θ (in the direction $x \to y$) from the x axis of the old coordinate system. For this, **R** has the simple form

$$\mathbf{R} = \begin{bmatrix} \mathbf{u}_1^* \cdot \mathbf{u}_x & \mathbf{u}_1^* \cdot \mathbf{u}_y \\ \mathbf{u}_2^* \cdot \mathbf{u}_x & \mathbf{u}_2^* \cdot \mathbf{u}_y \end{bmatrix} = \begin{bmatrix} \cos\theta & \sin\theta \\ -\sin\theta & \cos\theta \end{bmatrix} \quad (3.99)$$

3.12. RELATIONSHIP BETWEEN P AND q, AND THE MODIFIED POLARIZATION RATIO

The inverse circular polarization ratio of a plane wave can be found in terms of its linear polarization ratio by using (3.81) and the definitions for P and q, thus

$$q = \frac{E_x - jE_y}{E_x + jE_y} = \frac{1 - jP}{1 + jP} \tag{3.100}$$

Solving for P gives

$$P = -j\frac{1 - q}{1 + q} \tag{3.101}$$

The j multiplier can be removed from these equations if a *modified polarization ratio*, p, is defined as

$$p = jP \tag{3.102}$$

With this definition the polarization ratio relationships become

$$q = \frac{1 - p}{1 + p} \tag{3.103a}$$

$$p = \frac{1 - q}{1 + q} \tag{3.103b}$$

The symmetry is pleasing, but more important is that by using q and p (rather than P) the common Smith transmission line chart can be used to plot polarizations.

Some special values of the polarization ratios are:

Wave	Characteristics	P	p	q
Linear vertical	$E_x = 0$	∞	$j\infty$	-1
Linear horizontal	$E_y = 0$	0	0	$+1$
Right circular	$E_L = 0$	$-j$	$+1$	0
Left circular	$E_R = 0$	$+j$	-1	∞

3.13. ELLIPSE CHARACTERISTICS IN TERMS OF q

Equations (3.77) and (3.82) can be combined to give

$$\mathbf{E} = E_R(q\mathbf{u}_L + \mathbf{u}_R) = |E_R|e^{j\theta_R}(q\mathbf{u}_L + \mathbf{u}_R) \tag{3.104}$$

and the corresponding time-varying field is

$$\mathscr{E} = |E_R|\text{Re}\left[e^{j(\omega t - kz + \theta_R)}(q\mathbf{u}_L + \mathbf{u}_R)\right] \tag{3.105}$$

ELLIPSE CHARACTERISTICS IN TERMS OF q

It may be written as

$$\frac{\mathscr{E}}{|E_R|} = \frac{1}{2}\left[e^{j\beta}(q\mathbf{u}_L + \mathbf{u}_R) + e^{-j\beta}(q^*\mathbf{u}_L^* + \mathbf{u}_R^*)\right] \quad (3.106)$$

where

$$\beta = \omega t - kz + \theta_R \quad (3.107)$$

Now the unit vectors $\mathbf{u}_L$ and $\mathbf{u}_R$ are not only orthogonal but conjugate to each other, so that they obey

$$\mathbf{u}_L \cdot \mathbf{u}_L^* = \mathbf{u}_R \cdot \mathbf{u}_R^* = \mathbf{u}_L \cdot \mathbf{u}_R = \mathbf{u}_R \cdot \mathbf{u}_L = 1 \quad (3.108\text{a})$$
$$\mathbf{u}_L \cdot \mathbf{u}_R^* = \mathbf{u}_R \cdot \mathbf{u}_L^* = \mathbf{u}_L \cdot \mathbf{u}_L = \mathbf{u}_R \cdot \mathbf{u}_R = 0 \quad (3.108\text{b})$$

If these relations are used, the square of the time-varying field intensity is

$$\frac{\mathscr{E}^2}{|E_R|^2} = \frac{1}{2}\left[1 + |q|^2 + qe^{j2\beta} + q^*e^{-j2\beta}\right]$$
$$= \frac{1}{2}\left[1 + |q|^2 + 2|q|\cos(2\beta - \theta)\right] \quad (3.109)$$

The maximum and minimum values of this field are

$$\frac{1}{\sqrt{2}}(1 + |q|) \quad \text{at} \quad 2\beta - \theta = 0, 2\pi, \ldots$$

and

$$\frac{1}{\sqrt{2}}|1 - |q|| \quad \text{at} \quad 2\beta - \theta = \pi, 3\pi, \ldots$$

From these the axial ratio of the polarization ellipse,

$$\text{AR} = \left|\frac{1 + |q|}{1 - |q|}\right| \quad (3.110)$$

is immediately noted.

To find the tilt angle of the polarization ellipse, substitute the value of β that gives maximum field intensity magnitude into (3.106). Doing so and converting the resulting equation to rectangular coordinates gives the field,

$$\left.\frac{\mathscr{E}}{|E_R|}\right|_{\text{max}} = \frac{1}{\sqrt{2}}\left[(1 + |q|)\cos\left(\frac{\theta}{2}\right)\mathbf{u}_x + (1 + |q|)\sin\left(\frac{\theta}{2}\right)\mathbf{u}_y\right]$$

Since this is the field having maximum magnitude, its rotation angle is the ellipse tilt angle, or

$$\tau = \tan^{-1} \frac{\mathscr{E}_y}{\mathscr{E}_x} = \tan^{-1} \frac{\sin(\theta/2)}{\cos(\theta/2)} \qquad (3.111)$$

with solutions

$$\tau = \tfrac{1}{2}\theta \qquad \tfrac{1}{2}\theta \pm \pi \qquad (3.112)$$

The appropriate form can be chosen to keep τ in the desired range.

3.14. ELLIPSE CHARACTERISTICS IN TERMS OF P AND p

The axial ratio can be written in terms of the linear polarization ratio P with the use of (3.100), giving

$$\mathrm{AR} = \left| \frac{|1+jP| + |1-jP|}{|1+jP| - |1-jP|} \right| \qquad (3.113)$$

The tilt angle may be obtained from

$$e^{-2\tau} = \frac{q}{|q|} = \frac{(1-jP)/(1+jP)}{|(1-jP)/(1+jP)|} \qquad (3.114)$$

or from the more convenient form which may be obtained from it,

$$\tan(2\tau) = \frac{2\,\mathrm{Re}(P)}{1 - |P|^2} \qquad (3.115)$$

In terms of the modified linear polarization ratio p, these equations become

$$\mathrm{AR} = \left| \frac{|1+p| + |1-p|}{|1+p| - |1-p|} \right| \qquad (3.116)$$

and

$$\tan(2\tau) = \frac{2\,\mathrm{Im}(p)}{1 - |p|^2} \qquad (3.117)$$

3.15. POLARIZATION CHARACTERISTICS FOR RANGES OF P, p, AND q

It is not immediately clear from a knowledge of one of the polarization ratios just what the polarization ellipse characteristics (or more generally the physical polarization characteristics) are. For example, consider two waves with polarization ratios $P_1 = 2\exp(\pi/6)$ and $P_2 = 2\exp(-\pi/6)$. Are the polarization ellipses similar? Are the rotation senses the same? If two antennas transmit, respectively, waves with these polarizations, can they be used satisfactorily in a transmit–receive configuration? The answer to the last question is reserved to another chapter, but we can begin to examine the first two.

From

$$\frac{\partial \psi}{\partial t} = \frac{-\omega |E_x||E_y|\sin \phi}{|\mathscr{E}|^2} \tag{3.62}$$

which gives the rate of increase of the angle of $\mathscr{E}$ measured from the x axis, we see that the derivative is negative, corresponding to left elliptic rotation (LER) for $0 < \phi < \pi$ and positive, corresponding to right elliptic rotation (RER), for $\pi < \phi < 2\pi$. The end points of these ranges correspond to linear polarizations.

From

$$P = \frac{|E_y|}{|E_x|}e^{j\phi} = \frac{|E_y|}{|E_x|}(\cos \phi + j \sin \phi) \tag{3.118}$$

it is clear that left elliptic rotation requires

$$\begin{aligned}\text{Im}(P) &> 0 \\ \text{Re}(p) &< 0\end{aligned} \quad \text{LER} \tag{3.119}$$

and for right elliptic rotation

$$\begin{aligned}\text{Im}(P) &< 0 \\ \text{Re}(p) &> 0\end{aligned} \quad \text{RER} \tag{3.120}$$

From the relationship between P and q, it follows that

$$|q|^2 = \frac{1-jP}{1+jP}\frac{1+jP^*}{1-jP^*} = \frac{1 + 2\left|\frac{E_y}{E_x}\right|\sin \phi + \left|\frac{E_y}{E_x}\right|^2}{1 - 2\left|\frac{E_y}{E_x}\right|\sin \phi + \left|\frac{E_y}{E_x}\right|^2} \tag{3.121}$$

and recalling the ranges of ϕ for right- and left-handed rotation informs us that

$$|q| > 1 \quad \text{(LER)} \qquad |q| < 1 \quad \text{(RER)}$$

This result was of course expected from the definition of q as the ratio of the left circular component of a wave to the right circular component.

3.16. THE TRANSFORMATIONS $p(q)$ AND $q(p)$

In this section and some following sections, the modified polarization ratio p is used instead of the conventional linear polarization ratio P. This allows the conventional Smith transmission line chart to be used for plotting polarization states.

Before examining the transformations

$$q = \frac{1-p}{1+p} \qquad (3.103a)$$

$$p = \frac{1-q}{1+q} \qquad (3.103b)$$

consider a configuration that gives rise to similar equations, the electric transmission line shown in Figure 3.7. In terms of the current at the load or receiving end, the voltage and current on the line are given by

$$V = \frac{I_R}{2}\left[(Z_R + Z_0)e^{\gamma d} + (Z_R - Z_0)e^{-\gamma d}\right] \qquad (3.122a)$$

$$I = \frac{I_R}{2Z_0}\left[(Z_R + Z_0)e^{\gamma d} - (Z_R - Z_0)e^{-\gamma d}\right] \qquad (3.122b)$$

where Z_R is the load impedance, Z_0 is the characteristic line impedance, and γ is the propagation constant, which are given in terms of the line parameters by

$$Z_0 = \sqrt{\frac{R + j\omega L}{G + j\omega C}} \qquad \gamma = \sqrt{(R + j\omega L)(G + j\omega C)} \qquad (3.123)$$

The voltage reflection coefficient K is the ratio of the reflected voltage wave (second term in the voltage equation) to the incident wave (first term). Similarly, the current reflection coefficient is the ratio of the reflected current

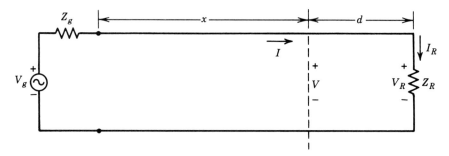

FIGURE 3.7 The transmission line.

THE TRANSFORMATIONS $p(q)$ AND $q(p)$ **141**

to the incident current. They are

$$K = \frac{Z_R - Z_0}{Z_R + Z_0} e^{-2\gamma d} \tag{3.124a}$$

$$K_I = -\frac{Z_R - Z_0}{Z_R + Z_0} e^{-2\gamma d} = -K \tag{3.124b}$$

If the impedance of the line at any point is defined as

$$Z = \frac{V}{I} \tag{3.125}$$

its normalized value is found from the voltage and current equations and the reflection coefficient definitions to be

$$z = \frac{Z}{Z_0} = \frac{1 + K}{1 - K} = \frac{1 - K_I}{1 + K_I} \tag{3.126}$$

Solving for K and K_I leads to

$$K = \frac{z - 1}{z + 1} \tag{3.127a}$$

$$K_I = \frac{1 - z}{1 + z} \tag{3.127b}$$

and we see that the transformation between z and K_I is the same as that between the polarization parameters p and q.

It is well known that if curves of constant $\text{Re}(z)$ and constant $\text{Im}(z)$ are plotted on the complex K plane, a widely available transmission line chart, the Smith chart, results. While the common usage of the Smith chart employs z and K (not K_I), it may also be used with K_I. The advantage of using the modified polarization ratio p, rather than P, is that the transformation between p and q allows a representation on the commercially available Smith chart.

The Smith chart is plotted on the K plane. If we let q be analogous to K and plot polarization states on the q plane, all right elliptical polarizations will fall within the unit circle $|q| < 1$. This unit circle is important since the common Smith chart is restricted (for convenience) to the unit circle. If the choice were made to let p be analogous to K and plot polarizations on the p plane, the left half plane would contain all LER points and the right half plane all RER points, thus requiring an infinite plane for all polarizations.

The q plane will therefore be used. Rather than use known equations for the Smith chart, new equations are developed here, and the confusion inherent in translating from z to p and $K_I (= -K)$ to q is avoided.

In (3.103) let

$$p = u + jv \tag{3.128a}$$
$$q = s + jt \tag{3.128b}$$

Then

$$p = u + jv = \frac{1-q}{1+q} = \frac{1-s-jt}{1+s+jt} = \frac{1-s^2-t^2-j2t}{(1+s)^2+t^2} \tag{3.129}$$

Equating real and imaginary parts of this expression, collecting terms, and completing the squares gives, if $u \neq -1$ and $v \neq 0$,

$$\left(s + \frac{u}{u+1}\right)^2 + t^2 = \left(\frac{1}{u+1}\right)^2 \tag{3.130a}$$

$$(s+1)^2 + \left(t + \frac{1}{v}\right)^2 = \left(\frac{1}{v}\right)^2 \tag{3.130b}$$

These equations describe two families of circles on the complex q plane. One family has centers at

$$s_c, t_c = -\frac{u}{u+1}, 0 \tag{3.131}$$

and radii

$$r = \left|\frac{1}{u+1}\right| \tag{3.132}$$

The second has centers at

$$s_c, t_c = -1, -\frac{1}{v} \tag{3.133}$$

and radii

$$r = \left|\frac{1}{v}\right| \tag{3.134}$$

Some of these circles are shown in Figure 3.8. It is obvious that the portion of the complex plane within the circle of unit radius, $|q| < 1$, is identical to the common Smith chart.

In developing the previous equations, the values $u = -1$ and $v = 0$ were excluded. As $u \to -1$, however, the first family of circles with centers and radii given by (3.131) and (3.132) merely degenerates to a straight line, the vertical line $s = -1$. As $v \to 0$, the second family of circles yields the horizontal axis. It is then not necessary to exclude these u and v values.

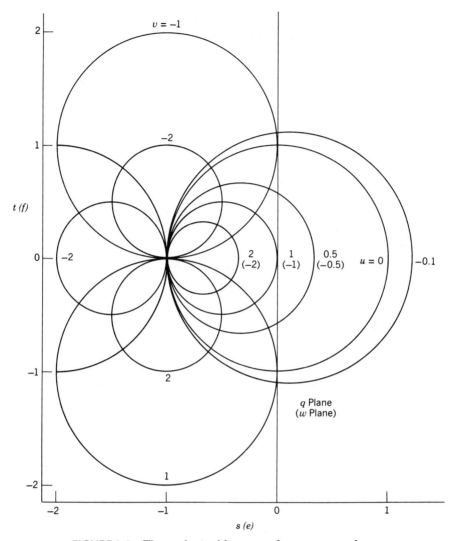

FIGURE 3.8 The q plane with curves of constant u and v.

3.17. THE TRANSFORMATION FOR $u < 0$

Equation (3.131) and Figure 3.8 show that for $0 > u > -1$ the center for the family of circles $u =$ constant is on the $+s$ axis. Further, all the circles pass through the point $(s = -1, t = 0)$ since the radius exceeds the location of the center by

$$\frac{1}{u+1} - \frac{|u|}{u+1} = \frac{1+u}{u+1} = 1$$

All the circles for $0 > u > -1$ therefore lie outside the unit circle.

For $u < -1$, the circle centers are on the $-s$ axis. Again the circles pass through $s = -1$, $t = 0$, since the magnitude of the distance from the origin to the circle center exceeds the circle radius by

$$\frac{|u|}{|u + 1|} - \frac{1}{|u + 1|} = 1$$

We see that all values of $u = \text{Re}(p) < 0$ transform to points outside the unit radius circle $|q| = 1$. This might have been expected since $|q| < 1$ and $\text{Re}(p) > 0$ both correspond to right elliptic rotation. This appears to limit the usefulness of the polarization chart developed here to right-handed polarizations, since it is important to stay within the unit circle. To eliminate this restriction, a new form of the circular polarization ratio is desirable.

Define a new phase angle γ (not to be confused with the transmission line propagation constant discussed earlier),

$$\gamma = -\theta \tag{3.135}$$

Then the inverse circular polarization ratio may be written

$$q = |q|e^{-j\theta} = Qe^{j\gamma} \tag{3.136}$$

with Q the magnitude of the inverse circular polarization ratio. If $Q > 1$, the polarization state point will lie outside the unit circle, and we define a new parameter to use instead of q, thus

$$w = \frac{1}{q^*} = \frac{1}{Q}e^{-j\theta} = We^{j\gamma} \tag{3.137}$$

It may be seen that w has the same angle as q and in fact is a reflection of q in the unit circle.

The transformations between p and w are

$$p = \frac{w^* - 1}{w^* + 1} \tag{3.138a}$$

$$w = \frac{1 + p^*}{1 - p^*} \tag{3.138b}$$

The transformations are not symmetric, as they were for p and q, but they will serve. Let

$$w = e + jf \tag{3.139}$$

and repeat the steps leading to the families of circles on the q plane. The result is

$$\left(e + \frac{u}{u-1}\right)^2 + f^2 = \left(\frac{1}{u-1}\right)^2 \qquad (3.140a)$$

$$(e+1)^2 + \left(f + \frac{1}{v}\right)^2 = \left(\frac{1}{v}\right)^2 \qquad (3.140b)$$

These equations represent families of circles on the complex w plane. The first family, for constant u, has centers and radii

$$e_c, f_c = -\frac{u}{u-1}, 0 \qquad r = \left|\frac{1}{u-1}\right| \qquad (3.141)$$

and the second family, for constant v, has centers and radii

$$e_c, f_c = -1, -\frac{1}{v} \qquad r = \left|\frac{1}{v}\right| \qquad (3.142)$$

A comparison of these equations is instructive,

$q = s + jt$ PLANE		$w = e + jf$ PLANE
$q = \dfrac{1-p}{1+p}$	$p = u + jv$	$w = \dfrac{1+p^*}{1-p^*}$
$\left(s + \dfrac{u}{u+1}\right)^2 + t^2 = \left(\dfrac{1}{u+1}\right)^2$		$\left(e + \dfrac{u}{u-1}\right)^2 + f^2 = \left(\dfrac{1}{u-1}\right)^2$
$(s+1)^2 + \left(t + \dfrac{1}{v}\right)^2 = \left(\dfrac{1}{v}\right)^2$		$(e+1)^2 + \left(f + \dfrac{1}{v}\right)^2 = \left(\dfrac{1}{v}\right)^2$

FAMILY OF CIRCLES, u = CONSTANT

Center at $\dfrac{-u}{u+1}, 0$ \qquad Center at $\dfrac{-u}{u-1}, 0$

Radius $= \left|\dfrac{1}{u+1}\right|$ \qquad Radius $= \left|\dfrac{1}{u-1}\right|$

FAMILY OF CIRCLES, v = CONSTANT

Center at $-1, -\dfrac{1}{v}$ \qquad Center at $-1, -\dfrac{1}{v}$

Radius $= \left|\dfrac{1}{v}\right|$ \qquad Radius $= \left|\dfrac{1}{v}\right|$

The comparison of the circle centers and radii for the two planes clearly shows two facts.

1. The circles on the q plane for some u, say $u = u_0$, are identical to the circles on the w plane for $u = -u_0$.
2. The circles on the q plane for some v, say $v = v_0$, are identical to the circles on the w plane for $v = v_0$.

It follows therefore that Figure 3.8 may be considered the q plane with all curves $u =$ constant inside the unit circle corresponding to positive values of u. All polarization points inside the unit circle on the q plane correspond to right-elliptical polarization. With equal justification, Figure 3.8 may be considered the w plane with all curves inside the unit circle corresponding to negative values of u. All polarization points inside the unit circle on the w plane correspond to left-elliptical polarization. A particular circle $u = u_0$ on the q plane would be labeled $-u_0$ on the w plane. For the two planes there is no difference in the constant v circles. In Figure 3.8 all values relating to the w plane are in parentheses.

The polarization ellipse characteristics in terms of w can be found by substituting the relation between q and w into the appropriate equations. The axial ratio is

$$\text{AR} = \left|\frac{1+|q|}{1-|q|}\right| = \left|\frac{1+|1/w^*|}{1-|1/w^*|}\right| = \left|\frac{1+|w|}{1-|w|}\right| \quad (3.143)$$

Since by definition the phase angle of w is the same as that of q, the tilt angle is still

$$\tau = \tfrac{1}{2}\theta = -\tfrac{1}{2}\gamma \quad (3.144)$$

The rotation sense is left handed if $|q| \geq 1$ and $|w| \leq 1$, and right handed if $|q| \leq 1$ and $|w| \geq 1$.

3.18. POLARIZATION CHART AS THE p PLANE

Since $p(q)$ and $q(p)$ have the same form, it is obvious that we can consider the polarization chart of Figure 3.8 as the p plane, with the circles being curves of constant $\text{Re}(q)$ and constant $\text{Im}(q)$. This use has limited value, however, since all polarizations of interest do not fall within the unit circle.

3.19. COINCIDENT POINTS ON THE q AND w PLANES

Figure 3.8 represents the q plane for right-elliptic and the w plane for left-elliptic polarizations. A point on the chart then represents two polarizations, one left handed and the other right handed. Consider the polarization described by a point q_0 and that described by the coincident point w_0 (not the transformed point $w = 1/q^*$). With $w_0 = q_0$ the axial ratios on the two planes are related by

$$\text{AR}|_{q_0} = \left|\frac{1 + |q_0|}{1 - |q_0|}\right| = \left|\frac{1 + |w_0|}{1 - |w_0|}\right| = \text{AR}|_{w_0}$$

and the tilt angles by

$$\tau|_{q_0} = \tfrac{1}{2}\theta = \tau|_{w_0}$$

Coincident points on the q and w planes represent waves having the same axial ratios and tilt angles, but opposite rotation senses.

3.20. CONTOURS OF CONSTANT AXIAL RATIO AND TILT ANGLE

It is useful to consider contours of constant axial ratio and tilt angle on the polarization chart. It is apparent from (3.143) that contours of constant axial ratio are also contours of constant $|q|$ or $|w|$. These are circles on the q or w planes with centers at the origin. On the q plane inside the unit circle, the circle radius is found from (3.143) to be

$$\text{Radius} = |q| = \frac{\text{AR} - 1}{\text{AR} + 1} \qquad (3.145)$$

On the w plane, the same equation is used, with q replaced by w. For a particular axial ratio, the circles on the q and w planes coincide.

Contours of constant tilt angle, for either the q or w plane, are found from

$$\gamma = \text{ang } q = \text{ang } w = -2\tau \qquad (3.146)$$

Figure 3.9 shows curves of constant axial ratio and tilt angle on the q and w plane. Left- and right-circular waves are represented by points at the origin and linear polarizations by points on the unit circle.

148 REPRESENTATIONS OF WAVE POLARIZATION

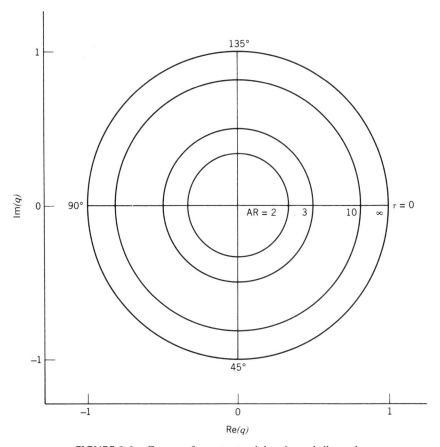

FIGURE 3.9 Curves of constant axial ratio and tilt angle.

3.21. CONTOURS OF CONSTANT $|p|$

Consider first the RER case and use

$$p = \frac{1-q}{1+q} \tag{3.103b}$$

which becomes

$$|p|e^{j(\phi+\pi/2)} = \frac{1 - Qe^{j\gamma}}{1 + Qe^{j\gamma}} \tag{3.147}$$

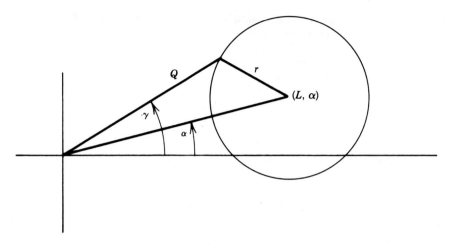

FIGURE 3.10 Circle with polar coordinates.

Multiplying both sides by the complex conjugate and rearranging leads to

$$Q^2 - 2\frac{1+|p|^2}{1-|p|^2}Q\cos\gamma + \left(\frac{1+|p|^2}{1-|p|^2}\right)^2 = \left(\frac{1+|p|^2}{1-|p|^2}\right)^2 - 1 \quad (3.148)$$

Now a circle in plane polar coordinates (Q, γ) of radius r and center at (L, α) is given by

$$Q^2 - 2LQ\cos(\gamma - \alpha) + L^2 = r^2 \quad (3.149)$$

where $0 < (\gamma - \alpha) < 2\pi$. This is readily seen by applying the law of cosines to Figure 3.10. Therefore, (3.148) represents a family of circles on the q plane with centers and radii

$$Q_c, \gamma_c = \frac{1+|p|^2}{1-|p|^2}, 0 \qquad r = \left[\left(\frac{1+|p|^2}{1-|p|^2}\right)^2 - 1\right]^{1/2} \qquad |p| < 1 \quad (3.150)$$

If $|p| > 1$, (3.148) is not in the correct form to represent a circle. For this case, it can be rewritten, dividing numerators and denominators by $|p|^2$, to obtain

$$Q^2 - 2\cos(\gamma \pm \pi)\frac{1+1/|p|^2}{1-1/|p|^2}Q + \left(\frac{1+1/|p|^2}{1-1/|p|^2}\right)^2 = \left(\frac{1+1/|p|^2}{1-1/|p|^2}\right)^2 - 1$$

$$(3.151)$$

150 REPRESENTATIONS OF WAVE POLARIZATION

Comparison to the standard equation shows this to represent circles on the q plane with centers and radii

$$Q_c, \gamma_c = \frac{1 + 1/|p|^2}{1 - 1/|p|^2}, \pi \qquad r = \left[\left(\frac{1 + 1/|p|^2}{1 - 1/|p|^2}\right)^2 - 1\right]^{1/2} \qquad |p| > 1 \tag{3.152}$$

For the left-elliptical case, with

$$p = \frac{w^* - 1}{w^* + 1} = \frac{We^{-j\gamma} - 1}{We^{-j\gamma} + 1} \tag{3.153}$$

if both sides are multiplied by the complex conjugate, the result is

$$|p|^2 = \frac{W^2 - 2W\cos\gamma + 1}{W^2 + 2W\cos\gamma + 1} \tag{3.154}$$

This equation may be put into the standard circle form, as was done with the RER case, and the results are

$$W_c, \gamma_c = \frac{1 + |p|^2}{1 - |p|^2}, 0 \qquad r = \left[\left(\frac{1 + |p|^2}{1 - |p|^2}\right)^2 - 1\right]^{1/2} \qquad |p| < 1 \quad (3.155)$$

$$W_c, \gamma_c = \frac{1 + 1/|p|^2}{1 - 1/|p|^2}, \pi \qquad r = \left[\left(\frac{1 + 1/|p|^2}{1 - 1/|p|^2}\right)^2 - 1\right]^{1/2} \qquad |p| > 1 \quad (3.156)$$

It is obvious from these results that circles of constant $|p|$ coincide for the q and w planes.

3.22. CONTOURS OF CONSTANT ϕ

Consider contours of constant ϕ, the phase difference between x and y components of the propagating wave, for the RER case. If the conjugate of

$$p = |p|e^{j(\phi + \pi/2)} = \frac{1 - Qe^{j\gamma}}{1 + Qe^{j\gamma}} \tag{3.147}$$

is first added to p and then subtracted, and if the difference is divided by the sum, the result is

$$\tan\left(\phi + \frac{\pi}{2}\right) = \frac{-2Q \sin \gamma}{1 - Q^2} \qquad (3.157)$$

If the substitution is made

$$C = \cot \phi = -\tan\left(\phi + \frac{\pi}{2}\right) \qquad (3.158)$$

then

$$Q^2 - \frac{2Q}{C}\cos\left(\gamma + \frac{\pi}{2}\right) + \frac{1}{C^2} = 1 + \frac{1}{C^2} \qquad (3.159)$$

is obtained. It is in the standard form to represent circles on the q plane with centers at

$$Q_c = \frac{1}{C} = \tan \phi \qquad \gamma_c = -\frac{\pi}{2} \qquad C > 0 \qquad (3.160)$$

and radii

$$r = \left(1 + \frac{1}{C^2}\right)^{1/2} = (1 + \tan^2 \phi)^{1/2} = |\sec \phi| \qquad C > 0 \qquad (3.161)$$

These equations are restricted to $C > 0$ since the circle equation is in a standard form only if $Q/C > 0$, and $Q > 0$ by definition. The condition $C = \cot \phi > 0$ leads to

$$0 < \phi < \frac{\pi}{2} \quad \text{LER} \qquad \pi < \phi < \frac{3\pi}{2} \quad \text{RER}$$

The range for left-elliptic rotation need not be considered, since the q plane is not used for that case.

For $C < 0$, (3.159) can be rewritten as

$$Q^2 - \frac{2Q}{|C|}\cos\left(\gamma - \frac{\pi}{2}\right) + \frac{1}{C^2} = 1 + \frac{1}{C^2} \qquad (3.162)$$

Now $C = \cot \phi < 0$ gives phase angle ranges

$$\frac{\pi}{2} < \phi < \pi \quad \text{LER} \qquad \frac{3\pi}{2} < \phi < 2\pi \quad \text{RER}$$

152 REPRESENTATIONS OF WAVE POLARIZATION

Again, only the right-elliptic case is considered, with circle centers and radii

$$Q_c, \gamma_c = \frac{1}{|C|}, \gamma_c = |\tan \phi|, \frac{\pi}{2} \qquad r = \sec \phi \tag{3.163}$$

Next, consider the LER case, for which

$$p = |p|e^{j(\phi + \pi/2)} = \frac{w^* - 1}{w^* + 1} = \frac{We^{-j\gamma} - 1}{We^{-j\gamma} + 1} \tag{3.164}$$

Proceeding in the same manner as for the RER case leads to

$$W^2 - \frac{2W}{C}\cos\left(\gamma - \frac{\pi}{2}\right) + \frac{1}{C^2} = 1 + \frac{1}{C^2} \tag{3.165}$$

For $C > 0$, if the RER case is excluded, ϕ has the range

$$0 < \phi < \frac{\pi}{2} \quad \text{LER}$$

Then (3.165) represents circles on the w plane with centers

$$W_c, \gamma_c = \frac{1}{C}, \gamma_c = \tan \phi, \frac{\pi}{2} \tag{3.166}$$

and radii

$$r = \left(1 + \frac{1}{C^2}\right)^{1/2} = (1 + \tan^2 \phi)^{1/2} = \sec \phi \tag{3.167}$$

For $C < 0$, excluding the RER case, the range of ϕ is

$$\frac{\pi}{2} < \phi < \pi \quad \text{LER}$$

and the circle equation should be rewritten as

$$W^2 - \frac{2W}{|C|}\cos\left(\gamma + \frac{\pi}{2}\right) + \frac{1}{C^2} = 1 + \frac{1}{C^2} \tag{3.168}$$

Then the circle centers and radii on the w plane are

$$W_c, \gamma_c = \frac{1}{|C|}, \gamma_c = |\tan \phi|, -\frac{\pi}{2} \quad r = |\sec \phi| \quad (3.169)$$

At this point a summary would be useful, and shown here are the centers, radii, and range of ϕ for the two polarization states.

$$\begin{array}{cc} \text{RER} & \text{LER} \\ Q_c = \tan \phi & W_c = |\tan \phi| \\ \gamma_c = -\dfrac{\pi}{2} & \gamma_c = -\dfrac{\pi}{2} \\ r = |\sec \phi| & r = |\sec \phi| \\ \pi < \phi < \dfrac{3\pi}{2} & \dfrac{\pi}{2} < \phi < \pi \end{array} \quad (3.170)$$

$$\begin{array}{cc} \text{RER} & \text{LER} \\ Q_c = |\tan \phi| & W_c = \tan \phi \\ \gamma_c = \dfrac{\pi}{2} & \gamma_c = \dfrac{\pi}{2} \\ r = \sec \phi & r = \sec \phi \\ \dfrac{3\pi}{2} < \phi < 2\pi & 0 < \phi < \dfrac{\pi}{2} \end{array} \quad (3.171)$$

This summary can be condensed further to

$$\begin{array}{cc} \text{RER} & \text{LER} \\ Q_c = \tan \phi & W_c = \tan \phi \\ \gamma_c = -\dfrac{\pi}{2} & \gamma_c = \dfrac{\pi}{2} \\ r = |\sec \phi| & r = |\sec \phi| \\ \pi < \phi < 2\pi & 0 < \phi < \pi \end{array} \quad (3.172)$$

if the convention is adopted that Q_c and W_c can take on negative values. When $\tan \phi$ becomes negative, γ is increased by π.

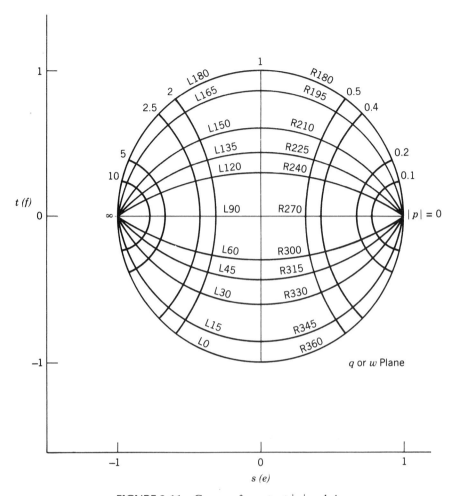

FIGURE 3.11 Curves of constant $|p|$ and ϕ.

It is seen from the summary that on the combined q and w plane the q-plane circles for $\pi < \phi < 3\pi/2$ coincide with the w-plane circles for $\pi/2 < \phi < \pi$. Likewise the q-plane circles for $3\pi/2 < \phi < 2\pi$ coincide with the w-plane circles for $0 < \phi < \pi/2$.

Figure 3.11 shows the circles of constant $|p|$ and constant phase angle ϕ on the q and w planes. The constant $|p|$ circles are the same for both planes. The constant ϕ circles are labeled with the value of ϕ and either R or L, meaning, respectively, right-elliptical polarization, in which case the chart is considered the q plane, or left-elliptical polarization, in which case it is the w plane. A large, carefully constructed chart like this one will allow one to find quickly any polarization parameter from a knowledge of others.

3.23. STOKES PARAMETERS

In his studies of partially polarized (quasi-monochromatic) light, Stokes introduced four quantities to characterize the amplitude and polarization of a wave. These are discussed more extensively in Chapter 7, but here they are presented for a strictly monochromatic wave, for which they are

$$G_0 = |E_x|^2 + |E_y|^2 \tag{3.173a}$$

$$G_1 = |E_x|^2 - |E_y|^2 \tag{3.173b}$$

$$G_2 = 2|E_x||E_y|\cos \phi \tag{3.173c}$$

$$G_3 = 2|E_x||E_y|\sin \phi \tag{3.173d}$$

where the terms of the equations are, as defined previously, the rectangular component amplitudes and phase difference of the wave.

It is obvious that the parameters are sufficient to describe both amplitude and polarization. Parameter G_0 gives the amplitude directly, while $|E_x|$ and $|E_y|$ can be found from G_0 and G_1. Then ϕ can be determined from either G_2 or G_3. Only three of the equations are independent for the monochromatic case (unlike the situation for partial polarization) since it is easily seen from (3.173) that

$$G_0^2 = G_1^2 + G_2^2 + G_3^2 \tag{3.174}$$

It is not difficult to relate the Stokes parameters to the terms used previously to describe the polarization ellipse. From (3.42) and the definitions of the Stokes parameters,

$$G_3 = G_0 \sin 2\epsilon \tag{3.175}$$

where ϵ is the ellipticity angle of the polarization ellipse.

From (3.40), giving the ellipse tilt angle, and (3.39),

$$\tan 2\tau = \frac{G_2}{G_1} \tag{3.176}$$

If this equation is combined with (3.174) and (3.175), the result is

$$G_1 = G_0 \cos 2\epsilon \cos 2\tau \tag{3.177}$$

Substitution of this equation into (3.176) in turn gives

$$G_2 = G_0 \cos 2\epsilon \sin 2\tau \tag{3.178}$$

3.24. THE POINCARÉ SPHERE

Collecting some of the previous equations, namely

$$G_0 = |E_x|^2 + |E_y|^2 \quad (3.173\text{a})$$

$$G_1 = G_0 \cos 2\epsilon \cos 2\tau \quad (3.177)$$

$$G_2 = G_0 \cos 2\epsilon \sin 2\tau \quad (3.178)$$

$$G_3 = G_0 \sin 2\epsilon \quad (3.175)$$

suggests a geometric interpretation of the Stokes parameters. G_1, G_2, and G_3 are the Cartesian coordinates of a point on a sphere of radius G_0. Then 2ϵ and 2τ are the latitude and azimuth angles measured to the point. This interpretation was introduced by Poincaré, and the sphere is called the Poincaré sphere. Such a sphere, with the Stokes parameters and tilt and ellipticity angles, is shown in Figure 3.12.

Since the amplitude of the wave can be described by G_0 and its polarization by G_1, G_2, and G_3, it is clear that any monochromatic wave can be described by a point on the Poincaré sphere. To every state of polarization there corresponds one point on the sphere and vice versa.

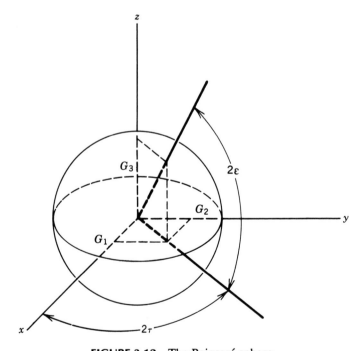

FIGURE 3.12 The Poincaré sphere.

3.25. SPECIAL POINTS ON THE POINCARÉ SPHERE

Left Circular

For this case $|E_x| = |E_y|$ and $\phi = \frac{1}{2}\pi$. Then the Stokes parameters become

$$G_0 = |E_x|^2 + |E_y|^2 = 2|E_x|^2$$
$$G_1 = |E_x|^2 - |E_y|^2 = 0$$
$$G_2 = 2|E_x||E_y|\cos\phi = 0$$
$$G_3 = 2|E_x||E_y|\sin\phi = 2|E_x|^2 = G_0$$

The point representing left-circular polarization is the north pole (the $+z$ axis) of the Poincaré sphere.

Right Circular

For this case

$$|E_x| = |E_y| \qquad \phi = -\tfrac{1}{2}\pi$$
$$G_0 = 2|E_x|^2 \qquad G_1 = G_2 = 0 \qquad G_3 = -G_0$$

which is the south pole of the sphere.

Left Elliptic

For this we have $0 < \phi < \pi$, and it follows from (3.173d) that $G_3 > 0$. All points for left-elliptic polarizations are plotted on the upper hemisphere.

Right Elliptic

For right-elliptic polarizations $\pi < \phi < 2\pi$, $G_3 < 0$, and the polarization points are in the lower hemisphere.

Linear

For linear polarizations, if $|E_x|$ and $|E_y|$ are nonzero, then $\phi = 0, \pi$; $G_3 = 0$; and the polarization points are at the equator. For linear vertical polarization the Poincaré sphere point is at the $-x$ axis intersection with the sphere and for linear horizontal it is at the $+x$ axis intersection. The $+y$ axis intersection corresponds to linear polarization with a tilt angle of $\pi/4$, and the $-y$ axis intersection to a tilt angle of $-\pi/4$.

3.26. OTHER RELATIONSHIPS BETWEEN THE VARIABLES

It may be seen from (3.173) that

$$\phi = \tan^{-1} \frac{G_3}{G_2} \tag{3.179}$$

and

$$|E_x| = \sqrt{\tfrac{1}{2}(G_0 + G_1)} \tag{3.180a}$$

$$|E_y| = \sqrt{\tfrac{1}{2}(G_0 - G_1)} \tag{3.180b}$$

From these equations we easily find that

$$P = \frac{G_2 + jG_3}{G_0 + G_1} \tag{3.181a}$$

$$p = \frac{-G_3 + jG_2}{G_0 + G_1} \tag{3.181b}$$

$$q = \frac{G_1 - jG_2}{G_0 - G_3} \tag{3.181c}$$

The polarization ratios can also be found in terms of the Poincaré sphere angles. If the Stokes parameters given by (3.175), (3.177), and (3.178) are substituted into the last equation, the polarization ratios become

$$P = \frac{\cos 2\epsilon \sin 2\tau + j \sin 2\epsilon}{1 + \cos 2\epsilon \cos 2\tau} \tag{3.182a}$$

$$p = \frac{-\sin 2\epsilon + j \cos 2\epsilon \sin 2\tau}{1 + \cos 2\epsilon \cos 2\tau} \tag{3.182b}$$

$$q = \frac{\cos 2\epsilon \cos 2\tau - j \cos 2\epsilon \sin 2\tau}{1 - \sin 2\epsilon} \tag{3.182c}$$

Finally, we note that the Poincaré sphere angles in terms of the Stokes parameters are

$$2\epsilon = \sin^{-1} \frac{G_3}{G_0} \tag{3.183a}$$

$$2\tau = \tan^{-1} \frac{G_2}{G_1} \tag{3.183b}$$

3.27. MAPPING THE POINCARÉ SPHERE ONTO A PLANE

Since the state of wave polarization can be represented by a point on the Poincaré sphere as well as by a point on the p, q, or w planes, it is not surprising that the Poincaré sphere may be mapped onto these complex planes. Before carrying out the mapping, however, let us find the Stokes parameters in terms of p. From (3.181b) and (3.174), we get

$$|p| = \sqrt{\frac{G_0 - G_1}{G_0 + G_1}} \tag{3.184}$$

which may be solved to give

$$\frac{G_1}{G_0} = \frac{1 - |p|^2}{1 + |p|^2} = \frac{1 - |P|^2}{1 + |P|^2} \tag{3.185}$$

From

$$p = jP = j\frac{|E_y|}{|E_x|}e^{j\phi}$$

and (3.173c) and (3.173d) for G_2 and G_3, it follows that

$$G_2 = 2|E_x|^2 \operatorname{Im}(p) \tag{3.186a}$$

$$G_3 = -2|E_x|^2 \operatorname{Re}(p) \tag{3.186b}$$

Substituting this equation into

$$\frac{G_2^2}{G_0^2} + \frac{G_3^2}{G_0^2} = 1 - \frac{G_1^2}{G_0^2} = \frac{4|p|^2}{\left(1 + |p|^2\right)^2} \tag{3.187}$$

which results from combining (3.174) and (3.185), we obtain

$$|E_x|^2 = \frac{G_0}{1 + |p|^2} \tag{3.188}$$

If the last equation is substituted into (3.186), the result is

$$\frac{G_2}{G_0} = \frac{2\operatorname{Im}(p)}{1 + |p|^2} = \frac{2\operatorname{Re}(P)}{1 + |P|^2} \tag{3.189a}$$

$$\frac{G_3}{G_0} = \frac{-2\operatorname{Re}(p)}{1 + |p|^2} = \frac{2\operatorname{Im}(P)}{1 + |P|^2} \tag{3.189b}$$

160 REPRESENTATIONS OF WAVE POLARIZATION

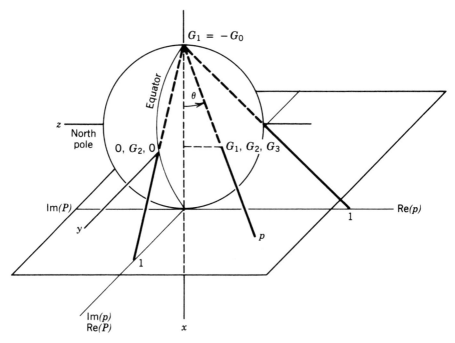

FIGURE 3.13 Stereographic projection of the Poincaré sphere onto the p and P planes.

The Poincaré sphere may be mapped onto the p and P planes by a stereographic projection. Proof of this assertion is deferred until the projection is described. First, as an aid in orienting the Poincaré sphere, the parameters for some special polarizations are tabulated.

Polarization	p	P	q	w	G_1	G_2	G_3
Right circular	1	$-j$	0	∞	0	0	$-G_0$
Left circular	-1	j	∞	0	0	0	G_0
Linear vertical	$j\infty$	∞	-1	-1	$-G_0$	0	0
Linear horizontal	0	0	1	1	G_0	0	0
Linear ($\tau = \pi/4$)	j	1	$-j$	$-j$	0	G_0	0
Linear ($\tau = -\pi/4$)	$-j$	-1	j	j	0	$-G_0$	0

The stereographic projection of a point on the Poincaré sphere onto the P or p plane is shown in Figure 3.13. The sphere is oriented so that its north pole–south pole (z) axis is parallel to the real axis of the p plane (imaginary axis of the P plane). Points are projected onto the plane by a ray from the sphere point farthest from the plane. Note that this projection point itself projects to infinity on the polarization plane. Recall from Figure 3.12 that G_1, G_2, and G_3 are rectangular coordinates of a point on the sphere measured, respectively, along the x, y, and z axes.

Now the Stokes parameters and the Poincaré sphere give both amplitude and polarization information about the wave, while points on the P or p plane describe only the wave polarization, so we expect something to be lost in the mapping. Since the projection of the south pole ($G_1 = G_2 = 0$; $G_3 = -G_0$) onto the plane gives the point $p = 1$, the sphere radius must be

$$G_0 = \tfrac{1}{2} \tag{3.190}$$

and we give up amplitude information, commonly not of great interest in polarization problems, in going from the sphere to the polarization plane.

It is easy to see from the preceding table and Figure 3.13 that all of the special points on the sphere project to the correct values of P and p on the polarization plane, and thus if the stereographic projection is valid, the sphere is oriented properly on the plane. To show that the stereographic projection is valid, we find from Figure 3.13 geometric relationships between the Stokes parameters and their projection on the p plane. These equations are then compared to previous equations for the Stokes parameters in terms of p to justify the mapping.

From Figure 3.13,

$$\sin\theta = \sqrt{\frac{G_2^2 + G_3^2}{(G_1 + G_0)^2 + G_2^2 + G_3^2}} = \sqrt{\frac{G_0^2 - G_1^2}{2G_0^2 + 2G_0 G_1}} = \sqrt{\frac{G_0 - G_1}{2G_0}} \tag{3.191}$$

Then

$$|p| = 2G_0 \tan\theta = 2G_0 \frac{\sin\theta}{\sqrt{1 - \sin^2\theta}} = 2G_0 \sqrt{\frac{G_0 - G_1}{G_0 + G_1}} \tag{3.192}$$

Since $G_0 = \tfrac{1}{2}$, this equation is identical to (3.184), and because (3.185) was derived from (3.184), it must be satisfied by the mapping.

From Figure 3.13 we note that for the point projected $G_1 > 0$, $G_2 > 0$, $G_3 < 0$, and for the corresponding polarization plane point $\mathrm{Re}(p) > 0$ and $\mathrm{Im}(p) > 0$. It follows that

$$G_2 = (\text{positive constant})\mathrm{Im}(p) = C\,\mathrm{Im}(p) \tag{3.193a}$$

$$G_3 = (\text{negative constant})\mathrm{Re}(p) = -C\,\mathrm{Re}(p) \tag{3.193b}$$

To determine the unknown constant in these equations, they and (3.185) are substituted into

$$G_1^2 + G_2^2 + G_3^2 = G_0^2 \tag{3.174}$$

giving

$$G_0^2 \left[\frac{1 - |p|^2}{1 + |p|^2}\right]^2 + C^2[\text{Im}(p)]^2 + C^2[\text{Re}(p)]^2 = G_0^2$$

from which it follows that

$$C = \frac{2G_0}{1 + |p|^2} \tag{3.194}$$

If this constant is used in (3.193), the result is

$$\frac{G_2}{G_0} = \frac{2\,\text{Im}(p)}{1 + |p|^2} \tag{3.195a}$$

$$\frac{G_3}{G_0} = \frac{-2\,\text{Re}(p)}{1 + |p|^2} \tag{3.195b}$$

which are the same as (3.189). We conclude then that the stereographic projection of the Poincaré sphere onto the p plane, with the sphere oriented

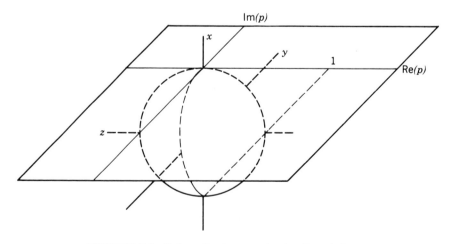

FIGURE 3.14 Poincaré sphere projection from below.

as in Figure 3.13, gives the correct correspondence between a polarization point on the P or p plane and a polarization point on the Poincaré sphere.

The P and p planes of Figure 3.13 are oriented awkwardly as we look down on them. The real part of P is plotted downward and $\text{Im}(P)$ to the left, while $\text{Re}(p)$ is plotted to the right and $\text{Im}(p)$ downward. This is not of great consequence, but if desired the orientation can be changed by using left-handed coordinates for the sphere (although this solution is not generally advisable) or by projecting the sphere onto the plane from below, as shown in Figure 3.14.

3.28. MAPPING ONTO THE q AND w PLANES

Since the transformation between p and q is linear, we might expect that the Poincaré sphere can be mapped onto the q plane. To obtain the mapping equations, substitute

$$p = p' + jp'' \qquad q = q' + q'' \qquad (3.196)$$

into (3.103) to obtain

$$p' = \frac{1 - |q|^2}{1 + 2q' + |q|^2} \qquad (3.197a)$$

$$p'' = \frac{-2q''}{1 + 2q' + |q|^2} \qquad (3.197b)$$

If these equations are substituted into (3.185) and (3.189), which give the Stokes parameters in terms of p, and if we note from (3.103) that

$$|p|^2 = \frac{1 - q' + |q|^2}{1 + 2q' + |q|^2} \qquad (3.198)$$

we find that

$$\frac{G_1}{G_0} = \frac{2q'}{1 + |q|^2} \qquad (3.199a)$$

$$\frac{G_2}{G_0} = \frac{-2q''}{1 + |q|^2} \qquad (3.199b)$$

$$\frac{G_3}{G_0} = -\frac{1 - |q|^2}{1 + |q|^2} \qquad (3.199c)$$

The w plane is the analog of the q plane for left-handed polarizations. If $w = 1/q^*$ is substituted into the previous equations, the mapping to the w plane is found to be

$$\frac{G_1}{G_0} = \frac{2w'}{1+|w|^2} \tag{3.200a}$$

$$\frac{G_2}{G_0} = \frac{-2w''}{1+|w|^2} \tag{3.200b}$$

$$\frac{G_3}{G_0} = \frac{1-|w|^2}{1+|w|^2} \tag{3.200c}$$

It is useful to collect all of the mapping equations together. They are

$$\frac{G_1}{G_0} = \frac{1-|P|^2}{1+|P|^2} = \frac{1-|p|^2}{1+|p|^2} = \frac{2\,\mathrm{Re}(q)}{1+|q|^2} = \frac{2\,\mathrm{Re}(w)}{1+|w|^2} \tag{3.201a}$$

$$\frac{G_2}{G_0} = \frac{2\,\mathrm{Re}(P)}{1+|P|^2} = \frac{2\,\mathrm{Im}(p)}{1+|p|^2} = \frac{-2\,\mathrm{Im}(q)}{1+|q|^2} = \frac{-2\,\mathrm{Im}(w)}{1+|w|^2} \tag{3.201b}$$

$$\frac{G_3}{G_0} = \frac{2\,\mathrm{Im}(P)}{1+|P|^2} = \frac{-2\,\mathrm{Re}(p)}{1+|p|^2} = -\frac{1-|q|^2}{1+|q|^2} = \frac{1-|w|^2}{1+|w|^2} \tag{3.201c}$$

This summary shows that the G_1 (or x) coordinate in the p-plane transformation is analogous to the $-G_3$ (or $-z$) coordinate in the q-plane transformation, and so on. The Poincaré sphere can therefore be projected onto the q or w plane by interchanging the axes of the Poincaré sphere of Figure 3.13. The appropriate interchanges are:

p Plane	q Plane	w Plane
x	$-z$	$-z$ or $+z$
y	$-y$	$-y$ or $+y$
z	$-x$	x or $-x$

The first coordinate interchange for the w plane gives a left-handed coordinate system for the sphere. Therefore, following a hint given at the end of the preceding section, we reverse the coordinates to give a right-handed system, and to compensate we reverse the direction in which $\mathrm{Im}(w)$ is plotted.

The projections of the Poincaré sphere onto the q and w planes are shown in Figure 3.15. Figure 3.15a shows that the lower hemisphere ($z < 0$) maps into the unit circle on the q plane. This is expected since the lower hemisphere contains all right elliptically polarized points, and so does the

MAPPING ONTO THE q AND w PLANES

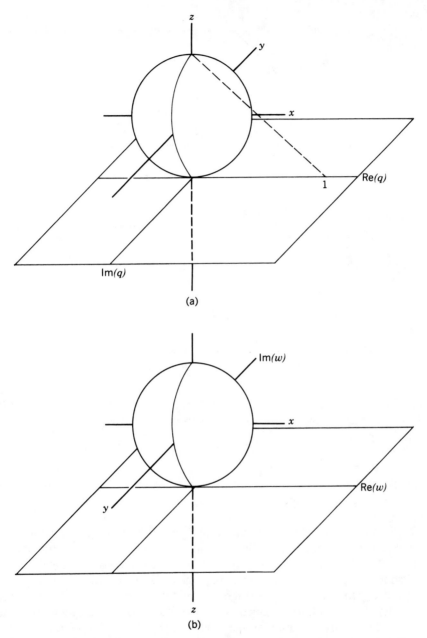

FIGURE 3.15 Projections onto the q and w planes.

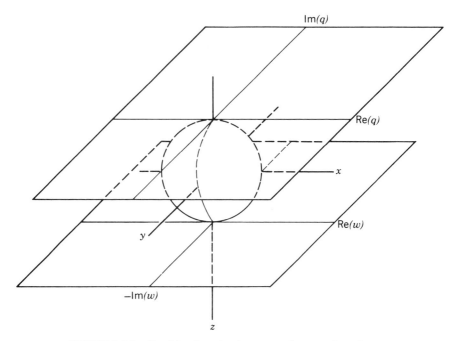

FIGURE 3.16 Combined projections onto the q and w planes.

unit circle on the q plane. Figure 3.15b shows that the hemisphere $z > 0$, which contains all left elliptically polarized points, maps into the unit circle on the w plane.

The two parts of Figure 3.15 may be combined into one drawing, and the combination is shown in Figure 3.16. In this figure the q plane is above the sphere and the w plane below, but the reverse is also possible. The origin of the projection ray for the q plane is $G_3 = +G_0$ and for the w plane it is $G_3 = -G_0$.

In Figure 3.17 the Poincaré sphere and the three projection planes p, q, and w are shown together. The drawing suggests that three other planes could be used in simple fashion to describe polarization states, since Figure 3.17 shows only three of a possible six planes. The transformations from G_1, G_2, and G_3 to the remaining three parameters may be found from Figure 3.17. Other polarization parameters can be found by interchanging the real and imaginary axes of one of the polarization planes (as in the relation $p = jP$). Of the large number of parameters that can be used it appears that the P (or p) and q (together with w for left-handed polarizations) parameters are the most useful. P is easily obtained from the rectangular wave components, and q is useful because *only* on the q plane will all right-handed polarizations be plotted in a finite region, and similarly for left-handed polarizations on the w plane.

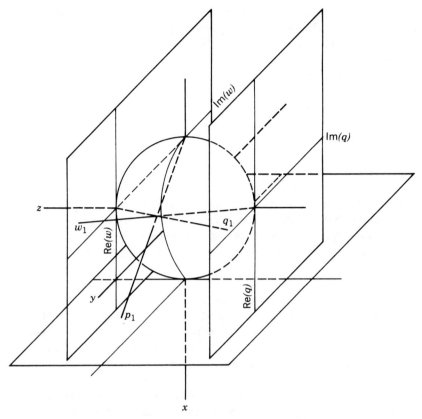

FIGURE 3.17 Poincaré sphere with three projection planes.

3.29. OTHER MAPS OF THE POINCARÉ SPHERE

The stereographic projection of the Poincaré sphere onto a tangent plane was covered in detail in the preceding section because if the projection point is chosen correctly, the projection plane is either the *P*-plane (polarization ratio) or the *q*-plane (circular polarization ratio). Maps of the Poincaré sphere onto a plane are desirable for other purposes also. In the next chapter we shall see that the distance on the sphere between the polarization value for a transmitting antenna and the point representing the receiving polarization ratio of a receiving antenna determines the polarization matching efficiency of the two antennas. In Chapter 9 we shall see that the locations on the Poincaré sphere of certain characteristic polarizations for a target and clutter can be of great value in separating targets and clutter or distinguishing between targets. It is easier to plot these polarization values on a plane than on a sphere, and of course essential to use a plane for computer plotting. Also, for wide dissemination of graphic information a plane representation is necessary. For these reasons, some other useful maps of the Poincaré sphere to a plane are discussed here. However, bear in mind that the Poincaré sphere is the natural surface to display polarization information if distances between polarization points are needed, and remember that *no* map of a sphere onto a plane is distortionless. A plane map can be quite useful, however, for visualizing polarization information if these reservations are kept in mind.

The procedures of map making for our earth have been developed to a high degree and will be useful for the polarization sphere. In some respects, however, mapping the Poincaré sphere onto a plane presents problems not always encountered by an earth map maker. Many earth maps need cover only relatively small regions (and that is of course true for some Poincaré sphere maps), but in many instances the wide variation in polarization parameters to be mapped requires a map covering very large regions of the sphere. Distortions will therefore be noticeable and perhaps significant. A critical factor is preserving distances between polarization points in the mapping, but no map can do this everywhere [5].

A *conformal* map of the earth preserves angles at all points. Lines of constant latitude (parallels) and longitude lines (meridians), which cross perpendicularly on earth, will cross perpendicularly on a conformal map (except for singular points). It follows that the shape of small areas (but not of large) is preserved in the mapping, although the area may depend on its map location. An *equal-area* map (*homolographic*, *homalographic*, or *authalic*) preserves area, so that a coin, for example, large or small, placed on any part of the map covers the same earth area as the coin when placed on another part of the map. In general, the shape of an earth region is not preserved. A rectangular area of the United States will not have the same shape as an equal rectangular area in Canada, for example, when projected onto an equal-area map, although it will have the same area. A map cannot be both conformal and equal-area, but it may be neither [5, 6]. An *equidistant* projection has the same distance scale between one or two points and every other point on the map or along certain lines, such as meridians. If these lines are chosen properly, distance (scale) errors in other regions may be minimized [5]. Again it must be noted that no map can have a constant scale throughout. Although no unqualified criteria for a map of the Poincaré sphere may be given, the significance of distances on the sphere argues for the value of an equidistant map or an equal-area (for a cluster of polarization points) projection, as compared to a conformal map.

It was noted earlier that the azimuth angle, corresponding to longitude on a map of the earth, is twice the tilt angle τ of the polarization ellipse, and the elevation angle, corresponding to earth latitude, is twice the ellipticity angle ϵ of the polarization ellipse. Equations given here will be in terms of tilt and ellipticity angles τ and ϵ, rather than azimuth and elevation angles. The radius of the Poincaré sphere is taken as one.

OTHER MAPS OF THE POINCARÉ SPHERE   169

On each map shown here are two triads of polarization points. In each triad one point, marked by an ×, is at the intersection of two perpendicular great circles. The other two points, marked by a O, are at an equal distance from the first point along the two intersecting great circles. For the first triad, the great circles are the equator and the 80° meridian (40° in τ). Spacing between points is 20° in both azimuth and elevation (10° in both τ and ϵ). For the second triad, the great circles are the 80° and 170° meridians. Spacing in elevation is again 20° (10° in ϵ). Thus on the Poincaré sphere, the points have the same spacing. Comparisons can be made on each map of spacing differences and distortion of the polar and equatorial triads. Comparisons between maps can be made of the shape distortion of the triads.

The Stereographic Projection

The lower hemisphere of the Poincaré sphere, containing all right-handed polarizations, can be projected into a circular region with a projection point at $G_3 = G_0$, or at the z-axis intersection with the sphere, as shown in Figure 3.15a. Similarly, all left-handed polarizations map into a circular region if the projection point is at $G_3 = -G_0$ and the projection plane is tangent to the sphere at $G_3 = G_0$.

The mapping equations for both hemispheres are

$$\rho = 2R \frac{\cos(2|\epsilon|)}{1 + \sin(2|\epsilon|)} \quad (3.202a)$$

$$\phi = 2(\tau - \tau_0) \quad (3.202b)$$

where ρ and ϕ are the radius and azimuth map values, and R is a scale multiplier to make the map the desired size. Figure 3.18 shows the resultant maps. Two are shown, although they are identical except for the sign of the

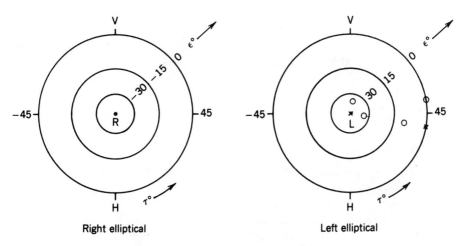

Right elliptical   Left elliptical

FIGURE 3.18   The stereographic projection.

ellipticity angle, because both are required to give full coverage of the Poincaré sphere.

Along a sphere meridian, which corresponds to a map radial, an incremental distance is $\Delta(2|\epsilon|)$ and on the map the corresponding incremental radial distance is

$$\Delta\rho = \frac{d\rho}{d(2|\epsilon|)}\Delta(2|\epsilon|) = -\frac{2R\Delta(2|\epsilon|)}{1+\sin(2|\epsilon|)}$$

The ratio of magnitudes is the radial scale factor

$$R_\rho = \frac{2R}{1+\sin(2|\epsilon|)} \qquad (3.203)$$

which varies from R at the map center to $2R$ at the edge.

In the azimuthal direction, an incremental distance on the Poincaré sphere is $\cos(2|\epsilon|)\Delta(2\tau)$, whereas on the map a corresponding change in azimuth is

$$\rho\Delta(2\tau) = \frac{2R\cos(2|\epsilon|)}{1+\sin(2|\epsilon|)}\Delta(2\tau)$$

The azimuthal scale factor is the ratio of these distance increments,

$$R_\phi = \frac{2R}{1+\sin(2|\epsilon|)} \qquad (3.204)$$

which is the same as the radial scale factor.

Relative to the map center, two polarization points near the outer edge of the map, corresponding to linear polarizations, appear to be twice as far apart as they actually are on the polarization sphere. One must also bear in mind, when using this stereographic projection, that near the outer edge two polarization points having the same tilt angle τ, with one point on the map for right-handed polarizations and the other on the left-handed map, are relatively close to each other.

Lambert Azimuthal Equal-Area Projection

On the stereographic map an area at the equator (linear polarization) will appear to be four times as large as an equal area at a pole (circular polarization). This can give a misleading conception to someone considering a group of characteristic polarization states. An improvement in the visualization may result if the Lambert azimuthal equal-area map in its polar aspect is used instead of a stereographic map. Unlike the stereographic

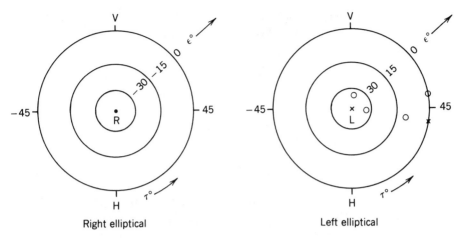

FIGURE 3.19 Lambert azimuthal equal-area projection.

projection, it is not perspective; that is, there is no single point from which rays are drawn to transfer sphere features to the map. Like the stereographic projection, two circular maps are required to show the complete Poincaré sphere.

The mapping equations are [5]

UPPER HEMISPHERE, LEFT-ELLIPTIC POLARIZATIONS:

$$\rho = 2R \sin(\pi/4 - \epsilon)$$
$$\phi = 2(\tau - \tau_0)$$
(3.205)

LOWER HEMISPHERE, RIGHT-ELLIPTIC POLARIZATIONS:

$$\rho = 2R \cos(\pi/4 - \epsilon)$$
$$\phi = 2(\tau - \tau_0)$$
(3.206)

The resulting maps are shown in Figure 3.19.*

The radial scale factor is readily found, using the upper hemisphere equation and proceeding as we did for the stereographic map, to be

$$R_\rho = \left|\frac{\Delta\rho}{2\Delta\epsilon}\right| = \frac{1}{2}\left|\frac{d\rho}{d\epsilon}\right| = R \cos\left(\frac{\pi}{4} - \epsilon\right)$$
(3.207)

It decreases from a value of R at the map center to $R/\sqrt{2}$ at the edge.

---

*Reference 5 reverses the map azimuth angle in the lower hemisphere. The direction is arbitrary, and the relationship between the upper- and lower-hemisphere maps is clearer in the form used here.

The azimuth scale factor is easily found as the ratio of an incremental azimuth distance on the map to the corresponding azimuth distance on the sphere. It is

$$R_\phi = \frac{\rho \Delta(2\tau)}{\cos 2\epsilon \Delta(2\tau)} = \frac{2R \sin(\pi/4 - \epsilon)}{\cos 2\epsilon} \quad (3.208)$$

Unlike the radial scale factor, this azimuth scale factor increases from center to edge, with a value of R at the center. It is easily shown that

$$R_\rho R_\phi = R^2 \quad (3.209)$$

Thus a unit area anywhere on the sphere is mapped to the plane as the area R^2.

Orthogonal Projection and Modified Form

Another projection that shows all points on the Poincaré sphere in two circular maps is the orthogonal projection of the points onto an equatorial plane (or a plane parallel to it) by rays perpendicular to the plane. Its use is widespread, and it is sometimes called *the* polarization chart [7, 8]. The mapping equations are

$$\begin{aligned} \rho &= R \cos 2\epsilon \\ \phi &= 2(\tau - \tau_0) \end{aligned} \quad (3.210)$$

with radial and azimuthal scale factors

$$\begin{aligned} R_\rho &= R \sin 2\epsilon \\ R_\phi &= R \end{aligned} \quad (3.211)$$

There is considerable distortion for linear polarizations, since the radial scale factor is zero at the equator.

Another form referred to by its primary users as a "modified polarization chart" [7] has mapping equations

$$\begin{aligned} \rho &= R(1 - |\tan \epsilon|) \\ \phi &= 2(\tau - \tau_0) \end{aligned} \quad (3.212)$$

with scale factors

$$\begin{aligned} R_\rho &= \frac{R}{2 \cos^2 \epsilon} \\ R_\phi &= R \frac{1 - |\tan \epsilon|}{\cos(2\epsilon)} \end{aligned} \quad (3.213)$$

There is considerably less distortion with this map than with the orthogonal projection.

The Mercator Projection

The conformal Mercator map of the earth is constructed symbolically by forming a cylinder around the earth tangent to it at the equator. After the earth features are projected onto the cylinder, it is unrolled to form a plane rectangular map. The process for the Poincaré sphere is the same. It is unlike the stereographic map in that there is not a single projection point from which rays are drawn to transfer sphere points to the cylinder.

The projection equations from Poincaré sphere angles to rectangular coordinates are [5]

$$x = 2R(\tau - \tau_0)$$
$$y = R \ln \tan(\pi/4 + \epsilon) \qquad (3.214)$$

where τ_0 is the tilt angle arbitrarily chosen to be the map center. The resulting map is shown in Figure 3.20. The vertical coordinate y becomes

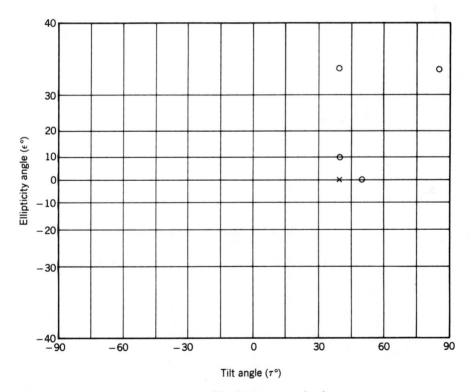

FIGURE 3.20 The Mercator projection.

174 REPRESENTATIONS OF WAVE POLARIZATION

infinite for $\epsilon = +\pi/4$, and the Mercator map cannot be used to display polarization states near circular.

Equidistant Cylindrical Projection: The τ–ϵ Map

The simplest map of the Poincaré sphere is a rectangular plane with tilt and ellipticity angles, as shown in Figure 3.21. As it is drawn, the horizontal dimension (for unit radius sphere) is $2\pi R$, with R a scale constant, and the vertical dimension is πR. The grid lines form squares. It is commonly referred to as a τ–ϵ map if the Poincaré sphere is mapped. It is neither equal-area nor conformal [5]. Rather, it falls into the category of equidistant projections mentioned previously. Along the meridians the map scale is constant. It is also constant for each parallel, but not the same for all parallels. Along a parallel an incremental map distance is $\Delta(2\tau)$, while on the sphere an incremental distance in the azimuth direction is $R \cos 2(|\epsilon|)\Delta(2\tau)$, giving a map scale factor in the horizontal or x direction of

$$R_x = \frac{R}{\cos 2(|\epsilon|)} \tag{3.215}$$

This scale factor is R for linear polarizations and approaches infinity for circular polarizations.

As the τ–ϵ map is drawn in Figure 3.21, with the horizontal, or azimuth, dimension twice that of the vertical, or elevation, dimension, the horizontal and vertical map scale factors are equal along the equator, or line of linear polarizations. Suppose, however, that it is drawn with horizontal extent D_x

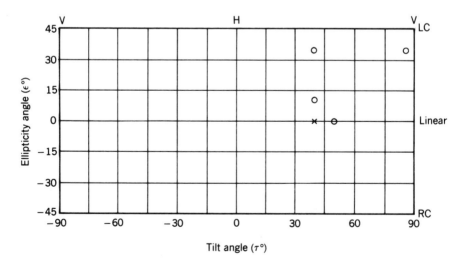

FIGURE 3.21 The τ–ϵ map with square grid.

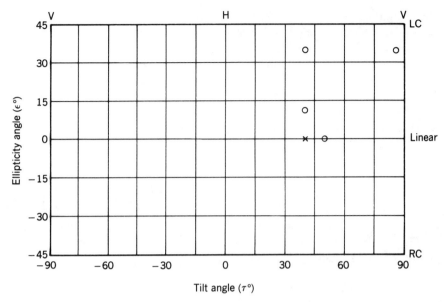

FIGURE 3.22 The τ–ϵ map with rectangular grid.

and vertical extent D_y instead of the values $2\pi R$ and πR of Figure 3.21. The horizontal and vertical scale factors then become

$$R_x = \frac{D_x}{2\pi \cos(2|\epsilon|)}$$
$$R_y = \frac{D_y}{\pi}$$
(3.216)

The scale factors may be set equal at some ellipticity angle other than zero by a choice of the ratio D_y/D_x. If the ratio is increased to 0.625 rather than 0.5 as in Figure 3.21, the map of Figure 3.22 results. For this map the horizontal and vertical scale factors are equal at $\epsilon = \pm 18.4°$, and for linear polarizations ($\epsilon = 0$) the vertical scale factor is 1.25 as great as the horizontal. Depending on the polarization points to be mapped, this choice may result in less distortion than the choice of equal scale factors for linear polarizations.

The τ–ϵ map can be particularly useful for presenting computer-plotted histograms of characteristic polarization states in a pseudo-three-dimensional form.

Mollweide Equal-Area Projection

The Mollweide map, like the τ–ϵ map, shows the entire polarization sphere, but with less distortion. The map form is easily constructed since the parallels

176 REPRESENTATIONS OF WAVE POLARIZATION

are straight lines and each meridian, together with a corresponding meridian on the other side of the central meridian, forms an ellipse. The map is shown in Figure 3.23. The equator, or line of linear polarizations, is the central horizontal parallel of the Mollweide map. The central meridian is arbitrarily chosen, and for this map is the line of zero tilt angle. The parallels are unequally spaced, and their spacing may be obtained from the mapping equations. The central parallel is twice the length of the central meridian, which is appropriate for a map representing the complete Poincaré sphere.

The mapping equations are [5]

$$x = \frac{4\sqrt{2}}{\pi} R(\tau - \tau_0) \cos \theta \tag{3.217}$$

$$y = \sqrt{2} R \sin \theta$$

where θ is a parametric angle given by

$$2\theta + \sin(2\theta) = \pi \sin(2\epsilon) \tag{3.218}$$

This must be solved by iteration. A suggested method is to use a Newton–Raphson iteration with this equation [5],

$$\Delta \theta' = -\frac{\theta' + \sin \theta' - \pi \sin(2\epsilon)}{1 + \cos \theta'} \tag{3.219}$$

that converges rapidly (but slowly at the poles). After it is solved,

$$\theta = \frac{\theta'}{2} \tag{3.220}$$

The scale factors are not as easily obtained for the Mollweide map as for the other maps considered, but examination of the map at the equator— bearing in mind that equal increments in azimuth and elevation angles at the equator of a sphere give equal distance increments on the surface—shows that sphere distances are increased by mapping along a meridian compared to mapping along the equator. This stretching in the north–south direction is about 23% at the equator, disappears at ellipticity angles of about $\pm 20.4°$, and becomes compression at greater ellipticity angles [5].

Aitoff–Hammer Equal-Area Projection

This map is variously called the Aitoff, Hammer, Aitoff–Hammer, and Hammer–Aitoff map. In essence it is a Lambert azimuthal equal-area projection in an equatorial aspect (rather than the polar aspect considered earlier)

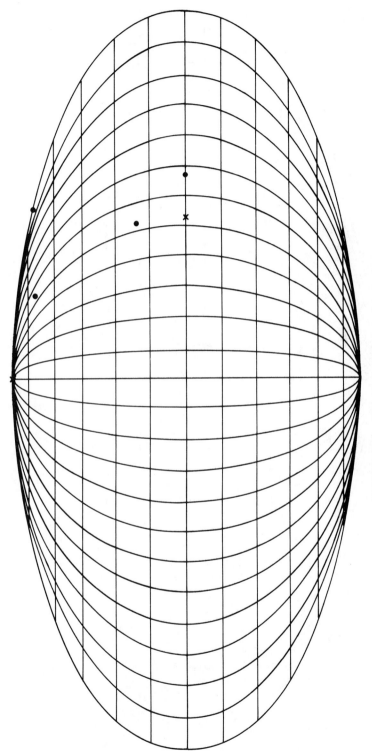

FIGURE 3.23 The Mollweide equal-area map.

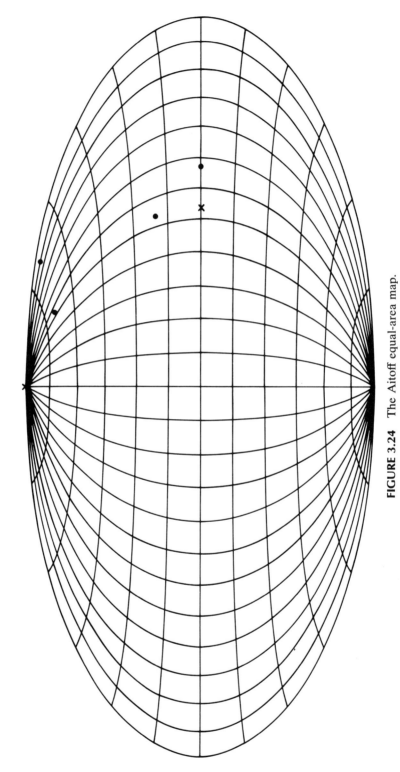

FIGURE 3.24 The Aitoff equal-area map.

with a modification suggested by Aitoff [6] that converts the projection to a full sphere. The projection was devised by Hammer in 1892 [5].

In appearance the map, shown in Figure 3.24, is remarkably similar to the Mollweide map. It too is constructed in an ellipse with the horizontal major axis, representing the sphere equator, twice the length of the minor axis. It also shares the equal-area characteristic.

The mapping equations, restricted to the equatorial aspect so that linear polarizations lie on the horizontal center line of the map, are [9]

$$U = \cos^{-1}[\cos(\tau - \tau_0)\cos 2\epsilon]$$
$$V = \sqrt{2}\,\sin(U/2) \tag{3.221}$$
$$W = \sin^{-1}[\sin 2\epsilon/\sin U]$$

where U, V, and W are intermediate parameters. The map coordinates are then found from

$$\begin{aligned} x &= 2RV\cos W \\ y &= RV\sin W \end{aligned} \tag{3.222}$$

with R a scale constant.

REFERENCES

1. M. Born and E. Wolf, *Principles of Optics*, 3rd ed., Pergamon, Oxford, 1965.
2. R. P. Feynman, R. B. Leighton, and M. Sands, *The Feynman Lectures on Physics*, Addison-Wesley, Reading, MA, 1973.
3. "IEEE Standard Definitions of Terms for Antennas," IEEE Standard 145-1983, The Institute of Electrical and Electronics Engineers, New York, 1983.
4. V. H. Rumsey, "Transmission Between Elliptically Polarized Antennas," *Proc. IRE*, **39**(5), 535–540, May, 1951.
5. J. P. Snyder, *Map Projections—A Working Manual*, US Geological Survey Professional Paper 1385, US Government Printing Office, Washington, DC, 1987.
6. C. H. Deetz and O. S. Adams, *Elements of Map Projection*, 5th ed., US Department of Commerce, Coast and Geodetic Survey, Special Publication No. 68, US Government Printing Office, Washington, DC, 1944.
7. A. J. Poelman and J. R. F. Guy, "Polarization Information Utilization in Primary Radar," *Inverse Methods in Electromagnetic Imaging*, W-M. Boerner, Ed., Part 1, D. Reidel, Dordrecht, Holland, 1985, pp. 521–572.
8. D. Giuli, "Polarization Diversity in Radars," *Proc. IEEE*, **74**(2), 245–269, February, 1986.
9. R. S. Raven, "On the Null Representation of Polarization Parameters," *Inverse Methods in Electromagnetic Imaging*, W-M. Boerner, Ed., Part 1, D. Reidel, Dordrecht, Holland, 1985, pp. 629–641.

180 REPRESENTATIONS OF WAVE POLARIZATION

PROBLEMS

3.1. A wave travels in the z direction with electric field

$$\mathscr{E} = 2\mathbf{u}_x \cos(\omega t - kz) + \mathbf{u}_y \cos(\omega t - kz - \pi/6)$$

Find its polarization ratio P and the tilt angle, axial ratio, and rotation sense of the polarization ellipse.

3.2. The field of a uniform plane wave is

$$\mathbf{E} = \left(2\mathbf{u}_y + 1e^{-j\pi/6}\mathbf{u}_z\right)e^{-jkx}$$

Find the tilt angle, axial ratio, and rotation sense of the polarization ellipse.

3.3. A left-handed elliptically polarized wave has a tilt angle of 30° and an axial ratio of 2. Find the polarization ratios P and q (or w).

3.4. An antenna radiates a wave in the z direction with a polarization ratio $P = 2\exp(j\pi/4)$. The antenna is rotated in the xy plane by 30° from the x axis toward the y axis. Find the new value of P.

3.5. Find the points on the q (or w) plane representing the two polarization states of Problem 3.4. Can you draw a general conclusion about the change in polarization state caused by rotating the antenna?

3.6. Consider the stereographic projection of the Poincaré sphere from the point $G_1 = G_0$ onto a complex plane. Find the equation for the complex variable on this plane in terms of the Stokes parameters.

3.7. Show that the polarization ellipse for the magnetic field of a uniform plane wave is identical to that of the electric field except that it is rotated by angle $\pi/2$ around the z axis.

3.8. Make the substitution outlined immediately after (3.188) to obtain (3.189).

3.9. Use (3.181a) and the relations between P, p, and q to obtain (3.181c).

3.10. Electric field intensity in rectangular component form is

$$\mathbf{E} = \left(2\mathbf{u}_x + 1e^{-j\pi/6}\mathbf{u}_y\right)e^{-jkz}$$

We wish to write it as the sum of two elliptical waves

$$\mathbf{E} = \mathbf{E}_1 + \mathbf{E}_2 = (E_1\mathbf{u}_1 + E_2\mathbf{u}_2)e^{-jkz}$$

where E_1 has polarization ratio

$$P_1 = 2e^{j\pi/4}$$

Find E_1 and E_2.

3.11. Write the wave of Problem 3.10 as the sum of two circular waves $\mathbf{E}_R$ and $\mathbf{E}_L$.

3.12. Construct an orthogonal projection of the Poincaré sphere onto the equatorial plane. Use the same ellipticity angle circles used for the Lambert projection of Figure 3.19. Is the orthogonal projection conformal? Equal area?

3.13. In xy coordinates, the wave incident on a target is

$$\mathbf{E}^i = \begin{bmatrix} E_x^i \\ E_y^i \end{bmatrix}$$

A new coordinate system is to be used, determined by rotating the old system by angle θ ($x \to y$). Find the incident wave $\mathbf{E}^{i'}$ in the new coordinates.

CHAPTER FOUR

Polarization Matching of Antennas

4.1. INTRODUCTION

It is obvious that when two antennas are used in a communication system they should be matched in polarization so that the available power at the receiving antenna can be fully utilized. In this chapter a polarization match factor is developed and is given in terms of the standard polarization parameters. The relationship between the effective length of a receiving antenna and the field components of an incident wave that yield maximum power is developed. In the final section a step-by-step process is outlined for obtaining the power received when two antennas are mismatched in polarization and do not have their main beam axes pointing at each other. This topic is not treated in most of the standard texts on antenna theory.

4.2. EFFECTIVE LENGTH OF AN ANTENNA

The electric field in the radiation zone of a dipole antenna that is short compared to a free-space wavelength, as shown in Figure 4.1, is given by

$$E_\theta(r, \theta, \phi) = \frac{jZ_0 IL}{2\lambda r} e^{-jkr} \sin \theta \qquad (4.1)$$

where Z_0 is the intrinsic impedance of free space, k the free space propagation constant, λ the wavelength, and I the current into the antenna terminals of Figure 4.1.

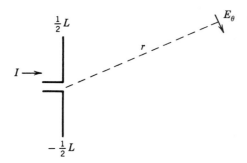

FIGURE 4.1 Short dipole antenna.

Equation (4.1) may be generalized to give the transmitted field of any antenna, thus [1]

$$\mathbf{E}^t(r,\theta,\phi) = \frac{jZ_0 I}{2\lambda r} e^{-jkr} \mathbf{h}(\theta,\phi) \quad (4.2)$$

where θ is the colatitude angle of Figure 4.1 and ϕ is the azimuth angle. The current I is an input current at an arbitrary pair of terminals. Equation (4.2) describes a general antenna in terms of its *effective length*, $\mathbf{h}(\theta,\phi)$. The effective length does not necessarily correspond to a physical length of the antenna, although there is a correspondence for the dipole. In fact, comparison of (4.1) and (4.2) shows that the effective length of the short dipole antenna is

$$\mathbf{h} = \mathbf{u}_\theta h_\theta = \mathbf{u}_\theta L \sin\theta \quad (4.3)$$

We see that $\mathbf{h}$ is not fixed for an antenna but depends on the angle θ (and more generally on ϕ) at which we measure the radiated field.

As mentioned, I is the current at an arbitrary pair of terminals, and it follows that the effective length $\mathbf{h}$ depends on the choice of terminal pair. Further, if $\mathbf{E}^t$ is to describe an elliptically polarized field, it must be complex, and therefore $\mathbf{h}$ is a complex vector. With the proper choice of coordinate system, $\mathbf{E}^t$ and $\mathbf{h}$ will have only two components since in the radiation zone $\mathbf{E}^t$ has no radial component.

4.3. RELATION BETWEEN EFFECTIVE LENGTH AND GAIN

The directivity of an antenna is, from Section 1.8, the ratio of the radiation intensity in a specified direction to the radiation intensity averaged over all

184 POLARIZATION MATCHING OF ANTENNAS

space. From

$$U(\theta,\phi) = r^2 \mathscr{P}(r,\theta,\phi) = \frac{r^2}{2Z_0} \mathbf{E} \cdot \mathbf{E}^* \quad (4.4)$$

and

$$U_{av} = \frac{1}{4\pi} \iint_{4\pi} U \, d\Omega \quad (4.5)$$

we can obtain the directivity

$$D(\theta,\phi) = \frac{\mathbf{E} \cdot \mathbf{E}^*}{\frac{1}{4\pi} \iint_{4\pi} \mathbf{E} \cdot \mathbf{E}^* \, d\Omega} = \frac{\mathbf{h} \cdot \mathbf{h}^*}{\frac{1}{4\pi} \iint_{4\pi} \mathbf{h} \cdot \mathbf{h}^* \, d\Omega} \quad (4.6)$$

and gain

$$G(\theta,\phi) = \frac{e \mathbf{h} \cdot \mathbf{h}^*}{\frac{1}{4\pi} \iint_{4\pi} \mathbf{h} \cdot \mathbf{h}^* \, d\Omega} \quad (4.7)$$

where e is the antenna efficiency.

The gain and effective area of the antenna are related by $\lambda^2/4\pi$, so that

$$A_e(\theta,\phi) = \frac{\lambda^2 e \mathbf{h} \cdot \mathbf{h}^*}{\iint_{4\pi} \mathbf{h} \cdot \mathbf{h}^* \, d\Omega} \quad (4.8)$$

Note that (θ,ϕ) now refers to the direction from which the wave comes to strike the receiving antenna.

Radiation resistance of an antenna, from Section 1.8, is the ratio of the power radiated to the square of the rms current at arbitrarily chosen terminals. Then

$$R_r = \frac{2}{I^2} \frac{r^2}{2Z_0} \iint_{4\pi} \mathbf{E} \cdot \mathbf{E}^* \, d\Omega \quad (4.9)$$

and if we use (4.2)

$$R_r = \frac{Z_0}{4\lambda^2} \iint_{4\pi} \mathbf{h} \cdot \mathbf{h}^* \, d\Omega \quad (4.10)$$

In terms of a total antenna resistance $R_a \, (= R_r + R_{\text{loss}})$, we can use (1.90) and obtain

$$R_a = \frac{Z_0}{4e\lambda^2} \iint_{4\pi} \mathbf{h} \cdot \mathbf{h}^* \, d\Omega \qquad (4.11)$$

4.4. RECEIVED VOLTAGE

We defined the effective length of an antenna in terms of the radiation field produced by it. In this section we show that the open-circuit voltage induced in the antenna by an externally produced field is proportional to this effective length; in fact, some authors *define* effective length in terms of the open-circuit voltage produced when the antenna is receiving a wave.

By the principle of reciprocity, if two antennas are fed by equal current sources the open-circuit voltage produced across the terminals of antenna 1 by the current source feeding antenna 2 is the same as the open-circuit voltage produced across the terminals of antenna 2 by the current source feeding antenna 1.

We apply this principle to determine the open-circuit voltage across the terminals of a general antenna whose transmitted field is assumed to interact with a short dipole, as shown in Figure 4.2 together with the coordinate system used and the assumed current directions and voltage polarities. Note that the same rectangular coordinate system is used for both antennas, although θ', ϕ' are not equal to θ, ϕ.

First we let the general antenna, fed by a current source of 1 A, transmit a wave toward the short dipole. Its field at the dipole is

$$\mathbf{E}^t = \frac{jZ_0}{2\lambda r} e^{-jkr} \mathbf{h} \qquad (4.12)$$

and the open-circuit voltage across the dipole terminals, with the polarity shown in Figure 4.2, is

$$V_2 = \mathbf{E}^t \cdot \mathbf{L} \qquad (4.13)$$

where $\mathbf{L}$ is the vector length of the dipole. Here we are considering that the dipole has infinitesimal length so that the incident field $\mathbf{E}^t$ is constant over the dipole length. The dipole may be arbitrarily oriented, but since $\mathbf{E}^t$ has no radial component, a radial component of the dipole length will not contribute to the received voltage. Then, still using the coordinate system of Figure 4.2,

$$V_2 = E_\theta^t L_\theta + E_\phi^t L_\phi \qquad (4.14)$$

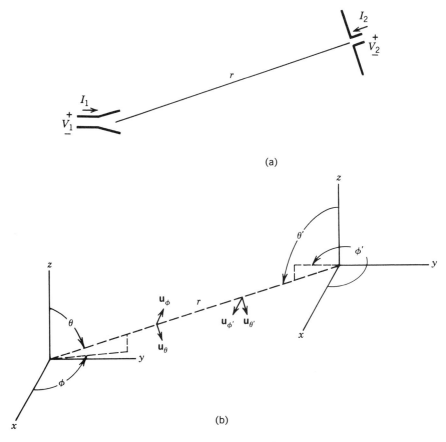

FIGURE 4.2 General antenna and short dipole: (*a*) antennas; (*b*) coordinates and unit vectors.

where the dipole components L_θ and L_ϕ are given by

$$L_\theta = \mathbf{u}_\theta \cdot \mathbf{L} \qquad L_\phi = \mathbf{u}_\phi \cdot \mathbf{L} \tag{4.15}$$

Combining (4.12) and (4.13) gives the voltage induced across the open dipole terminals by the incident wave from the general antenna. It is

$$V_2 = \frac{jZ_0}{2\lambda r} e^{-jkr} \mathbf{h} \cdot \mathbf{L} \tag{4.16}$$

Next, suppose the dipole, fed by a 1-A current source, is transmitting, and the general antenna, with open terminals, is receiving. The field produced at the general antenna, 1, is given by

$$E^i_{\theta'} = \frac{jZ_0}{2\lambda r} e^{-jkr} L_{\theta'} \tag{4.17a}$$

$$E^i_{\phi'} = \frac{jZ_0}{2\lambda r} e^{-jkr} L_{\phi'} \tag{4.17b}$$

where we continue to use the same coordinate system but note that θ', ϕ' differ from θ, ϕ. We note from Figure 4.2 that although the angles just mentioned are different we have

$$\mathbf{u}_\theta = \mathbf{u}_{\theta'} \qquad \mathbf{u}_\phi = -\mathbf{u}_{\phi'} \tag{4.18}$$

It follows that

$$E^i_{\theta'} = E^i_\theta \tag{4.19a}$$

$$E^i_{\phi'} = -E^i_\phi \tag{4.19b}$$

$$L_{\theta'} = L_\theta \tag{4.19c}$$

$$L_{\phi'} = -L_\phi \tag{4.19d}$$

and therefore the wave incident on antenna 1 is

$$E^i_\theta = \frac{jZ_0}{2\lambda r} e^{-jkr} L_\theta \tag{4.20a}$$

$$-E^i_\phi = \frac{jZ_0}{2\lambda r} e^{-jkr} (-L_\phi) \tag{4.20b}$$

or

$$\mathbf{E}^i = \frac{jZ_0}{2\lambda r} e^{-jkr} \mathbf{L} \tag{4.21}$$

The open-circuit voltage induced in the general antenna, 1, is V_1, which by the reciprocity theorem is equal to V_2, as given by (4.16). Then from (4.16) and the reciprocity theorem, we get

$$V_1 = V_2 = \frac{jZ_0}{2\lambda r} e^{-jkr} \mathbf{L} \cdot \mathbf{h} \tag{4.22}$$

and if we recognize that the first part of this expression is the incident wave of (4.21), we get an expression for the received voltage across the open terminals of antenna 1, in terms of the incident field on it, $\mathbf{E}^i$, and its effective length $\mathbf{h}$. It is

$$V_1 = \mathbf{E}^i \cdot \mathbf{h} \tag{4.23}$$

188 POLARIZATION MATCHING OF ANTENNAS

It should be noted that in this equation both $\mathbf{E}^i$ and $\mathbf{h}$ are measured in the same coordinate system (in contrast to a situation to be discussed later). The voltage V_1 is in general a complex phasor voltage, since both $\mathbf{E}^i$ and $\mathbf{h}$ are complex. Finally, in specifying the effective length $\mathbf{h}$ of an antenna, a terminal pair at which input current is to be measured must be specified. Then V_1 is the open-circuit voltage measured across those terminals.

4.5. MAXIMUM RECEIVED POWER

It is reasonable to believe, from looking at (4.23), that by proper selection of the effective length $\mathbf{h}$ of a receiving antenna we can increase the open-circuit voltage and hence the received power. If we neglect such extraneous problems as impedance mismatch, the power received by the general antenna is proportional to the square of the magnitude of the open-circuit voltage; thus, using an equality rather than a proportional symbol (an inconsequential action since we later consider a power ratio), we have

$$W = VV^* = |\mathbf{E}^i \cdot \mathbf{h}|^2 = |h_\theta E_\theta^i + h_\phi E_\phi^i|^2 \qquad (4.24)$$

where an appropriate coordinate system is used so that $\mathbf{h}$ has only two components and so only two are needed for $\mathbf{E}^i$.

Let us define

$$h_\theta = |h_\theta| e^{j\alpha} \qquad (4.25a)$$

$$\frac{h_\phi}{|h_\phi|} = \frac{h_\theta}{|h_\theta|} e^{j\delta_1} \qquad (4.25b)$$

$$\frac{E_\theta^i}{|E_\theta^i|} = \frac{h_\theta}{|h_\theta|} e^{j\beta} \qquad (4.25c)$$

$$\frac{E_\phi^i}{|E_\phi^i|} = \frac{E_\theta^i}{|E_\theta^i|} e^{j\delta_2} \qquad (4.25d)$$

with δ_1 the phase angle by which h_ϕ leads h_θ, β the angle by which E_θ^i leads h_θ, and δ_2 the angle by which E_ϕ^i leads E_θ^i. Using these equations, the received power, (4.24), becomes

$$W = \left||h_\theta| |E_\theta^i| + |h_\phi| |E_\phi^i| e^{j(\delta_1+\delta_2)}\right|^2 \qquad (4.26)$$

where the angle $2\alpha + \beta$ has been removed as common to both terms in the sum.

Clearly W is a maximum if

$$\delta_1 + \delta_2 = 0 \tag{4.27}$$

and has value

$$W_m = \left[|h_\theta|\,|E^i_\theta| + |h_\phi|\,|E^i_\phi|\right]^2 \tag{4.28}$$

Now W_m can be maximized further, for a fixed incident wave, $\mathbf{E}^i$, by varying $|h_\theta|$ or $|h_\phi|$. Certainly, however, there must be some constraint on the lengths; otherwise W_m could be made as great as we please by increasing $|h_\theta|$ and $|h_\phi|$ arbitrarily. To determine this constraint, return to (4.2), which gives the transmitted field of an antenna in terms of its effective length. The transmitted Poynting vector, from (4.2), is obviously proportional to $\mathbf{h}\cdot\mathbf{h}^*$. Then a reasonable constraint on an antenna is that this Poynting vector remain constant as $\mathbf{h}$ is varied. Therefore, we vary $\mathbf{h}$ to maximize W_m in (4.28) with the constraint

$$\mathbf{h}\cdot\mathbf{h}^* = |h_\theta|^2 + |h_\phi|^2 = C \tag{4.29}$$

Substituting this equation into the equation for W_m gives

$$W_m = \left[|h_\theta|\,|E^i_\theta| + \left(C - |h_\theta|^2\right)^{1/2}|E^i_\phi|\right]^2 \tag{4.30}$$

and differentiating with respect to $|h_\theta|$ in order to maximize W_m gives

$$\frac{\partial W_m}{\partial |h_\theta|} = 2\left[|h_\theta|\,|E^i_\theta| + \left(C^2 - |h_\theta|^2\right)^{1/2}|E^i_\phi|\right]\left[|E^i_\theta| - \frac{|h_\theta|\,|E^i_\phi|}{\left(C - |h_\theta|^2\right)^{1/2}}\right] = 0 \tag{4.31}$$

from which it is clear that

$$|E^i_\theta| - \frac{|h_\theta|\,|E^i_\phi|}{\left(C - |h_\theta|^2\right)^{1/2}} = |E^i_\theta| - \frac{|h_\theta|}{|h_\phi|}|E^i_\phi| = 0 \tag{4.32}$$

or

$$\frac{|h_\theta|}{|h_\phi|} = \frac{|E^i_\theta|}{|E^i_\phi|} \tag{4.33}$$

It seems quite reasonable that selecting $\mathbf{h}$ according to the last equation will give maximum received power, rather than minimum, since if $\mathbf{E}^i$ has a large θ component we expect a large $|h_\theta|$ to give best reception. This belief can be verified by substituting the last relationship into (4.28), which is rewritten as

$$W_m = |h_\theta|^2|E_\theta^i|^2 + |h_\phi|^2|E_\phi^i|^2 + |h_\theta||E_\theta^i||h_\phi||E_\phi^i| + |h_\theta||E_\theta^i||h_\phi||E_\phi^i| \tag{4.34}$$

In the third term of W_m, we make the substitution from (4.33) that

$$|E_\theta^i||h_\phi| = |h_\theta||E_\phi^i| \tag{4.35}$$

and in the fourth term it is made in reverse. This leads to the maximum value of W_m, which is the largest possible value of received power,

$$\begin{aligned}W_{mm} &= |h_\theta|^2|E_\theta^i|^2 + |h_\phi|^2|E_\phi^i|^2 + |h_\theta|^2|E_\phi^i|^2 + |h_\phi|^2|E_\theta^i|^2 \\ &= \left(|h_\theta|^2 + |h_\phi|^2\right)\left(|E_\theta^i|^2 + |E_\phi^i|^2\right)\end{aligned} \tag{4.36}$$

This is quite obviously maximum power rather than minimum. Finally, the last equation can be written as

$$W_{mm} = (\mathbf{h} \cdot \mathbf{h}^*)(\mathbf{E}^i \cdot \mathbf{E}^{i*}) = |\mathbf{h}|^2|\mathbf{E}^i|^2 \tag{4.37}$$

There may be some concern on the part of the reader that (4.29) is a legitimate constraint. While we vary $\mathbf{h}$ for maximum *received* power, why should we apply a constraint that is meaningful only for the *transmitting* case? For this reason, we return to (4.28), assume that $\mathbf{h}$ is fixed, and vary $\mathbf{E}^i$ to maximize the power. Now in this situation it is quite clear that only a fixed amount of power impinges on the receiving antenna. Therefore

$$\mathbf{E}^i \cdot \mathbf{E}^{i*} = C \tag{4.38}$$

Using this equation makes W_m become

$$W_m = \left[|h_\theta||E_\theta^i| + |h_\phi|\left(C - |E_\theta^i|^2\right)^{1/2}\right]^2 \tag{4.39}$$

and differentiation with respect to $|E_\theta^i|$ gives

$$\frac{\partial W_m}{\partial |E_\theta^i|} = 2\left[|h_\theta||E_\theta^i| + |h_\phi|\left(C - |E_\theta^i|^2\right)^{1/2}\right]\left[|h_\theta| - \frac{|h_\phi||E_\theta^i|}{\left(C - |E_\theta^i|^2\right)^{1/2}}\right] = 0 \tag{4.40}$$

from which it follows that

$$\frac{|h_\theta|}{|h_\phi|} = \frac{|E_\theta^i|}{|E_\phi^i|} \qquad (4.41)$$

which is the same condition we arrived at previously.

The last equation gives one condition on **h** for maximum power reception. The other is given by

$$\delta_1 + \delta_2 = 0 \qquad (4.27)$$

where δ_1 is the angle by which h_ϕ leads h_θ and δ_2 is the angle by which E_ϕ^i leads E_θ^i.

If (4.25b) and (4.27) are combined with (4.41) the result is

$$\frac{h_\phi}{h_\theta} = \frac{|E_\phi^i|}{|E_\theta^i|} e^{-j\delta_2} \qquad (4.42)$$

The term on the right in this equation is recognizable from (4.25d), and the equation may then be written as

$$\frac{h_\phi}{h_\theta} = \left(\frac{E_\phi^i}{E_\theta^i}\right)^* \qquad (4.43)$$

In rectangular coordinates, the equation for maximum received power is

$$\frac{h_y}{h_x} = \left(\frac{E_y^i}{E_x^i}\right)^* \qquad (4.44)$$

where **E** and **h** are given in the same coordinate system.

4.6. POLARIZATION EFFICIENCY — MATCH FACTOR

If we maintain the same degree of impedance matching for an antenna as we vary its polarization properties, then the ratio of actual power received to that received under the most favorable circumstances is, from (4.24) and (4.37),

$$\rho = \frac{|\mathbf{E}^i \cdot \mathbf{h}|^2}{|\mathbf{E}^i|^2 |\mathbf{h}|^2} \qquad (4.45)$$

We refer to ρ as the *polarization match factor* or the *polarization efficiency*. Its range is obviously $0 \leq \rho \leq 1$.

192 POLARIZATION MATCHING OF ANTENNAS

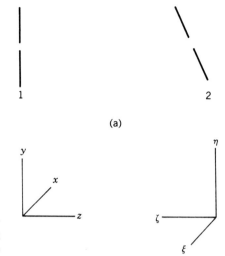

FIGURE 4.3 Antennas and coordinate systems used to develop the polarization match factor: (*a*) antennas; (*b*) coordinate systems.

The polarization match factor shows how well a receiving antenna of effective length **h** is matched in polarization to an incoming wave. Now let us recognize that the incoming wave was transmitted by another antenna and introduce the polarization properties of that antenna into the problem.

Figure 4.3 shows two antennas in a transmit–receive configuration. The polarization properties of the transmitting antenna 1 are described by the right-handed coordinate system *xyz* adjacent to it since the polarization of a wave is normally based on a right-handed coordinate system with one axis pointing in the direction of wave travel. The receiving antenna is described by the right-handed $\xi\eta\zeta$ system. The antennas need not have their main beams pointed at each other, but the z and ζ axes are antiparallel and each points at the other antenna.

The incident wave from antenna 1 may be written as

$$\mathbf{E}_1^i = E_{1x}\left(\mathbf{u}_x + \frac{|E_{1y}|}{|E_{1x}|}e^{j\phi_1}\mathbf{u}_y\right) = E_{1x}(\mathbf{u}_x + P_1\mathbf{u}_y) \qquad (4.46)$$

where P_1 is the polarization ratio of the incident wave produced by antenna 1.

The *polarization ratio of an antenna* is defined as the polarization ratio of the field that it transmits (far field). Therefore P_1 is the polarization ratio of antenna 1. It is a function of θ and ϕ, the colatitude and azimuth angles measured for the transmission direction.

If antenna 2 were transmitting, its radiated wave could be written as

$$\mathbf{E}_2 = E_{2\xi}\mathbf{u}_\xi + E_{2\eta}\mathbf{u}_\eta \qquad (4.47)$$

where appropriate coordinates $\xi\eta\zeta$ for the wave propagating in the ζ direction, toward the first antenna, are used. The equation may be written in terms of the polarization ratio of antenna 2 as

$$\mathbf{E}_2 = E_{2\xi}\left(\mathbf{u}_\xi + \frac{|E_{2\eta}|}{|E_{2\xi}|}e^{j\phi_2}\mathbf{u}_\eta\right) = E_{2\xi}(\mathbf{u}_\xi + P_2\mathbf{u}_\eta) \quad (4.48)$$

where P_2 is the polarization ratio of antenna 2 in the ζ direction, using the right-handed coordinates at antenna 2.

Now the transmitted field (4.48) is related to the vector effective length of antenna 2 by

$$\mathbf{E}_2 = \frac{jZ_0 I_2}{2\lambda r}e^{-jkr}\mathbf{h}_2 \quad (4.49)$$

The last two equations can be combined to give

$$\mathbf{h}_2 = h_{2\xi}(\mathbf{u}_\xi + P_2\mathbf{u}_\eta) \quad (4.50)$$

where

$$h_{2\xi} = \frac{2\lambda r}{jZ_0 I_2}e^{jkr}E_{2\xi} \quad (4.51)$$

Return now to the situation where antenna 1 transmits and 2 receives. The open-circuit voltage across the terminals of 2 is

$$V_2 = \mathbf{E}_1^i \cdot \mathbf{h}_2 = E_{1\xi}^i h_{2\xi} + E_{1\eta}^i h_{2\eta} \quad (4.52)$$

If we note from Figure 4.3 that

$$E_{1\xi}^i = -E_{1x}^i \quad (4.53a)$$

$$E_{1\eta}^i = E_{1y}^i \quad (4.53b)$$

and use the field and effective length components from (4.46) and (4.50), the open-circuit voltage is

$$V_2 = -E_{1x}h_{2\xi}(1 - P_1 P_2) \quad (4.54)$$

It may be noted also, from (4.46) and (4.50), that

$$|\mathbf{E}^i|^2 = \mathbf{E}^i \cdot \mathbf{E}^{i*} = |E_{1x}|^2(1 + P_1 P_1^*) \quad (4.55)$$

and

$$|\mathbf{h}_2|^2 = |h_{2\xi}|^2(1 + P_2 P_2^*) \quad (4.56)$$

194 POLARIZATION MATCHING OF ANTENNAS

If the last three equations and (4.52) are substituted into the polarization match factor equation (4.45) it becomes

$$\rho = \frac{(1 - P_1 P_2)(1 - P_1^* P_2^*)}{(1 + P_1 P_1^*)(1 + P_2 P_2^*)} \tag{4.57}$$

It is worthwhile to repeat that the definition of P_1 uses wave components measured in a right-handed system with the z axis pointing away from antenna 1 and toward 2. In defining P_2 a right-handed system with the η axis parallel to and in the same direction as the y axis and the ζ axis pointing toward antenna 1 was used.

In (4.45), if it is noted that $\mathbf{h}$ is $\mathbf{h}_2$ and the incident field $\mathbf{E}^i$ is proportional to the effective length $\mathbf{h}_1$ of the transmitting antenna, the polarization efficiency can be written as

$$\rho = \frac{|\mathbf{h}_1 \cdot \mathbf{h}_2|^2}{|\mathbf{h}_1|^2 |\mathbf{h}_2|^2} \tag{4.58}$$

which is a useful form. In this equation, $\mathbf{h}_1$ and $\mathbf{h}_2$ may be interpreted as the effective lengths or normalized to be the polarization states of the two antennas.

4.7. THE MODIFIED FRIIS TRANSMISSION EQUATION

In Chapter 1 an equation was given for the power received in a transmitting–receiving antenna configuration,

$$W_r = \frac{W_{at} G_t(\theta_t, \phi_t) A_{er}(\theta_r, \phi_r)}{4\pi r^2} \tag{1.155}$$

with W_{at} the transmitted power, G_t the gain of the transmitting antenna, A_{er} the effective area of the receiving antenna, and r the distance between them. This is commonly called the Friis equation. It was pointed out that it must be modified by a polarization efficiency, but it is not clear that the correct multiplier is that given by (4.45) or (4.58), since the gains and effective areas were not used in developing those equations. It will be seen here that ρ, given by (4.58), is indeed the correct multiplier in the Friis equation to obtain the correct power for polarization-mismatched antennas.

The Friis equation, modified to account for polarization mismatch, is

$$W_r = \frac{W_{at} G_t A_{er} \rho}{4\pi r^2} \tag{4.59}$$

where we wish to determine if ρ in this equation agrees with (4.58). The received power is also given, if the receiving antenna is impedance-matched, by

$$W_r = \frac{|\mathbf{E}^i \cdot \mathbf{h}_r|^2}{8R_r} \tag{4.60}$$

where $\mathbf{h}_r$ and R_r are effective length and resistance of the receiving antenna. The two forms are equated, noting that the power accepted by the transmitting antenna is

$$W_{at} = \tfrac{1}{2} R_t |I|^2 \tag{4.61}$$

and that the field incident on the receiving antenna is

$$\mathbf{E}^i = \frac{jZ_0 I}{2\lambda r} \mathbf{h}_t e^{-jkr} \tag{4.62}$$

This leads to

$$R_t G_t A_{er} R_r \rho = \frac{\pi Z_0^2 |\mathbf{h}_t \cdot \mathbf{h}_r|^2}{4\lambda^2} \tag{4.63}$$

But from (4.7), (4.11), and (1.154),

$$R_t G_t = \frac{\pi Z_0 \mathbf{h}_t \cdot \mathbf{h}_t^*}{\lambda^2}$$

and

$$A_{er} R_r = \frac{Z_0 \mathbf{h}_r \cdot \mathbf{h}_r^*}{4}$$

Substitution in (4.63) gives

$$\rho = \frac{|\mathbf{h}_t \cdot \mathbf{h}_r|^2}{|\mathbf{h}_t|^2 |\mathbf{h}_r|^2} \tag{4.64}$$

which agrees with (4.58).

We see then that the polarization efficiency developed in this chapter should be used as a multiplier in the Friis transmission formula to give the received power with polarization-mismatched antennas. However, since the effective lengths (or polarization states) of both antennas must be known to determine the polarization efficiency, a more convenient form than (4.59) that does not use gain and effective area can be found from (4.60). Substituting (4.61) and (4.62) in (4.60) gives

$$W_r = \frac{W_{at} Z_0^2 |\mathbf{h}_t \cdot \mathbf{h}_r|^2}{16\lambda^2 R_t R_r r^2} \qquad (4.65)$$

Note that in this form the effective lengths, not the normalized antenna polarization states, must be used.

4.8. POLARIZATION MATCH FACTOR — SPECIAL CASES

Polarization-Matched Antennas

If two polarization-matched antennas are in a transmit–receive configuration, the polarization match factor is equal to 1. (It may change if the orientation of one of the antennas is changed.) Then

$$\rho = \frac{(1 - P_1 P_2)(1 - P_1^* P_2^*)}{(1 + P_1 P_1^*)(1 + P_2 P_2^*)} = 1 \qquad (4.66)$$

Cross-multiplying, expanding, and canceling terms gives $(P_1^* + P_2)(P_2^* + P_1) = 0$ which has a solution

$$P_2 = -P_1^* \qquad (4.67)$$

In terms of the inverse circular polarization ratio defined in Chapter 3, the matched-polarization condition becomes

$$q_2 = \frac{1 - jP_2}{1 + jP_2} = \frac{1 + jP_1^*}{1 - jP_1^*} = q_1^* \qquad (4.68)$$

and the axial ratios and tilt angles of the polarization ellipses of the two antennas are related by

$$AR_1 = \left|\frac{1 + |q_1|}{1 - |q_1|}\right| = \left|\frac{1 + |q_2|}{1 - |q_2|}\right| = AR_2 \qquad (4.69a)$$

$$\tau_1 = \theta_1/2 = -\gamma_1/2 = +\gamma_2/2 = -\tau_2 \qquad (4.69b)$$

In this equation τ_1 and τ_2 are described in different coordinate systems, as shown in Figure 4.3. It is obvious from the figure that the condition $\tau_1 = -\tau_2$ means that the major axes of the two polarization ellipses coincide. Equation (4.68) shows that the rotation senses of the two polarization ellipses are the same, when described in the appropriate coordinate systems. Having the

Cross-Polarized Antennas

Two antennas in a transmit–receive configuration that are so polarized that no signal is received are said to be cross-polarized. For this situation

$$\rho = \frac{(1 - P_1 P_2)(1 - P_1^* P_2^*)}{(1 + P_1 P_1^*)(1 + P_2 P_2^*)} = 0 \tag{4.70}$$

from which it follows that

$$P_1 = \frac{1}{P_2} \tag{4.71}$$

The inverse circular polarization ratios for cross-polarized antennas are readily found to be related by

$$q_1 = -\frac{1}{q_2} \tag{4.72}$$

We see immediately that the rotation senses of the polarization ellipses of the antennas are opposite, so that if both antennas transmitted simultaneously their field vectors would appear to rotate in the same direction.

The axial ratios are

$$AR_1 = \left| \frac{1 + |q_1|}{1 - |q_1|} \right| = \left| \frac{|q_2| + 1}{|q_2| - 1} \right| = AR_2 \tag{4.73}$$

Also, from (4.72),

$$Q_1 e^{j\gamma_1} = -\frac{1}{Q_2 e^{j\gamma_2}} = \frac{1}{Q_2} e^{-j(\gamma_2 + \pi)}$$

so that $\gamma_1 = -\gamma_2 + \pi$ and

$$\tau_1 = -\frac{\gamma_1}{2} = -\tau_2 \mp \frac{\pi}{2} \tag{4.74}$$

Bearing in mind that τ_1 is measured from the x axis toward the y axis and τ_2 is measured from the ξ axis toward the η axis in Figure 4.3, we see that the

198 POLARIZATION MATCHING OF ANTENNAS

major axis of one polarization ellipse coincides with the minor axis of the other.

Identical, Polarization-Matched Antennas

It would seem quite easy to define identical antennas, but surprisingly there is a degree of arbitrariness involved. When placed side by side and oriented similarly, identical antennas are indistinguishable except by position. Although not overly precise, this definition is quite clear. Now we make the assumption that they are placed into a transmit–receive configuration by rotating one of them by π radians around a *vertical* axis (the y axis of Fig. 4.3). We might also consider a rotation about a horizontal axis, or the major or minor axis of the polarization ellipse—hence the arbitrariness mentioned—but we rotate first about the vertical axis.

For identical antennas, before one is rotated to put it into a receiving position,

$$h_{x1} = h'_{x2} \qquad h_{y1} = h'_{y2}$$
$$P_1 = \frac{h_{y1}}{h_{x1}} = P'_2 = \frac{h'_{y2}}{h'_{x2}} \tag{4.75}$$

where the primes are used with the parameters of the antenna to be rotated. After antenna 2 is rotated 180° about the y axis of Figure 4.3, its new length components are

$$h_{x2} = -h'_{x2} \qquad h_{y2} = h'_{y2} \tag{4.76}$$

Changing these components to $\xi\eta\zeta$ coordinates, which are now appropriate for antenna 2, we have

$$h_{\xi 2} = -h_{x2} = h'_{x2} \tag{4.77a}$$
$$h_{\eta 2} = h_{y2} = h'_{y2} \tag{4.77b}$$

Then the new value for the polarization ratio P_2 is

$$P_2 = \frac{h_{\eta 2}}{h_{\xi 2}} = \frac{h'_{y2}}{h'_{x2}} = P'_2 \tag{4.78}$$

and thus P_2 is unchanged by rotation about a vertical axis. A little thought will show that, in general, the major axes of the ellipses no longer coincide.

Let the antennas be not only identical but polarization-matched. Then they must satisfy (4.67), (4.75), and (4.78), or

$$P_1 = -P_2^* \tag{4.79a}$$
$$P_1 = P_2 \tag{4.79b}$$

and also

$$q_1 = q_2^* \quad (4.80a)$$
$$q_1 = q_2 \quad (4.80b)$$

We conclude then that identical, polarization-matched antennas must have polarization ratios that are imaginary and circular polarization ratios that are real, from which it follows that

$$\gamma_1 = \gamma_2 = 0, \pi \qquad \tau_1 = \tau_2 = 0, \pm \pi/2 \quad (4.81)$$

We see that the major axes of the polarization ellipse must be either vertical or horizontal if the antennas are to be identical (in our sense of rotation about a vertical axis) and matched. This does not include circularly polarized antennas for which the concept of major axis is not meaningful.

Antennas that are identical and cross-polarized must satisfy

$$P_1 = \frac{1}{P_2} \qquad q_1 = -\frac{1}{q_2}$$
$$P_1 = P_2 \qquad q_1 = q_2 \quad (4.82)$$

which gives

$$P_1 = \pm 1 \qquad q_1 = \pm j1 \quad (4.83)$$

from which we find that

$$AR \to \infty$$
$$\tau = \pm \tfrac{1}{4}\pi \quad (4.84)$$

which describe linearly polarized antennas with tilt angles of $\pm 45°$.

When we rotated one of the antennas about a vertical axis we found that the two antennas are matched if their major axes are vertical. Perhaps if we started with identical, side-by-side antennas and rotated one of them about its major axis we would obtain polarization matching.

Let the polarization parameters *before* rotation be P_1, P_2', and so on, where the primes are used with the parameters of the antenna to be rotated. Then for identical antennas,

$$q_1 = q_2' \qquad AR_1 = AR_2' \qquad \tau_1 = \tau_2' \quad (4.85)$$

After antenna 2 is rotated about its major axis, the axial ratio and rotation sense are unchanged, so that

$$AR_2 = AR_2' \qquad |q_2| = |q_2'| \quad (4.86)$$

200 POLARIZATION MATCHING OF ANTENNAS

Since the rotation takes place about the major axis, obviously the major axis does not change, but as Figure 4.3 shows, the tilt angle in the $\xi\eta\zeta$ system is measured opposite that in xyz. Therefore, after rotation the new tilt angle is given by

$$\tau_2 = -\tau_2' \tag{4.87}$$

The last two equations lead to

$$q_2 = q_2'^* = q_1^* \tag{4.88}$$

Now, from (4.68), this is the condition for polarization matching. Therefore, identical antennas (indistinguishable when placed side by side and similarly oriented) will be matched in polarization if one of them is rotated 180° about its major polarization axis to bring it to a receive position.

Orthogonal Wave Components

In Chapter 3 it was shown that a plane harmonic wave can be expressed as the sum of two orthogonal elliptically polarized waves, thus

$$\mathbf{E} = \mathbf{E}_1 + \mathbf{E}_2 = E_1 \mathbf{u}_1 + E_2 \mathbf{u}_2 \tag{4.89}$$

where the base vectors obey the orthonormal conditions

$$\begin{aligned} \mathbf{u}_1 \cdot \mathbf{u}_1^* = \mathbf{u}_2 \cdot \mathbf{u}_2^* = 1 \\ \mathbf{u}_1 \cdot \mathbf{u}_2^* = \mathbf{u}_1^* \cdot \mathbf{u}_2 = 0 \end{aligned} \tag{4.90}$$

It was seen there that $\mathbf{E}_1$ and $\mathbf{E}_2$ have the linear polarization ratios P_1 and $-1/P_1^*$ respectively.

A physical interpretation of orthogonality as it characterizes the two component waves was not given, but we shall see now that if a receiving antenna has a polarization that causes it to receive one of the component waves without loss, it cannot receive the other component at all. Let the field $\mathbf{E}$ with its two components be incident on a receiving antenna with linear polarization ratio P_2. If the antenna is matched to receive component $\mathbf{E}_1$ without loss, then we must have

$$P_2 = -P_1^* \tag{4.67}$$

Consider next the match between the receiving antenna and the wave component $\mathbf{E}_2$ with polarization ratio $-1/P_1^*$. Substitution in (4.57), using $-1/P_1^*$ instead of P_1 and $P_2 = -P_1^*$, gives $\rho = 0$.

4.9. MATCH FACTOR IN OTHER FORMS

The linear polarization ratio P is not always the most convenient parameter for an antenna, and an equation for the polarization efficiency in terms of other parameters is needed. If we make the substitutions

$$P = -jp \qquad p = \frac{1-q}{1+q}$$

into (4.57) the match factor can be found in terms of the modified polarization ratio p and the inverse circular polarization ratio q. The equations are symmetric, thus

$$\rho = \frac{(1+p_1 p_2)(1+p_1^* p_2^*)}{(1+p_1 p_1^*)(1+p_2 p_2^*)} = \frac{(1+q_1 q_2)(1+q_1^* q_2^*)}{(1+q_1 q_1^*)(1+q_2 q_2^*)} \qquad (4.91)$$

The symmetry is not surprising, since the transformations between p and q were found to be symmetric in Chapter 3.

This equation is valid for any value of q, but nonetheless if left-elliptic polarizations are treated by means of the parameter w, where $w = 1/q^*$, it is desirable to have the efficiency in terms of w. It is

$$\rho = \frac{(1+w_1 w_2)(1+w_1^* w_2^*)}{(1+w_1 w_1^*)(1+w_2 w_2^*)} \qquad (4.92)$$

Mixed forms in terms of q and w might also be useful and are easily found by substitution to be

$$\rho = \frac{(w_2^* + q_1)(w_2 + q_1^*)}{(1+q_1 q_1^*)(1+w_2 w_2^*)} = \frac{(w_1^* + q_2)(w_1 + q_2^*)}{(1+q_2 q_2^*)(1+w_1 w_1^*)} \qquad (4.93)$$

All of these forms are valid for any q or w, but it is natural to use the one in terms of q for both antennas right handed, the one with w for both left handed, and one of the mixed forms for one antenna right handed and the other left handed.

The efficiency may be found in terms of axial ratios and tilt angles of the polarization ellipses, but here we must be careful about the rotation sense of the ellipses, since axial ratio and tilt angle alone are not sufficient to describe the antenna polarization. Consider first that both antennas are right handed, with $|q| < 1$. Then, from (3.110),

$$AR = \frac{1+|q|}{1-|q|} \qquad (4.94)$$

If this equation and

$$q = |q|e^{-j2\tau} \tag{4.95}$$

are substituted into the equation for polarization efficiency in terms of q,

$$\rho = \frac{(AR_1 AR_2 + 1)^2 + (AR_1 + AR_2)^2 + (AR_1^2 - 1)(AR_2^2 - 1)\cos 2(\tau_1 + \tau_2)}{2(AR_1^2 + 1)(AR_2^2 + 1)} \tag{4.96}$$

If both antennas are left handed, with $|w| < 1$, (3.143) gives

$$AR = \frac{1 + |w|}{1 - |w|} \tag{4.97}$$

Because of the manner in which w was defined, we also have

$$w = |w|e^{-j2\tau} \tag{4.98}$$

These equations have the same forms as corresponding equations in q. If they are substituted into (4.92), which has the same form as (4.91), it is obvious that (4.96) will result. Therefore (4.96) holds if both antennas are right handed or if both are left handed.

If antenna 1 is right handed and antenna 2 is left handed, we substitute

$$AR_1 = \frac{1 + |q_1|}{1 - |q_1|} \quad q_1 = |q_1|e^{-j2\tau_1} \tag{4.99a}$$

$$AR_2 = \frac{1 + |w_2|}{1 - |w_2|} \quad w_2 = |w_2|e^{-j2\tau_2} \tag{4.99b}$$

into the appropriate mixed form for the polarization efficiency and obtain

$$\rho = \frac{(AR_1 AR_2 - 1)^2 + (AR_1 - AR_2)^2 + (AR_1^2 - 1)(AR_2^2 - 1)\cos 2(\tau_1 + \tau_2)}{2(AR_1^2 + 1)(AR_2^2 + 1)} \tag{4.100}$$

for the match factor in terms of axial ratios and tilt angles.

If antenna 1 is left handed and antenna 2 right handed, the appropriate substitutions lead to the same equation for the polarization efficiency. Then (4.96) is valid if both antennas have the same rotation sense and (4.100) holds if they are of the opposite sense.

Finally, we note that ρ may be written in terms of the rectangular field components as

$$\rho = \frac{1 - 2(|E_{1y}|/|E_{1x}|)(|E_{2\eta}|/|E_{2\xi}|)\cos(\phi_1 + \phi_2) + [(|E_{1y}|/|E_{1x}|)(|E_{2\eta}|/|E_{2\xi}|)]^2}{\left[1 + (|E_{1y}|/|E_{1x}|)^2\right]\left[1 + (|E_{2\eta}|/|E_{2\xi}|)^2\right]} \quad (4.101)$$

and in terms of right- and left-circular components as

$$\rho = \frac{1 + 2(|E_{1L}|/|E_{1R}|)(|E_{2L}|/|E_{2R}|)\cos(\theta_1 + \theta_2) + [(|E_{1L}|/|E_{1R}|)(|E_{2L}|/|E_{2R}|)]^2}{\left[1 + (|E_{1L}|/|E_{1R}|)^2\right]\left[1 + (|E_{2L}|/|E_{2R}|)^2\right]} \quad (4.102)$$

Receiving Polarization

The polarization efficiency is maximum if the polarization ratio of the receiving antenna, P_2, and that of the incident wave are related by

$$P_2 = -P_1^*$$

where P_1 is defined in xy coordinates and P_2 in $\xi\eta$. The *receiving polarization* of an antenna is by definition the polarization of a plane wave that results in maximum available power to the antenna [3]. Then the receiving polarization ratio of an antenna that has polarization ratio P is

$$P_R = -P^* \quad (4.103)$$

Note that P_R is still measured in coordinates appropriate to the antenna as a transmitter.

4.10. CONTOURS OF CONSTANT MATCH FACTOR

Examination of the equation for polarization efficiency in terms of q leads one to suspect that, for a given value of q_1, a range of q_2 values might give the same match factor ρ. It is desirable to consider this problem in terms of q because all polarization states can be plotted inside unit circles on the q- and w-planes. Hold ρ constant and use

$$q = Qe^{j\gamma} \quad (3.136)$$

204 POLARIZATION MATCHING OF ANTENNAS

If this is substituted into (4.91), the resulting equation may be written as

$$Q_2^2 - \frac{2Q_1 \cos(\gamma_1 + \gamma_2)}{(1 + Q_1^2)\rho - Q_1^2} Q_2 + \frac{Q_1^2}{[(1 + Q_1^2)\rho - Q_1^2]^2} = \frac{(1 - \rho)\rho}{[\rho - Q_1^2/(1 + Q_1^2)]^2} \quad (4.104)$$

This equation represents a family of circles on the q plane (actually the q_2 plane) with center and radius

$$Q_{2c}, \gamma_{2c} = \frac{Q_1}{(1 + Q_1^2)\rho - Q_1^2}, -\gamma_1 \quad (4.105a)$$

$$r = \left| \frac{[(1 - \rho)\rho]^{1/2}}{\rho - Q_1^2/(1 + Q_1^2)} \right| \quad (4.105b)$$

Now, (4.104) is in the correct form of a circle only if $(1 + Q_1^2)\rho - Q_1^2 > 0$ or

$$\rho > \frac{Q_1^2}{1 + Q_1^2} \quad (4.106)$$

If this condition is not met, (4.104) may be rewritten as

$$Q_2^2 - \frac{2Q_1 \cos(\gamma_1 + \gamma_2 + \pi)}{|(1 + Q_1^2)\rho - Q_1^2|} Q_2 + \frac{Q_1^2}{[(1 + Q_1^2)\rho - Q_1^2]^2}$$

$$= \frac{(1 - \rho)\rho}{[\rho - Q_1^2/(1 + Q_1^2)]^2} \quad (4.107)$$

which represents a family of circles with center and radius

$$Q_{2c}, \gamma_{2c} = \frac{Q_1}{|(1 + Q_1^2)\rho - Q_1^2|}, -\gamma_1 + \pi \quad (4.108a)$$

$$r = \left| \frac{[(1 - \rho)\rho]^{1/2}}{\rho - Q_1^2/(1 + Q_1^2)} \right| \quad (4.108b)$$

if

$$\rho < \frac{Q_1^2}{1 + Q_1^2} \quad (4.109)$$

CONTOURS OF CONSTANT MATCH FACTOR   205

These equations represent contours of constant match factor on the q plane representing antenna 2 in terms of given polarization characteristics for antenna 1, the other part of the communication system. Given a transmitting antenna with polarization q_1, this family of circles on the q_2 plane allows us to determine quickly the effect of varying the receiving antenna polarization. The words *transmitting* and *receiving* were used above for clarity. Of course, it makes no difference which antenna transmits and which receives.

It is tempting at this point to draw these constant ρ contours on the q plane, but we used an equation, (4.91), in which both antennas are represented by q, which prevents our considering a left-handed antenna if we wish to remain in the unit circle on the q_2 plane. We therefore go on to consider, before drawing these contours, antennas described by w_1 and w_2, q_1 and w_2, and q_2 and w_1.

For antennas described by w_1 and w_2 we should substitute

$$w = We^{j\gamma} \tag{4.110}$$

into (4.92), but this equation for w has the same form as the equation for q, and the equation for efficiency in terms of w has the same form as that in terms of q. The result is that contours of constant ρ on the w_2 plane are circles with centers and radii given by (4.105) and (4.108) using W instead of Q.

$$W_{2c}, \gamma_{2c} = \frac{W_1}{(1 + W_1^2)\rho - W_1^2}, -\gamma_1 \tag{4.111a}$$

$$r = \left| \frac{[(1-\rho)\rho]^{1/2}}{\rho - W_1^2/(1 + W_1^2)} \right| \tag{4.111b}$$

if

$$\rho > \frac{W_1^2}{1 + W_1^2} \tag{4.112}$$

and

$$W_{2c}, \gamma_{2c} = \frac{W_1}{|(1 + W_1^2)\rho - W_1^2|}, -\gamma_1 + \pi \tag{4.113a}$$

$$r = \left| \frac{[(1-\rho)\rho]^{1/2}}{\rho - W_1^2/(1 + W_1^2)} \right| \tag{4.113b}$$

if

$$\rho < \frac{W_1^2}{1 + W_1^2} \tag{4.114}$$

Now we take the situation of antenna 1 described by q_1 and antenna 2 by w_2. Substitute (3.136) and (4.110) into the appropriate part of (4.93) and obtain

$$\rho = \frac{W_2^2 + 2Q_1 W_2 \cos(\gamma_1 + \gamma_2) + Q_1^2}{(1 + Q_1^2)(1 + W_2^2)} \tag{4.115}$$

This equation may be put into the standard circle form

$$W_2^2 - \frac{2Q_1 \cos(\gamma_1 + \gamma_2)}{(1 + Q_1^2)\rho - 1} W_2 + \frac{Q_1^2}{[(1 + Q_1^2)\rho - 1]^2} = \frac{(1 - \rho)\rho}{[\rho - 1/(1 + Q_1^2)]^2} \tag{4.116}$$

if the denominator of the second term is positive. If the denominator is negative, the correct form is

$$W_2^2 - \frac{2Q_1 \cos(\gamma_1 + \gamma_2 + \pi)}{|(1 + Q_1^2)\rho - 1|} W_2 + \frac{Q_1^2}{[(1 + Q_1^2)\rho - 1]^2}$$

$$= \frac{(1 - \rho)\rho}{[\rho - 1/(1 + Q_1^2)]^2} \tag{4.117}$$

These equations represent circles on the w_2 plane with centers and radii

$$W_{2c}, \gamma_{2c} = \frac{Q_1}{(1 + Q_1^2)\rho - 1}, -\gamma_1 \tag{4.118a}$$

$$r = \left| \frac{[(1 - \rho)\rho]^{1/2}}{\rho - 1/(1 + Q_1^2)} \right| \tag{4.118b}$$

if

$$\rho > \frac{1}{1 + Q_1^2} \tag{4.119}$$

and

$$W_{2c}, \gamma_{2c} = \frac{Q_1}{|(1 + Q_1^2)\rho - 1|}, -\gamma_1 + \pi \tag{4.120a}$$

$$r = \left| \frac{[(1 - \rho)\rho]^{1/2}}{\rho - 1/(1 + Q_1^2)} \right| \tag{4.120b}$$

if

$$\rho < \frac{1}{1 + Q_1^2} \quad (4.121)$$

Finally we take the sense of antenna 1 being described by w_1 and antenna 2 by q_2. Substituting (3.136) and (4.110) into (4.93) gives

$$\rho = \frac{W_1^2 + 2Q_2W_1 \cos(\gamma_1 + \gamma_2) + Q_2^2}{(1 + W_1^2)(1 + Q_2^2)} \quad (4.122)$$

an equation that is the same as (4.115) with Q and W interchanged. We may therefore use the circle equations (4.118) and (4.120) with the same replacements, giving

$$Q_{2c}, \gamma_{2c} = \frac{W_1}{(1 + W_1^2)\rho - 1}, \; -\gamma_1 \quad (4.123a)$$

$$r = \left| \frac{[(1-\rho)\rho]^{1/2}}{\rho - 1/(1 + W_1^2)} \right| \quad (4.123b)$$

if

$$\rho > \frac{1}{1 + W_1^2} \quad (4.124)$$

and

$$Q_{2c}, \gamma_{2c} = \frac{W_1}{|(1 + W_1^2)\rho - 1|}, \; -\gamma_1 + \pi \quad (4.125a)$$

$$r = \left| \frac{[(1-\rho)\rho]^{1/2}}{\rho - 1/(1 + W_1^2)} \right| \quad (4.125b)$$

if

$$\rho < \frac{1}{1 + W_1^2} \quad (4.126)$$

All of the preceding equations for constant ρ were given in terms of circles on the q_2 and w_2 planes. The designations 1 and 2 are arbitrary, so the subscripts may be interchanged. Also, it makes no difference which antenna transmits and which receives.

We have not restricted the antennas to a particular rotation sense. However, to stay in the unit circle, if antenna 2 is right handed, we normally

208 POLARIZATION MATCHING OF ANTENNAS

use equations (4.105) and (4.108), or (4.123) and (4.125), which represent circles on the q plane, on which all polarizations fall within the unit circle.

We can combine these eight equation sets into two if we note that (4.105) and (4.108) differ only by the magnitude sign in the form for Q_{2c} and in the value of γ_{2c} (and similarly for the other pairs of equations) and if we note that if we use $W_1 = 1/Q_1$ in (4.105), it becomes

$$Q_{2c} = \frac{W_1}{(1 + W_1^2)\rho - 1}$$

which is the same equation as (4.123a).

Combining the equations for the constant ρ circles appropriately leads to

$$Q_{2c} = \frac{Q_1}{|(1 + Q_1^2)\rho - Q_1^2|} = \frac{W_1}{|(1 + W_1^2)\rho - 1|} \qquad (4.127a)$$

$$\gamma_{2c} = -\gamma_1 \qquad \rho > \frac{Q_1^2}{1 + Q_1^2} = \frac{1}{1 + W_1^2} \qquad (4.127b)$$

$$\gamma_{2c} = -\gamma_1 + \pi \qquad \rho < \frac{Q_1^2}{1 + Q_1^2} = \frac{1}{1 + W_1^2} \qquad (4.127c)$$

$$r = \left|\frac{[(1-\rho)\rho]^{1/2}}{\rho - Q_1^2/(1 + Q_1^2)}\right| = \left|\frac{[(1-\rho)\rho]^{1/2}}{\rho - 1/(1 + W_1^2)}\right| \qquad (4.127d)$$

and

$$W_{2c} = \frac{W_1}{|(1 + W_1^2)\rho - W_1^2|} = \frac{Q_1}{|(1 + Q_1^2)\rho - 1|} \qquad (4.128a)$$

$$\gamma_{2c} = -\gamma_1 \qquad \rho > \frac{W_1^2}{1 + W_1^2} = \frac{1}{1 + Q_1^2} \qquad (4.128b)$$

$$\gamma_{2c} = -\gamma_1 + \pi \qquad \rho < \frac{W_1^2}{1 + W_1^2} = \frac{1}{1 + Q_1^2} \qquad (4.128c)$$

$$r = \left|\frac{[(1-\rho)\rho]^{1/2}}{\rho - W_1^2/(1 + W_1^2)}\right| = \left|\frac{[(1-\rho)\rho]^{1/2}}{\rho - 1/(1 + Q_1^2)}\right| \qquad (4.128d)$$

Example

To understand these equations and the constant match factor curves, consider an example of antenna 1 right handed with $q_1 = 1/2$. Substituting into both sets of circle equations gives

$$Q_{2c} = \frac{2}{|5\rho - 1|} \qquad W_{2c} = \frac{2}{|5\rho - 4|}$$

$$\gamma_{2c} = \begin{cases} 0, & \rho > 0.2 \\ \pi, & \rho < 0.2 \end{cases} \qquad \gamma_{2c} = \begin{cases} 0, & \rho > 0.8 \\ \pi, & \rho < 0.8 \end{cases}$$

$$r = \frac{[(1-\rho)\rho]^{1/2}}{|\rho - 0.2|} \qquad r = \frac{[(1-\rho)\rho]^{1/2}}{|\rho - 0.8|}$$

Figure 4.4 shows the circles of constant ρ determined from these equations. The contours are labeled in decibels, with the negative sign omitted.

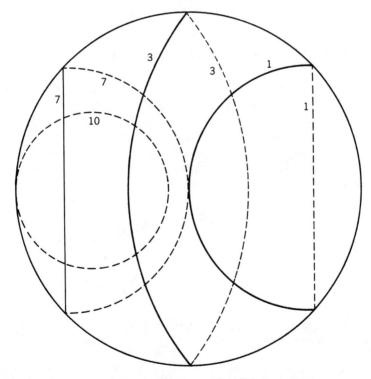

FIGURE 4.4 The q_2 (or w_2) plane with curves of constant polarization match factor: antenna 1—right elliptic, $q_1 = 0.5$; antenna 2—right elliptic (solid curves) and left elliptic (dashed curves).

210 POLARIZATION MATCHING OF ANTENNAS

If antenna 2 is right handed, it is only necessary to show the circles from the equation set in the left column above. However, it is possible that a left-elliptic antenna might be selected for antenna 2, so from the equations in the right column the constant ρ circles are also drawn on the w_2 plane. When considered as the q_2 plane (antenna 2 right handed), the contours are solid in Figure 4.4, and when considered as the w_2 plane (antenna 2 left handed), the contours are dashed.

A study of Figure 4.4 illustrates points worth considering:

1. A chart such as Figure 4.4 applies for only one value of Q_1 or W_1, but remains valid if the angle γ_1 of q_1 changes. The line of circle centers is rotated to $-\gamma_1$ by rotating the chart.

2. A wide range of antenna polarizations will result in the same polarization match, since in general one of the constant ρ circles spans a wide range of q_2.

3. For the constant ρ circles that intersect the unit circle, right- and left-elliptic antennas exist with the same polarization match.

4. The polarization match *may* be better between right- and left-elliptic antennas than between two right-handed antennas.

4.11. THE POINCARÉ SPHERE AND POLARIZATION EFFICIENCY

Since contours of constant polarization match factor exist on the q and w planes, and since those planes are stereographic projections of polarization points on the Poincaré sphere, constant polarization match curves might be expected on the Poincaré sphere itself.

As a matter of notation, we use P_a and P_b for antennas a and b to avoid confusion with the numerical subscripts of the Stokes parameters.

The polarization match between the two antennas is

$$\rho = \frac{(1 - P_a P_b)(1 - P_a^* P_b^*)}{(1 + |P_a|^2)(1 + |P_b|^2)} \quad (4.129)$$

where

$$P_a = \frac{|E_y^a|}{|E_x^a|} e^{j\phi_a} \qquad P_b = \frac{|E_\eta^b|}{|E_\xi^b|} e^{j\phi_b} \quad (4.130)$$

with all quantities measured in the coordinate systems of Figure 4.3.

Combining these equations gives

$$\rho = \frac{|E_x^a|^2 |E_\xi^b|^2 + |E_y^a|^2 |E_\eta^b|^2 - 2|E_x^a| \, |E_y^a| \, |E_\xi^b| \, |E_\eta^b|(\cos\phi_a \cos\phi_b - \sin\phi_a \sin\phi_b)}{(|E_x^a|^2 + |E_y^a|^2)(|E_\xi^b|^2 + |E_\eta^b|^2)}$$

$$(4.131)$$

The Stokes parameters of the waves are

$$G_0^a = |E_x^a|^2 + |E_y^a|^2 \qquad G_0^b = |E_\xi^b|^2 + |E_\eta^b|^2$$
$$G_1^a = |E_x^a|^2 - |E_y^a|^2 \qquad G_1^b = |E_\xi^b|^2 - |E_\eta^b|^2$$
$$G_2^a = 2|E_x^a|\,|E_y^a|\cos\phi_a \qquad G_2^b = 2|E_\xi^b|\,|E_\eta^b|\cos\phi_b \qquad (4.132)$$
$$G_3^a = 2|E_x^a|\,|E_y^a|\sin\phi_a \qquad G_3^b = 2|E_\xi^b|\,|E_\eta^b|\sin\phi_b$$

and if these are substituted into the equation for polarization match factor, the result is

$$\rho = \frac{1}{2}\left(1 + \frac{G_1^a G_1^b}{G_0^a G_0^b} - \frac{G_2^a G_2^b}{G_0^a G_0^b} + \frac{G_3^a G_3^b}{G_0^a G_0^b}\right) \qquad (4.133)$$

Recall that the receiving polarization of an antenna with polarization ratio P was defined previously as $P_R = -P^*$. The Stokes parameters corresponding to the receiving polarization of antenna b are, from Chapter 3, $G_1^b, -G_2^b,$ and G_3^b.

If the two points corresponding to the polarization of antenna a and the receiving polarization of antenna b, with Stokes parameters G_1^a, G_2^a, G_3^a, and $G_1^b, -G_2^b, G_3^b$, are plotted on the Poincaré sphere and two rays are drawn from the origin to these points, the angle between the rays is given by

$$\cos\beta = \frac{G_1^a G_1^b}{G_0^a G_0^b} - \frac{G_2^a G_2^b}{G_0^a G_0^b} + \frac{G_3^a G_3^b}{G_0^a G_0^b} \qquad (4.134)$$

Then

$$\cos\frac{\beta}{2} = \left(\frac{1 + \cos\beta}{2}\right)^{1/2} = \frac{1}{\sqrt{2}}\left(1 + \frac{G_1^a G_1^b}{G_0^a G_0^b} - \frac{G_2^a G_2^b}{G_0^a G_0^b} + \frac{G_3^a G_3^b}{G_0^a G_3^b}\right)^{1/2}$$
$$(4.135)$$

Comparing this result to the equation for polarization match factor shows that

$$\rho = \cos^2(\tfrac{1}{2}\beta) \qquad (4.136)$$

For a transmit–receive antenna configuration, if the polarization ratio of one antenna and the receiving polarization ratio of the other are plotted on the Poincaré sphere via their Stokes parameters, the polarization efficiency of the antenna pair may be found, using the last equation, from the angle between rays from the sphere center to the plotted points.

212 POLARIZATION MATCHING OF ANTENNAS

It is evident that curves of constant match factor for a two-antenna system can be drawn on the Poincaré sphere. These curves are circles with center at the plotted polarization point of one of the antennas.

4.12. MATCH FACTOR USING ONE COORDINATE SYSTEM

Throughout this chapter the polarization ratio of an antenna has been consistently defined in a right-handed coordinate system with the propagation axis pointing away from the antenna. When two antennas in a transmit–receive arrangement are considered, as in Figure 4.3, two right-handed coordinate systems are used, with the z and ζ axes pointing at each other. Both polarization ratios can be defined in one coordinate system if desired, although that is not the practice in this book (with the exception of this section). Consider that the field components radiated by antenna 2 of Figure 4.3 are given by their x and y components. The polarization ratio of antenna 2 may then be defined in x and y coordinates as $P_2' = E_y/E_x$.

But, from Figure 4.3, $E_y = E_\eta$, $E_x = -E_\xi$, and, obviously,

$$P_2' = -P_2 \tag{4.137}$$

If this reversed coordinate system definition of the polarization ratio is used for antenna 2, the polarization efficiency equation becomes

$$\rho = \frac{(1 + P_1 P_2')(1 + P_1^* P_2'^*)}{(1 + P_1 P_1^*)(1 + P_2' P_2'^*)} \tag{4.138}$$

where both polarization ratios are defined in the same coordinate system.

4.13. POLARIZATION MATCH FACTOR — MISALIGNED ANTENNAS

In this chapter we have generally considered two antennas in transmit–receive configuration to have aligned axes (see Fig. 4.3). The orientation of the antennas was arbitrary, but the fields and effective length components were known in the particular coordinate systems. In the general case the radiated field components and effective length components are known in some appropriate coordinate system for the antennas, and these coordinate systems are not aligned. The antenna locations and field components must then be transformed to other coordinate systems before determining the polarization match factor.

Figure 4.5 shows the coordinate systems to be considered. The system without subscripts is a ground or reference system. The a system is appropriate to the transmitting antenna, with the radiated fields known in that system.

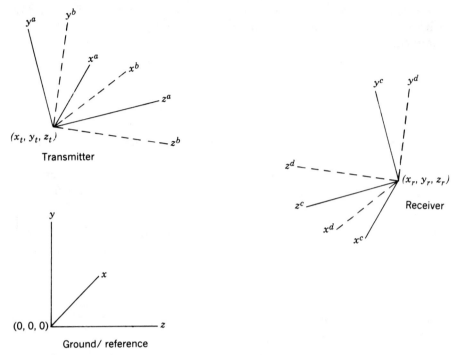

FIGURE 4.5 Coordinate systems for misaligned antennas.

The b system is rotated so that its z axis points toward the receiving antenna. Likewise, the c system is the natural one for the receiving antenna, the one in which its radiated field (or equivalently its effective length $\mathbf{h}$) is known. The d system is rotated so that its z axis points to the transmitting antenna.

Economy and conciseness of notation are essential to clarity when the number of coordinate systems is considered. We use $\mathbf{E}$ to represent the field of the transmitting antenna and $\mathbf{h}$ for the effective length of the receiving antenna. A letter superscript refers to the coordinate system in which a quantity is measured. Vector fields are represented by the usual boldface letters and are treated in this section as column matrices; thus

$$\mathbf{E} = \mathrm{col}(E_x, E_y, E_z) = \begin{bmatrix} E_x \\ E_y \\ E_z \end{bmatrix} = \begin{bmatrix} E_x & E_y & E_z \end{bmatrix}^T \qquad (4.139)$$

A column matrix also represents the coordinates of a point; thus

$$\mathbf{X} = \begin{bmatrix} x & y & z \end{bmatrix}^T \qquad (4.140)$$

Products may be represented in a vector form, as by a dot, or as a product of matrices, as appropriate. A 3 × 3 matrix is also represented by a boldface letter; thus

$$\mathbf{A} = \begin{bmatrix} A_{xx} & A_{xy} & A_{xz} \\ A_{yx} & A_{yy} & A_{yz} \\ A_{zx} & A_{zy} & A_{zz} \end{bmatrix} \quad (4.141)$$

The distinction between a vector and such a square matrix should present no difficulty.

The transformation, by rotations, of a point from coordinate system 1 to system 2, having the same origin, is carried out by the Euler angle matrix,

$$\mathbf{U} = \begin{bmatrix} \cos\beta\cos\gamma & \cos\beta\sin\gamma & -\sin\beta \\ \sin\alpha\sin\beta\cos\gamma - \cos\alpha\sin\gamma & \sin\alpha\sin\beta\sin\gamma + \cos\alpha\cos\gamma & \sin\alpha\cos\beta \\ \cos\alpha\sin\beta\cos\gamma + \sin\alpha\sin\gamma & \cos\alpha\sin\beta\sin\gamma - \sin\alpha\cos\gamma & \cos\alpha\cos\beta \end{bmatrix}$$
$$(4.142)$$

The angles α, β, and γ are measured from an axis in the old system (1) toward the corresponding axis in the new (2). The rotations are taken in order:

1. γ around the z axis in the direction $x \to y$.
2. β around the y axis in the direction $z \to x$.
3. α around the x axis in the direction $y \to z$.

Not all authors define the Euler angle transformations in the same manner [4].

The location of point $\mathbf{X}^{(2)}$ in the new system is related to its location $\mathbf{X}^{(1)}$ in the old system by

$$\mathbf{X}^{(2)} = \mathbf{U}\mathbf{X}^{(1)} \quad (4.143)$$

where $\mathbf{X}$ is the vector of (4.140) and (4.143) is the matrix product. Transformation of vector functions is carried out by the same matrix; thus

$$\mathbf{F}^{(2)} = \mathbf{U}\mathbf{F}^{(1)} \quad (4.144)$$

where

$$\mathbf{F} = \begin{bmatrix} F_x & F_y & F_z \end{bmatrix}^T \quad (4.145)$$

The transformation from system 2 to 1 is carried out with the inverse matrix

$$\mathbf{X}^{(1)} = \mathbf{U}^{-1}\mathbf{X}^{(2)} \quad (4.146)$$

But the Euler angle matrix is unitary, so that its inverse is equal to its transpose. Then

$$\mathbf{X}^{(1)} = \mathbf{U}^T \mathbf{X}^{(2)} \qquad (4.147)$$

The following Euler angle matrices are used for the coordinate systems of Figure 4.5:

A from the ground–reference system (translated to x_t, y_t, z_t) to system a.
B from the ground–reference system to system b.
C from the ground–reference system to system c.
D from the ground–reference system to system d.

Note that in many cases the geometry is simpler than this general case. The transmitter, for example, may also be the reference system, and z^a may already point to the receiver, making two transformations unnecessary.

Let us now consider the polarization-matching problem for the two antennas. More generally, we shall obtain the received power at the receiving antenna and separate the effects of polarization mismatch from the antenna gains. The positions and orientations of the transmitting and receiving antenna systems with respect to the reference system are known, that is, the matrices **A** and **C** are known. Further, we know the far fields of the transmitting antenna, E_θ^a and E_ϕ^a, in its natural coordinate system. The effective length components h_θ^c and h_ϕ^c of the receiving antenna are also known in its natural coordinate system. We may proceed using one of two methods, both of which are given here.

Method 1

Step 1. Translate the reference system to the transmitter position. Obtain the receiver position $\mathbf{X}'_r$ in this translated system,

$$\mathbf{X}'_r = \mathbf{X}_r - \mathbf{X}_t \qquad (4.148)$$

Step 2. Use the Euler angle matrix **A** to find the receiver position in the natural system (system a) of the transmitter.

$$\mathbf{X}_r^a = \mathbf{A}\mathbf{X}'_r \qquad (4.149)$$

Determine the colatitude and azimuth angles of the receiver in system a.

Step 3. From the known properties of the transmitter, find E_θ^a and E_ϕ^a at the receiver. The absolute values of the fields must be found if the receiver power is needed. This requires a knowledge of transmitted power and the

216 POLARIZATION MATCHING OF ANTENNAS

distance from transmitter to receiver, $|\mathbf{X}_r - \mathbf{X}_t|$. If relative values are sufficient, the transmitter–receiver distance and the transmitter power may be neglected. In fact, only the effective length $\mathbf{h}(\theta, \phi)$ of the transmitting antenna is needed. We shall continue to use $\mathbf{E}$, however, since it is more general and so that it can be distinguished easily from the $\mathbf{h}$ value used for the receiving antenna.

Step 4. Convert E_θ^a and E_ϕ^a at the receiver to rectangular form.

$$E_x^a = E_\theta^a \cos\theta \cos\phi - E_\phi^a \sin\phi \qquad (4.150\text{a})$$

$$E_y^a = E_\theta^a \cos\theta \sin\phi + E_\phi^a \cos\phi \qquad (4.150\text{b})$$

$$E_z^a = -E_\theta^a \sin\theta \qquad (4.150\text{c})$$

where θ and ϕ are the known values at the receiver found in step 2. The subscripts refer to axes in the a system, specifically x^a, y^a, z^a.

Step 5. Transform the field components to the receiving antenna system, going to the reference system as an intermediate step and then to the receiving antenna system (system c) using the known matrix $\mathbf{C}$.

$$\mathbf{E} = \mathbf{A}^T \mathbf{E}^a \qquad (4.151)$$

$$\mathbf{E}^c = \mathbf{C}\mathbf{E} = \mathbf{C}\mathbf{A}^T \mathbf{E}^a \qquad (4.152)$$

Step 6. Determine the colatitude and azimuth angles of the transmitter in system c. Use these with the known receiver effective length $\mathbf{h}^c(\theta_c, \phi_c)$ to find the receiver length $\mathbf{h}^c$ in the direction of the transmitter. Convert to rectangular form using the equations of step 4.

Step 7. If the absolute value of $\mathbf{E}^c$ is known, find the receiver open-circuit voltage from

$$V = \mathbf{E}^c \cdot \mathbf{h}^c = \mathbf{E}^{cT}\mathbf{h}^c \qquad (4.153)$$

where the multiplication can be considered either the scalar product (dot product) of two vectors or as the matrix product of a 1×3 and a 3×1 matrix. The received power is easily found if antenna and load impedances are known.

If the power is not needed, or if only a relative value of $\mathbf{E}^c$ has been obtained, find the polarization match factor from

$$\rho = \frac{|\mathbf{E}^c \cdot \mathbf{h}^c|}{|\mathbf{E}^c|^2 |\mathbf{h}^c|^2} \qquad (4.154)$$

Method 2

Note that Method 1 does not utilize the polarization ratios of the antennas, nor does it use the coordinate systems b and d. An alternate method to develop the polarization match factor between the antennas does lead naturally to the use of polarization ratios.

Steps 1–4. Same as steps 1–4 of Method 1.

Step 5. Create two new coordinate systems, b and d of Figure 4.5, and obtain the Euler angle matrices **B** and **D**. The z axes are to be antiparallel and so are the x axes. The y axes are parallel. These systems will then correspond to those of Figure 4.3, and equations developed using that figure will be valid. The requirements on systems b and d are not yet sufficient to yield unique coordinates. It is convenient to require further that the axes x^b and x^d lie in the xz plane of the reference system. In the Euler angle matrices, this leads to the requirement that $\gamma = 0$. The Euler angle matrices then become

$$\mathbf{B} = \begin{bmatrix} \cos\beta_b & 0 & -\sin\beta_b \\ \sin\alpha_b \sin\beta_b & \cos\alpha_b & \sin\alpha_b \cos\beta_b \\ \cos\alpha_b \sin\beta_b & -\sin\alpha_b & \cos\alpha_b \cos\beta_b \end{bmatrix} \quad (4.155)$$

$$\mathbf{D} = \begin{bmatrix} \cos\beta_d & 0 & -\sin\beta_d \\ \sin\alpha_d \sin\beta_d & \cos\alpha_d & \sin\alpha_d \cos\beta_d \\ \cos\alpha_d \sin\beta_d & -\sin\alpha_d & \cos\alpha_d \cos\beta_d \end{bmatrix} \quad (4.156)$$

Consider again the reference system translated to the transmitter position so that the receiver position in the translated system is

$$\mathbf{X}'_r = \mathbf{X}_r - \mathbf{X}_t \quad (4.148)$$

The receiver position in system b is then

$$\mathbf{X}^b_r = \mathbf{B}\mathbf{X}'_r \quad (4.157)$$

or

$$\begin{bmatrix} x^b_r \\ y^b_r \\ z^b_r \end{bmatrix} = \mathbf{B} \begin{bmatrix} x'_r \\ y'_r \\ z'_r \end{bmatrix} \quad (4.158)$$

But in system b, the x and y coordinates x^b_r and y^b_r of the receiver are zero. Then

$$\cos\beta_b x'_r - \sin\beta_b z'_r = 0 \quad (4.159a)$$
$$\sin\alpha_b \sin\beta_b x'_r + \cos\alpha_b y'_r + \sin\alpha_b \cos\beta_b z'_r = 0 \quad (4.159b)$$

These equations may be solved for the angles in terms of the known positional values. The solutions are

$$\tan \beta_b = \frac{x'_r}{z'_r} \qquad (4.160a)$$

$$\tan \alpha_b = -\frac{1}{\cos \beta_b} \frac{y'_r z'_r}{(x'_r)^2 + (z'_r)^2} \qquad (4.160b)$$

The matrix **B** is thus completely specified.

To find **D**, the reference system is first translated to the position of the receiving antenna. The location of the transmitter in this translated system is

$$\mathbf{X}''_t = \mathbf{X}_t - \mathbf{X}_r \qquad (4.161)$$

and the transmitter in system d is at

$$\mathbf{X}^d_t = \mathbf{D}\mathbf{X}''_t \qquad (4.162)$$

The x^d_t and y^d_t coordinates are zero, which leads to

$$\cos \beta_d x''_t - \sin \beta_d z''_t = 0 \qquad (4.163a)$$
$$\sin \alpha_d \sin \beta_d x''_t + \cos \alpha_d y''_t + \sin \alpha_d \cos \beta_d z''_t = 0 \qquad (4.163b)$$

with solutions

$$\tan \beta_d = \frac{x''_t}{z''_t} \qquad (4.164a)$$

$$\tan \alpha_d = -\frac{1}{\cos \beta_d} \frac{y''_t z''_t}{(x''_t)^2 + (z''_t)^2} \qquad (4.164b)$$

Step 6. Transform the field components obtained in step 4 to the b system, using the reference system as an intermediate step.

$$\mathbf{E}^b = \mathbf{B}\mathbf{A}^T\mathbf{E}^a \qquad (4.165)$$

Note that $\mathbf{E}^b$ will not have a z^b component but only transverse components.

Step 7. Use the Euler angle matrix **C** to find the transmitter position in the natural system (system c) of the receiving antenna,

$$\mathbf{X}^c_t = \mathbf{C}\mathbf{X}''_t \qquad (4.166)$$

Determine the colatitude and azimuth angles of the transmitter in system c.

Step 8. From knowledge of the receiver effective length $\mathbf{h}^c(\theta_c, \phi_c)$ in its natural coordinate system, find the effective length (rectangular components) in the direction of the transmitter.

Step 9. Transform the effective length to system d using the reference system as an intermediate step.

$$\mathbf{h}^d = \mathbf{DC}^T\mathbf{h}^c \tag{4.167}$$

Note that in system d, $\mathbf{h}^d$ has only transverse components.

Step 10. Define polarization ratios P_t and P_r for the transmitting and receiving antennas,

$$P_t = \frac{E_y^b}{E_x^b} \tag{4.168a}$$

$$P_r = \frac{h_y^d}{h_y^d} \tag{4.168b}$$

The subscripts refer to axes in the proper coordinate system. Thus in defining P_t, the components are those along the y^b and x^b axes, while in defining P_r, the components are along the y^d and x^d axes.

Step 11. Find the polarization match factor from

$$\rho = \frac{(1 - P_t P_r)(1 - P_t^* P_r^*)}{(1 + P_t P_t^*)(1 + P_r P_r^*)} \tag{4.169}$$

It is obvious that the second method is more cumbersome than the first. It has the advantage, however, that the polarization ratios are obtained, and the aids developed for understanding polarization problems, such as the Poincaré sphere and the complex plane charts, can be readily applied.

REFERENCES

1. G. Sinclair, "The Transmission and Reception of Elliptically Polarized Waves," *Proc. IRE*, **38**(2), 148–151, February 1950.
2. V. H. Rumsey, "Transmission between Elliptically Polarized Antennas," *Proc. IRE*, **39**(5), 535–540, May 1951.
3. "IEEE Standard Definitions of Terms for Antennas," IEEE Standard 145-1983, The Institute of Electrical and Electronics Engineers, New York, 1983.
4. H. Goldstein, *Classical Mechanics*, Addison-Wesley, Reading, MA, 1950.

PROBLEMS

4.1. In a reference coordinate system located at the ground (see Fig. 4.5) a transmitting antenna is located at (500, 1000, 2000) and a receiving antenna is located at (0, 500, 5000). Develop a coordinate system at the transmitting antenna with its z axis pointing at the receiving antenna and its x axis parallel to the xz plane of the reference system. Find the Euler angle matrix of this system.

4.2. Two short dipoles are used as transmitter and receiver in a communications link. The first dipole lies along the y axis of the reference coordinate system of Figure 4.5. The xz plane of the reference system is parallel to the ground. The z axis of the reference system points to the east. The second dipole is located at (400, 400, 2000). It leans toward the northeast and makes an angle of 75° with the ground. Find the polarization match factor between the antennas.

4.3. Show that the polarization match factor between two antennas can be written as

$$\rho = \frac{|\mathbf{h}_1 \cdot \mathbf{h}_2|^2}{|\mathbf{h}_1|^2 |\mathbf{h}_2|^2}$$

where $\mathbf{h}_1$ and $\mathbf{h}_2$ are the effective lengths of the two antennas measured in the same coordinate system.

4.4. Show that, for two antennas in a communication link, if the axial ratio of one antenna is much greater than that of the other, it is of little concern that the antennas have the same or opposite rotation sense.

4.5. Verify the statement following (4.78) that if two identical antennas are first placed side by side and then moved into a transmit–receive configuration by rotating one of them 180° around a vertical axis, the major axes of their polarization ellipses no longer coincide.

4.6. Find the Euler angle matrix if the first rotation is around the x axis by angle α, in the direction $y \to z$, the second around the new y axis by angle β, in the direction $z \to x$, and the third around the newest z axis by angle γ, in the direction $x \to y$.

4.7. A wave incident on an antenna of effective length

$$\mathbf{h} = 0.5\mathbf{u}_x + 0.4\mathbf{u}_y \quad \text{m}$$

has components

$$E_x = 1 \qquad E_y = 0.5e^{j\pi/4} \quad \mu\text{V/m}$$

(a) Find the open-circuit received voltage.
(b) The receiving antenna is impedance matched and the load resistance is 50 Ω. Find the received power.

4.8. In Figure 4.5 the transmitter is located at $x_t, y_t, z_t = 0, 2, 10$ km with respect to the reference coordinate origin. The receiver is at $-1, 8, 40$ km. The transmitter coordinate system is unrotated from the reference system, that is, the Euler angle matrix **A** is the identity matrix. At the transmitter find the polar and azimuth angles measured to the receiver.

4.9. Two short dipoles in a transmit–receive configuration are correctly oriented so that maximum power is received. Then the receiving antenna is rotated 15° while remaining in the plane containing the transmitting antenna and the line of sight. Find the loss in power, in decibels, caused by the rotation. Is this power loss a loss in gain, or is it a polarization loss?

4.10. In Problem 4.9, if the receiving antenna is rotated 15° around the line of sight between the antennas, rather than as specified in that problem, find the power loss. Is it a gain loss or a polarization loss?

4.11. The effective length of an antenna is

$$\mathbf{h} = h_0 \sin\theta \sin\phi \, \mathbf{u}_\theta \qquad \begin{cases} 0 \le \theta \le \pi \\ 0 \le \phi \le \pi \end{cases}$$

If the radiation efficiency is 0.98, find the gain $G(\theta, \phi)$.

CHAPTER FIVE

Polarization Characteristics of Some Antennas

5.1. INTRODUCTION

In this chapter we shall obtain the polarization parameters of several common antennas. We shall also obtain polarization match factors when these antennas are paired with standard antennas in a transmit–receive configuration. In this way we can compare the off-axis performance of the antennas of interest to their polarization performance on-axis (or in some design direction).

In the previous chapters the polarization parameters were defined in the context of a wave traveling in the z direction, as in Figure 5.1a. Specifically, the polarization ratio P was defined as

$$P = \frac{E_y}{E_x} \qquad (3.73)$$

The coordinates of Figure 5.1a are common in a discussion of polarization [1]. The value of P clearly depends on the coordinates used. For example, if we use the ratio $P = E_y/E_x$ for the rotated system of Figure 5.1b, it will not be the same as (3.73). A somewhat more general definition of the polarization ratio is

$$P = \frac{E_{\text{vertical}}}{E_{\text{horizontal}}} \qquad (5.1)$$

with unit vectors chosen so that

$$\mathbf{u}_{\text{horizontal}} \times \mathbf{u}_{\text{vertical}} = \mathbf{u}_{\text{propagation direction}} \qquad (5.2)$$

INTRODUCTION 223

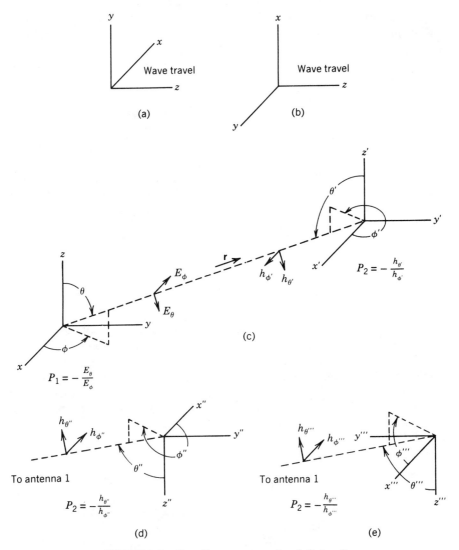

FIGURE 5.1 Coordinate systems for defining P.

Note that the first coordinate system of Figure 5.1 satisfies the requirement (5.2) if x is taken as the "horizontal" component and y the "vertical," but the coordinates of Figure 5.1b do not satisfy (5.2) if y is treated as horizontal.

Near the earth's surface we may define horizontal somewhat loosely as parallel to the surface. More precisely, if a line is drawn from the coordinate origin to the earth's center and if a tangent plane is drawn at the intersection of this line and the earth's surface, then the horizontal axis of the coordinate system is parallel to the tangent plane. The direction of wave propagation may or may not be parallel to this plane, but it either case, vertical is defined as perpendicular to the horizontal axis and to the direction of propagation.

At points far from the earth the coordinate system for defining the polarization ratio is essentially arbitrary.

The electric field radiated by an antenna is commonly defined by a spherical coordinate system as in Figure 5.1c. The wave, having only E_θ and E_ϕ components (in the far field), travels in the radial direction. If the xy plane is parallel to the earth's surface, then E_ϕ is the horizontal component of the wave (it is always parallel to the xy plane and hence to the earth's surface) and $-E_\theta$ is the vertical component. [To establish this, either lay the coordinates of Fig. 5.1a over those of Fig. 5.1c with the z axis coincident with **r** or take the cross product $\mathbf{u}_\phi \times (-\mathbf{u}_\theta)$.] The appropriate definition of the polarization ratio is then

$$P = \frac{-E_\theta}{E_\phi} \qquad (5.3)$$

Some difficulty arises if we establish a coordinate system whose xy plane is not parallel to the earth's surface. Neither E_θ nor E_ϕ is in general horizontal (parallel to the earth's surface), and we cannot define P in terms of vertical and horizontal components. In this text, therefore, we use coordinate systems, if possible, with the z axis perpendicular to the earth's surface. If that is inappropriate, we continue to use (5.3) to define P and consider horizontal to mean parallel to the xy plane.

5.2. TEST ANTENNAS FOR DETERMINING EFFECT OF POLARIZATION

One purpose for obtaining the polarization parameters of an antenna is to find the polarization match factor between the antenna used, for example, as a transmitter and some other antenna used as a receiver. What antenna should we use as a receiver, and how should it be oriented? It is reasonable that if the antenna being examined is intended to produce a circularly polarized wave, for example, we should see how faithfully it does so by using a circularly polarized receiving antenna. Since most antennas are meant to produce either linear or circular polarization (and, with one exception, we consider only these in this chapter), we shall use a linearly or circularly polarized antenna, as appropriate, to receive the wave.

The polarization ratio was redefined in Section 5.1, and we must determine the effect on the equations for the polarization match factor developed in Chapter 4. To do this, we consider the coordinate systems of Figure 5.1c. Antenna 1 is the transmitter and antenna 2 is the receiver. The rectangular

TEST ANTENNAS FOR DETERMINING EFFECT OF POLARIZATION

coordinate system at antenna 2 is translated from antenna 1. The polarization match factor developed in Chapter 4 is

$$\rho = \frac{|\mathbf{E}^i \cdot \mathbf{h}|^2}{|\mathbf{E}^i|^2 |\mathbf{h}|^2} \tag{4.45}$$

where $\mathbf{E}^i$ is the field at the receiving antenna caused by the transmitter and $\mathbf{h}$ is the effective length of the receiving antenna. Both $\mathbf{E}^i$ and $\mathbf{h}$ are measured in the same coordinate system. Let us arbitrarily choose to do so in the receiver system of Figure 5.1c. Then

$$\mathbf{E}^i \cdot \mathbf{h} = E^i_{\theta'} h_{\theta'} + E^i_{\phi'} h_{\phi'} \tag{5.4}$$

From Figure 5.1c we see that

$$E^i_\theta = E^i_{\theta'} \qquad E^i_\phi = -E^i_{\phi'} \tag{5.5}$$

and therefore

$$\mathbf{E}^i \cdot \mathbf{h} = E^i_\theta h_{\theta'} - E^i_\phi h_{\phi'} = E^i_\phi h_{\phi'} \left(\frac{E^i_\theta}{E^i_\phi} \frac{h_{\theta'}}{h_{\phi'}} - 1 \right)$$

It is clear that for the transmitting antenna, 1,

$$P_1 = \frac{-E^i_\theta}{E^i_\phi} \tag{5.6}$$

and if we think for a moment of antenna 2 as transmitting a wave $\mathbf{E}$ toward 1, its field would be proportional to its effective length. Then

$$P_2 = \frac{-E_{\theta'}}{E_{\phi'}} = -\frac{h_{\theta'}}{h_{\phi'}} \tag{5.7}$$

If (5.6) and (5.7) are used in the component parts of (4.45),

$$\mathbf{E}^i \cdot \mathbf{h} = E^i_\phi h_{\phi'} (P_1 P_2 - 1) \tag{5.8a}$$

$$|\mathbf{E}^i|^2 = |E^i_\phi|^2 (1 + P_1 P_1^*) \tag{5.8b}$$

$$|\mathbf{h}|^2 = |h_{\phi'}|^2 (1 + P_2 P_2^*) \tag{5.8c}$$

then substitution into (4.45) gives for the polarization match factor

$$\rho = \frac{(1 - P_1 P_2)(1 - P_1^* P_2^*)}{(1 + P_1 P_1^*)(1 + P_2 P_2^*)} \tag{5.9}$$

This equation is identical to (4.57), which we developed using the definition (3.73) for polarization ratio. We may therefore use (4.57) or any equation for ρ involving polarization parameters p, q, w, and so on developed in Chapter 4.

It is not always convenient to define P_2 in the translated coordinate system of Figure 5.1c. If we have, for example, two identical antennas pointing at each other, a reversed coordinate system is appropriate for one of them. Two such systems are shown in Figure 5.1d and e. If P_2 is defined in the manner shown, (5.9) remains valid. Although we define P_2 in one of the coordinate systems of Figure 5.1c, d, or e, it is nonetheless convenient to express P_2 in terms of θ and ϕ, the direction of the receiving antenna from the transmitter. This will be clearer as we discuss the polarization of the receiving antenna necessary to give a polarization match with the transmitter.

To select the receiving antenna and orient it correctly, we start by recognizing that a dipole antenna produces a wave that is everywhere linearly polarized. If we use it as a transmitter, then we should use another linearly polarized antenna (perhaps another dipole) as a receiver. If the receiver is correctly oriented, there is no polarization mismatch in any direction from

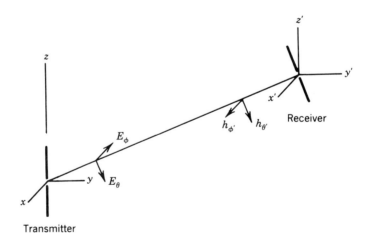

FIGURE 5.2 Orientation of receiving dipole for testing z-directed transmitting dipole antenna.

the transmitter. Figure 5.2 shows the correct orientation for the receiving antenna if the antenna undergoing examination is a z-directed dipole. The receiving dipole must lie in the plane containing the transmitting dipole, and the line from transmitter to receiver must be perpendicular to the receiving dipole. We anticipate the results of Section 5.3 and note that the z-directed transmitter has a polarization ratio

$$P_z = \infty \tag{5.24}$$

except at $\theta = 0$ where the field is zero. From Figure 5.2 it is clear that on a line perpendicular to the receiving dipole $h_{\phi'} = 0$ and the receiver has a polarization ratio

$$P_r = \infty \tag{5.10}$$

Substitution in (5.9) shows that the polarization matching requirement is met for these two antennas.

Let us now suppose that the transmitting antenna is a dipole lying on the x axis. Again we anticipate the results of Section 5.3 and note that the polarization ratio in the direction θ, ϕ is

$$P_x = \cos \theta \cot \phi \tag{5.22}$$

For polarization matching the receiving antenna should be a dipole lying in a plane that contains the x axis, and a perpendicular from the receiving antenna should point to the transmitter. Since we know that these antennas will be polarization matched, the polarization ratio of the receiving antenna must be

$$P_r = -P_x^* = -\cos \theta \cot \phi \tag{5.11}$$

An earlier statement may now be clearer. In this expression, P_r is defined as $-h_{\theta'}/h_{\phi'}$, using the translated coordinate system of Figure 5.1c (or the translated and reversed systems of Fig. 5.1d or e). Nonetheless, it is convenient to give P_r not in terms of the angles θ', ϕ' of Figure 5.1c, but in terms of the direction θ, ϕ of the receiving antenna from the transmitting antenna.

In Section 5.3 we shall find that the polarization ratio of a y-directed dipole is

$$P_y = -\cos \theta \tan \phi \tag{5.23}$$

For test purposes we use a dipole receiver with polarization ratio

$$P_r = -P_y^* = \cos\theta \tan\phi \qquad (5.12)$$

Now suppose that the transmitting antenna is intended to produce a wave with polarization characteristics similar to one of these x-, y-, or z-oriented dipoles. The open waveguide antenna of Section 5.8 is an example, with a polarization similar to that of the y-oriented dipole. For such an antenna we continue to use as a test (receiving) antenna a dipole with polarization ratio given by (5.12), even though there is a polarization mismatch in some directions. The antenna is intended to produce a linearly polarized wave, and it is appropriate to examine how well it does so. We are not measuring the polarization of the antenna, although we could certainly do so; rather we are comparing its polarization to a standard, and the dipole is the standard.

It should be noted that the receiving antenna changes only in orientation as it is moved from one point to another. The change in polarization ratio is caused by this change in orientation.

Right and left circularly polarized antennas have polarization ratios of $-j$ and $+j$ respectively. If an antenna is intended to produce a right-circular wave, whether it does so in all directions or not, we test it by obtaining its polarization match factor with a receiving antenna having

$$P_r = -(-j)^* = -j \qquad (5.13)$$

Similarly, to test a left-circular antenna, we use a receiver with

$$P_r = +j \qquad (5.14)$$

In this chapter we consider relative radiation intensity

$$|E_\theta(\theta,\phi)|^2 + |E_\phi(\theta,\phi)|^2$$

and polarization match factor

$$\rho(\theta,\phi)$$

We refer to a radiation intensity pattern and to the *radiation intensity beamwidth* as the angle between the two directions in which the radiation intensity drops to half (-3 dB) of its maximum value.

In the same way the *polarization beamwidth* is the angle between two directions for which ρ is one half.

If both radiation intensity and polarization effects are used in determining the 3-dB points for an antenna, *overall beamwidth* is used as the angle between two directions for which received power in an appropriate receiving antenna drops to half the possible received power.

5.3. THE SHORT DIPOLE

The far fields produced by short dipoles oriented along the coordinate axes are obtained in this section. The fields of a dipole with any orientation may then be written as the sum of the fields produced by these.

The magnetic vector potential of a short dipole directed along one of the coordinate axes is given by (1.33) and similar forms,

$$A_{x,y,z} = \frac{\mu IL}{4\pi r} e^{-jkr} \tag{5.15}$$

where I is the value of the current at all points in the dipole of length L. A more realistic model for a short antenna is a center-fed dipole having a triangular current distribution with maximum current I_0 at the center and zero current at the ends. The vector potential for this antenna is (see Problem 1.5)

$$A_{x,y,z} = \frac{\mu I_0 L}{8\pi r} e^{-jkr} \tag{5.16}$$

There is no essential difference between these equations, and we shall continue to use (5.15).

The fields of the dipoles can be found by the transformations

$$\begin{aligned}
A_r &= A_x \sin\theta \cos\phi + A_y \sin\theta \sin\phi + A_z \cos\theta \\
A_\theta &= A_x \cos\theta \cos\phi + A_y \cos\theta \sin\phi - A_z \sin\theta \\
A_\phi &= -A_x \sin\phi + A_y \cos\phi
\end{aligned} \tag{5.17}$$

and the far-field equations of (1.58), repeated here for convenience,

$$E_r = 0 \qquad E_\theta = -j\omega A_\theta \qquad E_\phi = -j\omega A_\phi \tag{5.18}$$

Fields

The fields that result from these equations are

x-DIRECTED DIPOLE

$$E_r = 0 \tag{5.19a}$$

$$E_\theta = -\frac{j\omega\mu IL}{4\pi r} \cos\theta \cos\phi \, e^{-jkr} \tag{5.19b}$$

$$E_\phi = \frac{j\omega\mu IL}{4\pi r} \sin\phi \, e^{-jkr} \tag{5.19c}$$

y-DIRECTED DIPOLE

$$E_r = 0 \tag{5.20a}$$

$$E_\theta = -\frac{j\omega\mu IL}{4\pi r} \cos\theta \sin\phi \, e^{-jkr} \tag{5.20b}$$

$$E_\phi = -\frac{j\omega\mu IL}{4\pi r} \cos\phi \, e^{-jkr} \tag{5.20c}$$

z-DIRECTED DIPOLE

$$E_r = 0 \tag{5.21a}$$

$$E_\theta = \frac{j\omega\mu IL}{4\pi r} \sin\theta \, e^{-jkr} \tag{5.21b}$$

$$E_\phi = 0 \tag{5.21c}$$

Polarization Ratios

x-DIRECTED DIPOLE

$$P_x = -\frac{E_\theta}{E_\phi} = \cos\theta \cot\phi \tag{5.22}$$

y-DIRECTED DIPOLE

$$P_y = -\cos\theta \tan\phi \tag{5.23}$$

z-DIRECTED DIPOLE

$$P_z = \infty \tag{5.24}$$

Polarization Match Factors

This is essentially trivial since we have already postulated that the dipole in question is to be compared to a correctly oriented receiving antenna, which is also a dipole with

$$P_r = -P^*_{x,y,z} \tag{5.25}$$

Nevertheless, let us see the process for the x-directed dipole. We substitute

$$P_x = \cos\theta \cot\phi \tag{5.22}$$

and

$$P_r = -\cos\theta \cot\phi \tag{5.11}$$

into the match factor equation

$$\rho = \frac{(1 - P_x P_r)(1 - P^*_x P^*_r)}{(1 + P_x P^*_x)(1 + P_r P^*_r)} \tag{5.26}$$

and immediately obtain $\rho = 1$. The test antenna does not receive the same power at all points, but the variation is due to the dipole directive gain, not to its polarization properties.

Received Power

The received power density is quickly found from the fields to be

x-DIRECTED DIPOLE

$$\mathcal{P}_x = \frac{1}{2}\sqrt{\frac{\epsilon}{\mu}}\left(|E_\theta|^2 + |E_\phi|^2\right) = \frac{1}{2}\sqrt{\frac{\epsilon}{\mu}}\left(\frac{\omega\mu IL}{4\pi r}\right)^2 (1 - \sin^2\theta \cos^2\phi) \tag{5.27}$$

y-DIRECTED DIPOLE

$$\mathcal{P}_y = \frac{1}{2}\sqrt{\frac{\epsilon}{\mu}}\left(\frac{\omega\mu IL}{4\pi r}\right)^2 (1 - \sin^2\theta \sin^2\phi) \tag{5.28}$$

z-DIRECTED DIPOLE

$$\mathcal{P}_z = \frac{1}{2}\sqrt{\frac{\epsilon}{\mu}}\left(\frac{\omega\mu IL}{4\pi r}\right)^2 \sin^2\phi \tag{5.29}$$

5.4. CROSSED DIPOLES (TURNSTILE ANTENNA)

An antenna used to produce a circularly polarized wave is shown in Figure 5.3. If the vertical (*y*-directed) and horizontal (*x*-directed) dipoles are identical and are fed with currents having the same amplitudes and $\pi/2$ phase difference, the radiated wave is circularly polarized on the *z* axis.

Take the feed current or voltage to the *x*-directed dipole as reference, and let the feed to the *y* dipole lead it by $\pi/2$. The resulting fields are the sum of (5.19) and (5.20) multiplied by *j*. The result is

$$E_\theta = -\frac{j\omega\mu IL}{4\pi r}(\cos\theta\cos\phi + j\cos\theta\sin\phi)e^{-jkr} \qquad (5.30a)$$

$$E_\phi = \frac{j\omega\mu IL}{4\pi r}(\sin\phi - j\cos\phi)e^{-jkr} \qquad (5.30b)$$

The polarization ratio is

$$P_t = \frac{\cos\theta\cos\phi + j\cos\theta\sin\phi}{\sin\phi - j\cos\phi} = j\cos\theta \qquad (5.31)$$

On the *z* axis, $\theta = 0$ and $P_t = j1$, which corresponds to a left-circular wave propagating in the *z* direction. Had the *y*-dipole feed lagged in phase by $\pi/2$, the wave would have been right circular along the *z* axis.

In Section 5.2 we saw that it is appropriate to examine the polarization loss as a function of propagation direction by allowing the antenna to radiate toward a left-circular receiving antenna with $P_r = +j$. Then for any angle θ the polarization match factor for this antenna pair is

$$\rho(\theta,\phi) = \frac{(1 - P_t P_r)(1 - P_t^* P_r^*)}{(1 + P_t P_t^*)(1 + P_r P_r^*)} = \frac{(1 + \cos\theta)^2}{2(1 + \cos^2\theta)} = \frac{1}{2} + \frac{\cos\theta}{1 + \cos^2\theta} \qquad (5.32)$$

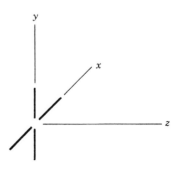

FIGURE 5.3 Crossed dipoles or turnstile antenna.

CROSSED DIPOLES WITH GROUND PLANE

This equation shows that the power received drops to half its maximum value at $\theta = \pi/2$.

The antenna gain, neglecting polarization effects, is proportional to

$$|E_\theta|^2 + |E_\phi|^2 = |\cos\theta\cos\phi + j\cos\theta\sin\phi|^2 + |\sin\phi - j\cos\phi|^2$$
$$= 1 + \cos^2\theta$$

and the gain relative to the maximum gain (in the direction $\theta = 0$) is

$$G_r = \tfrac{1}{2}(1 + \cos^2\theta) \tag{5.33}$$

Note that the half-power angle for polarization is $\theta = \pi/2$, and the half-power beamwidth for polarization is π. The half-power beamwidth, neglecting polarization, is also π, from (5.33). When ρ and G_r are combined to determine the actual power received by a circularly polarized antenna, we obtain

$$G_r\rho = \tfrac{1}{4}(1 + \cos\theta)^2 \tag{5.34}$$

Setting $G_r\rho$ to $\tfrac{1}{2}$ gives

$$\theta_{3\,\mathrm{dB}} = 65.5°$$
$$\text{Half-power beamwidth} = 2\theta_{3\,\mathrm{dB}} = 131°$$

A note of caution is in order. The polarization ratio for the crossed dipoles was found to be

$$P_t = j\cos\theta \tag{5.31}$$

In the xz plane, for large θ (measured from the z axis), the electric field is primarily y directed, whereas in the yz plane, for large θ, it is primarily x directed. The reader who thinks of y as the "vertical" axis and x as "horizontal" in Figure 5.3 will be concerned that the polarization ratio is the same in both planes. As noted in Section 5.1, the xy plane must be treated as the horizontal plane in defining polarization ratio in terms of E_θ and E_ϕ, even though it need not be parallel to the earth. In that context, both the x and y axes in Figure 5.3 are horizontal, and it is reasonable that the polarization ratio has the same form in the two cases described above.

5.5. CROSSED DIPOLES WITH GROUND PLANE

The radiation intensity pattern of the crossed dipoles can be sharpened by placing the dipoles in front of an infinite conducting plane, as in Figure 5.4. By image theory the fields in front of the plane remain the same if the screen

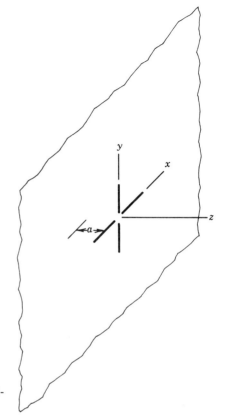

FIGURE 5.4 Crossed dipoles near an infinite conducting plane.

is removed and image dipoles, fed by currents differing in phase from the real dipoles by π radians, are placed at a distance $2a$ from the real dipoles.

From the pattern multiplication principle of array theory the far-zone field of a uniform array of identical elements is the product of the field of a single element and the *array factor*. The array factor is a function of geometry and the excitation phases of the elements and is essentially the pattern of an array of isotropic radiators located at the real antennas [2]. The array factor of two elements on the z axis separated by distance d is [2]

$$F(\theta, \phi) = 2\cos\left[\tfrac{1}{2}(kd\cos\theta + \beta)\right] \tag{5.35}$$

where β is the phase of the excitation of the element at the greater z value compared to that of the element at the lesser value of z.

If we let the dipoles be a quarter wavelength from the plane, and note that $\beta = \pi$ and $d = 2a = \lambda/2$, the array factor for the crossed dipoles in front of the conducting screen becomes

$$F(\theta, \phi) = 2\cos\left[\tfrac{1}{2}\pi(\cos\theta + 1)\right] \qquad (5.36)$$

Both E_θ and E_ϕ of (5.30) are multiplied by this array factor to give the new fields, and since both are altered by the same factor, it is clear that P_t, as given by (5.31), and ρ, as given by (5.32), are unchanged. On the other hand, the radiation intensity is multiplied by the square of the array factor, and the relative gain becomes

$$G_r = \tfrac{1}{2}(1 + \cos^2\theta)\cos^2\left[\tfrac{1}{2}\pi(\cos\theta + 1)\right] \qquad (5.37)$$

The product of G_r and ρ is now

$$G_r\rho = \tfrac{1}{4}(1 + \cos\theta)^2\cos^2\left[\tfrac{1}{2}\pi(\cos\theta + 1)\right] \qquad (5.38)$$

and if this is set equal to $\tfrac{1}{2}$, the half-power beamwidth, considering both radiation intensity and polarization match, is

$$\text{Beamwidth} = 2\theta_{3\,\text{dB}} = 98.4°$$

The value of ρ, the polarization match factor, is 0.854 at the overall 3-dB angle for the crossed dipoles without a screen and 0.958 for the crossed dipoles with a screen. Using the screen produces a narrower beam and one that is still almost circularly polarized at the beam edge.

We saw in Chapter 2 and again in this example that the array factor has no polarization properties, since it is the pattern of an array of isotropic radiators. Both E_θ and E_ϕ produced by one element are multiplied by the same factor. Then the polarization ratio of an array of identical elements whose fields are not altered by the presence of other elements in the array is the same as that of one of the elements. This may be advantageous in some applications since the radiation intensity pattern can be primarily controlled by the array geometry, whereas the polarization of the radiated wave is completely established by the choice of array element.

5.6. THE LOOP ANTENNA

The far electric field of a small circular loop antenna with uniform in-phase current lying in the xy plane, as shown in Figure 5.5, is [2]

$$E_r = E_\theta = 0 \qquad (5.39a)$$

$$E_\phi = \frac{\omega\mu ka^2 I \sin\theta}{4r} e^{-jkr} \qquad (5.39b)$$

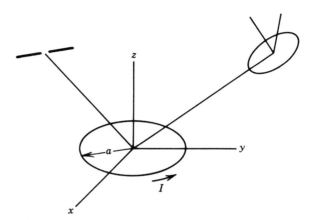

FIGURE 5.5 Circular loop antenna and test antennas.

It is obvious that the field is everywhere linearly polarized and horizontal, as discussed in Section 5.1. Then another circular loop antenna or a horizontal dipole can be used to receive a field radiated by the transmitting loop without polarization loss. The receiving loop is oriented so that the radial line from the transmitting loop center is in the receiving loop plane. This will keep the receiving loop gain constant as it is moved from one location to another. A horizontal dipole receiving antenna must be tangent to a circle drawn with the z axis as center. With either receiving antenna, $\rho = 1$. We now have two additional test antennas to use with an antenna intended to produce a field linearly polarized in a horizontal direction.

The loop may be considered small, and (5.39) is valid if $a \ll \lambda$. If that is not the case, more general equations for the loop fields are [1]

$$E_r = E_\theta = 0 \tag{5.40a}$$

$$E_\phi = \frac{\omega \mu a I_L J_1(ka \sin \theta)}{2r} e^{-jkr} \tag{5.40b}$$

where J_1 is the Bessel function of the first kind and first order. For loops with circumference $\lambda/4$ or greater, phase shifters must be inserted at intervals around the loop to maintain a uniform in-phase current in the loop [1].

It should be noted that the radiated field of the large loop is still linearly polarized in the azimuth direction.

5.7. LOOP AND DIPOLE

We saw in Section 5.6 that the field of a small loop is azimuthal and varies as $\sin \theta$. In Section 5.3 we noted that the field of a z-directed short dipole is directed wholly in the θ direction and also varies as $\sin \theta$. It is obvious then that a combination of the two antennas with proper current amplitudes and phases can produce a wave that is everywhere circular. Figure 5.6 shows such a combination.

From a comparison of the E_θ field of the short dipole and the E_ϕ field of a small loop,

$$E_\theta = \frac{j\omega\mu I_d L \sin \theta}{4\pi r} e^{-jkr} \tag{5.41a}$$

$$E_\phi = \frac{\omega\mu k a^2 I_L \sin \theta}{4r} e^{-jkr} \tag{5.41b}$$

it is obvious that if the wave is to be, for example, right circular, so that

$$P = -\frac{E_\theta}{E_\phi} = -j$$

then we must require that

$$\frac{I_d}{I_L} = \frac{\pi k a^2}{L} \tag{5.42}$$

Reversing either current will give a left-circular wave.

The use of the loop and dipole antenna with a circularly polarized receiving antenna, of the correct sense, obviously gives a polarization match

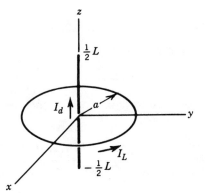

FIGURE 5.6 Loop and dipole.

238   POLARIZATION CHARACTERISTICS OF SOME ANTENNAS

factor of unity. The relative gain is also clearly

$$G_r = \sin^2 \theta \tag{5.43}$$

and the pattern is omnidirectional in azimuth with a half-power beamwidth of $\pi/2$ in a constant-azimuth plane.

Consider now a longer dipole with a field

$$E_\theta = \frac{jZ_0 I_m}{2\pi r} \frac{\cos[(kL/2)\cos\theta] - \cos(kL/2)}{\sin\theta} e^{-jkr} \tag{1.104b}$$

and the larger loop of Section 5.6 with field

$$E_\phi = \frac{\omega\mu a I_L J_1(ka\sin\theta)}{2r} e^{-jkr} \tag{5.40b}$$

This combination no longer is circularly polarized for all values of θ. By a choice of the relative feed currents of loop and dipole, the antenna can be made to radiate a circularly polarized wave in one direction, θ. Choose the wave to be right circular at $\theta = \pi/2$ (in the xy plane).

The polarization ratio is, from the equations for E_θ and E_ϕ,

$$P_t = C \frac{\cos[(kL/2)\cos\theta] - \cos(kL/2)}{\sin\theta J_1(ka\sin\theta)} \tag{5.44}$$

where C is a constant. If the wave is to be right circular at $\theta = \pi/2$,

$$C \frac{1 - \cos(kL/2)}{J_1(ka)} = -j$$

which gives

$$C = \frac{-jJ_1(ka)}{1 - \cos(kL/2)} \tag{5.45}$$

Then for general θ

$$P_t = \frac{-jJ_1(ka)}{1 - \cos(kL/2)} \frac{\cos[(kL/2)\cos\theta] - \cos(kL/2)}{\sin\theta J_1(ka\sin\theta)} \tag{5.46}$$

If this transmitting antenna is matched with a right-circular antenna having $P_r = -j$, the match factor can be written as

$$\rho = \frac{1}{2} \frac{(\sin\theta[1 - \cos(kL/2)]J_1(ka\sin\theta) + J_1(ka)\{\cos[(kL/2)\cos\theta] - \cos(kL/2)\})^2}{\sin^2\theta[1 - \cos(kL/2)]^2 J_1^2(ka\sin\theta) + J_1^2(ka)\{\cos[(kL/2)\cos\theta] - \cos(kL/2)\}^2} \tag{5.47}$$

This is a rather complicated equation, but matters can be simplified if the product of ρ and the relative gain G_r is considered. Using (1.104b) and (5.40b), we note that

$$|E_\theta|^2 + |E_\phi|^2 = \left(\frac{Z_0 I_m}{2\pi r}\right)^2 \frac{\{\cos[(kL/2)\cos\theta] - \cos(kL/2)\}^2}{\sin^2\theta}$$

$$+ \left(\frac{\omega\mu a I_L}{2r}\right)^2 J_1^2(ka\sin\theta) \tag{5.48}$$

At $\theta = \pi/2$, the two terms given above are equal because of the choice of circular polarization at $\theta = \pi/2$. In addition, we use the intensity there to normalize the intensity at any angle. It follows from these two facts that

$$G_r = \frac{\{\cos[(kL/2)\cos\theta] - \cos(kL/2)\}^2}{2\sin^2\theta[1 - \cos(kL/2)]^2} + \frac{J_1^2(ka\sin\theta)}{2J_1^2(ka)} \tag{5.49}$$

The product of G_r and ρ is

$$G_r\rho = \frac{1}{4}\frac{(\sin\theta[1 - \cos(kL/2)]J_1(ka\sin\theta) + J_1(ka)\{\cos[(kL/2)\cos\theta] - \cos(kL/2)\})^2}{\sin^2\theta[1 - \cos(kL/2)]^2 J_1^2(ka)} \tag{5.50}$$

It may be verified that this is unity in the plane $\theta = \pi/2$. For a half-wave dipole the product is simpler, namely,

$$G_r\rho\big|_{L=\lambda/2} = \frac{1}{4}\left[\frac{J_1(ka\sin\theta)}{J_1(ka)} + \frac{\cos[(\pi/2)\cos\theta]}{\sin\theta}\right]^2 \tag{5.51}$$

5.8. WAVEGUIDE OPENING INTO INFINITE GROUND PLANE

In Section 1.13 we developed the equations for the far fields of a rectangular waveguide carrying the TE_{10} mode and opening into an infinite ground plane. With the ground plane taken as the xy plane and with the long dimension of the waveguide directed along x, as shown in Figure 5.7, the far

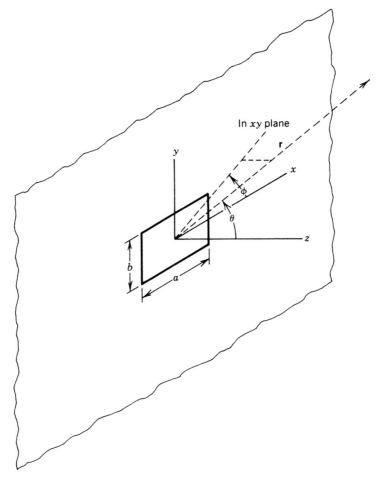

FIGURE 5.7 Waveguide opening into a plane.

fields are

$$E_\theta = \frac{\omega ab E_0}{cr} \sin\phi \frac{\cos[(\pi a/\lambda)\sin\theta\cos\phi]}{\pi^2 - 4[(\pi a/\lambda)\sin\theta\cos\phi]^2} \frac{\sin[(\pi b/\lambda)\sin\theta\sin\phi]}{(\pi b/\lambda)\sin\theta\sin\phi} \quad (1.135a)$$

$$E_\phi = \frac{\omega ab E_0}{cr} \cos\theta\cos\phi \frac{\cos[(\pi a/\lambda)\sin\theta\cos\phi]}{\pi^2 - 4[(\pi a/\lambda)\sin\theta\cos\phi]^2}$$

$$\times \frac{\sin[(\pi b/\lambda)\sin\theta\sin\phi]}{(\pi b/\lambda)\sin\theta\sin\phi} \quad (1.135b)$$

where a and b are the waveguide dimensions in the x and y directions, respectively.

WAVEGUIDE OPENING INTO INFINITE GROUND PLANE

In spite of the complexity of the field components, the polarization ratio for this antenna is quite simple. It is

$$P_t = -\frac{\tan \phi}{\cos \theta} \qquad (5.52)$$

On the z axis the wave is polarized in the y direction. (The word *vertical* is ambiguous and is avoided.)

We earlier considered an antenna that also produces a y-polarized wave on the z axis, namely, the y-directed dipole of Section 5.3, with a polarization ratio

$$P_y = -\cos \theta \tan \phi \qquad (5.23)$$

Since the polarization is the same in at least one direction, an important direction at that, it is interesting to compare polarizations in other directions.

It should be noted first that both antennas radiate a wave that is everywhere linearly polarized and that may be received everywhere without polarization loss by a correctly oriented, linearly polarized antenna. Since the polarization ratios are different, we might expect, correctly, that the receiving antenna orientation will be different for the dipole and the waveguide. Let us consider first the principal E and H planes and the xy plane. The polarization ratios and the field components are compared here:

	Polarization		Fields	
Plane	Dipole	Waveguide	Dipole	Waveguide
Principal E plane $\phi = \pi/2$	∞	∞	$E_\theta(E_y, E_z)$	$E_\theta(E_y, E_z)$
Principal H plane $\phi = 0$	0	0	$E_\phi(E_y)$	$E_\phi(E_y)$
xy plane $\theta = \pi/2$	0	∞	$E_\phi(E_x, E_y)$	$E_\theta(E_z)$

This table shows that the polarization behavior of the two antennas is the same in the principal E and H planes but differs markedly in the xy plane (which is generally of little interest for the waveguide opening).

Now consider the reception of the wave transmitted by the waveguide antenna. It is worthwhile to repeat that the radiated wave is everywhere linear and can be received without polarization loss by any linearly polarized antenna that is correctly oriented. We saw in Section 5.2 and Figure 5.2 that if the transmitter is a dipole, the receiver can be a dipole that lies in the same plane as the transmitter and is perpendicular to a line drawn from it. That natural orientation will not do for the waveguide antenna, however. In theory, the field components of the waveguide antenna can be calculated and a receiving antenna (dipole) oriented parallel to the field. This means nothing, however, since, in theory, $\rho = 1$. In making gain measurements of the waveguide antenna, a receiving dipole can be oriented perpendicular to a line from the waveguide antenna and rotated around that axis to maximize received power. This eliminates any polarization mismatch and allows a correct measurement of the gain.

If we wish specifically to consider the polarization behavior of the waveguide opening into a plane, it is appropriate to use as a receiving antenna a dipole oriented as it would be if the transmitting antenna itself were a y-directed dipole. The polarization ratio of the receiver is then

$$P_r = \cos\theta \tan\phi \tag{5.12}$$

and the polarization match factor between this dipole and the waveguide antenna is

$$\rho = \frac{\cos^2\theta(1+\tan^2\phi)^2}{(\cos^2\theta + \tan^2\phi)(1+\cos^2\theta\tan^2\phi)} \tag{5.53}$$

It is quickly noted that $\rho = 1$ in the principal E and H planes. It may also be determined without difficulty that the maximum rate of change of ρ with angle θ, near $\theta = 0$, occurs for $\phi = \pi/4, 3\pi/4,\ldots$ (which is not surprising since ρ is independent of θ for $\phi = 0, \pi/2, \ldots$).

Along a line giving maximum rate of change of ρ with θ ($\tan\phi = 1$), the value of ρ drops 3 dB where $\theta_{3\,\mathrm{dB}} = 65.53°$. Then the minimum 3-dB beamwidth is

$$2\theta_{3\,\mathrm{dB}} = 131.1°$$

for polarization effects alone.

The radiation intensity beamwidth can be found from the fields of (1.135). The 3-dB beamwidths in the principal E and H planes are $50.6\lambda/b$ and $68.8\lambda/a$ respectively [2]. These beamwidths are on the order of the polarization beamwidth for standard rectangular waveguides used in their designed frequency ranges. Polarization effects are therefore important within the radiation intensity beamwidth, and they decrease the overall beamwidth of this antenna significantly.

5.9. HORNS

Horns are among the most widely used microwave antennas. In this section we consider primarily the polarization properties of a pyramidal horn fed by

HORNS   243

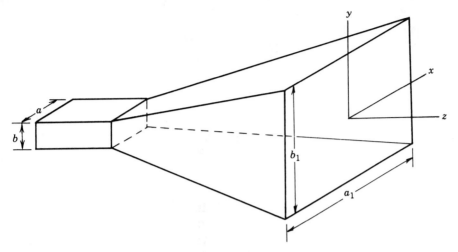

FIGURE 5.8 Pyramidal horn antenna.

a rectangular waveguide carrying the TE_{10} mode with a y-directed field. The geometry is shown in Figure 5.8. The radiation fields are relatively complex, and their development is widely available in the literature and thus is not repeated here.

The far fields of the pyramidal horn of Figure 5.8 are [2]

$$E_\theta = \frac{jkE_0 e^{-jkr}}{4\pi r} \sin\phi(1+\cos\theta) I_1 I_2 \quad (5.54a)$$

$$E_\phi = \frac{jkE_0 e^{-jkr}}{4\pi r} \cos\phi(1+\cos\theta) I_1 I_2 \quad (5.54b)$$

where I_1 and I_2 are given by the rather complicated expressions

$$I_1 = \frac{1}{2}\sqrt{\frac{\pi\rho_2}{k}} \left(e^{jk_x'^2 \rho_2/2k} \{[C(t_2') - C(t_1')] - j[S(t_2') - S(t_1')]\} \right.$$

$$\left. + e^{jk_x''^2 \rho_2/2k} \{[C(t_2'') - C(t_1'')] - j[S(t_2'') - S(t_1'')]\} \right) \quad (5.55a)$$

$$I_2 = \sqrt{\frac{\pi\rho_1}{k}} e^{jk_y^2 \rho_1/2k} \{[C(t_2) - C(t_1)] - j[S(t_2) - S(t_1)]\} \quad (5.55b)$$

where

$$k = \frac{2\pi}{\lambda} \tag{5.56a}$$

$$k'_x = k \sin\theta \cos\phi + \frac{\pi}{a_1} \tag{5.56b}$$

$$k''_x = k \sin\theta \cos\phi - \frac{\pi}{a_1} \tag{5.56c}$$

$$k_y = k \sin\theta \sin\phi \tag{5.56d}$$

$$t'_1 = \sqrt{\frac{1}{\pi k \rho_2}} \left(-\frac{k a_1}{2} - k'_x \rho_2 \right) \tag{5.56e}$$

$$t'_2 = \sqrt{\frac{1}{\pi k \rho_2}} \left(\frac{k a_1}{2} - k'_x \rho_2 \right) \tag{5.56f}$$

$$t''_1 = \sqrt{\frac{1}{\pi k \rho_2}} \left(-\frac{k a_1}{2} - k''_x \rho_2 \right) \tag{5.56g}$$

$$t''_2 = \sqrt{\frac{1}{\pi k \rho_2}} \left(\frac{k a_1}{2} - k''_x \rho_2 \right) \tag{5.56h}$$

$$t_1 = \sqrt{\frac{1}{\pi k \rho_1}} \left(-\frac{k b_1}{2} - k_y \rho_1 \right) \tag{5.56i}$$

$$t_2 = \sqrt{\frac{1}{\pi k \rho_1}} \left(\frac{k b_1}{2} - k_y \rho_1 \right) \tag{5.56j}$$

$$C(x) = \int_0^x \cos\left(\tfrac{1}{2}\pi t^2\right) dt \tag{5.56k}$$

$$S(x) = \int_0^x \sin\left(\tfrac{1}{2}\pi t^2\right) dt \tag{5.56l}$$

If one looks at the horn in the x direction of Figure 5.8, the upper and lower horn surfaces, if extended, meet inside the waveguide. The distance from this line to the aperture plane $z = 0$ is ρ_1. Similarly, the two side surfaces of the horn, if extended, meet in a line, and the distance to the aperture plane is ρ_2.

It is evident from the field equations that numerical computation of the radiation pattern of the pyramidal horn is necessary for the best understanding of its radiation characteristics. The reader is referred to Balanis for three-dimensional patterns [2].

In contrast to the radiation intensity, the polarization ratio of the pyramidal horn is quite simple. From (5.54) it is

$$P_t = -\frac{E_\theta}{E_\phi} = -\tan\phi \qquad (5.57)$$

It is noteworthy that the E-plane sectoral horn (flared in the E plane, the y direction, but not in the x direction) and the H-plane sectoral horn (flared in the H plane, the x direction, but not in the y direction) have fields that are quite different from those of the pyramidal horn [2] and yet have the same polarization ratio. Note that the fields are everywhere linearly polarized, but not in the same direction.

Since these horn antennas are intended to produce a y-polarized linear field in the main beam, it is appropriate to use as a test receiving antenna the dipole that was used for the waveguide opening into a plane. The polarization ratio of the test dipole is

$$P_r = \cos\theta \tan\phi \qquad (5.12)$$

Then the polarization match factor between this dipole and any one of the horn antennas is

$$\rho = \frac{(1 + \cos\theta \tan^2\phi)^2}{(1 + \tan^2\phi)(1 + \cos^2\theta \tan^2\phi)} \qquad (5.58)$$

Look first at the principal E and H planes. In the principal E plane, $\phi = \pi/2$, and in the principal H plane, $\phi = 0$, it is immediately seen from (5.58) that $\rho = 1$. In those planes the field radiated from a horn is indistinguishable from that of a dipole in its polarization characteristics.

By differentiating (5.58) with respect to $\tan^2\phi$, it can be determined that the greatest rate of change of ρ with θ near the direction $\theta = 0$ occurs, as it did for the waveguide opening into a plane, where $\tan^2\phi = 1$ or $\phi = \pi/4, 3\pi/4, \ldots$. If this value is substituted into the polarization match factor equation, it becomes

$$\rho = \frac{(1 + \cos\theta)^2}{2(1 + \cos^2\phi)} \qquad \phi = \frac{\pi}{4} \qquad (5.59)$$

It is quickly ascertained from this equation that the polarization beamwidth in a plane tilted at 45° with respect to the principal E and H planes is

$$2\theta_{3\,\text{dB}} = 120°$$

Furthermore, this is the minimum polarization beamwidth.

246 POLARIZATION CHARACTERISTICS OF SOME ANTENNAS

Typically, a pyramidal horn antenna will have E- and H-plane beamwidths determined by radiation intensity that are much smaller than the minimum polarization beamwidth. In many situations we may therefore neglect polarization in the main beam. It is a different matter for other horns however; the E-plane sectoral horn generally has a large H-plane radiation intensity beamwidth, and the H-plane sectoral horn has a large E-plane radiation intensity beamwidth. It is necessary therefore to consider polarization effects in the main beam of the E- and H-plane sectoral horns.

5.10. PARABOLOIDAL REFLECTOR

The surface formed by rotating a parabola about its axis is perhaps the most frequently used reflector antenna. Geometric optics, valid for vanishingly small wavelengths, shows that rays from the focal point are reflected in a beam parallel to the axis. Analyses using finite wavelengths show that the beam diverges, but the beamwidth is small for a paraboloid whose dimensions are large compared to a wavelength.

In this section we carry out an analysis to find the fields produced by a source at the focal point, using the aperture distribution technique. The field reflected by the paraboloid is found over a plane passing through the focal point and perpendicular to the paraboloid axis using geometric optics techniques. Equivalent sources are formed over that plane, and the far fields are obtained by integration over these sources using the procedures of Section 1.13. This method and others for finding the fields of a paraboloid are discussed in standard texts [2–5].

The geometry of the reflection problem is shown in Figure 5.9, and a cross-sectional view of the reflector in a plane ϕ' = constant appears as Figure 5.10. From the equation for the paraboloid surfaces the unit normal **n** may be found [2],

$$\mathbf{n} = -\mathbf{u}_{r'} \cos(\tfrac{1}{2}\theta') + \mathbf{u}_{\theta'} \sin(\tfrac{1}{2}\theta') \tag{5.60}$$

and from **n** the angles α and β are easily shown to be equal to each other, and

$$\alpha = \beta = \tfrac{1}{2}\theta' \tag{5.61}$$

Since α is the angle from the surface normal measured to an incident ray from the focal point, and β is the angle from **n** to a ray parallel to the paraboloid axis, it is clear that all rays from the focal point are reflected parallel to each other.

A source at the origin produces a wave that is incident on the reflector surface, inducing a surface current with density

$$\mathbf{J}_s = \mathbf{n} \times \mathbf{H} = \mathbf{n} \times (\mathbf{H}_i + \mathbf{H}_r) \tag{5.62}$$

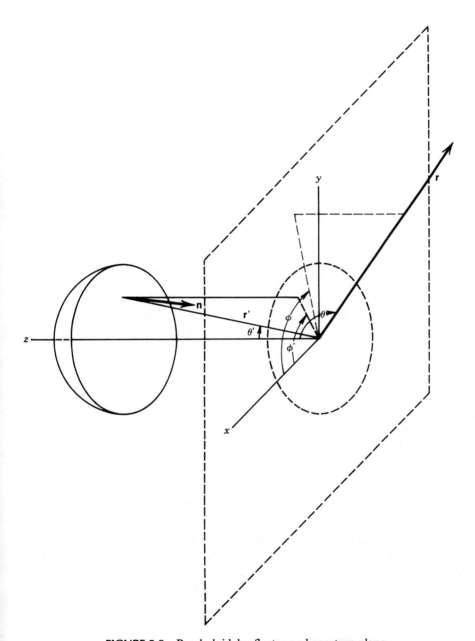

FIGURE 5.9 Paraboloidal reflector and aperture plane.

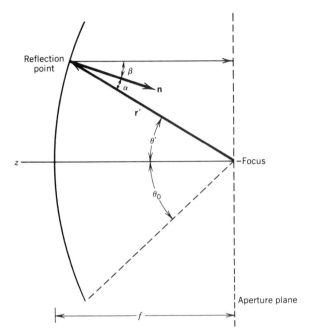

FIGURE 5.10 Cross section of reflector.

where $\mathbf{H}_i$ and $\mathbf{H}_r$ are incident and reflected fields. If we can approximate the surface in the vicinity of the reflection point by an infinite plane conductor, the tangential components of incident and reflected magnetic fields are equal, and

$$\mathbf{n} \times \mathbf{H}_i = \mathbf{n} \times \mathbf{H}_r \qquad (5.63)$$

which allows $\mathbf{J}_s$ to be written as

$$\mathbf{J}_s = 2\mathbf{n} \times \mathbf{H}_i = 2\mathbf{n} \times \mathbf{H}_r \qquad (5.64)$$

Next require all points on the reflecting surface to be in the far field of the source so that the electric field of the incident wave is transverse to the magnetic field and the direction of propagation and is related to $\mathbf{H}_i$ by the intrinsic impedance of free space, Z_0. Then

$$\mathbf{J}_s = \frac{2}{Z_0}[\mathbf{n} \times (\mathbf{u}_{r'} \times \mathbf{E}_i)] \qquad (5.65)$$

where $\mathbf{u}_{r'}$ is a unit vector directed from the focal point to the point of reflection. In the same way the electric field of the reflected wave is related to the reflected magnetic field, and at the reflector surface

$$\mathbf{J}_s = \frac{2}{Z_0}[\mathbf{n} \times (-\mathbf{u}_z \times \mathbf{E}_r)] \qquad (5.66)$$

with $-\mathbf{u}_z$ a unit vector from the reflection point and antiparallel to the z axis.

Let a source at the focal point with gain G accept power W_t from a generator. The power density is then

$$\mathscr{P}(r', \theta', \phi') = \frac{W_t G(\theta', \phi')}{4\pi r'^2} \qquad (5.67)$$

From the relationship

$$\mathscr{P} = \frac{1}{2Z_0}|\mathbf{E}|^2 \qquad (5.68)$$

the field of the incident wave at the reflector surface is

$$\mathbf{E}_i(r', \theta', \phi') = \mathbf{e}_i\left[Z_0 \frac{W_t}{2\pi} G(\theta', \phi')\right]^{1/2} \frac{e^{-jkr'}}{r'} \qquad (5.69)$$

where $\mathbf{e}_i$ is a unit vector perpendicular to $\mathbf{u}_{r'}$. We extract the constants from this equation and define

$$C = \sqrt{Z_0 W_t / 2\pi} \qquad (5.70)$$

We may then write

$$\mathbf{E}_i(r', \theta', \phi') = \mathbf{e}_i C \sqrt{G(\theta', \phi')}\, \frac{e^{-jkr'}}{r'} \qquad (5.71)$$

If this expression for $\mathbf{E}_i$ is substituted into (5.65), the surface current density in terms of the source parameters can be written as

$$\mathbf{J}_s = \frac{2}{Z_0} C\sqrt{G(\theta', \phi')}\, \frac{e^{-jkr'}}{r'}\, \mathbf{a} \qquad (5.72)$$

where

$$\mathbf{a} = \mathbf{n} \times (\mathbf{u}_{r'} \times \mathbf{e}_i) \qquad (5.73)$$

We may also write the second expression for $\mathbf{J}_s$, (5.66), as

$$\mathbf{J}_s = \frac{2}{Z_0} E_r [\mathbf{n} \times (-\mathbf{u}_z \times \mathbf{e}_r)] \tag{5.74}$$

where E_r is the reflected field value at the surface and $\mathbf{e}_r$ is a unit vector expressing its polarization (considered here a real vector). Since both

$$\mathbf{n} \times (\mathbf{u}_{r'} \times \mathbf{e}_i)$$

and

$$\mathbf{n} \times (-\mathbf{u}_z \times \mathbf{e}_r)$$

are unit vectors, and since both give the direction of $\mathbf{J}_s$, it is apparent that they are equal and that at the surface

$$\mathbf{E}_r = C\sqrt{G(\theta', \phi')} \, \frac{e^{-jkr'}}{r'} \mathbf{e}_r \tag{5.75}$$

Source Polarized in the y Direction

It is difficult to proceed further without assuming a polarization for the source. If it is taken in the y direction, $\mathbf{e}_i$ may be written as

$$\mathbf{e}_i = \frac{\mathbf{u}_{r'} \times (\mathbf{u}_y \times \mathbf{u}_{r'})}{|\mathbf{u}_{r'} \times (\mathbf{u}_y \times \mathbf{u}_{r'})|} \tag{5.76}$$

and if $\mathbf{u}_y$, the unit vector in the y direction, is expanded as

$$\mathbf{u}_y = \sin\theta' \sin\phi' \, \mathbf{u}_{r'} + \cos\theta' \sin\phi' \, \mathbf{u}_{\theta'} + \cos\phi' \, \mathbf{u}_{\phi'} \tag{5.77}$$

it is straightforward to show that

$$\mathbf{e}_i = \frac{\cos\theta' \sin\phi' \, \mathbf{u}_{\theta'} + \cos\phi' \, \mathbf{u}_{\phi'}}{\sqrt{1 - \sin^2\theta' \sin^2\phi'}} \tag{5.78}$$

It is tedious, but not difficult, to obtain $\mathbf{a}$ in rectangular coordinates by substituting $\mathbf{e}_i$ into (5.73) and transforming all vectors to rectangular coordinates. The result is

$$\mathbf{a} = \frac{-\sin\theta' \sin(\theta'/2) \sin\phi' \cos\phi'}{\sqrt{1 - \sin^2\theta' \sin^2\phi'}} \mathbf{u}_x + \frac{\cos(\theta'/2)(\cos\theta' \sin^2\phi' + \cos^2\phi')}{\sqrt{1 - \sin^2\theta' \sin^2\phi'}} \mathbf{u}_y$$

$$- \frac{\sin(\theta'/2) \cos\theta' \sin\phi'}{\sqrt{1 - \sin^2\theta' \sin^2\phi'}} \mathbf{u}_z \tag{5.79}$$

We may next find $\mathbf{e}_r$ by writing $\mathbf{a}$ as

$$\mathbf{a} = \mathbf{n} \times (-\mathbf{u}_z \times \mathbf{e}_r) = -(\mathbf{n} \cdot \mathbf{e}_r)\mathbf{u}_z + (\mathbf{n} \cdot \mathbf{u}_z)\mathbf{e}_r$$
$$= -(\mathbf{n} \cdot \mathbf{e}_r)\mathbf{u}_z - \cos(\tfrac{1}{2}\theta')\mathbf{e}_r \qquad (5.80)$$

If these expressions for $\mathbf{a}$ are equated, the components of $\mathbf{e}_r$ transverse to z can be found. (We need not be concerned with ascertaining if $\mathbf{e}_r$ has a z component since it will not contribute to equivalent surface currents over the aperture plane.) The result is

$$\mathbf{e}_r = \frac{\sin \phi' \cos \phi'(1 - \cos \theta')\mathbf{u}_x - (\cos \theta' \sin^2 \phi' + \cos^2 \phi')\mathbf{u}_y}{\sqrt{1 - \sin^2 \theta' \sin^2 \phi'}} \qquad (5.81)$$

Now make the assumption that the electric field intensity at a point on the aperture plane is given by the field intensity transverse to z at a corresponding point (same x and y coordinates) on the reflector, except for the phase retardation $kr' \cos \theta'$ caused by the path from reflector to aperture [3]. The aperture field is then given by

$$\mathbf{E}_{ap} = C\sqrt{G_y(\theta', \phi')} \, \frac{e^{-jkr'(1+\cos \theta')}}{r'} \mathbf{e}_r \qquad (5.82)$$

where G_y is used as a reminder that the source is y directed.

If $\mathbf{E}_{ap}$ is written as

$$\mathbf{E}_{ap} = \mathbf{u}_x E_{ax} + \mathbf{u}_y E_{ay} \qquad (5.83)$$

where

$$E_{ax} = C\sqrt{G_y(\theta', \phi')} \, \frac{e^{-jkr'(1+\cos \theta')}}{r'} \, \frac{\sin \phi' \cos \phi'(1 - \cos \theta')}{\sqrt{1 - \sin^2 \theta' \sin^2 \phi'}} \qquad (5.84a)$$

$$E_{ay} = -C\sqrt{G_y(\theta', \phi')} \, \frac{e^{-jkr'(1+\cos \theta')}}{r'} \, \frac{\cos \theta' \sin^2 \phi' + \cos^2 \phi'}{\sqrt{1 - \sin^2 \theta' \sin^2 \phi'}} \qquad (5.84b)$$

we can find equivalent surface currents on the aperture plane by

$$\mathbf{M}_{sa} = -\mathbf{u}_z \times \mathbf{E}_{ap} = \mathbf{u}_x E_{ay} - \mathbf{u}_y E_{ax} \qquad (5.85a)$$

$$\mathbf{J}_{sa} = \mathbf{u}_z \times \mathbf{H}_{ap} = -\mathbf{u}_x H_{ay} + \mathbf{u}_y H_{ax} = -\mathbf{u}_x \frac{E_{ax}}{Z_0} - \mathbf{u}_y \frac{E_{ay}}{Z_0} \qquad (5.85b)$$

We fill the region on the reflector side of the aperture plane with a perfect electric conductor, as discussed in Section 1.11, and apply image theory, as in

Section 1.13. The result is that we double $\mathbf{M}_{sa}$ over the aperture plane and let $\mathbf{J}_{sa}$ be zero. Further, assume that $\mathbf{M}_{sa}$ has value only over the circle on the aperture plane, which is the projection of the paraboloidal reflector on the plane. Since we are concerned only with the far fields, we can use the approximation

$$\mathbf{F}(\mathbf{r}) = \frac{\epsilon e^{-jkr}}{4\pi r} \int\int 2\mathbf{M}_{sa}(\mathbf{r}') e^{jk\mathbf{u}_r \cdot \mathbf{r}'} \, dA' \tag{5.86}$$

where

$$\mathbf{u}_r \cdot \mathbf{r}' = x' \sin\theta \cos\phi + y' \sin\theta \sin\phi \tag{5.87}$$

Substitution of the magnetic surface current value of (5.85) into the equation for $\mathbf{F}(\mathbf{r})$ gives

$$F_x = \frac{\epsilon e^{-jkr}}{2\pi r} \int\int E_{ay} e^{jk(x' \sin\theta \cos\phi + y' \sin\theta \sin\phi)} \, dx' \, dy' \tag{5.88a}$$

$$F_y = -\frac{\epsilon e^{-jkr}}{2\pi r} \int\int E_{ax} e^{jk(x' \sin\theta \cos\phi + y' \sin\theta \sin\phi)} \, dx' \, dy' \tag{5.88b}$$

where the integration is carried out over the projection of the reflector onto the aperture plane.

The electric field components may be found from

$$E_\theta = -j\omega Z_0 F_\phi = -j\omega Z_0 (-F_x \sin\phi + F_y \cos\phi) \tag{5.89a}$$

$$E_\phi = j\omega Z_0 F_\theta = j\omega Z_0 (F_x \cos\theta \cos\phi + F_y \cos\theta \sin\phi) \tag{5.89b}$$

Substitution of the components of $\mathbf{F}$ leads to

$$E_\theta = \frac{j\omega Z_0 \epsilon e^{-jkr}}{2\pi r} \int\int (E_{ax} \cos\phi + E_{ay} \sin\phi)$$
$$\times e^{jk(x' \sin\theta \cos\phi + y' \sin\theta \sin\phi)} \, dx' \, dy' \tag{5.90a}$$

$$E_\phi = \frac{j\omega Z_0 \epsilon e^{-jkr}}{2\pi r} \cos\theta \int\int (-E_{ax} \sin\phi + E_{ay} \cos\phi)$$
$$\times e^{jk(x' \sin\theta \cos\phi + y' \sin\theta \sin\phi)} \, dx' \, dy' \tag{5.90b}$$

Using the geometry of the reflector, a change of variables in the integrals achieves forms that are easier to evaluate. Express r' in Figure 5.10 as [2]

$$r' = \frac{2f}{1 + \cos\theta'} \tag{5.91}$$

and also

$$x' = r' \sin\theta' \cos\phi' \tag{5.92a}$$
$$y' = r' \sin\theta' \sin\phi' \tag{5.92b}$$
$$z' = r' \cos\theta' \tag{5.92c}$$

With these changes, the aperture fields become

$$E_{ax} = C\sqrt{G_y(\theta',\phi')}\,\frac{(1+\cos\theta')e^{-j2kf}}{2f}\,\frac{\sin\phi'\cos\phi'(1-\cos\theta')}{\sqrt{1-\sin^2\theta'\sin^2\phi'}} \tag{5.93a}$$

$$E_{ay} = -C\sqrt{G_y(\theta',\phi')}\,\frac{(1+\cos\theta')e^{-j2kf}}{2f}\,\frac{\cos\theta'\sin^2\phi' + \cos^2\phi'}{\sqrt{1-\sin^2\theta'\sin^2\phi'}} \tag{5.93b}$$

and the radiated fields are

$$E_\theta = \frac{j\omega Z_0 \epsilon e^{-jkr}}{2\pi r}\iint (E_{ax}\cos\phi + E_{ay}\sin\phi)$$

$$\times \exp\left(j2kf\,\frac{\sin\theta'\sin\theta\cos(\phi'-\phi)}{1+\cos\theta'}\right)\frac{(2f)^2 \sin\theta'}{(1+\cos\theta')^2}\,d\theta'\,d\phi' \tag{5.94a}$$

$$E_\phi = \frac{j\omega Z_0 \epsilon e^{-jkr}}{2\pi r}\cos\theta \iint (-E_{ax}\sin\phi + E_{ay}\cos\phi)$$

$$\times \exp\left(j2kf\,\frac{\sin\theta'\sin\theta\cos(\phi'-\phi)}{1+\cos\theta'}\right)\frac{(2f)^2 \sin\theta'}{(1+\cos\theta')^2}\,d\theta'\,d\phi' \tag{5.94b}$$

Principal Plane Fields: y-Polarized Source

The expressions for the fields in general require numerical integration, but they can be simplified if we consider the principal E and H planes only, given respectively, for the y-polarized source, by $\phi = \pi/2$ and $\phi = 0$.

In the principal E plane E_ϕ reduces to

$$E_\phi = \frac{j\omega Z_0 \epsilon e^{-jkr}}{2\pi r}\cos\theta \iint E_{ax}e^{jky'\sin\theta}\,dx'\,dy' \tag{5.95}$$

where E_{ax} may be written from (5.84) and (5.92) as

$$E_{ax} = C\sqrt{G_y(\theta',\phi')}\,\frac{e^{-jkr'(1+\cos\theta')}}{r'}\,\frac{x'y'(1-\cos\theta')}{r'^2 \sin^2\theta'\sqrt{1-\sin^2\theta'\sin^2\phi'}} \tag{5.96}$$

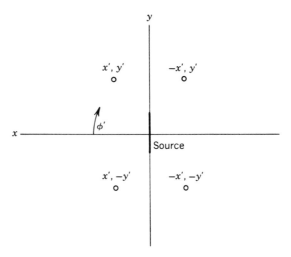

FIGURE 5.11 Aperture plane for a y-directed source.

We have previously restricted our development to a y-directed primary source. Now make the assumption that the gain $G_y(\theta', \phi')$ is symmetric so that the x-directed fields over the aperture plane have the same magnitudes at symmetrically located points, as shown in Figure 5.11. Then

$$G_y(x', y') = G_y(-x', y') = G_y(x', -y') = G_y(-x', -y') \quad (5.97)$$

It follows that

$$E_{ax}(x', y') = -E_{ax}(-x', y') = -E_{ax}(x', -y') = E_{ax}(-x', -y') \quad (5.98)$$

and the contributions to the integral for E_ϕ are

$$E_{ax}(x', y')e^{jk|y'|\sin\theta} \qquad -E_{ax}(x', y')e^{-jk|y'|\sin\theta}$$
$$-E_{ax}(x', y')e^{jk|y'|\sin\theta} \qquad E_{ax}(x', y')e^{-jk|y'|\sin\theta}$$

These contributions cancel in pairs. It follows that in the principal E plane $E_\phi = 0$, and the radiated field has only an E_θ component.

In the principal H plane, $\phi = 0$, E_θ becomes

$$E_\theta = \frac{j\omega Z_0 \epsilon e^{-jkr}}{2\pi r} \int\int E_{ax} e^{jkx'\sin\theta} \, dx' \, dy' \quad (5.99)$$

where E_{ax} is given by (5.96).

If the analysis carried out for the principal E plane is repeated, we find that $E_\theta = 0$ in the principal H plane.

If the fields in the principal planes are converted to rectangular coordinates, $E_x = 0$ in both planes, and in the H plane the field has only an E_y component. We recognize that both statements would be correct for a y-directed dipole, for example, without the parabolic reflector, and it is interesting to note that the conditions carry over to the reflector antenna if the required symmetry conditions are met.

Source Polarized in the x Direction

If the source is x polarized instead of y, the vector (5.76) for the field incident on the reflector should be altered to

$$\mathbf{e}'_i = \frac{\mathbf{u}_{r'} \times (\mathbf{u}_x \times \mathbf{u}_{r'})}{|\mathbf{u}_{r'} \times (\mathbf{u}_x \times \mathbf{u}_{r'})|} \tag{5.100}$$

which becomes

$$\mathbf{e}'_i = \frac{\cos\theta' \cos\phi' \, \mathbf{u}_{\theta'} - \sin\phi' \, \mathbf{u}_{\phi'}}{\sqrt{1 - \sin^2\theta' \cos^2\phi'}} \tag{5.101}$$

The vector $\mathbf{a}$, given by (5.79) for the y-directed source, becomes, for the x-polarized case,

$$\mathbf{a}' = \frac{\cos(\theta'/2)(\sin^2\phi' + \cos\theta' \cos^2\phi')}{\sqrt{1 - \sin^2\theta' \cos^2\phi'}} \mathbf{u}_x - \frac{\sin\theta' \sin(\theta'/2)\sin\phi' \cos\phi'}{\sqrt{1 - \sin^2\theta' \cos^2\phi'}} \mathbf{u}_y$$

$$- \frac{\cos\theta' \sin(\theta'/2)\cos\phi'}{\sqrt{1 - \sin^2\theta' \cos^2\phi'}} \mathbf{u}_z \tag{5.102}$$

The vector $\mathbf{e}'_r$, representing the polarization of the reflected field is found by the process used earlier and is

$$\mathbf{e}'_r = -\frac{\sin^2\phi' + \cos\theta' \cos^2\phi'}{\sqrt{1 - \sin^2\theta' \cos^2\phi'}} \mathbf{u}_x + \frac{\sin\phi' \cos\phi'(1 - \cos\theta')}{\sqrt{1 - \sin^2\theta' \cos^2\phi'}} \mathbf{u}_y \tag{5.103}$$

This equation can be verified, or in fact could have been derived, by noting that the x-polarized source is the y-polarized source rotated in azimuth by $-90°$. Then we should have

$$e'_{rx}(\phi' + \tfrac{1}{2}\pi) = e_{ry}(\phi') \tag{5.104a}$$
$$e'_{ry}(\phi' + \tfrac{1}{2}\pi) = -e_{rx}(\phi') \tag{5.104b}$$

These equalities are easily verified.

POLARIZATION CHARACTERISTICS OF SOME ANTENNAS

The aperture fields may now be found from

$$E'_{ax} = -C\sqrt{G_x(\theta', \phi')} \, \frac{e^{-jkr'(1+\cos\theta')}}{r'} \, \frac{\sin^2\phi' + \cos\theta'\cos^2\phi'}{\sqrt{1 - \sin^2\theta'\cos^2\phi'}} \quad (5.105a)$$

$$E'_{ay} = C\sqrt{G_x(\theta', \phi')} \, \frac{e^{-jkr'(1+\cos\theta')}}{r'} \, \frac{\sin\phi'\cos\phi'(1 - \cos\theta')}{\sqrt{1 - \sin^2\theta'\cos^2\phi'}} \quad (5.105b)$$

and the radiated fields found from (5.90) as before.

Principal Plane Fields: x-Polarized Source

Earlier we considered the fields in the principal planes for a y-polarized source with gain symmetry and found E_ϕ to be zero in the E plane, $\phi = \pi/2$, and E_θ to be zero in the H plane, $\phi = 0$. If this development is repeated for an x-polarized source with symmetric gain,

$$G_x(x', y') = G_x(-x', y') = G_x(x', -y') = G_x(-x', -y') \quad (5.106)$$

a result is that in the E plane, $\phi = 0$, of the x-polarized source, $E_\phi = 0$. In the H plane, $\phi = \pi/2$, the component $E_\theta = 0$. In both planes the rectangular component E_y is zero, an unsurprising result for an x-polarized source. Note that the E plane for an x-polarized source is the H plane for a y-polarized source, and vice versa.

Dipole Feed Antenna: y Directed

A commonly used feed antenna is a center-fed dipole. We found in Section 5.3 that the field components of a y-directed short dipole are

$$E_r = 0 \quad (5.107a)$$

$$E_\theta = -\frac{j\omega\mu IL}{4\pi r'} e^{-jkr'} \cos\theta' \sin\phi' \quad (5.107b)$$

$$E_\phi = \frac{j\omega\mu IL}{4\pi r'} e^{-jkr'} \cos\phi' \quad (5.107c)$$

where the factor 4 in the denominators is doubled if a triangular current distribution is assumed. With this feed the incident field magnitude at the reflector surface is

$$|E_i|^2 = \left(\frac{\omega\mu IL}{4\pi r'}\right)^2 (1 - \sin^2\theta' \sin^2\phi') \quad (5.108)$$

Comparison of this equation with that for the incident field, (5.71), shows that for the y-directed dipole, we should use

$$C\sqrt{G_y(\theta', \phi')} = \frac{j\omega\mu IL}{4\pi}\sqrt{1 - \sin^2\theta' \sin^2\phi'} \qquad (5.109)$$

Substitution into the aperture fields simplifies them to

$$E_{ax} = \frac{j\omega\mu IL}{4\pi r'} \sin\phi' \cos\phi' (1 - \cos\theta') e^{-jkr'(1+\cos\theta')} \qquad (5.110a)$$

$$E_{ay} = -\frac{j\omega\mu IL}{4\pi r'} (\cos\theta' \sin^2\phi' + \cos^2\phi') e^{-jkr'(1+\cos\theta')} \qquad (5.110b)$$

Principal Plane Polarization: y-Directed Dipole

We saw earlier that for a y-directed source with symmetric gain, $E_\phi = 0$ in the principal E plane and $E_\theta = 0$ in the principal H plane. Using this finding, the fields produced by a y-directed dipole feed become

E PLANE, $\phi = \frac{1}{2}\pi$

$$E_\theta = j\omega Z_0 \frac{\epsilon e^{-jkr}}{2\pi r} \int\int E_{ay} e^{jky' \sin\theta} dx' dy'$$

$$= j\omega Z_0 \frac{\epsilon e^{-jkr}}{2\pi r} \int\int \left(-\frac{j\omega\mu IL}{4\pi r'}\right)(\cos\theta' \sin^2\phi' + \cos^2\phi')$$

$$\times e^{-jkr'(1+\cos\theta')} e^{jky' \sin\theta} dx' dy' \qquad (5.111a)$$

$$E_\phi = 0 \qquad (5.111b)$$

The polarization ratio is

$$P = \infty \qquad (5.112)$$

H PLANE, $\phi = 0$

$$E_\theta = 0$$

$$E_\phi = j\omega Z_0 \frac{\epsilon e^{-jkr}}{2\pi r} \cos\theta \int\int E_{ay} e^{jkx' \sin\theta} dx' dy' \qquad (5.113a)$$

$$= j\omega Z_0 \frac{\epsilon e^{-jkr}}{2\pi r} \cos\theta \int\int \left(-\frac{j\omega\mu IL}{4\pi r'}\right)(\cos\theta' \sin^2\phi' + \cos^2\phi')$$

$$\times e^{-jkr'(1+\cos\theta')} e^{jkx' \sin\theta} dx' dy' \qquad (5.113b)$$

258   POLARIZATION CHARACTERISTICS OF SOME ANTENNAS

The polarization ratio is

$$P = 0 \tag{5.114}$$

Dipole Feed Antenna: x Directed

We found in Section 5.3 the field components of a short x-directed dipole to be

$$E_\theta = -\frac{j\omega\mu IL}{4\pi r'} e^{-jkr'} \cos\theta' \cos\phi' \tag{5.115a}$$

$$E_\phi = \frac{j\omega\mu IL}{4\pi r'} e^{-jkr'} \sin\phi' \tag{5.115b}$$

It follows from the same reasoning used previously that we should let in (5.105)

$$C\sqrt{G_x(\theta',\phi')} = \frac{j\omega\mu IL}{4\pi} \sqrt{1 - \sin^2\theta' \cos^2\phi'} \tag{5.116}$$

The aperture fields for the x-directed dipole reduce to

$$E'_{ax} = -\frac{j\omega\mu IL}{4\pi r'}(\sin^2\phi' + \cos\theta' \cos^2\phi')e^{-jkr'(1+\cos\theta')} \tag{5.117a}$$

$$E'_{ay} = \frac{j\omega\mu IL}{4\pi r'} \sin\phi' \cos\phi'(1 - \cos\theta')e^{-jkr'(1+\cos\theta')} \tag{5.117b}$$

Principal Plane Polarization: x-Directed Dipole

We saw earlier that for an x-polarized source, $E_\phi = 0$ in the principal E plane, $\phi = 0$, and $E_\theta = 0$ in the H plane, $\phi = \pi/2$. Using this information and the fields of the x-directed dipole in (5.90), we find in the principal planes

E PLANE, $\phi = 0$

$$E'_\theta = j\omega Z_0 \frac{\epsilon e^{-jkr}}{2\pi r} \int\int E'_{ax} e^{jkx' \sin\theta} \, dx' \, dy'$$

$$= j\omega Z_0 \frac{\epsilon e^{-jkr}}{2\pi r} \int\int \left(-\frac{j\omega\mu IL}{4\pi r'}\right)(\sin^2\phi' + \cos\theta' \cos^2\phi')$$

$$\times e^{-jkr'(1+\cos\theta')} e^{jkx' \sin\theta} \, dx' \, dy' \tag{5.118a}$$

$$E'_\phi = 0 \tag{5.118b}$$

$$P = \infty \tag{5.118c}$$

H PLANE, $\phi = \frac{1}{2}\pi$

$$E'_\theta = 0 \tag{5.119a}$$

$$E'_\phi = j\omega Z_0 \frac{\epsilon e^{-jkr}}{2\pi r} \cos\theta \iint -E'_{ax} e^{jky'\sin\theta}\, dx'\, dy'$$

$$= j\omega Z_0 \frac{\epsilon e^{-jkr}}{2\pi r} \cos\theta \iint \frac{j\omega\mu IL}{4\pi r'}(\sin^2\phi' + \cos\theta'\cos^2\phi')$$
$$\times e^{-jkr'(1+\cos\theta')} e^{jky'\sin\theta}\, dx'\, dy' \tag{5.119b}$$

$$P = 0 \tag{5.119c}$$

Crossed-Dipole Feed

A circularly polarized wave can be produced on the paraboloid axis if crossed dipoles with a $\pi/2$ phase difference are used as the feed antenna. The fields in a general direction must be computed numerically, but the polarization ratio in two planes is relatively simple. We therefore restrict our consideration to the planes $\phi = 0$ and $\phi = \pi/2$.

In the equations for the dipole feeds previously developed let the y-directed dipole lead in phase by $\pi/2$ and use currents I and jI, respectively, in the x- and y-directed dipoles. Then, in general, the fields produced by the crossed-dipole feed are

$$\hat{E}_\theta = jE_\theta + E'_\theta \tag{5.120a}$$

$$\hat{E}_\phi = jE_\phi + E'_\phi \tag{5.120b}$$

where E_θ, E_ϕ, E'_θ, and E'_ϕ are given by (5.111), (5.113), (5.118), and (5.119). In the plane $\phi = 0$ these equations give

$$\hat{E}_\theta = E'_\theta \tag{5.121a}$$

$$\hat{E}_\phi = jE_\phi \tag{5.121b}$$

From these equations the polarization ratio may be written as

$$P = -\frac{\hat{E}_\theta}{\hat{E}_\phi} = \frac{jE'_\theta}{E_\phi}$$

$$= \frac{j}{\cos\theta} \frac{\iint (1/r')(\sin^2\phi' + \cos\theta'\cos^2\phi') e^{-jkr'(1+\cos\theta')} e^{jkx'\sin\theta}\, dx'\, dy'}{\iint (1/r')(\cos\theta'\sin^2\phi' + \cos^2\phi') e^{-jkr'(1+\cos\theta')} e^{jkx'\sin\theta}\, dx'\, dy'}$$

$$\tag{5.122}$$

Now if (5.92) is used, portions of the integrands in the numerator and denominator of this expression can be written as

$$\sin^2 \phi' + \cos \theta' \cos^2 \phi' = \frac{1}{r'^2 \sin^2 \theta'}\left(y'^2 + \frac{z'}{r'}x'^2\right) \tag{5.123a}$$

$$\cos \theta' \sin^2 \phi' + \cos^2 \phi' = \frac{1}{r'^2 \sin^2 \theta'}\left(\frac{z'}{r'}y'^2 + x'^2\right) \tag{5.123b}$$

The variables x' and y' can be interchanged without having any effect on other terms in the integrands, and the limits on x' and y' are the same. Therefore the numerator and denominator integrals in (5.122) are equal, and in the plane $\phi = 0$ the paraboloidal reflector with crossed dipole feed has polarization ratio

$$P = \frac{j}{\cos \theta} \qquad \frac{\pi}{2} \le \theta \le \pi \tag{5.124}$$

In the plane $\phi = \pi/2$ the fields are

$$\hat{E}_\theta = jE_\theta \tag{5.125a}$$

$$\hat{E}_\phi = E'_\phi \tag{5.125b}$$

and the polarization ratio becomes

$$P = \frac{-jE_\theta}{E'_\phi}$$

$$= \frac{j}{\cos \theta} \frac{\iint (1/r')(\cos \theta' \sin^2 \phi' + \cos^2 \phi')e^{-jkr'(1+\cos \theta')}e^{jky' \sin \theta}\,dx'\,dy'}{\iint (1/r')(\sin^2 \phi' + \cos \theta' \cos^2 \phi')e^{-jkr'(1+\cos \theta')}e^{jky' \sin \theta}\,dx'\,dy'} \tag{5.126}$$

As before, it is easily shown that the integrands are equal, and the polarization ratio in the plane $\phi = \pi/2$ becomes

$$P = \frac{j}{\cos \theta} \qquad \frac{\pi}{2} \le \theta \le \pi \tag{5.127}$$

which is what it was for the $\phi = 0$ plane.

The polarization match factor in either plane is obtained by using a circularly polarized receiver with

$$P_r = -j \tag{5.128}$$

since in the region of interest $\cos \theta$ is negative. The polarization match factor then becomes

$$\rho = \frac{1}{2} - \frac{\cos \theta}{1 + \cos^2 \theta} \qquad (5.129)$$

If this value of ρ is compared to that of (5.32) for the crossed dipoles without the parabolic reflector, it may be seen that they are the same if the different reference for the measurement of polar angle θ is considered. Another difference is that for the reflector antenna, the equation (5.129) for ρ is valid only in the planes $\phi = 0$ and $\phi = \pi/2$, whereas the crossed dipoles alone (5.32) is valid everywhere.

It is obvious from (5.129) that the 3-dB polarization beamwidth for the crossed-dipole feed is π in the planes $\phi = 0$ and $\phi = \pi/2$. On the other hand, the 3-dB radiation intensity beamwidth of a parabolic reflector is approximately λ/D for a uniformly illuminated aperture [1]. For a tapered illumination, which would occur with a crossed-dipole feed, the beamwidth is somewhat greater. In addition, the polarization beamwidth obtained here is valid only in two planes. Nevertheless, it is clear that for large aperture diameter, polarization effects are small in the main beam of the parabolic reflector if received power is the quantity of interest.

5.11. NARROW-POLARIZATION-BEAMWIDTH ARRAY

In our examination of the polarization characteristics of various antennas we have not yet encountered one with a small polarization beamwidth, even though some of them have small radiation intensity beamwidths. In this section we examine an array, shown in Figure 5.12, that can produce narrow beams in both radiation intensity and polarization. The array elements are short dipoles, although other linearly polarized elements can be used. The array is intended to produce a circularly polarized wave in the main beam, so we assume that the phases of the elements along the y axis lead those of the x-axis elements by $\pi/2$ when the beam is broadside. An even number of elements is shown for each linear array, but an odd number can be used without changing the equations of this section.

For simplicity, a uniform array, with equal feed amplitudes for all elements and a constant difference between the phases of adjacent elements, is assumed on both x and y axes. Furthermore, the same number of elements, with the same spacing, is assumed for the x- and y-axis arrays.

The array factor for a linear array of isotropic elements along the axes is [2]

$$F(\theta, \phi) = \frac{1}{N} \frac{\sin[(N/2)\Psi]}{\sin[(1/2)\Psi]} \qquad (5.130)$$

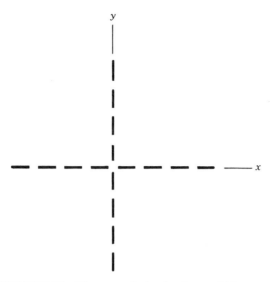

FIGURE 5.12 Narrow-polarization-beamwidth array.

where

$$\Psi = \Psi_x = kd \sin\theta \cos\phi + \delta_x \quad x \text{ axis} \quad (5.131a)$$

$$\Psi = \Psi_y = kd \sin\theta \sin\phi + \delta_y \quad y \text{ axis} \quad (5.131b)$$

with δ_x and δ_y the feed phase differences.

If the pattern multiplication principle of array theory and the fields of x- and y-directed short dipoles of (5.19) and (5.20) are used, the fields produced by the x-directed dipoles along the x axis are

$$E_r = 0 \tag{5.132a}$$

$$E_\theta = -\frac{j\omega\mu IL}{4\pi r} \cos\theta \cos\phi \frac{\sin[(N/2)(kd \sin\theta \cos\phi + \delta_x)]}{N \sin[(1/2)(kd \sin\theta \cos\phi + \delta_x)]} e^{-jkr} \tag{5.132b}$$

$$E_\phi = \frac{j\omega\mu IL}{4\pi r} \sin\phi \frac{\sin[(N/2)(kd \sin\theta \cos\phi + \delta_x)]}{N \sin[(1/2)(kd \sin\theta \cos\phi + \delta_x)]} e^{-jkr} \tag{5.132c}$$

and the fields produced by the array of y-directed dipoles along the y axis are

$$E_r = 0 \tag{5.133a}$$

$$E_\theta = -\frac{j\omega\mu IL}{4\pi r} \cos\theta \sin\phi \frac{\sin[(N/2)(kd\sin\theta\sin\phi + \delta_y)]}{N\sin[(1/2)(kd\sin\theta\sin\phi + \delta_y)]} e^{-jkr} \tag{5.133b}$$

$$E_\phi = -\frac{j\omega\mu IL}{4\pi r} \cos\phi \frac{\sin[(N/2)(kd\sin\theta\sin\phi + \delta_y)]}{N\sin[(1/2)(kd\sin\theta\sin\phi + \delta_y)]} e^{-jkr} \tag{5.133c}$$

For convenience, consider only the broadside case, $\delta_x = \delta_y = 0$, and group several factors in the field equations as constant C. If the dipoles along the y axis lead those along the x axis by phase difference $\pi/2$, the total fields are

$$E_\theta = -C\cos\theta\cos\phi \frac{\sin[(N/2)kd\sin\theta\cos\phi]}{N\sin[(1/2)kd\sin\theta\cos\phi]}$$

$$-jC\cos\theta\sin\phi \frac{\sin[(N/2)kd\sin\theta\sin\phi]}{N\sin[(1/2)kd\sin\theta\sin\phi]} \tag{5.134a}$$

$$E_\phi = C\sin\phi \frac{\sin[(N/2)kd\sin\theta\cos\phi]}{N\sin[(1/2)kd\sin\theta\cos\phi]}$$

$$-jC\cos\phi \frac{\sin[(N/2)kd\sin\theta\sin\phi]}{N\sin[(1/2)kd\sin\theta\sin\phi]} \tag{5.134b}$$

Radiation intensity and polarization ratio are readily found from (5.134). It is sufficient here to consider these quantities only in the yz plane, $\phi = \pi/2$. The xz plane is obviously like the yz plane. In other planes the beamwidths are more difficult to determine but still may be found from (5.134).

In the yz plane, the radiation intensity, normalized to its maximum value at $\theta = 0$, is

$$G_r = \frac{U}{U_{max}} = \frac{1}{2}\left(\cos^2\theta \frac{\sin^2[(N/2)kd\sin\theta]}{N^2\sin^2[(1/2)kd\sin\theta]} + 1\right) \tag{5.135}$$

and the radiation intensity half-power beamwidth, obtained by setting this equal to $1/2$, occurs where

$$\sin^2(\tfrac{1}{2}Nkd\sin\theta) = 0 \tag{5.136}$$

The value of θ for which this holds is readily recognized as the first array factor null of the linear array on the y axis.

Still in the yz plane, the polarization ratio of the wave is

$$P = -\frac{E_\theta}{E_\phi} = j \cos\theta \frac{\sin[(N/2)kd\sin\theta]}{N\sin[(1/2)kd\sin\theta]} \quad (5.137)$$

from which it is readily seen that the wave is circularly polarized on the z axis.

If a circularly polarized receiving antenna is used in conjunction with this array, the polarization match factor in the yz plane is

$$\rho = \frac{1}{2} + \cos\theta \frac{\sin[(N/2)kd\sin\theta]}{N\sin[(1/2)kd\sin\theta]} \bigg/ \left(1 + \cos^2\theta \frac{\sin^2[(N/2)kd\sin\theta]}{N^2\sin^2[(1/2)kd\sin\theta]}\right) \quad (5.138)$$

If this expression is set equal to $1/2$ to obtain the polarization beamwidth, we again obtain (5.136) and conclude that the polarization beamwidth is the same as the radiation intensity beamwidth and may be quite small for a large number of array elements.

The product of G_r and ρ in the yz plane is simpler than either factor alone. It is

$$G_r\rho = \frac{1}{4}\left(1 + \cos\theta \frac{\sin[(N/2)kd\sin\theta]}{N\sin[(1/2)kd\sin\theta]}\right)^2 \quad (5.139)$$

If we set this to $1/2$ to find the overall beamwidth in the yz plane, we find, not unexpectedly, that the overall beamwidth is smaller than that for the radiation intensity or polarization alone.

It is instructive to study the radiation intensity and polarization if all of the dipole antenna elements are rotated by $\pi/2$, so that the dipoles on the y axis are oriented in the x direction, and vice versa. On the z axis the wave is still circularly polarized, but the off-axis behavior is different. This is left to the reader as an exercise. One of the problems at the end of this chapter also asks for the polarization behavior if each element in Figure 5.12 is replaced by crossed dipoles, with each element producing a circularly polarized wave on the z axis.

5.12. TRAVELING-WAVE LOOP ANTENNA

Let a traveling wave of current on the circular loop antenna of radius a lying in the xy plane, Figure 5.13, be given by

$$I(r') = I_0 e^{-jka\phi'} \quad (5.140)$$

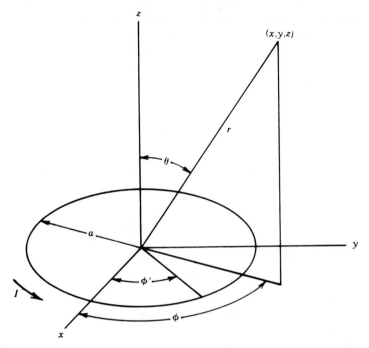

FIGURE 5.13 Traveling-wave-loop antenna.

where the current is assumed to propagate unattenuated on the loop at the velocity of light in free space. The vector magnetic potential far from the loop is given by

$$\mathbf{A}(\mathbf{r}) = \frac{\mu}{4\pi} \int \frac{I(\mathbf{r}')}{|\mathbf{r} - \mathbf{r}'|} e^{-jk|\mathbf{r}-\mathbf{r}'|} \, d\mathbf{L}' \tag{5.141}$$

where

$$\begin{aligned}
\mathbf{r}' &= \mathbf{u}_x x' + \mathbf{u}_y y' = \mathbf{u}_x a \cos \phi' + \mathbf{u}_y a \sin \phi' \\
|\mathbf{r} - \mathbf{r}'| &= [(x - x')^2 + (y - y')^2 + (z - 0)^2]^{1/2} \\
&= [(r \sin \theta \cos \phi - x')^2 + (r \sin \theta \sin \phi - y')^2 + z^2]^{1/2} \\
&\approx r - (x' \sin \theta \cos \phi + y' \sin \theta \sin \phi) \\
d\mathbf{L}' &= \mathbf{u}_{\phi'} a \, d\phi'
\end{aligned}$$

If these values are used in the integral for $\mathbf{A}(\mathbf{r})$, it becomes

$$\mathbf{A}(\mathbf{r}) = \frac{\mu I_0 a}{4\pi r} e^{-jkr} \int_0^{2\pi} e^{jka[\sin \theta \cos(\phi'-\phi)-\phi']} \mathbf{u}_{\phi'} \, d\phi' \tag{5.142}$$

POLARIZATION CHARACTERISTICS OF SOME ANTENNAS

To simplify the integration, replace the unit vector by

$$\mathbf{u}_{\phi'} = -\sin\phi' \,\mathbf{u}_x + \cos\phi' \,\mathbf{u}_y$$

use

$$A_\theta = A_x \cos\theta \cos\phi + A_y \cos\theta \sin\phi - A_z \sin\theta$$
$$A_\phi = -A_x \sin\phi + A_y \cos\phi$$

and move the terms involving θ and ϕ inside the integral as appropriate. The result is

$$A_\theta = -\frac{\mu I_0 a \cos\theta}{4\pi r} e^{-jkr} \int_0^{2\pi} \sin(\phi' - \phi) e^{jka[\sin\theta \cos(\phi'-\phi)-\phi']} \, d\phi' \quad (5.143a)$$

$$A_\phi = \frac{\mu I_0 a}{4\pi r} e^{-jkr} \int_0^{2\pi} \cos(\phi' - \phi) e^{jka[\sin\theta \cos(\phi'-\phi)-\phi']} \, d\phi' \quad (5.143b)$$

Changing the variable of integration and noting that for integration all ranges of 2π are equivalent gives

$$A_\theta = -\frac{\mu I_0 a \cos\theta}{4\pi r} e^{-jkr} e^{-jka\phi} \int_0^{2\pi} \sin\phi' e^{jka(\sin\theta \cos\phi' - \phi')} \, d\phi' \quad (5.144a)$$

$$A_\phi = \frac{\mu I_0 a}{4\pi r} e^{-jkr} e^{-jka\phi} \int_0^{2\pi} \cos\phi' e^{jka(\sin\theta \cos\phi' - \phi')} \, d\phi' \quad (5.144b)$$

If the integrand symmetries are used, these potentials become

$$A_\theta = -\frac{\mu I_0 a \cos\theta}{2\pi r} e^{-jkr} e^{-jka\phi}$$
$$\times \int_0^\pi \sin\phi' \sin(ka\phi')[\sin(ka \sin\theta \cos\phi') - j\cos(ka \sin\theta \cos\phi')] \, d\phi'$$
(5.145a)

$$A_\phi = \frac{\mu I_0 a}{2\pi r} e^{-jkr} e^{-jka\phi}$$
$$\times \int_0^\pi \cos\phi' \cos(ka\phi')[\cos(ka \sin\theta \cos\phi') + j\sin(ka \sin\theta \cos\phi')] \, d\phi'$$
(5.145b)

The potentials simplify even more if the loop circumference is made equal to a free-space wavelength or $ka = 1$. Then the potentials become

$$A_\theta = j\frac{\mu I_0 a \cos\theta}{\pi r} e^{-jkr} e^{-j\phi} \int_0^{\pi/2} \sin^2\phi' \cos(\sin\theta \cos\phi') \, d\phi' \quad (5.146a)$$

$$A_\phi = \frac{\mu I_0 a}{\pi r} e^{-jkr} e^{-j\phi} \int_0^{\pi/2} \cos^2\phi' \cos(\sin\theta \cos\phi') \, d\phi' \quad (5.146b)$$

The far electric field, from (1.58), is

$$\mathbf{E} = -j\omega \mathbf{A} \quad (5.147)$$

It is of interest to find the electric field on the z axis, transverse to the loop, for which it becomes

$$\mathbf{E}(\theta = 0) = \frac{\omega \mu I_0 a}{4r} e^{-j(kr+\phi)} (\mathbf{u}_\theta - j\mathbf{u}_\phi) \quad (5.148)$$

This is the maximum field intensity. Further, it is seen that the wave on an axis transverse to the loop is circularly polarized. The polarization ratio is

$$P = -\frac{E_\theta}{E_\phi} = -j$$

which corresponds to right-circular polarization. On the negative z axis, with $\theta = \pi$, the wave polarization is left circular. To an observer at the coordinate origin of Figure 5.13, looking in the direction of the positive z axis, the loop current travels in a clockwise direction and the electric field vector rotates clockwise.

The magnitude of E_θ forms a radiation pattern axially symmetric about the z axis, with a maximum on the z axis, mirror symmetry about the plane of the loop, and a null at $\theta = \pi/2$. It is relatively broad. The magnitude of E_ϕ is also maximum on the $\pm z$ axis, but has no null and does not differ greatly from isotropicity. Since the field component magnitudes do not vary in the same way with angle θ, the wave is circularly polarized only in two directions, along the $\pm z$ axis [6].

5.13. THE AXIAL-MODE HELIX*

A widely used antenna for radiating and receiving circularly polarized waves is a wire helix operating in the so-called "axial mode" [1, 6]. A diversity of feed arrangements is possible, but Figure 5.14 shows a commonly used feed

*Some of the material in this section is adapted from Elliot [6].

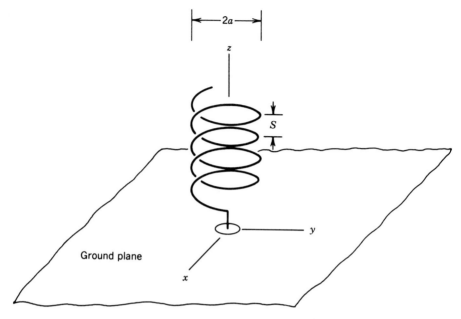

FIGURE 5.14 The axial-mode helix.

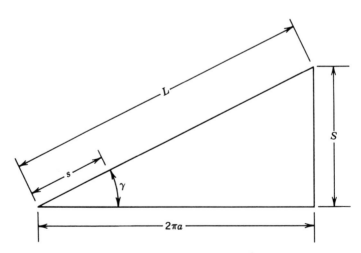

FIGURE 5.15 Helix turn unrolled.

by the extended conductor of a coaxial line, with a ground plane. Figure 5.15 shows one turn of the helix unrolled from the imaginary cylinder about which the helix is formed. The length of one turn is L and when wrapped around an imaginary cylinder of radius a at helix angle γ it advances distance S, determined by helix angle γ, in the z direction.

It has been found experimentally that if the helix circumference, $2\pi a$, is approximately one free-space wavelength and the helix angle γ is about $12°$, the current along much of the helix length is a traveling wave moving circumferentially and outward along the helix from the feed point. The current decays with distance near the feed point, and a standing wave exists near the outer end, but over the central portion of the helix, the VSWR is relatively small, indicating that the current is primarily a traveling wave. From the results of the previous section on the traveling wave loop, we expect each "loop" of the helix to radiate, and it is found to do so.

The parametric equations of a source point on a helix are

$$x' = a \cos \phi' \qquad y' = a \sin \phi' \qquad z' = \frac{S}{2\pi}\phi' = b\phi' \qquad (5.149)$$

where ϕ' is the azimuth angle measured in the xy plane, as in Figure 5.13. From these, the distance term in the magnetic potential integral is

$$|\mathbf{r} - \mathbf{r}'| = r - (a \cos \phi' \sin \theta \cos \phi + a \sin \phi' \sin \theta \sin \phi + b\phi' \cos \theta)$$

and the vector length element is

$$d\mathbf{L}' = \mathbf{u}_{\phi'} a\, d\phi' + \mathbf{u}_z b\, d\phi' = (-\mathbf{u}_x a \sin \phi' + \mathbf{u}_y a \cos \phi' + \mathbf{u}_z b)\, d\phi'$$

If the current near the helix ends, which is unlikely to contribute significantly to the far field, is neglected, the helix current can be written as

$$I(s) = I_0 e^{-(\alpha + j\beta)s} \qquad (5.150)$$

where s is the distance measured along the helix conductor. The inclusion of the attenuation term α allows for a decay in the helix current, although measurements show that such a decay is not great. Note from Figure 5.15 that when ϕ' changes by 2π, s changes by helix turn length L, so that

$$\frac{s}{L} = \frac{\phi'}{2\pi} \qquad (5.151)$$

Then the current on the helix is

$$I(\phi') = I_0 e^{-(\alpha + j\beta)(L/2\pi)\phi'} \qquad (5.152)$$

270   POLARIZATION CHARACTERISTICS OF SOME ANTENNAS

If the distance term and vector length element are used in the magnetic vector potential integral and the relations between rectangular and spherical components of the potential are used, the result is

$$A_\theta = \frac{\mu}{4\pi r} e^{-jkr} \int_0^{2\pi N} I(\phi')[-a\cos\theta \sin(\phi' - \phi) - b\sin\theta]$$
$$\times e^{jk[a\sin\theta \cos(\phi' - \phi) + b\phi' \cos\theta]} d\phi' \qquad (5.153a)$$

$$A_\phi = \frac{\mu}{4\pi r} e^{-jkr}$$
$$\times \int_0^{2\pi N} I(\phi')[a\cos(\phi' - \phi)] e^{jk[a\sin\theta \cos(\phi' - \phi) + b\phi' \cos\theta]} d\phi' \qquad (5.153b)$$

for a helix of N turns.

The integration over the full helix length can be altered to an integration over one helix turn if the periodic nature of quantities in the integrand is recognized. Consider a point x_1', y_1', z_1' corresponding to angle ϕ_1' on the first helix turn and another point on the nth turn x_n', y_n', z_n' corresponding to angle

$$\phi_n' = \phi_1' + 2\pi n \qquad (5.154)$$

It is clear that

$$x_n' = x_1' \qquad y_n' = y_1' \qquad z_n' = z_1' + 2\pi bn \qquad (5.155)$$

Note also that distance s_1 measured along the conductor of the first turn has a corresponding point on the nth turn,

$$s_n = s_1 + nL \qquad (5.156)$$

Then the helix current for the nth turn is

$$I(s_n) = I_0 e^{-(\alpha + j\beta)s_n} = I_0 e^{-(\alpha + j\beta)s_1} e^{-(\alpha + j\beta)nL}$$
$$= I(\phi_1')e^{-(\alpha + j\beta)nL} \qquad (5.157)$$

The terms

$$-a\cos\theta \sin(\phi' - \phi) - b\sin\theta \qquad a\cos(\phi' - \phi)$$

of the integrals for the potentials come from the vector length element and are the same for all helix turns. The phase terms, however, involve distances x_n', y_n', z_n'. The term

$$a\sin\theta \cos(\phi' - \phi)$$

originates in x' and y' and is unchanged. However, for the nth turn, the phase term arising from z' is altered, thus

$$b\phi' = z'_n = z'_1 + 2\pi bn = b\phi'_1 + 2\pi bn \qquad (5.158)$$

If this information is used for the potential, A_θ, the integral for the nth turn becomes

$$I_n = \int_{2\pi(n-1)}^{2\pi n} \{I(\phi'_1) e^{-(\alpha+j\beta)nL} [-a \cos\theta \sin(\phi'_1 - \phi) - b \sin\theta]$$
$$\times e^{jk[a \sin\theta \cos(\phi'_1 - \phi) + \cos\theta(b\phi'_1 + 2\pi bn)]} \} \, d\phi' \qquad (5.159)$$

The integral limits can now be 0 and 2π, and the subscript designating the first term can be dropped, yielding

$$I_n = e^{-\alpha nL - jn(\beta L - 2k\pi b \cos\theta)} \int_0^{2\pi} \{I(\phi')[-a \cos\theta \sin(\phi' - \phi) - b \sin\theta]$$
$$\times e^{jk[a \sin\theta \cos(\phi' - \phi) + b\phi' \cos\theta]} \} \, d\phi' \qquad (5.160)$$

When all the terms are summed and multiplied by $-j\omega$, the electric field E_θ is

$$E_\theta = -\frac{j\omega\mu I_0}{4\pi r} e^{-jkr} f(\theta) g_\theta(\theta, \phi) \qquad (5.161)$$

Similarly

$$E_\phi = -\frac{j\omega\mu I_0}{4\pi r} e^{-jkr} f(\theta) g_\phi(\theta, \phi) \qquad (5.162)$$

where

$$f(\theta) = \sum_{n=1}^{N} e^{-\alpha nL - jn(\beta L - 2\pi kb \cos\theta)} \qquad (5.163a)$$

$$g_\theta(\theta, \phi) = -\int_0^{2\pi} \{e^{-(\alpha+j\beta)(L/2\pi)\phi'} [a \cos\theta \sin(\phi' - \theta) + b \sin\theta]$$
$$\times e^{jk[a \sin\theta \cos(\phi' - \phi) + b\phi' \cos\theta]} \} \, d\phi' \qquad (5.163b)$$

$$g_\phi(\theta, \phi) = \int_0^{2\pi} \{e^{-(\alpha+j\beta)(L/2\pi)\phi'} [a \cos(\phi' - \phi)]$$
$$\times e^{jk[a \sin\theta \cos(\phi' - \phi) + b\phi' \cos\theta]} \} \, d\phi' \qquad (5.163c)$$

The terms g_θ and g_ϕ represent the field components produced by one helix turn, and the effect of $f(\theta)$ is to sum the fields of the N turns with the correct phase. The helix may be looked at as an array of N single turns, and $f(\theta)$ is thus the array factor of the helix "array." In the design of a helix for a specified frequency, only a and b (and thus L) may be selected, and β may be constrained or selected by the chosen a and b. Suppose it is desired to have a beam maximum at $\theta = 0°$. Now g_θ and g_ϕ will probably be maximum at or near $\theta = 0°$, particularly if $b \ll a$ and the helix circumference is about one free-space wavelength, since then the helix turn approximates the traveling wave loop antenna examined in the previous section. Then both E_θ and E_ϕ will be maximum at $\theta = 0°$ if all the terms in the array factor add in phase, or if

$$\beta L - 2\pi k b \cos\theta |_{\theta=0°} = 2m\pi \qquad m = 0, 1, 2, \ldots \qquad (5.164)$$

The phase velocity of the traveling wave of current on the helix is

$$v = \frac{\omega}{\beta} = pc = p\frac{\omega}{k} \qquad (5.165)$$

where p is the ratio of the current phase velocity to the free-space wave velocity. The last two equations can be solved for p, giving

$$p = \frac{L/\lambda}{(2\pi b/\lambda) + m} \qquad (5.166)$$

where λ is the free-space wavelength. From the experimental findings mentioned earlier, that for radiation in the direction of the helix axis the circumference is about one wavelength and the helix angle is about 12°, the numerator is approximately one and

$$\frac{2\pi b}{\lambda} \approx \tan(12°) = 0.21$$

Then $m = 1$ gives a value of v somewhat less than the free-space velocity, a value confirmed experimentally for a radiating helix [1]. The array factor simplifies somewhat to be

$$f(\theta) = \sum_{n=1}^{N} e^{-\alpha nL - j2\pi nkb(1-\cos\theta)} \qquad (5.167)$$

Specializing the helix element factors g_θ and g_ϕ to the helix axis and letting $ka = 1$ gives

$$g_\theta(\theta = 0, \phi = 0) = -\int_0^{2\pi} a \sin\phi' e^{-\alpha L\phi'/2\pi} e^{-j(\beta L/2\pi - kb)\phi'} d\phi' \qquad (5.168a)$$

$$g_\phi(\theta = 0, \phi = 0) = \int_0^{2\pi} a \cos\phi' e^{-\alpha L\phi'/2\pi} e^{-j(\beta L/2\pi - kb)\phi'} d\phi' \qquad (5.168b)$$

But the phase term is equal to the integer m that we found to be 1. Then

$$g_\theta(\theta = 0, \phi = 0) = -\int_0^{2\pi} a \sin \phi' e^{-\alpha L \phi'/2\pi} e^{-j\phi'} d\phi' \qquad (5.169a)$$

$$g_\phi(\theta = 0, \phi = 0) = \int_0^{2\pi} a \cos \phi' e^{-\alpha L \phi'/2\pi} e^{-j\phi'} d\phi' \qquad (5.169b)$$

If these expressions are compared to the field components of the traveling wave loop, also specialized to the z axis, they are found to be identical except for the current decay included in the helix turn but neglected for the loop. It follows that, on the helix axis, the helix may be considered as an end-fire array of traveling-wave loops if the amplitude of each loop current is reduced according to its distance from the feed point and the attenuation constant α. The radiated field on the axis is then circularly polarized. Here a word of caution is in order. The helix was idealized by treating only the traveling wave current component and neglecting the helix ends. It might be expected, then, that the longer the helix the more nearly circular the polarization will be. Kraus gives the axial ratio, on the z axis, as

$$\mathrm{AR} = \frac{2N + 1}{2N} \qquad (5.170)$$

where N is the number of turns.

Off the $\theta = 0$ axis, the wave is neither quite circularly polarized nor is the field exactly that produced by an array of circular loops. The shape of the overall pattern is determined primarily by the array factor $f(\theta)$, however, so the use of a loop to find the element pattern is justified, particularly if b is small. Just as the loop field remains close to circular polarization in a useful range of θ about the axis, so is the helix field circular over its useful beamwidth.

5.14. SIMPLE WAVEGUIDE SYSTEM FOR ELLIPTICAL POLARIZATION

It is desirable, when working with elliptically polarized waves, to have an antenna system that can generate any desired polarization and to know, from attenuator and phase shifter settings, what that polarization is. Conversely, we need a receiving antenna that can measure the polarization of an incoming wave. Figure 5.16 shows a waveguide and antenna system capable of transmitting a wave with any desired polarization [7]. It consists of a circular horn fed by a circular guide loaded with a quarter-wave plate, a circular-to-circular waveguide rotary joint, a rectangular-to-circular waveguide (TE_{11} mode) transducer, and a rectangular-to-rectangular waveguide collinear rotary joint.

274 POLARIZATION CHARACTERISTICS OF SOME ANTENNAS

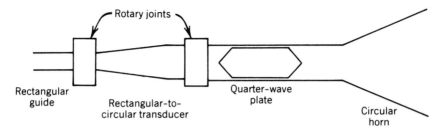

FIGURE 5.16 Waveguide system for generating a wave with general polarization.

The first rectangular guide establishes a reference frame, and the output of the first rotary joint is a rectangular waveguide mode (TE_{10}) with fields rotated from the reference field direction. The rectangular-to-circular transducer establishes the circular TE_{11} mode with a plane of symmetry dependent on the rotation of the first joint. The second rotary joint allows the quarter-wave plate in the circular guide feeding the horn to be oriented independently of the mode plane of symmetry.

For reference purposes the broadwall of the input waveguide is parallel to the horizontal plane and considered the x axis of the fixed coordinate system. The broadwall of the rotatable rectangular waveguide serves as a reference for the angular displaced axis x' with angle of displacement β. The angle between the y' axis and the plane of the quarter-wave plate (plane in which the component parallel to the plate is delayed in phase by $\pi/2$) is denoted by δ. The unit vectors $\mathbf{u}_{\parallel}$ and $\mathbf{u}_{\perp}$ are, respectively, in the plane and perpendicular to the plane of the quarter-wave plate, and both are transverse to the axis of revolution of the circular horn. Figure 5.17 shows these coordinates.

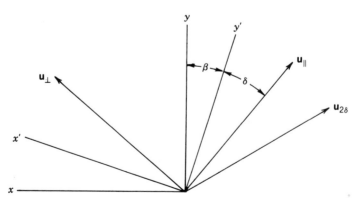

FIGURE 5.17 Coordinates for waveguide system.

SIMPLE WAVEGUIDE SYSTEM FOR ELLIPTICAL POLARIZATION 275

The initial reference frame may be established independently of the first waveguide section and the rectangular-to-rectangular rotary joint replaced, for example, by a movable coaxial-to-rectangular transducer. Also, any antenna with circular symmetry, such as a cylindrical polyrod, may be used in place of the horn.

The far field transmitted by the antenna is

$$\mathbf{E}(\beta, \delta) = E_\| \mathbf{u}_\| + jE_\perp \mathbf{u}_\perp \tag{5.171}$$

where $E_\|$ and $E_\perp$ are the relative field strengths, in the plane and perpendicular to the plane of the quarter-wave plate, respectively, on the axis of the horn at the far-field point. From Figure 5.17 it is seen that

$$\mathbf{u}_\| = \cos(\beta + \delta)\mathbf{u}_y - \sin(\beta + \delta)\mathbf{u}_x \tag{5.172a}$$
$$\mathbf{u}_\perp = \sin(\beta + \delta)\mathbf{u}_y + \cos(\beta + \delta)\mathbf{u}_x \tag{5.172b}$$

and that

$$E_\| = E_{y'} \cos \delta \tag{5.173a}$$
$$E_\perp = E_{y'} \sin \delta \tag{5.173b}$$

For convenience, (5.171) can be normalized by requiring that $|E_{y'}| = 1$. Using (5.172) and (5.173), it becomes

$$\mathbf{E}(\beta, \delta) = \cos \delta \left[\cos(\beta + \delta)\mathbf{u}_y - \sin(\beta + \delta)\mathbf{u}_x\right]$$
$$+ e^{j\pi/2} \sin \delta \left[\sin(\beta + \delta)\mathbf{u}_y + \cos(\beta + \delta)\mathbf{u}_x\right] \tag{5.174}$$

which by trigonometric manipulation becomes

$$\mathbf{E}(\beta, \delta) = \tfrac{1}{2}\{[\mathbf{u}_y \cos \beta - \mathbf{u}_x \sin \beta] + [\mathbf{u}_y \cos(\beta + 2\delta) - \mathbf{u}_x \sin(\beta + 2\delta)]$$
$$+ e^{j\pi/2}[\mathbf{u}_y \cos \beta - \mathbf{u}_x \sin \beta]$$
$$+ e^{-j\pi/2}[\mathbf{u}_y \cos(\beta + 2\delta) - \mathbf{u}_x \sin(\beta + 2\delta)]\} \tag{5.175}$$

The first and third bracketed terms in this equation are unit vectors in the y' direction ($\mathbf{u}_{y'}$). The second and fourth terms describe a unit vector leading $\mathbf{u}_{y'}$ by angle 2δ, ($\mathbf{u}_{2\delta}$), as shown in Figure 5.17.

The linearly polarized field components of (5.175) may be expressed in terms of right-circular and left-circular rotating components. Substituting

$$\mathbf{u}_x = \frac{1}{\sqrt{2}}(\mathbf{u}_R + \mathbf{u}_L) \tag{5.176a}$$

$$\mathbf{u}_y = \frac{1}{\sqrt{2}}(\mathbf{u}_R - \mathbf{u}_L) \tag{5.176b}$$

where $\mathbf{u}_R$ and $\mathbf{u}_L$ are the unit vectors representing right and left circularly polarized waves discussed in Section 3.10, into (5.175) leads to the radiated field in the form

$$\mathbf{E}(\beta,\delta) = \tfrac{1}{2}e^{j\pi/4}\{[e^{j(\beta+\pi/2)} + e^{j(\beta+2\delta)}]\mathbf{u}_R \\ + [e^{-j(\beta+\pi/2)} - e^{-j(\beta+2\delta)}]\mathbf{u}_L\} \qquad (5.177)$$

The inverse circular polarization ratio is found from this field to be

$$q = \frac{E_L}{E_R} = e^{-j(2\beta+2\delta+\pi)}\cot\left(\delta + \frac{\pi}{4}\right) \qquad (5.178)$$

The axial ratio, tilt angle, and rotation sense, found from q, are

$$AR = \begin{cases} |\cot\delta| \\ |\tan\delta| \end{cases} \quad AR \geq 1$$

$$\tau = \beta + \delta \pm \pi/2 \qquad (5.179)$$

$$\text{Rotation sense} \begin{cases} \text{LER} & -\pi/2 \leq \delta \leq 0 \\ \text{RER} & 0 \leq \delta \leq \pi/2 \end{cases}$$

A logical change in this system is the replacement of the quarter-wave plate with a "variable-wave" plate (such as a ferrite slab biased transverse to the direction of propagation and parallel to the slab) that introduces a phase delay ψ into the field component parallel to the plate. Physical considerations fix the angle of polarization inclination δ at $\pi/4$ radians. This configuration is amenable to the same analysis applied to the quarter-wave plate, with the result

$$q = e^{-j(\pi+2\beta)}\cot\left(\frac{\psi}{2} + \frac{\pi}{4}\right) \qquad (5.180)$$

leading to

$$AR = \begin{cases} \left|\cot\left(\dfrac{\psi}{2}\right)\right| \\ \left|\tan\left(\dfrac{\psi}{2}\right)\right| \end{cases} \quad AR \geq 1$$

$$\tau = \beta \pm \pi/2 \qquad (5.181)$$

$$\text{Rotation sense} \begin{cases} \text{LER} & -\pi \leq \psi \leq 0 \\ \text{RER} & 0 \leq \psi \leq \pi \end{cases}$$

This arrangement allows the replacement of a mechanical rotation by a bias current.

Although this discussion has been concerned with the transmission of an elliptically polarized wave, it is obvious that the system can also be used to measure the polarization of an incoming wave. It is left as an exercise to develop the required equations.

5.15. ANOTHER WAVEGUIDE SYSTEM

Figure 5.18 shows a second waveguide system for radiating an elliptical wave with any desired polarization ratio. It has been constructed and found to perform well. The two inputs are fed from a common source using a power splitter. Placing an attenuator before each input and a phase shifter before one input allows two orthogonal waveguide modes to be established with relative amplitude and phase controllable over any desired ranges.

The input signal at port 1 establishes a TE_{11} circular guide mode that travels toward the dielectric rod radiating element. The electric field of this mode is horizontal (in the plane of the paper) on the axis of the guide. The input at port 2 establishes the TE_{10} mode in the rectangular guide, with a vertical electric field. At the junction of the rectangular and circular guides, this TE_{10} rectangular mode excites a TE_{11} mode in the circular guide with a vertical electric field on the guide axis. A vertical post placed in the circular guide serves to prevent this vertical TE_{11} mode from traveling to the left, toward port 1. A grid of horizontal wires at the junction of the guides similarly serves as a mode filter for the rectangular guide.

Two orthogonal TE_{11} modes are thus set up in the circular guide with relative amplitudes and phase difference independently controlled. Off-axis the fields produced by a dielectric rod antenna excited with a TE_{11} mode are complex and will not be discussed here, but on-axis, because of mode symmetry, the vertical TE_{11} mode will produce a vertical linearly polarized wave in the far field. The orthogonal TE_{11} mode will produce a horizontal far

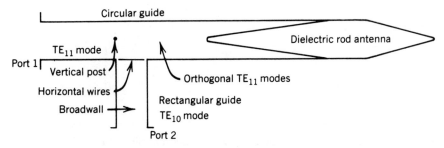

FIGURE 5.18 Two-port waveguide system for elliptical waves.

278 POLARIZATION CHARACTERISTICS OF SOME ANTENNAS

field. Since the amplitudes and phase difference can be set at will, it is clear that on the axis a wave of any desired polarization can be radiated.

It is clear that this system can also be used to measure the polarization ratio of an incoming wave. The two paths from the rod antenna to ports 1 and 2 are not equivalent, however, so the system must be calibrated in order to measure polarization by comparing outputs at ports 1 and 2. In this respect it is not as convenient as the waveguide system described in Section 5.14 or the one discussed in Section 5.16.

5.16. LOSSLESS POWER COMBINER AND DIVIDER SYSTEM

Figure 5.19 shows a lossless power combiner and divider system that is well suited for generating a wave with arbitrary polarization or measuring the

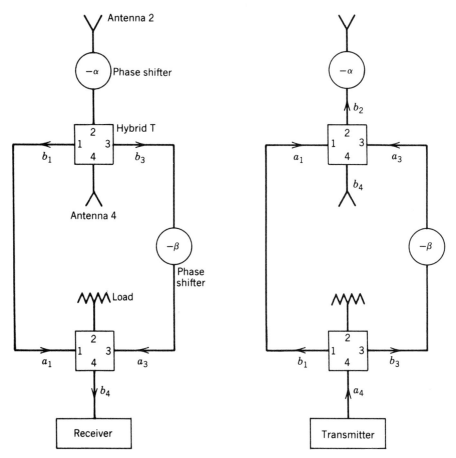

FIGURE 5.19 Power combiner and divider for transmitting and receiving arbitrarily polarized waves.

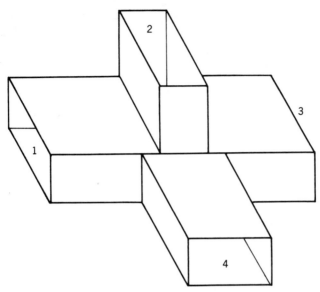

FIGURE 5.20 Port designations for the hybrid tee.

polarization of an incident elliptically polarized wave [8]. It is based on the variable-ratio power divider of Teeter and Bushore [9, 10]. The system may be set up either in waveguide or transmission line. The hybrid tees may be replaced by circulators, and in fact other variations are possible [9, 11]. One antenna is linear horizontal and the other is linear vertical. They are placed adjacent to each other and pointed in the same direction. Either antenna may be the vertically polarized one, but for definiteness it is the one marked 4.

The scattered signal from the waveguide hybrid tee, with the ports numbered as in Figure 5.20, is given by [12]

$$\mathbf{b} = \mathbf{Sa} = \frac{1}{\sqrt{2}} \begin{bmatrix} 0 & 1 & 0 & 1 \\ 1 & 0 & -1 & 0 \\ 0 & -1 & 0 & 1 \\ 1 & 0 & 1 & 0 \end{bmatrix} \mathbf{a} \qquad (5.182)$$

Consider first the system as used in receiving, Figure 5.19a. We assume that there is at least one direction in which the effective lengths h_2 and h_4 of the two antennas used are equal. For example, crossed dipoles with equal length transmission paths to the ports of the upper tee of Figure 5.19a have equal effective lengths along a direction perpendicular to the dipoles. For simplicity in the development we take this direction to be the z axis of Figure 5.21. The incident wave travels in the ζ direction in Figure 5.21.

280 POLARIZATION CHARACTERISTICS OF SOME ANTENNAS

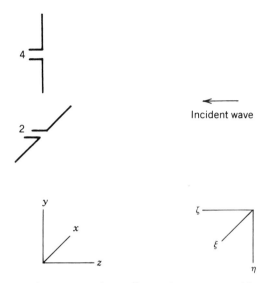

FIGURE 5.21 Antennas and coordinates for power combiner system.

Let the incident wave, in xyz coordinates, be

$$\mathbf{E}^i = E_0(\mathbf{u}_x a + \mathbf{u}_y b e^{j\phi}) \tag{5.183}$$

Without loss of generality we drop any common amplitude coefficients and neglect phase shifts common to both input arms of the upper tee of Figure 5.19a. Then the inputs to the upper tee are

$$a_2 = ae^{-j\alpha} \tag{5.184a}$$
$$a_4 = be^{j\phi} \tag{5.184b}$$

The outputs of the top tee, from the scattering matrix of the tee, are

$$b_1 = \frac{1}{\sqrt{2}}(a_2 + a_4) = \frac{1}{\sqrt{2}}(ae^{-j\alpha} + be^{j\phi}) \tag{5.185a}$$

$$b_3 = \frac{1}{\sqrt{2}}(-a_2 + a_4) = \frac{1}{\sqrt{2}}(-ae^{-j\alpha} + be^{j\phi}) \tag{5.185b}$$

Phase shifts common to both lines or waveguides connecting top and bottom tees can be neglected. Then if β is the differential phase shift, the inputs to the bottom tee are

$$a_1 = \frac{1}{\sqrt{2}}(ae^{-j\alpha} + be^{j\phi}) \tag{5.186a}$$

$$a_3 = \frac{1}{\sqrt{2}}(-ae^{-j\alpha} + be^{j\phi})e^{-j\beta} \tag{5.186b}$$

LOSSLESS POWER COMBINER AND DIVIDER SYSTEM

and the outputs from the bottom tee are

$$b_2 = \frac{1}{\sqrt{2}}(a_1 - a_3) = \frac{1}{2}e^{-j\alpha}\left[a + be^{j(\phi+\alpha)} + e^{-j\beta}(a - be^{j(\phi+\alpha)})\right] \quad (5.187a)$$

$$b_4 = \frac{1}{\sqrt{2}}(a_1 + a_3) = \frac{1}{2}e^{-j\alpha}\left[a + be^{j(\phi+\alpha)} - e^{-j\beta}(a - be^{j(\phi+\alpha)})\right] \quad (5.187b)$$

Now let

$$\phi + \alpha = \pm \tfrac{1}{2}\pi \quad (5.188)$$

and require that

$$b_2 = 0 \quad (5.189)$$

It follows from (5.187) that

$$e^{-j\beta} = -\frac{a \pm jb}{a \mp jb} \quad (5.190)$$

Using the upper and lower signs, respectively, in this equation gives

$$\beta = \pi \mp 2\tan^{-1}\frac{b}{a} \quad (5.191)$$

For these values, which make $b_2 = 0$, b_4 is

$$b_4 = \frac{1}{2}e^{-j\alpha}\left[a \pm jb + \frac{a \pm jb}{a \mp jb}(a \mp jb)\right] = e^{-j\alpha}(a \pm jb) \quad (5.192)$$

and it follows that

$$|b_4|^2 = a^2 + b^2 = 1 \quad (5.193)$$

All of the incident energy is therefore directed to port 4 of the bottom tee by the choice of phase delays given by (5.188) and (5.191).

We could just as well direct all of the energy to port 2 of the lower tee by the choices

$$\phi + \alpha = \pm \tfrac{1}{2}\pi \quad (5.188)$$

$$e^{-j\beta} = \frac{a \pm jb}{a \mp jb} \quad (5.194)$$

$$\beta = \mp 2\tan^{-1}\frac{b}{a} \quad (5.195)$$

282 POLARIZATION CHARACTERISTICS OF SOME ANTENNAS

from which it follows that

$$b_4 = 0 \tag{5.196a}$$

$$b_2 = e^{-j\alpha}(a \pm jb) \tag{5.196b}$$

$$|b_2|^2 = 1 \tag{5.196c}$$

This system has several uses. It can extract maximum power from an incident wave of any polarization by appropriate choice of the phase shifts α and β. It may also be used to measure the polarization of an incident wave. Finally we shall see that it allows the formation of a polarization-adaptive two-way communication system.

Consider that the system is set up for maximum output at port 4 of the lower tee, using phase shifts given by (5.188) and (5.191), and is used for transmission, with an input to arm 4 of the lower tee and a matched load at port 2, as in Figure 5.19b. Then at the bottom tee

$$a_2 = 0 \tag{5.197a}$$

$$a_4 = 1 \tag{5.197b}$$

The outputs from the bottom tee are

$$b_1 = \frac{1}{\sqrt{2}}(a_2 + a_4) = \frac{1}{\sqrt{2}} \tag{5.198a}$$

$$b_3 = \frac{1}{\sqrt{2}}(-a_2 + a_4) = \frac{1}{\sqrt{2}} \tag{5.198b}$$

and the inputs to the top tee are

$$a_1 = \frac{1}{\sqrt{2}} \tag{5.199a}$$

$$a_3 = \frac{1}{\sqrt{2}} e^{-j\beta} = -\frac{1}{\sqrt{2}} \frac{a \pm jb}{a \mp jb} \tag{5.199b}$$

where the special value of (5.190) is used for β.

The top tee outputs are then

$$b_2 = \frac{1}{\sqrt{2}}(a_1 - a_3) = \frac{1}{2}\left(1 + \frac{a \pm jb}{a \mp jb}\right) = \frac{a}{a \mp jb} \tag{5.200a}$$

$$b_4 = \frac{1}{\sqrt{r}}(a_1 + a_3) = \frac{1}{2}\left(1 - \frac{a \pm jb}{a \mp jb}\right) = \frac{\mp jb}{a \mp jb} \tag{5.200b}$$

The waves transmitted from the antennas, if $h_2 = h_4$ and if we neglect common amplitude coefficients, are

$$E_x^t = b_2 e^{-j\alpha} = \frac{a}{a \mp jb} e^{j(\phi \mp \pi/2)} \tag{5.201a}$$

$$E_y^t = b_4 = \frac{\mp jb}{a \mp jb} \tag{5.201b}$$

The total power radiated is

$$E_x^t E_x^{t*} + E_y^t E_y^{t*} = 1 \tag{5.202}$$

and the polarization ratio of the transmitted wave, in the directions for which $|h_2| = |h_4|$, is

$$P^t = \frac{E_y^t}{E_x^t} = \frac{\mp jb}{a e^{j(\phi \mp \pi/2)}} = \frac{b}{a} e^{-j\phi} \tag{5.203}$$

Now the polarization of the incoming wave is

$$P^i = \frac{E_\eta}{E_\xi} = -\frac{E_y}{E_x} = -\frac{b}{a} e^{j\phi} \tag{5.204}$$

and it may be seen that

$$P^t = -P^{i*} \tag{5.205}$$

This is the condition for complete polarization matching. The significance of this result is evident when this variable-polarization antenna is taken as one of the antennas in a two-way communication system.

If the phase shifters remain set so that on reception b_2 is zero and b_4 is maximum at the lower tee, and the generator is connected to port 2 of the bottom tee for transmission, the bottom tee inputs are

$$a_2 = 1 \tag{5.206a}$$

$$a_4 = 0 \tag{5.206b}$$

An analysis similar to the preceding one gives the transmitted signals as

$$E_x^t = -\frac{b}{a \mp jb} e^{j\phi} \tag{5.207a}$$

$$E_y^t = \frac{a}{a \mp jb} \tag{5.207b}$$

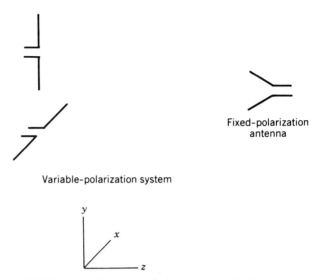

FIGURE 5.22 Power-maximized communications link.

leading to a polarization ratio relationship

$$P^t = -\frac{a}{b}e^{-j\phi} = \frac{1}{P^i} \qquad (5.208)$$

which is the relation for a total polarization mismatch.

Consider now a communication configuration with this variable-polarization system at one end, as in Figure 5.22, and a fixed arbitrarily polarized antenna at the other. The fixed-polarization antenna transmits a wave with polarization P^i toward the variable-polarization system, which is then set to receive maximum power. In turn, when it is used to transmit, from the port at which maximum power is received, the variable-polarization system transmits a wave with polarization $P^t = -P^{i*}$. This is the polarization that the fixed-polarization antenna receives best. Thus adjustment of the variable-polarization antenna until it receives maximum signal causes it, on transmission, to transmit a wave from which the fixed-polarization antenna receives maximum power. This offers the opportunity for an adaptive (in polarization) two-way communication system.

The condition (5.208) may be recognized as the cross-polarization condition in a communication link. Thus, in the link of Figure 5.22, if the variable-polarization system on reception is set for maximum power out at port 4 of the lower tee, connecting the generator to arm 2 on transmission would cause the fixed-polarization antenna in the link to receive *no* power.

In the preceding development the effective lengths of the two antennas were equal for propagation in the z direction. In general the effective lengths are complex functions of direction, and if the reference points for the effective lengths are taken at ports 2 and 4 of the upper tee of Figure 5.19, the effective lengths also depend on the lengths of the transmission paths to the tee. We therefore use

$$h_2(\theta, \phi) = |h_2|e^{j\delta_2} \tag{5.209a}$$

$$h_4(\theta, \phi) = |h_4|e^{j\delta_4} = |h_4|e^{j(\delta_2+\delta)} \tag{5.209b}$$

where δ_2 and δ_4 are sums of phase shifts inherent in the antennas and those occurring between the radiating elements and ports 2 and 4 of the upper tee.

Neglecting the common phase shift δ_2, the inputs to the top tee of Figure 5.29a change from (5.184) to

$$a_2 = |h_2|ae^{-j\alpha} \tag{5.210a}$$

$$a_4 = |h_4|e^{j\delta}be^{j\phi} \tag{5.210b}$$

The outputs of the bottom tee can be found by replacing a by $|h_2|a$ and b by $|h_4|\exp(j\delta)b$ in (5.187). The result is

$$b_2 = \tfrac{1}{2}e^{-j\alpha}\big[|h_2|a + |h_4|be^{j(\phi+\alpha+\delta)} + e^{-j\beta}(|h_2|a - |h_4|be^{j(\phi+\alpha+\delta)})\big] \tag{5.211a}$$

$$b_4 = \tfrac{1}{2}e^{-j\alpha}\big[|h_2|a + |h_4|be^{j(\phi+\alpha+\delta)} - e^{-j\beta}(|h_2|a - |h_4|be^{j(\phi+\alpha+\delta)})\big] \tag{5.211b}$$

If now we set

$$\phi + \alpha + \delta = \pm \tfrac{1}{2}\pi \tag{5.212}$$

and require

$$b_2 = 0 \tag{5.213}$$

it follows from (5.211) that

$$e^{-j\beta} = -\frac{|h_2|a \pm j|h_4|b}{|h_2|a \mp j|h_4|b} \tag{5.214}$$

which gives

$$\beta = \pi \mp 2\tan^{-1}\frac{|h_4|b}{|h_2|a} \tag{5.215}$$

Then

$$b_4 = \tfrac{1}{2}e^{-j\alpha}\left[|h_2|a \pm j|h_4|b + \frac{|h_2|a \pm j|h_4|b}{|h_2|a \mp j|h_4|b}(|h_2|a \mp j|h_4|b)\right]$$

$$= e^{-j\alpha}(|h_2|a \pm j|h_4|b) \tag{5.216}$$

which leads to the equation

$$|b_4|^2 = |h_2|^2 a^2 + |h_4|^2 b^2 \tag{5.217}$$

This again is all of the incident power, since $b_2 = 0$.

If this system is used for transmission by connecting a generator to arm 4 of the lower hybrid tee and a matched load to arm 2, while leaving phase shifts α and β set for maximum reception at port 4, the lower tee inputs are

$$a_2 = 0 \tag{5.197a}$$

$$a_4 = 1 \tag{5.197b}$$

The inputs to the top tee may be found by using (5.198) for b_1 and b_3 at the lower tee and (5.214) for $\exp(-j\beta)$. They are

$$a_1 = \frac{1}{\sqrt{2}} \tag{5.218a}$$

$$a_3 = -\frac{1}{\sqrt{2}}\frac{|h_2|a \pm j|h_4|b}{|h_2|a \mp j|h_4|b} \tag{5.218b}$$

The outputs from the top tee are

$$b_2 = \frac{1}{\sqrt{2}}(a_1 - a_3) = \frac{|h_2|a}{|h_2|a \mp j|h_4|b} \tag{5.219a}$$

$$b_4 = \frac{1}{\sqrt{2}}(a_1 + a_3) = \frac{\mp j|h_4|b}{|h_2|a \mp j|h_4|b} \tag{5.219b}$$

The transmitted wave has components

$$E_x^t = b_2|h_2|e^{-j\alpha} = \frac{|h_2|^2 a}{|h_2|a \mp j|h_4|b}e^{j(\phi + \delta \mp \pi/2)} \tag{5.220a}$$

$$E_y^t = b_4|h_4|e^{j\delta} = \frac{\mp j|h_4|^2 b}{|h_2|a \mp j|h_4|b}e^{j\delta} \tag{5.220b}$$

and polarization ratio

$$P^t = \frac{E_y^t}{E_x^t} = \frac{|h_4|^2}{|h_2|^2}\frac{b}{a}e^{-j\phi} = -\frac{|h_4|^2}{|h_2|^2}P^{i*} \qquad (5.221)$$

We conclude that in this more general case for which $h_2 \neq h_4$ in the direction of the incoming wave due perhaps to the use of nonidentical antennas, to improper orientation of the antennas, or to unequal transmission path lengths between antennas and hybrid tee inputs), if the system is set up for maximum power reception and then used for transmitting, the transmitted signal *in the direction from which the original signal was received* is modified in its polarization characteristics by the factor $|h_4|^2/|h_2|^2$.

REFERENCES

1. J. D. Kraus, *Antennas*, 2nd ed., McGraw-Hill, New York, 1988.
2. C. A. Balanis, *Antenna Theory*, Harper & Row, New York, 1982.
3. S. Silver, *Microwave Antenna Theory and Design*, MIT Radiation Laboratory Series, Vol. 12, McGraw-Hill, New York, 1949.
4. R. E. Collin and F. J. Zucker, *Antenna Theory*, Vol. 2, McGraw-Hill, New York, 1969.
5. W. L. Weeks, *Antenna Engineering*, McGraw-Hill, New York, 1968.
6. R. S. Elliott, *Antenna Theory and Design*, Prentice-Hall, Englewood Cliffs, NJ, 1981.
7. H. Mott and D. N. McQuiddy, "A Simple Waveguide System for Radiating Elliptically Polarized Waves," *IEEE Trans. on Antennas and Propagation*, **AP-16**(1), 134–135, January 1968.
8. H. Mott and D. N. McQuiddy, "A Polarization-Adaptive Antenna System," *IEEE Region 3 Convention Record*, 27.4.1–27.4.5, April 1968.
9. W. L. Teeter and K. R. Bushore, "A Variable-Ratio Microwave Power Divider and Multiplexer," *IRE Trans. Microwave Theory and Tech.*, **MTT-5**(4), 227–229, October 1957.
10. R. M. Vaillancourt, "Analysis of the Variable-Ratio Microwave Power Divider," *IRE Trans. Microwave Theory and Tech.*, **MTT-6**(2), 238–239, April 1958.
11. H. J. Riblet, "The Short-Slot Hybrid Junction," *Proc. IRE*, **40**(2), 180–184, February 1952.
12. R. N. Ghose, *Microwave Circuit Theory and Analysis*, McGraw-Hill, New York, 1963.
13. R. E. Ziemer and W. H. Tranter, *Principles of Communications*, Houghton Mifflin, Boston, MA, 1976.

PROBLEMS

5.1. Find the equation of a parabola, in rectangular coordinates, whose focus is the origin, if the x axis is transverse to the parabola and if the parabola intersects the x axis at 1 m. Show that a line drawn from the focus to a point on the parabola, a normal to the parabola, and a line parallel to the x axis form two equal angles.

5.2. Find the fields of a small circular loop antenna, with uniform current, lying in the xz plane. *Hint*: Compare the fields of y-oriented and z-oriented short dipoles.

5.3. Verify the text statement in Section 5.8 that the maximum rate of change of ρ with angle θ occurs, for example, for $\phi = \pi/4$.

5.4. Find the 3-dB polarization beamwidth in the plane $\phi = \pi/8$ of the waveguide opening into a plane (Section 5.8).

5.5. Plot the relative radiation intensity as a function of θ in the plane $\phi = \pi/8$ for the waveguide opening into a plane (Section 5.8). Find the 3-dB beamwidth and compare to the polarization beamwidth of Problem 5.4. Assume standard X-band waveguide ($a = 0.9$ in., $b = 0.4$ in.) and a frequency of 10 GHz.

5.6. Find the 3-dB polarization beamwidth of the pyramidal horn antenna as a function of the azimuth angle ϕ.

5.7. If each dipole element in Figure 5.12 is rotated by 90° so that the y axis array consists of x-directed dipoles, and vice versa, find the normalized radiation intensity in the yz plane. The other conditions of Section 5.11 remain the same. The number of elements on each axis and the element spacings are the same. All feed amplitudes are equal and the phases of all element feeds in each linear array are the same. The feed phases of the elements on the y axis lead those on the x axis by $\pi/2$. Compare radiation intensity beamwidth and polarization beamwidth in the yz plane. The receiving antenna is to be circularly polarized.

5.8. If each element in Figure 5.12 is replaced by crossed dipoles, with the y-directed dipole leading the x-directed dipole in phase by $\pi/2$, and if no phase difference exists between crossed dipoles on the x axis and those on the y axis, compare the radiation intensity and polarization beamwidths.

5.9. Suppose the turnstile antenna of Section 5.4 is used to transmit from an unstabilized satellite that rolls and tumbles. Let the earth-based receiving antenna be circularly polarized and always pointing at the transmitter. If all values of angle θ in (5.34) are equally probable, find the expected value of $G_r \rho$ [13].

PROBLEMS 289

5.10. Find the electric field components of a short y-directed dipole placed a quarter wavelength from a conducting plane, $z =$ constant.

5.11. The helix of Figure 5.14 is right-handed. If the helix wire is considered a screw thread, the screw would move in the z direction if rotated clockwise as one looks in the z direction. Is the radiated wave right- or left-circular? A receiving antenna identical to the transmitting antenna of Figure 5.14 is placed so that the z axes coincide (but are opposite in direction). Is the received power maximum or minimum?

CHAPTER SIX

Polarization Changes by Reflection and Transmission

6.1. REFLECTION AT AN INTERFACE: LINEAR POLARIZATION

In this section we consider the reflection and transmission of a linearly polarized plane wave at the plane interface between two media. It is convenient to use a rectangular coordinate system with two axes lying in the plane of the interface, but since the incident wave strikes the interface at some angle other than perpendicular we must change its components to the appropriate coordinates for the interface. The transformations are readily apparent from Figure 6.1. The plane wave being considered travels in the ζ direction. The x and ξ axes coincide and are into the plane of the page. From the figure,

$$\zeta = z \cos \theta + y \sin \theta$$
$$\eta = y \cos \theta - z \sin \theta \qquad (6.1)$$
$$\xi = x$$

and since space vector components transform like the coordinates of a point,

$$\mathbf{u}_\zeta = \mathbf{u}_z \cos \theta + \mathbf{u}_y \sin \theta$$
$$\mathbf{u}_\eta = \mathbf{u}_y \cos \theta - \mathbf{u}_z \sin \theta \qquad (6.2)$$
$$\mathbf{u}_\xi = \mathbf{u}_x$$

The *plane of incidence* is defined as that plane containing a vector in the direction of wave travel, $\mathbf{u}_\zeta$, and a vector normal to the interface, $\mathbf{u}_z$. It is convenient to consider linearly polarized waves by two cases, **E** lying in the

REFLECTION AT AN INTERFACE: LINEAR POLARIZATION 291

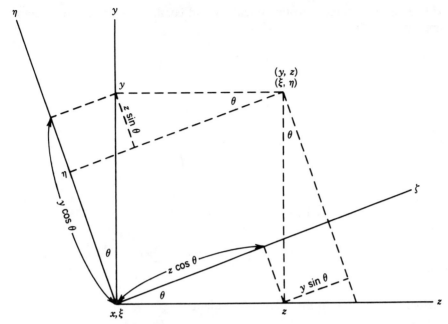

FIGURE 6.1 Coordinate transformations.

plane of incidence (**H** perpendicular to the plane) and **E** perpendicular to the plane of incidence (**H** in the plane).

Fields: Polarization Normal to the Plane of Incidence

For this case

$$E_\xi = E_0 e^{-jk\zeta} \qquad H_\eta = \frac{E_0}{Z} e^{-jk\zeta} \qquad (6.3)$$

where Z is the characteristic impedance of the medium.

In the general case, for lossy media, k and Z are given by

$$k = \omega\sqrt{\mu\epsilon}\sqrt{1 - j\frac{\sigma}{\omega\epsilon}} \qquad Z = \sqrt{\frac{j\omega\mu}{\sigma + j\omega\epsilon}} \qquad (6.4)$$

and are specialized to the lossless case by setting conductivity $\sigma = 0$.

Transforming to x, y, and z coordinates gives

$$\begin{aligned}\mathbf{E} &= \mathbf{u}_\xi E_\xi = \mathbf{u}_x E_x + \mathbf{u}_y E_y + \mathbf{u}_z E_z \\ \mathbf{H} &= \mathbf{u}_\eta H_\eta = \mathbf{u}_x H_x + \mathbf{u}_y H_y + \mathbf{u}_z H_z\end{aligned} \qquad (6.5)$$

292 POLARIZATION CHANGES BY REFLECTION AND TRANSMISSION

If (6.2) is substituted into this equation and coefficients of like unit vectors are equated, then

$$E_x = E_\xi$$
$$H_y = \cos\theta\, H_\eta \qquad (6.6)$$
$$H_z = -\sin\theta\, H_\eta$$

Combining these equations with (6.1) and (6.3) leads to

$$E_x = E_0 e^{-jk(z\cos\theta + y\sin\theta)}$$
$$H_y = \cos\theta\, \frac{E_0}{Z} e^{-jk(z\cos\theta + y\sin\theta)} \qquad (6.7)$$
$$H_z = -\sin\theta\, \frac{E_0}{Z} e^{-jk(z\cos\theta + y\sin\theta)}$$

Fields: Polarization in the Plane of Incidence

For this case, **H** has only a ξ component and it is appropriate to write

$$H_\xi = H_0 e^{-jk\zeta} \qquad E_\eta = -ZH_0 e^{-jk\zeta} \qquad (6.8)$$

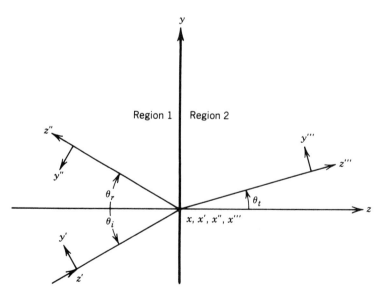

FIGURE 6.2 Coordinate systems for wave reflection.

REFLECTION AT AN INTERFACE: LINEAR POLARIZATION   293

where the negative sign for E_η is necessary to give wave travel in the ζ direction. The transformation to new coordinates is the same as for the electric field normal to the plane of incidence, with the roles of E_ξ and H_ξ interchanges. The resulting fields are

$$H_x = H_0 e^{-jk(z\cos\theta + y\sin\theta)}$$

$$E_y = -\cos\theta\, ZH_0 e^{-jk(z\cos\theta + y\sin\theta)} \qquad (6.9)$$

$$E_z = \sin\theta\, ZH_0 e^{-jk(z\cos\theta + y\sin\theta)}$$

Figure 6.2 is a superposition of all the rotated coordinate systems needed to study the reflection problem. The incident wave strikes the plane boundary at angle θ_i, a part is reflected at angle θ_r, and a part is transmitted at angle θ_t. Our examination is restricted to θ_i real so that (6.7) and (6.9) represent uniform plane incident waves.

Snell's Laws

Boundary conditions at the interface require the sum of two tangential fields to be equal to a third. It is obvious from the form of the waves [Eqs. (6.7) and (6.9) may represent incident, reflected, or transmitted waves with a proper choice of k and Z] that the boundary conditions can be met only if the phase variation with y in both media is the same for all field terms. These phase terms are (using the appropriate angles from Fig. 6.2 in the equations for the fields)

Incident wave	$k_1 y \sin\theta_i$
Reflected wave	$k_1 y \sin(\pi - \theta_r) = k_1 y \sin\theta_r$
Transmitted wave	$k_2 y \sin\theta_t$

It follows that

$$k_1 \sin\theta_i = k_1 \sin\theta_r = k_2 \sin\theta_t \qquad (6.10)$$

from which we obtain Snell's laws,

$$\theta_r = \theta_i \qquad (6.11a)$$

$$\frac{\sin\theta_t}{\sin\theta_i} = \frac{k_1}{k_2} \qquad (6.11b)$$

The equations to this point are valid for lossy media, for which k and Z are complex. In a commonly encountered situation the first medium is air and the second is the lossy earth. Snell's laws are valid, but the transmitted wave is nonuniform. For this case, we are normally interested in the reflected wave and restrict our attention to it.

Reflection and Transmission Coefficients: Perpendicular Polarization

We consider the waves to be composed of incident, reflected, and transmitted terms and identify them by superscripts. *Reflection* and *transmission coefficients*, or Fresnel coefficients, are defined as

$$\Gamma_\perp = \left.\frac{E_x^r}{E_x^i}\right|_{z=0} \qquad T_\perp = \left.\frac{E_x^t}{E_x^i}\right|_{z=0} \qquad (6.12)$$

The components of the incident wave are

$$E_x^i = E_0 e^{-jk_1(z\cos\theta_i + y\sin\theta_i)}$$

$$H_{y'}^i = \frac{E_0}{Z_1} e^{-jk_1(z\cos\theta_i + y\sin\theta_i)}$$

$$H_y^i = \cos\theta_i H_{y'}^i = \cos\theta_i \frac{E_0}{Z_1} e^{-jk_1(z\cos\theta_i + y\sin\theta_i)} \qquad (6.13)$$

$$H_z^i = -\sin\theta_i H_{y'}^i = -\sin\theta_i \frac{E_0}{Z_1} e^{-jk_1(z\cos\theta_i + y\sin\theta_i)}$$

For the reflected wave the appropriate angle to use in (6.7) is $\pi - \theta_r$. Then (6.7) and (6.12), with the first of Snell's laws, give the fields

$$E_x^r = \Gamma_\perp E_0 e^{-jk_1(-z\cos\theta_i + y\sin\theta_i)}$$

$$H_{y''}^r = \frac{E_x^r}{Z_1} = \Gamma_\perp \frac{E_0}{Z_1} e^{-jk_1(-z\cos\theta_i + y\sin\theta_i)}$$

$$H_y^r = -\cos\theta_i H_{y''}^r = -\cos\theta_i \Gamma_\perp \frac{E_0}{Z_1} e^{-jk_1(-z\cos\theta_i + y\sin\theta_i)} \qquad (6.14)$$

$$H_z^r = -\sin\theta_i H_{y''}^r = -\sin\theta_i \Gamma_\perp \frac{E_0}{Z_1} e^{-jk_1(-z\cos\theta_i + y\sin\theta_i)}$$

REFLECTION AT AN INTERFACE: LINEAR POLARIZATION 295

For the transmitted wave,

$$E_x^t = T_\perp E_0 e^{-jk_2(z\cos\theta_t + y\sin\theta_t)}$$

$$H_{y'''}^t = \frac{E_x^t}{Z_2} = T_\perp \frac{E_0}{Z_2} e^{-jk_2(z\cos\theta_t + y\sin\theta_t)}$$

$$H_y^t = \cos\theta_t H_{y'''}^t = \cos\theta_t T_\perp \frac{E_0}{Z_2} e^{-jk_2(z\cos\theta_t + y\sin\theta_t)} \quad (6.15)$$

$$H_z^t = -\sin\theta_t H_{y'''}^t = -\sin\theta_t T_\perp \frac{E_0}{Z_2} e^{-jk_2(z\cos\theta_t + y\sin\theta_t)}$$

The boundary condition on the tangential electric field components that must be met is

$$\left. E_x^i \right|_{z=0} + \left. E_x^r \right|_{z=0} = \left. E_x^t \right|_{z=0} \quad (6.16)$$

Using the appropriate field components from (6.13)–(6.15), and noting that the phase terms are equal because of Snell's laws, leads to

$$1 + \Gamma_\perp = T_\perp \quad (6.17)$$

Since the magnetic field components are also continuous across the boundary, we have

$$\left. H_y^i \right|_{z=0} + \left. H_y^r \right|_{z=0} = \left. H_y^t \right|_{z=0} \quad (6.18)$$

which becomes, using the field components and Snell's laws,

$$\frac{\cos\theta_i}{Z_1}(1 - \Gamma_\perp) = \frac{\cos\theta_t}{Z_2} T_\perp \quad (6.19)$$

If (6.17) is used, the reflection coefficient is

$$\Gamma_\perp = \frac{Z_2 \sec\theta_t - Z_1 \sec\theta_i}{Z_2 \sec\theta_t + Z_1 \sec\theta_i} \quad (6.20)$$

Reflection and Transmission Coefficients: Parallel Polarization

For this case the reflection and transmission coefficients are defined as

$$\Gamma_\| = \left. \frac{H_x^r}{H_x^i} \right|_{z=0} \qquad T_\| = \left. \frac{H_x^t}{H_x^i} \right|_{z=0} \quad (6.21)$$

Some authors define the reflection coefficient for parallel polarization as E_y^r/E_y^i rather than as above. Such a definition does not utilize the symmetry of the Maxwell equations, and it chooses one of two electric field components in preference to using the only magnetic field component. The choice made here agrees with Stratton [1].

From (6.9) and (6.21) the wave components are

$$H_x^i = H_0 e^{-jk_1(z\cos\theta_i + y\sin\theta_i)}$$
$$E_{y'}^i = -Z_1 H_x^i = -Z_1 H_0 e^{-jk_1(z\cos\theta_i + y\sin\theta_i)}$$
$$E_y^i = \cos\theta_i E_{y'}^i = -\cos\theta_i Z_1 H_0 e^{-jk_1(z\cos\theta_i + y\sin\theta_i)} \quad (6.22)$$
$$E_z^i = -\sin\theta_i E_{y'}^i = \sin\theta_i Z_1 H_0 e^{-jk_1(z\cos\theta_i + y\sin\theta_i)}$$

$$H_x^r = \Gamma_\| H_0 e^{-jk_1(-z\cos\theta_i + y\sin\theta_i)}$$
$$E_{y''}^r = -Z_1 H_x^r = -Z_1 \Gamma_\| H_0 e^{-jk_1(-z\cos\theta_i + y\sin\theta_i)}$$
$$E_y^r = -\cos\theta_i E_{y''}^r = \cos\theta_i Z_1 \Gamma_\| H_0 e^{-jk_1(-z\cos\theta_i + y\sin\theta_i)} \quad (6.23)$$
$$E_z^r = -\sin\theta_i E_{y''}^r = \sin\theta_i Z_1 \Gamma_\| H_0 e^{-jk_1(-z\cos\theta_i + y\sin\theta_i)}$$

$$H_x^t = T_\| H_0 e^{-jk_2(z\cos\theta_t + y\sin\theta_t)}$$
$$E_{y'''}^t = -Z_2 H_x^t = -Z_2 T_\| H_0 e^{-jk_2(z\cos\theta_t + y\sin\theta_t)}$$
$$E_y^t = \cos\theta_t E_{y'''}^t = -\cos\theta_t Z_2 T_\| H_0 e^{-jk_2(z\cos\theta_t + y\sin\theta_t)} \quad (6.24)$$
$$E_z^t = -\sin\theta_t E_{y'''}^t = \sin\theta_t Z_2 T_\| H_0 e^{-jk_2(z\cos\theta_t + y\sin\theta_t)}$$

Using these fields, the boundary condition

$$H_x^i\big|_{z=0} + H_x^r\big|_{z=0} = H_x^t\big|_{z=0} \quad (6.25)$$

and Snell's laws give

$$1 + \Gamma_\| = T_\| \quad (6.26)$$

The boundary condition

$$E_y^i\big|_{z=0} + E_y^r\big|_{z=0} = E_y^t\big|_{z=0} \quad (6.27)$$

and Snell's laws lead to

$$Z_1 \cos\theta_i (\Gamma_\| - 1) = -Z_2 T_\| \cos\theta_t \quad (6.28)$$

Use of (6.26) then gives for the reflection coefficient

$$\Gamma_\| = -\frac{Z_2 \cos\theta_t - Z_1 \cos\theta_i}{Z_2 \cos\theta_t + Z_1 \cos\theta_i} \qquad (6.29)$$

Alternate Forms for the Fresnel Coefficients

It is obvious that Snell's laws can be used to remove θ_t from the reflection and transmission coefficients. At the same time, the grazing angle

$$\alpha = \tfrac{1}{2}\pi - \theta_i \qquad (6.30)$$

is introduced and the reflection problem is specialized to the interface between air and a lossy medium such as earth. The parameters μ_1 and ϵ_1 are replaced by μ_0 and ϵ_0, and for the lossy medium $\mu_2 = \mu_0$, $\epsilon_2 = \epsilon$, and $\sigma_2 = \sigma$.

Since ϵ always occurs in the combination $\sigma + j\omega\epsilon$, we define a complex dielectric constant

$$\sigma + j\omega\epsilon = j\omega\epsilon\left(1 - j\frac{\sigma}{\omega\epsilon}\right) = j\omega\hat{\epsilon} \qquad (6.31)$$

where

$$\hat{\epsilon} = \epsilon\left(1 - j\frac{\sigma}{\omega\epsilon}\right) \qquad (6.32)$$

The propagation constant and characteristic impedance of the lossy medium may also be written as

$$k_2 = \omega\sqrt{\mu_0\epsilon}\sqrt{1 - j\sigma/\omega\epsilon} = \omega\sqrt{\mu_0\hat{\epsilon}}$$

$$Z_2 = \sqrt{\frac{j\omega\mu_0}{\sigma + j\omega\epsilon}} = \sqrt{\frac{\mu_0}{\hat{\epsilon}}} \qquad (6.33)$$

Substitution of these values into the reflection coefficients, and use of the grazing angle, causes them to become

$$\Gamma_\perp = \frac{\cos\theta_i - (\hat{\epsilon}/\epsilon_0 - \sin^2\theta_i)^{1/2}}{\cos\theta_i + (\hat{\epsilon}/\epsilon_0 - \sin^2\theta_i)^{1/2}} = \frac{\sin\alpha - (\hat{\epsilon}/\epsilon_0 - \cos^2\alpha)^{1/2}}{\sin\alpha + (\hat{\epsilon}/\epsilon_0 - \cos^2\alpha)^{1/2}} \qquad (6.34)$$

$$\Gamma_\| = \frac{(\hat{\epsilon}/\epsilon_0)\cos\theta_i - (\hat{\epsilon}/\epsilon_0 - \sin^2\theta_i)^{1/2}}{(\hat{\epsilon}/\epsilon_0)\cos\theta_i + (\hat{\epsilon}/\epsilon_0 - \sin^2\theta_i)^{1/2}}$$

$$= \frac{(\hat{\epsilon}/\epsilon_0)\sin\alpha - (\hat{\epsilon}/\epsilon_0 - \cos^2\alpha)^{1/2}}{(\hat{\epsilon}/\epsilon_0)\sin\alpha + (\hat{\epsilon}/\epsilon_0 - \cos^2\alpha)^{1/2}} \qquad (6.35)$$

Power: Perpendicular Polarization

This discussion is restricted to lossless media, although the extension to lossy media is simple. A distinction is made between Poynting vectors and the proportion of incident power that is reflected and transmitted.

The Poynting vectors, found from the fields, are

$$\mathcal{P}_i = \frac{1}{2}\frac{E_x^i E_x^{i*}}{Z_1} = \frac{1}{2}\frac{|E_0|^2}{Z_1}$$

$$\mathcal{P}_r = \frac{1}{2}\frac{E_x^r E_x^{r*}}{Z_1} = \frac{1}{2}\frac{|E_0|^2}{Z_1}|\Gamma_\perp|^2 \quad (6.36)$$

$$\mathcal{P}_t = \frac{1}{2}\frac{E_x^t E_x^{t*}}{Z_2} = \frac{1}{2}\frac{|E_0|^2}{Z_2}|T_\perp|^2$$

The ratio of reflected to incident power is the same as the ratio of their Poynting vectors, thus

$$\frac{W_r}{W_i} = \frac{\mathcal{P}_r}{\mathcal{P}_i} = |\Gamma_\perp|^2 \quad (6.37)$$

The ratio of transmitted to incident power is not equal to the Poynting vector ratio, however, as may be seen in Figure 6.3. The power incident on the interface within the confines of some arbitrary channel is partly reflected in a channel of equal cross section, and partly transmitted in a channel of different cross section. Equation (6.37) follows immediately, and the ratio of

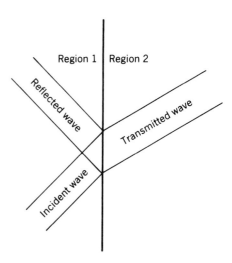

FIGURE 6.3 Channel widths for reflection and transmission.

REFLECTION AT AN INTERFACE: LINEAR POLARIZATION 299

transmitted to incident power is

$$\frac{W_t}{W_i} = 1 - \frac{W_r}{W_i} = 1 - \frac{\mathcal{P}_r}{\mathcal{P}_i} = 1 - |\Gamma_\perp|^2 \qquad (6.38)$$

Power: Parallel Polarization

The Poynting vectors for this case are

$$\begin{aligned}\mathcal{P}_i &= \tfrac{1}{2} Z_1 H_x^i H_x^{i*} = \tfrac{1}{2} Z_1 |H_0|^2 \\ \mathcal{P}_r &= \tfrac{1}{2} Z_1 H_x^r H_x^{r*} = \tfrac{1}{2} Z_1 |H_0|^2 |\Gamma_\parallel|^2 \\ \mathcal{P}_t &= \tfrac{1}{2} Z_2 H_x^t H_x^{t*} = \tfrac{1}{2} Z_2 |H_0|^2 |T_\parallel|^2 \end{aligned} \qquad (6.39)$$

The proportions of reflected and transmitted power are

$$\frac{W_r}{W_i} = \frac{\mathcal{P}_r}{\mathcal{P}_i} = |\Gamma_\parallel|^2$$

$$\frac{W_t}{W_i} = 1 - \frac{\mathcal{P}_r}{\mathcal{P}_i} = 1 - |\Gamma_\parallel|^2 \qquad (6.40)$$

Total Transmission

For parallel polarization an incidence angle, called the Brewster angle, can be found for which all of the incident power is transmitted across the interface into the second medium. The parallel reflection coefficient is zero if

$$Z_2 \cos \theta_t = Z_1 \cos \theta_i$$

For lossless dielectrics, for which

$$\frac{Z_2}{Z_1} = \sqrt{\frac{\epsilon_1}{\epsilon_2}} \qquad (6.41a)$$

$$\frac{\sin \theta_t}{\sin \theta_i} = \frac{k_1}{k_2} = \sqrt{\frac{\epsilon_1}{\epsilon_2}} \qquad (6.41b)$$

the reflection coefficient is zero if

$$\tan \theta_i = \sqrt{\frac{\epsilon_2}{\epsilon_1}} \qquad (6.42)$$

At this angle of incidence, all of the wave power is transmitted and none reflected. There is no comparable solution for perpendicular polarization, as

300 POLARIZATION CHANGES BY REFLECTION AND TRANSMISSION

may be seen by setting $\Gamma_\perp = 0$ in (6.20). This phenomenon allows the production of a linearly polarized wave by reflection of a wave with a general polarization.

Total Reflection

For lossless media with equal permeabilities, Snell's law for transmission is

$$\frac{\sin \theta_t}{\sin \theta_i} = \sqrt{\frac{\epsilon_1}{\epsilon_2}} \tag{6.43}$$

If $\sqrt{\epsilon_1/\epsilon_2} > 1$, then $\sin \theta_t > 1$ for a range of incidence angles θ_i. Then the exponential term for the transmitted fields, with either perpendicular or parallel polarization, from (6.15) or (6.24), becomes

$$e^{-jk_2(z\sqrt{1-\sin^2\theta_t}\, + y\sin\theta_t)} = e^{-k_2 z\sqrt{\sin^2\theta_t - 1}}\, e^{-jk_2 y \sin\theta_t} \tag{6.44}$$

which no longer represents a uniform plane wave in region 2. Examination of the fields shows that no power propagates in the z direction in region 2, and therefore no wave propagates across the interface. It follows that the magnitude of the reflection coefficients is unity, $|\Gamma_\perp| = |\Gamma_\parallel| = 1$ and all incident power is reflected. The angle

$$\theta_i = \sin^{-1}\sqrt{\frac{\epsilon_2}{\epsilon_1}} \tag{6.45}$$

which gives $\sin \theta_t = 1$, is called the *critical angle*. All greater angles of incidence lead to complete reflection at a boundary.

Note from (6.44) that in region 2 a nonuniform wave is set up which appears to propagate in the y direction and falls off in amplitude with z. Obviously this field could have been set up only by waves propagating across the interface, but this is not predicted by the steady-state solution.

6.2. ELLIPTICAL WAVES

A linearly polarized wave that is neither perpendicular to nor parallel to the plane of incidence can be broken into perpendicular and parallel components and each component multiplied by the appropriate reflection and transmission coefficients to obtain the complete reflected and transmitted fields. This procedure can also be applied to a generally polarized incident wave if the phase difference between perpendicular and parallel components is taken into account.

ELLIPTICAL WAVES

An incident wave of general polarization may be written, using the primed coordinate system of Figure 6.2, as

$$\mathbf{E}^i = \left(\mathbf{u}_{x'}|E_{x'}| + \mathbf{u}_{y'}|E_{y'}|e^{j\phi}\right)e^{-jk_1z'} \tag{6.46}$$

The magnetic field corresponding to the x' electric field component is

$$H_{y'} = \frac{|E_{x'}|}{Z_1}e^{-jk_1z'} \tag{6.47}$$

and that corresponding to the y' electric field component is

$$H_{x'} = -\frac{|E_{y'}|}{Z_1}e^{j\phi}e^{-jk_1z'} \tag{6.48}$$

The reflected and transmitted fields arising from the incident $E_{x'}$ are

$$E^r_{x''} = \Gamma_\perp |E_{x'}|e^{-jk_1z''} \tag{6.49a}$$

$$E^t_{x'''} = T_\perp |E_{x'}|e^{-jk_2z'''} \tag{6.49b}$$

Those arising from the incident $H_{x'}$ are

$$H^r_{x''} = -\Gamma_\| \frac{|E_{y'}|}{Z_1}e^{j\phi}e^{-jk_1z''} \tag{6.50a}$$

$$H^t_{x'''} = -T_\| \frac{|E_{y'}|}{Z_1}e^{j\phi}e^{-jk_2z'''} \tag{6.50b}$$

with associated electric fields

$$E^r_{y''} = -Z_1 H^r_{x''} = \Gamma_\||E_{y'}|e^{j\phi}e^{-jk_1z''} \tag{6.51a}$$

$$E^t_{y'''} = -Z_2 H^t_{x'''} = \frac{Z_2}{Z_1}T_\||E_{y'}|e^{j\phi}e^{-jk_2z'''} \tag{6.51b}$$

The total fields, incident, reflected, and transmitted, are then

$$\mathbf{E}^i = \left(\mathbf{u}_{x'}|E_{x'}| + \mathbf{u}_{y'}|E_{y'}|e^{j\phi}\right)e^{-jk_1z'} \tag{6.52a}$$

$$\mathbf{E}^r = \left(\mathbf{u}_{x''}\Gamma_\perp|E_{x'}| + \mathbf{u}_{y''}\Gamma_\||E_{y'}|e^{j\phi}\right)e^{-jk_1z''} \tag{6.52b}$$

$$\mathbf{E}^t = \left(\mathbf{u}_{x'''}T_\perp|E_{x'}| + \mathbf{u}_{y'''}\frac{Z_2}{Z_1}T_\||E_{y'}|e^{j\phi}\right)e^{-jk_2z'''} \tag{6.52c}$$

Polarization ratios are readily obtained from the fields. They are

$$P^i = \frac{|E_{y'}|}{|E_{x'}|} e^{j\phi} \tag{6.53a}$$

$$P^r = \frac{\Gamma_\| |E_{y'}|}{\Gamma_\perp |E_{x'}|} e^{j\phi} = \frac{\Gamma_\|}{\Gamma_\perp} P^i \tag{6.53b}$$

$$P^t = \frac{Z_2 T_\| |E_{y'}|}{Z_1 T_\perp |E_{x'}|} e^{j\phi} = \frac{Z_2}{Z_1} \frac{T_\|}{T_\perp} P^i \tag{6.53c}$$

Special Cases

Some physical insights into the reflection of elliptical waves can be discovered by looking at special cases.

1. Let θ_i, the angle of incidence, be 0. Then θ_t is also 0, and the equations for reflection and transmission coefficients simplify to

$$\Gamma_\perp = \frac{Z_2 - Z_1}{Z_2 + Z_1} \qquad \Gamma_\| = -\frac{Z_2 - Z_1}{Z_2 + Z_1} = -\Gamma_\perp$$

$$T_\perp = 1 + \Gamma_\perp = \frac{2Z_2}{Z_1 + Z_2} \qquad T_\| = 1 + \Gamma_\| = \frac{2Z_1}{Z_1 + Z_2}$$

Then, from (6.53)

$$P^r = -P^i \qquad P^t = P^i$$

The transmitted wave has the same polarization as the incident wave, but in general the reflected wave is neither matched to the incident wave nor cross polarized.

At this point coordinate systems must be considered. In Chapter 4, when the equations for polarization match of two antennas were developed, coordinate systems with coincident vertical axes were used, as shown in Figure 6.4a. It was convenient to use coordinate systems with coincident horizontal axes to study reflections at an interface, as in

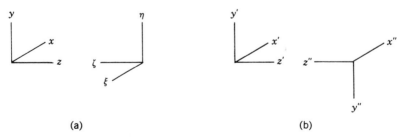

FIGURE 6.4 Coordinate systems for (a) transmission between antennas and (b) reflection at normal incidence.

Figure 6.4*b*. Examination of Figure 6.4 shows that since

$$\frac{E_\eta}{E_\xi} = \frac{-E_{y''}}{-E_{x''}} = \frac{E_{y''}}{E_{x''}}$$

the polarization parameter P is the same in the $\xi\eta\zeta$ system and the $x''y''z''$ system. Then the equations using P developed in Chapter 4 remain valid.

2. Let $\theta_i = 0$ and $P^i = \pm j$, where the positive sign corresponds to left circular and the negative to right circular polarization. Then

$$P^r = \mp j \qquad P^t = \pm j$$

The reflected wave is circularly polarized, but in the opposite sense to the incident wave. The incident and reflected waves are related by

$$P^r = 1/P^i$$

which is the condition for cross polarization. Thus, if a circularly polarized wave is transmitted normally toward a plane interface, the reflected wave cannot be received by the transmitting antenna.

3. Let the incident wave be linearly polarized. Then $\phi = 0$ and

$$P^i = \frac{|E_{y'}|}{|E_{x'}|} \qquad P^r = \frac{\Gamma_\parallel}{\Gamma_\perp}\frac{|E_{y'}|}{|E_{x'}|} \qquad P^t = \frac{Z_2}{Z_1}\frac{T_\parallel}{T_\perp}\frac{|E_{y'}|}{|E_{x'}|}$$

The reflected and transmitted waves are elliptically polarized unless both media are lossless or the wave incidence is normal.

4. Let the incident wave be linearly polarized and "vertical." Note that vertical here means in the plane of incidence. Then

$$P^i \to \infty \qquad P^r \to \infty \qquad P^t \to \infty$$

so the reflected and transmitted waves are also linear vertical.

5. Let the incident wave be linearly polarized and horizontal. Then

$$P^i = P^r = P^t = 0$$

and all waves are linear horizontal.

6. At the Brewster angle, with $\Gamma_\parallel = 0$,

$$P^r = 0 \qquad P^t = \frac{Z_2}{Z_1}\frac{1}{T_\perp}P^i$$

304 POLARIZATION CHANGES BY REFLECTION AND TRANSMISSION

The reflected wave is linear horizontal, no matter what the polarization of the incident wave (except for a linear vertical incident wave, which would not be reflected at all). The transmitted wave is in general different from the incident wave in polarization. This characteristic can be used to produce a linearly polarized wave. Its use is rare at the lower frequencies, but more common for light.

Reflections from a Conductor

Let a wave in air be reflected from a plane surface that is a good conductor. Equation (6.33) shows that Z_2 is 0 and the reflection coefficients become

$$\Gamma_\perp = -1 \qquad \Gamma_\parallel = +1 \tag{6.54}$$

independent of the angle of incidence. The polarization ratios of incident and reflected waves are then related by

$$P^r = -P^i \tag{6.55}$$

Note again that this is independent of the angle of incidence.

The Flat Plate Consider an infinite (so that edge effects are unimportant) flat plate (Fig. 6.5). Let the incident wave be linearly polarized perpendicular to the plane of incidence, so that $E_y^i = 0$ and $P^i = 0$. (The primes on the

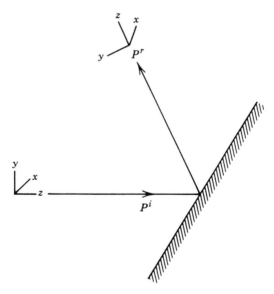

FIGURE 6.5 Polarization changes by reflection from a flat conducting plane.

coordinate subscripts that were used to distinguish incident, reflected, and transmitted waves will be omitted when confusion is not likely to arise.) This is a linear horizontal wave according to the previous definition and usage. From (6.55)

$$P^r = 0$$

which is also a linear horizontal wave.

For a linear vertical incident wave (polarized in the plane of incidence), with $E_x^i = 0$,

$$P^i \to \infty \qquad P^r \to -\infty$$

and the reflected wave is also linear vertical.

Let the incident wave be right circular so that

$$P^i = -j$$

Then (6.55) shows that

$$P^r = +j$$

which represents a left-circular wave. The reverse is also true; a left-circular wave will be reflected by a plane conductor as a right-circular wave, regardless of the angle of incidence. This leads to the well-known result that a monostatic radar transmitting a circular wave and receiving a wave of the same sense will be blind to a flat plate (at normal incidence of course for a monostatic radar). It is clear that this is true also for a bistatic radar.

Dihedral Corner Reflector A dihedral corner reflector that can be used for cross-section enhancement of a radar target is shown in Figure 6.6. The plane conducting surfaces form a right angle. It is apparent that a ray striking one

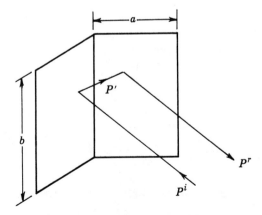

FIGURE 6.6 Dihedral corner reflector.

of the surfaces from a direction perpendicular to the line of intersection of the two planes will be reflected in the direction from which it came. The scattering matrix for this reflector is developed in another section. We consider now only the polarization ratios.

Application of (6.55) twice for a dihedral corner oriented so that its fold line coincides with either the x or y axes gives

$$P' = -P^i$$
$$P^r = -P' = P^i \qquad (6.56)$$

and thus the reflected wave has the same polarization as the incident wave, provided that the plates are large and edge effects are neglected.

From the developments of Section 4.8 on polarization-matched antennas and this section on reflections it is clear that for the reflected wave to be matched to the same antenna used to transmit the incident wave, it is necessary that

$$P^r = -P^{i*} \qquad (6.57)$$

Since this condition is not met by the dihedral corner in general, it follows that a monostatic radar may be blind to a dihedral corner reflector for some polarizations of the radar. Obviously, linear vertical and linear horizontal (with respect to the line of intersection of the plates) represent the polarization-matched cases.

If the incident wave is right circular,

$$P^i = -j$$

then the reflected wave is also right circular, since

$$P^r = P^i = -P^{i*} = -j$$

which represents the polarization-matched case. These cases should not mislead one into thinking that any wave reflected from a dihedral corner is polarization matched to the transmitter. Consider an incident wave that is linear and tilted at 45°, so that $E_y^i = E_x^i$ and

$$P^i = 1$$

Then

$$P^r = 1$$

which represents a linear wave tilted at 45° in a reversed coordinate system appropriate to the reflected wave, according to Figure 6.7. The reflected wave is cross polarized and the radar is blind to the dihedral corner.

ELLIPTICAL WAVES   307

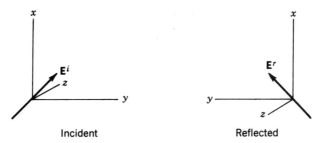

FIGURE 6.7 Incident wave with 45° tilt and reflected wave for a dihedral corner reflector.

Trihedral Corner Reflector The dihedral corner reflector has a serious defect as a cross-section enhancement device. The incident ray must lie in a plane perpendicular to the line of intersection of the planes that form the corner if the reflected ray is to be directed back to the radar. This deficiency can be overcome by using a trihedral corner reflector. Figure 6.8 shows a triangular trihedral corner reflector, although other shapes are possible.

In general, rays that strike an interior surface of the trihedral corner will undergo three reflections, as indicated by Figure 6.8, and will be returned parallel to the incident ray. There are exceptions to this rule, however. If the incident ray is at a sufficiently large angle from the axis of symmetry of the corner reflector, it will undergo only two reflections. The ray may, for example, strike plane AOB and be reflected to plane AOC. If plane BOC is

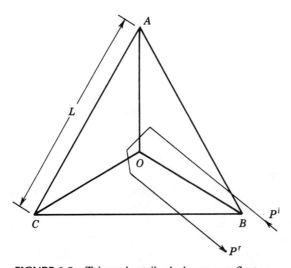

FIGURE 6.8 Triangular trihedral corner reflector.

not sufficiently extended, the ray from AOC will not strike it and thus will not be returned parallel to the incident ray.

It is also obvious that if the incident ray is parallel to one of the reflecting planes forming the trihedral corner it will be doubly reflected back to the source with polarization

$$P^r = P^i$$

If (6.55) is applied three times for the triply reflecting case, it is clear for the trihedral corner reflector that the polarizations of reflected and incident waves are related by

$$P^r = -P^i \tag{6.58}$$

From this it is seen that linear vertical waves are reflected as linear vertical and linear horizontal as linear horizontal. In fact, any linearly polarized wave is reflected so that it is polarization matched to the transmitting antenna. To see this, consider the polarization match factor

$$\rho = \frac{(1 - P_1 P_2)(1 - P_1^* P_2^*)}{(1 + P_1 P_1^*)(1 + P_2 P_2^*)} \tag{4.57}$$

The incident wave is 1 and the reflected wave is 2. If the incident wave is linearly polarized, the polarization ratio

$$P^i = \frac{|E_y|}{|E_x|} e^{j\phi} = \frac{|E_y|}{|E_x|}$$

is real. For the trihedral reflector

$$P^r = -P^i = -\frac{|E_y|}{|E_x|}$$

Substitution into the match factor equation produces $\rho = 1$.

This result is not unexpected because, although "vertical" polarization was used as an example, this reflector has no natural vertical axis for an incident ray directed along the symmetry axis of the trihedral corner. Note, however, that if the wave is reflected from only two planes of the trihedral corner, it may be significantly cross polarized.

Finally, note that if the incident wave is circularly polarized, the reflected wave is also circularly polarized with the opposite sense, and thus the trihedral corner reflector is invisible to a circularly polarized radar using the same antenna for transmitting and receiving.

The developments for the dihedral and trihedral corner reflectors can be extended to obtain the polarization of a wave reflected by n plane surfaces (large compared to a wavelength). It is

$$P^r = (-1)^n P^i \qquad (6.59)$$

6.3. REFLECTION AND TRANSMISSION MATRICES

In reflection problems we sometimes need to know field magnitudes of the reflected and transmitted waves in addition to their polarizations. In the field equations, the distance exponentials can be dropped since they do not affect either the power in a wave or its polarization, and they can be written as

$$\mathbf{E}^i = \mathbf{u}_{x'}E^i_{x'} + \mathbf{u}_{y'}E^i_{y'} = \mathbf{u}_{x'}|E_{x'}| + \mathbf{u}_{y'}|E_{y'}|e^{j\phi} \qquad (6.60a)$$

$$\mathbf{E}^r = \mathbf{u}_{x''}E^r_{x''} + \mathbf{u}_{y''}E^r_{y''} = \mathbf{u}_{x''}\Gamma_\perp|E_{x'}| + \mathbf{u}_{y''}\Gamma_\parallel|E_{y'}|e^{j\phi} \qquad (6.60b)$$

$$\mathbf{E}^t = \mathbf{u}_{x'''}E^t_{x'''} + \mathbf{u}_{y'''}E^t_{y'''} = \mathbf{u}_{x'''}T_\perp|E_{x'}| + \mathbf{u}_{y'''}\frac{Z_2}{Z_1}T_\parallel|E_{y'}|e^{j\phi} \qquad (6.60c)$$

It is clear from this equation that relationships between the field components can be written as

$$\begin{bmatrix} E^r_{x''} \\ E^r_{y''} \end{bmatrix} = \begin{bmatrix} \Gamma_\perp & 0 \\ 0 & \Gamma_\parallel \end{bmatrix} \begin{bmatrix} E^i_{x'} \\ E^i_{y'} \end{bmatrix} \qquad (6.61)$$

and

$$\begin{bmatrix} E^t_{x'''} \\ E^t_{y'''} \end{bmatrix} = \begin{bmatrix} T_\perp & 0 \\ 0 & \frac{Z_2}{Z_1}T_\parallel \end{bmatrix} \begin{bmatrix} E^i_{x'} \\ E^i_{y'} \end{bmatrix} \qquad (6.62)$$

The coefficient matrices may be called the reflection and transmission matrices. They are not the scattering matrices commonly encountered in radar (to be discussed in the next section) because they are concerned with reflections from an infinite plane.

The fields may also be written in terms of left- and right-circular components using the relationships developed in Chapter 3. They are

$$\mathbf{E}^i = E^i_{x'}\mathbf{u}_{x'} + E^i_{y'}\mathbf{u}_{y'} = E^i_L\mathbf{u}^i_L + E^i_R\mathbf{u}^i_R \qquad (6.63a)$$

$$\mathbf{E}^r = \Gamma_\perp E^i_{x'}\mathbf{u}_{x''} + \Gamma_\parallel E^i_{y'}\mathbf{u}_{y''} = E^r_L\mathbf{u}^r_L + E^r_R\mathbf{u}^r_R \qquad (6.63b)$$

$$\mathbf{E}^t = T_\perp E^i_{x'}\mathbf{u}_{x'''} + \frac{Z_2}{Z_1}T_\parallel E^i_{y'}\mathbf{u}_{y'''} = E^t_L\mathbf{u}^t_L + E^t_R\mathbf{u}^t_R \qquad (6.63c)$$

In these equations the vectors $\mathbf{u}_L$ and $\mathbf{u}_R$ are identified as to their corresponding coordinate system by their superscripts rather than by the primes used with the rectangular component unit vectors. For example $\mathbf{u}_L^i \neq \mathbf{u}_L^r$.

The circular wave components are readily found from these equations and the relationships developed in Chapter 3 between the circular and rectangular unit vectors. They are

$$E_L^i = \frac{1}{\sqrt{2}}\left(E_{x'}^i - jE_{y'}^i\right) \tag{6.64a}$$

$$E_R^i = \frac{1}{\sqrt{2}}\left(E_{x'}^i + jE_{y'}^i\right) \tag{6.64b}$$

$$E_L^r = \frac{1}{\sqrt{2}}\left(\Gamma_\perp E_{x'}^i - j\Gamma_\| E_{y'}^i\right) \tag{6.65a}$$

$$E_R^r = \frac{1}{\sqrt{2}}\left(\Gamma_\perp E_{x'}^i + j\Gamma_\| E_{y'}^i\right) \tag{6.65b}$$

$$E_L^t = \frac{1}{\sqrt{2}}\left(T_\perp E_{x'}^i - j\frac{Z_2}{Z_1}T_\| E_{y'}^i\right) \tag{6.66a}$$

$$E_R^t = \frac{1}{\sqrt{2}}\left(T_\perp E_{x'}^i + j\frac{Z_2}{Z_1}T_\| E_{y'}^i\right) \tag{6.66b}$$

The equations for reflected and transmitted fields may be put into matrix form without difficulty. The result is

$$\begin{bmatrix} E_L^r \\ E_R^r \end{bmatrix} = \begin{bmatrix} \Gamma_{LL} & \Gamma_{LR} \\ \Gamma_{RL} & \Gamma_{RR} \end{bmatrix}\begin{bmatrix} E_L^i \\ E_R^i \end{bmatrix} = \frac{1}{2}\begin{bmatrix} \Gamma_\perp + \Gamma_\| & \Gamma_\perp - \Gamma_\| \\ \Gamma_\perp - \Gamma_\| & \Gamma_\perp + \Gamma_\| \end{bmatrix}\begin{bmatrix} E_L^i \\ E_R^i \end{bmatrix} \tag{6.67}$$

$$\begin{bmatrix} E_L^t \\ E_R^t \end{bmatrix} = \begin{bmatrix} T_{LL} & T_{LR} \\ T_{RL} & T_{RR} \end{bmatrix}\begin{bmatrix} E_L^i \\ E_R^i \end{bmatrix} = \frac{1}{2}\begin{bmatrix} T_\perp + \frac{Z_2}{Z_1}T_\| & T_\perp - \frac{Z_2}{Z_1}T_\| \\ T_\perp - \frac{Z_2}{Z_1}T_\| & T_\perp + \frac{Z_2}{Z_1}T_\| \end{bmatrix}\begin{bmatrix} E_L^i \\ E_R^i \end{bmatrix} \tag{6.68}$$

Note that since the off-diagonal terms of (6.61) and (6.62) are zero, an incident linear vertical wave cannot give rise to a reflected or transmitted horizontal wave. Since the off-diagonal terms of (6.67) and (6.68) are not zero in general, an incident circular wave of one sense can create a circular wave of the other sense, either reflected or transmitted.

It may be seen from (6.67) and (6.68) that an incident left-circular wave gives the same reflected and transmitted powers as an incident right-circular wave.

6.4. SCATTERING BY A TARGET: SINCLAIR AND JONES MATRICES

The scattered power from a radar target is dependent on the polarization of the wave transmitted toward it. Further, many targets reflect a wave that differs in polarization from the incident wave. For example, consider the incident wave to be linear vertical and the target to be a thin wire transverse to a line from radar to target and neither vertical nor horizontal in the transverse plane. It is clear that a current will be induced in the wire by the incident vertical field and that the current will reradiate a wave with a horizontal component. Then the relationship between the incident and scattered fields must be given by a matrix,

$$\begin{bmatrix} E_x^s \\ E_y^s \end{bmatrix} = \begin{bmatrix} A_{xx} & A_{xy} \\ A_{yx} & A_{yy} \end{bmatrix} \begin{bmatrix} E_x^i \\ E_y^i \end{bmatrix} \tag{6.69}$$

or by

$$\mathbf{E}^s = \mathbf{A}\mathbf{E}^i \tag{6.70}$$

where boldface symbols are used both for the vector fields, considered as column matrices, and for the 2 × 2 matrix multiplier.

Coordinate Systems

Figure 6.9 shows the coordinate systems used to describe scattering by a target and transmission in a forward direction, as by an optical instrument. A

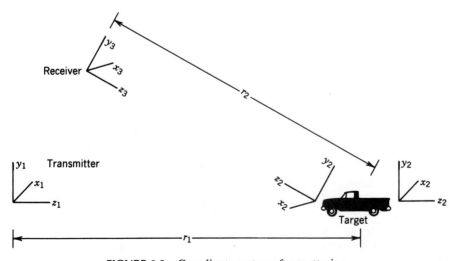

FIGURE 6.9 Coordinate systems for scattering.

wave incident on the target from the transmitter is appropriately given by its transverse fields E_{x1} and E_{y1} in the right-handed coordinate system $x_1 y_1 z_1$ of the transmitter, with z_1 pointing from transmitter to target. The coordinate systems $x_2 y_2 z_2$ and $x_3 y_3 z_3$ are both right-handed, but z_2 points away from the target and z_3 is directed toward it. There is a choice for the wave reflected from the target. It may be specified by its components E_{x2} and E_{y2} or by the reversed set E_{x3} and E_{y3}. Both conventions are used, leading to different matrix formulations.

The Sinclair Matrix

If the scattered wave components are given in $x_3 y_3 z_3$ coordinates, the field incident on the target and that scattered to a receiver at distance r are related by the *Sinclair matrix*,

$$\begin{bmatrix} E^s_{x3} \\ E^s_{y3} \end{bmatrix} = \frac{1}{\sqrt{4\pi} r} \begin{bmatrix} S_{x3x1} & S_{x3y1} \\ S_{y3x1} & S_{y3y1} \end{bmatrix} \begin{bmatrix} E^i_{x1} \\ E^i_{y1} \end{bmatrix} e^{-jkr} \tag{6.71}$$

Both the Sinclair matrix and the Jones matrix, to be discussed shortly, can be called by the more general term of *scattering matrix*. In this book, if the context is clear, "scattering matrix" may be used without specifying which is considered. If transmitter and receiver are separated, the matrix is the *bistatic* scattering matrix, and if they are colocated, it is called the *backscattering* matrix. The use of E_{x3} and E_{y3} is convenient for backscattering since then the $x_1 y_1 z_1$ and $x_3 y_3 z_3$ systems coincide and only one coordinate system is needed. If the coordinate systems being used are kept in mind, the numerical subscripts can be dropped and the scattering matrix for the scattered fields written as

$$\begin{bmatrix} E^s_x \\ E^s_y \end{bmatrix} = \frac{1}{\sqrt{4\pi} r} \begin{bmatrix} S_{xx} & S_{xy} \\ S_{yx} & S_{yy} \end{bmatrix} \begin{bmatrix} E^i_x \\ E^i_y \end{bmatrix} e^{-jkr} \tag{6.72}$$

It is appropriate to use the Sinclair matrix with the scattered fields expressed in $x_3 y_3 z_3$ (or $x_1 y_1 z_1$) only for rectangular field components. Descriptors for the scattered wave such as the polarization ratio and the Stokes parameters are customarily given in a right-handed coordinate system with an axis pointing in the direction of wave propagation. If they are to be used, the $x_2 y_2 z_2$ coordinates are more convenient for the scattered wave.

In general, all the elements of the scattering matrix are complex and unrelated to each other (except through the physics of a particular scatterer). For backscattering, however, the matrix is symmetric. To see this, write the wave incident on the target at distance r from the transmitter as

$$\begin{bmatrix} E_x^i \\ E_y^i \end{bmatrix} = \frac{jZ_0 I}{2\lambda r} \begin{bmatrix} h_{tx} \\ h_{ty} \end{bmatrix} e^{-jkr} \qquad (6.73)$$

where subscript t identifies the effective length components of the transmitting antenna. Then the scattered fields at a receiver colocated with the transmitter are

$$\begin{bmatrix} E_x^s \\ E_y^s \end{bmatrix} = \frac{jZ_0 I}{\sqrt{4\pi}\,(2\lambda r^2)} \begin{bmatrix} S_{xx} & S_{xy} \\ S_{yx} & S_{yy} \end{bmatrix} \begin{bmatrix} h_{tx} \\ h_{ty} \end{bmatrix} e^{-j2kr} \qquad (6.74)$$

The open-circuit voltage induced in a receiving antenna of effective length $\mathbf{h}_r$ is

$$V = \mathbf{h}_r \cdot \mathbf{E}^s = \frac{jZ_0 I}{\sqrt{4\pi}\,(2\lambda r^2)} [h_{rx} \quad h_{ry}] \begin{bmatrix} S_{xx} & S_{xy} \\ S_{yx} & S_{yy} \end{bmatrix} \begin{bmatrix} h_{tx} \\ h_{ty} \end{bmatrix} e^{-j2kr} \qquad (6.75)$$

The use of the same coordinate system for scattered and incident fields causes $\mathbf{h}_r$ and $\mathbf{E}^s$ to be in the same coordinates and their scalar product to be the open-circuit voltage without concern about component direction.

If the roles of transmitting and receiving antennas are interchanged and the feed current is kept constant, the reciprocity theorem requires that the same open-circuit voltage be induced in the new receiving antenna as in the old. With the interchange, the voltage is

$$V = \frac{jZ_0 I}{\sqrt{4\pi}\,(2\lambda r^2)} [h_{tx} \quad h_{ty}] \begin{bmatrix} S_{xx} & S_{xy} \\ S_{yx} & S_{yy} \end{bmatrix} \begin{bmatrix} h_{rx} \\ h_{ry} \end{bmatrix} e^{-j2kr} \qquad (6.76)$$

The requirement then exists that

$$[h_{rx} \quad h_{ry}] \begin{bmatrix} S_{xx} & S_{xy} \\ S_{yx} & S_{yy} \end{bmatrix} \begin{bmatrix} h_{tx} \\ h_{ty} \end{bmatrix} = [h_{tx} \quad h_{ty}] \begin{bmatrix} S_{xx} & S_{xy} \\ S_{yx} & S_{yy} \end{bmatrix} \begin{bmatrix} h_{rx} \\ h_{ry} \end{bmatrix} \qquad (6.77)$$

Multiplication of these matrices shows that the equality can be satisfied only if

$$S_{xy} = S_{yx} \qquad (6.78)$$

Note that the reciprocity theorem is invalid in a medium with Faraday rotation. In general, in such a medium $S_{xy} \neq S_{yx}$.

FIGURE 6.10 Simple target with symmetry plane.

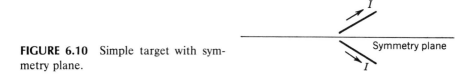

If the target is symmetric about a plane containing a ray from antenna to target in a monostatic radar system, the coordinate system can be chosen so that $S_{xy} = 0$. To see this, consider a simple target such as two diagonal wires, as shown in Figure 6.10. A ray from the transmitter to the target lies in the symmetry plane and is perpendicular to the wires. We choose the coordinate system so that the x axis lies in the symmetry plane and transmit a wave with only an x component, E_x^i in (6.72). At some instant the incident wave will set up currents, as shown in Figure 6.10. Obviously when these currents reradiate, the vertical components will cancel. Then in (6.72), E_y^s must be zero, and therefore $S_{xy} = 0$. The requirement that the ray from transmitter to target be perpendicular to the wires is unnecessary and was used only to make the physical process clearer. The wires of Figure 6.10 can have a component parallel to the transmitter–target ray, as long as symmetry about a plane is maintained. All targets that are axially symmetric are symmetric in the scattering matrix sense ($S_{xy} = 0$ for a properly chosen coordinate system) at all aspect angles [2]. Such targets include cylinders, ellipsoids, and cones. Other targets such as the two-wire target just considered and dihedral corner reflectors are symmetric at some aspect angles.

In the form used to this point, with all elements of the scattering matrix complex and $\exp(-jkr)$ in the field equations, the phase of the scattered wave may be related to the phase of the transmitted wave. The matrix in this form is called the *absolute scattering matrix* (SMA). More often than not, the absolute phase is not of interest, and the $\exp(-jkr)$ multiplier can be omitted. Further, one of the elements of the scattering matrix may be taken as real and used as a phase reference for the other elements. In this form the matrix is the *relative scattering matrix* (SMR). For backscattering, two elements are real if S_{xy} is used as phase reference, and the relative backscattering matrix is completely specified by three magnitudes and two phases. In the general usage of this section and the remaining text, the $\exp(-jkr)$ term is omitted and the relative scattering matrix is used. We then write

$$\mathbf{E}^s = \frac{1}{\sqrt{4\pi}\,r}\mathbf{S}\mathbf{E}^i \qquad (6.79)$$

The Jones Matrix

If the scattered field is described by its polarization ratio or Stokes parameters, its wave components are most conveniently expressed in a right-handed coordinate system with an axis pointed in the direction of wave propagation. This is the $x_2 y_2 z_2$ system of Figure 6.9. The relationship between the scattered field at distance r and the incident field at the target is then

$$\begin{bmatrix} E_{x2}^s \\ E_{y2}^s \end{bmatrix} = \frac{1}{\sqrt{4\pi}\, r} \begin{bmatrix} T_{x2x1} & T_{x2y1} \\ T_{y2x1} & T_{y2y1} \end{bmatrix} \begin{bmatrix} E_{x1}^i \\ E_{y1}^i \end{bmatrix} \tag{6.80}$$

where an $\exp(-jkr)$ term is omitted. The 2×2 matrix relating the field components is commonly called the *Jones matrix*, particularly in optics. Like the Sinclair matrix, it may be in absolute form or in relative form, with one element a real phase reference for the remaining elements. If the coordinate systems are kept in mind, the numerical subscripts may be omitted and the equation for the fields written as

$$\begin{bmatrix} E_x^s \\ E_y^s \end{bmatrix} = \frac{1}{\sqrt{4\pi}\, r} \begin{bmatrix} T_{xx} & T_{xy} \\ T_{yx} & T_{yy} \end{bmatrix} \begin{bmatrix} E_x^i \\ E_y^i \end{bmatrix} \tag{6.81}$$

or as

$$\mathbf{E}^s = \frac{1}{\sqrt{4\pi}\, r} \mathbf{T} \mathbf{E}^i \tag{6.82}$$

If transmission through some optical device in the forward direction (or forward scattering from a radar target) is considered, the $x_2 y_2 z_2$ and $x_1 y_1 z_1$ systems coincide. For an optical instrument it is also appropriate to omit the distance term and write

$$\mathbf{E}^s = \mathbf{T} \mathbf{E}^i \tag{6.83}$$

The elements of the Jones matrix may be related to those of the Sinclair matrix by noting that

$$E_{x3}^s = -E_{x2}^s \tag{6.84a}$$

$$E_{y3}^s = E_{y2}^s \tag{6.84b}$$

If the matrices of (6.71) and (6.80) are multiplied and substituted into this equation, the result is

$$T_{x2x1} E_{x1}^i + T_{x2y1} E_{y1}^i = -S_{x3x1} E_{x1}^i - S_{x3y1} E_{y1}^i \tag{6.85a}$$

$$T_{y2x1} E_{x1}^i + T_{y2y1} E_{y1}^i = S_{y3x1} E_{x1}^i + S_{y3y1} E_{y1}^i \tag{6.85b}$$

From these it follows that

$$T_{x2x1} = -S_{x3x1} \tag{6.86a}$$
$$T_{x2y1} = -S_{x3y1} \tag{6.86b}$$
$$T_{y2x1} = S_{y3x1} \tag{6.86c}$$
$$T_{y2y1} = S_{y3y1} \tag{6.86d}$$

If the coordinate systems to be used with $\mathbf{S}$ and $\mathbf{T}$ are kept in mind, the numerical subscripts may be omitted and the relationship written as

$$\begin{bmatrix} T_{xx} & T_{xy} \\ T_{yx} & T_{yy} \end{bmatrix} = \begin{bmatrix} -S_{xx} & -S_{xy} \\ S_{yx} & S_{yy} \end{bmatrix} \tag{6.87}$$

or as

$$\mathbf{T} = \begin{bmatrix} -1 & 0 \\ 0 & 1 \end{bmatrix} \mathbf{S} \tag{6.88}$$

This relation is valid for both absolute and relative Sinclair and Jones matrices and for both bistatic and backscattering. It may be noted that for backscattering

$$T_{xy} = -T_{yx} \tag{6.89}$$

6.5. SCATTERING WITH CIRCULAR WAVE COMPONENTS

It is often convenient to use incident and scattered waves in left- and right-circular component form with the relationship between incident and scattered waves given by

$$\begin{bmatrix} E_L^s \\ E_R^s \end{bmatrix} = \frac{1}{\sqrt{4\pi}\, r} \begin{bmatrix} S_{LL} & S_{LR} \\ S_{RL} & S_{RR} \end{bmatrix} \begin{bmatrix} E_L^i \\ E_R^i \end{bmatrix} \tag{6.90}$$

For this equation it is appropriate to write the incident wave in the $x_1 y_1 z_1$ coordinates of Figure 6.9 and the scattered fields in $x_2 y_2 z_2$ coordinates. The development of (6.64) then allows the incident wave to be written as

$$E_L^i = \frac{1}{\sqrt{2}} (E_{x1}^i - jE_{y1}^i)$$
$$E_R^i = \frac{1}{\sqrt{2}} (E_{x1}^i + jE_{y1}^i) \tag{6.91}$$

and the reflected wave as

$$E_L^s = \frac{1}{\sqrt{2}}(E_{x2}^s - jE_{y2}^s)$$
$$E_R^s = \frac{1}{\sqrt{2}}(E_{x2}^s + jE_{y2}^s)$$
(6.92)

If these equation sets are substituted into (6.90) and the matrices multiplied, the scattered fields in $x_2 y_2 z_2$ coordinates are obtained in terms of the incident wave in $x_1 y_1 z_1$ coordinates and the circular scattering parameters S_{LL}, and so forth. If the resulting fields are compared to (6.80) the circular scattering parameters in terms of the elements of the Jones matrix are

$$S_{LL} = \tfrac{1}{2}(T_{xx} + jT_{xy} - jT_{yx} + T_{yy})$$
$$S_{LR} = \tfrac{1}{2}(T_{xx} - jT_{xy} - jT_{yx} - T_{yy})$$
$$S_{RL} = \tfrac{1}{2}(T_{xx} + jT_{xy} + jT_{yx} - T_{yy})$$
$$S_{RR} = \tfrac{1}{2}(T_{xx} + jT_{xy} + jT_{yx} + T_{yy})$$
(6.93)

In these equations, the numerical part of the subscripts in the Jones matrix was omitted as unnecessary if the coordinate systems are kept in mind.

By using the relation between the Jones matrix and the Sinclair matrix the circular scattering parameters can be written in terms of the Sinclair matrix elements,

$$S_{LL} = \tfrac{1}{2}(-S_{xx} - jS_{xy} - jS_{yx} + S_{yy})$$
$$S_{LR} = \tfrac{1}{2}(-S_{xx} + jS_{xy} - jS_{yx} - S_{yy})$$
$$S_{RL} = \tfrac{1}{2}(-S_{xx} - jS_{xy} + jS_{yx} - S_{yy})$$
$$S_{RR} = \tfrac{1}{2}(-S_{xx} + jS_{xy} + jS_{yx} + S_{yy})$$
(6.94)

but it must be remembered that the Sinclair matrix parameters are defined using $x_3 y_3 z_3$ coordinates while the circular matrix parameters are defined in $x_2 y_2 z_2$.

For backscattering, the equations simplify to

$$S_{LL} = S_{RR}^* = \tfrac{1}{2}(-S_{xx} - j2S_{xy} + S_{yy})$$
$$S_{LR} = S_{RL} = \tfrac{1}{2}(-S_{xx} - S_{yy})$$
(6.95)

If the target has a plane of symmetry so that

$$S_{xy} = S_{yx} = 0$$

then

$$S_{LL} = S_{RR} = \tfrac{1}{2}(-S_{xx} + S_{yy})$$
$$S_{LR} = S_{RL} = \tfrac{1}{2}(-S_{xx} - S_{yy})$$

and the returns for left- and right-circular incident waves are equivalent.

For a sphere target (or more generally for a target with a plane of symmetry, which also appears to be unaltered by a rotation through 90°),

$$S_{xx} = S_{yy}$$
$$S_{LL} = S_{RR} = 0$$
$$S_{LR} = S_{RL} = -S_{xx}$$

The polarization sense is reversed for a circular wave incident on this target.

6.6. SINCLAIR MATRIX AND POLARIZATION RATIO

The polarization ratio of the reflected wave may be obtained in terms of the incident wave and the Sinclair (or Jones) matrix. We begin with

$$\begin{bmatrix} E_x^s \\ E_y^s \end{bmatrix} = \frac{1}{\sqrt{4\pi}\, r} \begin{bmatrix} S_{xx} & S_{xy} \\ S_{yx} & S_{yy} \end{bmatrix} \begin{bmatrix} E_x^i \\ E_y^i \end{bmatrix} \qquad (6.96)$$

where the incident wave is in $x_1 y_1 z_1$ coordinates and the scattered wave in $x_3 y_3 z_3$ (or $x_1 y_1 z_1$ for backscattering). The polarization ratio for the incident wave can be written directly, but $x_2 y_2 z_2$ coordinates are needed to define the polarization ratio for the scattered wave. The two polarization ratios are then

$$P^i = \frac{E_{y1}^i}{E_{x1}^i} = \frac{E_y^i}{E_x^i} \qquad P^s = \frac{E_{y2}^s}{E_{x2}^s} = \frac{E_y^s}{-E_x^s}$$

and if they are substituted into the scattering equation it becomes

$$\begin{bmatrix} 1 \\ -P^s \end{bmatrix} = \frac{1}{\sqrt{4\pi}\, r} \frac{E_x^i}{E_x^s} \begin{bmatrix} S_{xx} & S_{xy} \\ S_{yx} & S_{yy} \end{bmatrix} \begin{bmatrix} 1 \\ P^i \end{bmatrix} \qquad (6.97)$$

which is easily solved for P^s,

$$P^s = -\frac{S_{yy} P^i + S_{yx}}{S_{xy} P^i + S_{xx}} \qquad (6.98)$$

where we assume $E_x^s \neq 0$. The equation simplifies slightly for backscattering.

For a target with a plane of symmetry so that $S_{xy} = S_{yx} = 0$,

$$P^s = -\frac{S_{yy}}{S_{xx}} P^i$$

and further, if the target is unaltered by a 90° rotation,

$$P^s = -P^i$$

a result found earlier for the infinite conducting plane.

The inverse circular polarization ratio q^s can also be found from the elements of the circular scattering matrix. The scattered and incident fields are related by

$$\begin{bmatrix} E_L^s \\ E_R^s \end{bmatrix} = \frac{1}{\sqrt{4\pi} r} \begin{bmatrix} S_{LL} & S_{LR} \\ S_{RL} & S_{RR} \end{bmatrix} \begin{bmatrix} E_L^i \\ E_R^i \end{bmatrix} \quad (6.90)$$

Substitute

$$q = \frac{E_L}{E_R} \quad (6.99)$$

into the field equation, noting that the definition of inverse circular polarization ratio applies to both incident and reflected waves without any concern about coordinate systems; the circular components of incident and reflected waves were defined in a right-handed system with the z axis in the direction of wave travel, and such a system is proper for defining q. This gives

$$\begin{bmatrix} q^s \\ 1 \end{bmatrix} = \frac{1}{\sqrt{4\pi} r} \frac{E_R^i}{E_R^s} \begin{bmatrix} S_{LL} & S_{LR} \\ S_{RL} & S_{RR} \end{bmatrix} \begin{bmatrix} q^i \\ 1 \end{bmatrix} \quad (6.100)$$

from which q^s is found to be

$$q^s = \frac{S_{LL} q^i + S_{LR}}{S_{RL} q^i + S_{RR}} \quad (6.101)$$

6.7. CHANGE OF POLARIZATION BASE: THE SCATTERING MATRIX

In the preceding sections the Jones and Sinclair matrices of a target were developed for incident and scattered waves given in rectangular and circular component form. It is desirable to find the matrices if the waves are expressed in orthogonal elliptic components.*

---
*Note: In this section the distance variation of the scattered field, in both amplitude and phase, is omitted for convenience.

320 POLARIZATION CHANGES BY REFLECTION AND TRANSMISSION

The incident wave is given in $x_1 y_1 z_1$ coordinates of Figure 6.9. If the scattered field is expressed in $x_2 y_2 z_2$ coordinates it is appropriate to use the Jones matrix **T** and write

$$\mathbf{E}^s = \mathbf{T}\mathbf{E}^i \tag{6.102}$$

and if the scattered wave is in $x_3 y_3 z_3$ coordinates, the Sinclair matrix **S** is used, thus

$$\mathbf{E}^s = \mathbf{S}\mathbf{E}^i \tag{6.103}$$

In this book only the rectangular field components have been written using $x_3 y_3 z_3$ coordinates. All other polarization descriptors have used a right-handed coordinate system with an axis in the direction of wave propagation. To be consistent, when the scattered wave is considered in a polarization basis system other than rectangular, we shall use the $x_2 y_2 z_2$ system as a starting point and therefore work primarily with the Jones matrix **T**. Conversion to Sinclair matrix elements is straightforward.

Since there are two fields to consider, two basis transformations can be used. We can change the incident field $\mathbf{E}^i$ to its form $\mathbf{E}^i_a$ in the new basis system by the transformation

$$\mathbf{E}^i_a = \mathbf{R}_a \mathbf{E}^i \tag{6.104}$$

where $\mathbf{R}_a$ is a unitary matrix as discussed in Section 3.11. The scattered field can be transformed by

$$\mathbf{E}^s_b = \mathbf{R}_b \mathbf{E}^s \tag{6.105}$$

where $\mathbf{R}_b$ is also unitary. Combining these equations gives the relation between scattered and incident waves in the new basis systems,

$$\mathbf{E}^s_b = \mathbf{R}_b \mathbf{T} \mathbf{E}^i = \mathbf{R}_b \mathbf{T} \mathbf{R}_a^{-1} \mathbf{E}^i_a \tag{6.106}$$

Clearly

$$\mathbf{T}' = \mathbf{R}_b \mathbf{T} \mathbf{R}_a^{-1} \tag{6.107}$$

is a new Jones matrix relating incident and scattered fields.

Care must be exercised in forming the transformation matrices $\mathbf{R}_a$ and $\mathbf{R}_b$. In Chapter 3 it was noted that to transform the field vector from, for example, a rectangular coordinate system $\mathbf{u}_x, \mathbf{u}_y$ to a general polarization base system $\mathbf{u}_1, \mathbf{u}_2$, **R** is formed by

$$\mathbf{R} = \begin{bmatrix} \mathbf{u}_1^* \cdot \mathbf{u}_x & \mathbf{u}_1^* \cdot \mathbf{u}_y \\ \mathbf{u}_2^* \cdot \mathbf{u}_x & \mathbf{u}_2^* \cdot \mathbf{u}_y \end{bmatrix} \tag{6.108}$$

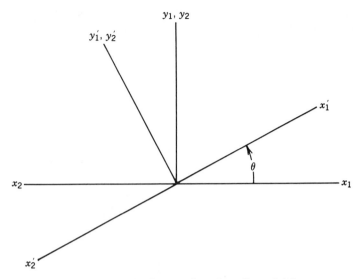

FIGURE 6.11 Radar rotation about line of sight.

In the Jones vector formulation, $\mathbf{R}_a$ for the incident wave is constructed using $\mathbf{u}_{x1}, \mathbf{u}_{y1}$, but $\mathbf{R}_b$ for the scattered wave is formed by using $\mathbf{u}_{x2}, \mathbf{u}_{y2}$. An example may clarify this:

Example A monostatic radar is rotated about its line of sight to the target by angle θ ($x \to y$) from its old coordinates x_1, y_1 to new x'_1, y'_1. Find the new Jones matrix $\mathbf{T}'$ in terms of θ and the old matrix $\mathbf{T}$.

Solution For the incident wave, find $\mathbf{R}_a$ to transform the field to x'_1, y'_1 of Figure 6.11. It is

$$\mathbf{R}_a = \mathbf{R}(\theta) = \begin{bmatrix} \cos\theta & \sin\theta \\ -\sin\theta & \cos\theta \end{bmatrix}$$

An observer at the target must transform the scattered field from x_2, y_2 coordinates to x'_2, y'_2 coordinates, which requires a rotation $-\theta$. Then $\mathbf{R}_b$ is given by

$$\mathbf{R}_b = \mathbf{R}(-\theta) = \begin{bmatrix} \cos\theta & -\sin\theta \\ \sin\theta & \cos\theta \end{bmatrix}$$

Substitution in (6.107) gives the new Jones matrix

$$\mathbf{T}' = \mathbf{R}_b \mathbf{T} \mathbf{R}_a^{-1} = \mathbf{R}(-\theta)\mathbf{T}\mathbf{R}^{-1}(\theta) = \mathbf{R}^{-1}(\theta)\mathbf{T}\mathbf{R}^{-1}(\theta)$$

For backscattering it is desirable to have available the transformation for the Sinclair matrix S. Equations (6.104) and (6.105) remain valid, and (6.103) can be used to write

$$\mathbf{E}_b^s = \mathbf{R}_b \mathbf{E}^s = \mathbf{R}_b \mathbf{S}\mathbf{E}^i = \mathbf{R}_b \mathbf{S}\mathbf{R}_a^{-1} \mathbf{E}_a^i \tag{6.109}$$

Then the transformed scattering matrix is

$$\mathbf{S}' = \mathbf{R}_b \mathbf{S} \mathbf{R}_a^{-1} \tag{6.110}$$

which is the same transformation used for the Jones matrix $\mathbf{T}$. The matrices $\mathbf{R}_a$ and $\mathbf{R}_b$ may be different, however.

Consider the example of the monostatic radar rotated by angle θ for which we transformed the Jones matrix. $\mathbf{R}_a$ is unchanged and is given by

$$\mathbf{R}_a = \mathbf{R}(\theta) = \begin{bmatrix} \cos\theta & \sin\theta \\ -\sin\theta & \cos\theta \end{bmatrix}$$

An observer at the target must transform the scattering field from x_1, y_1 to x_1', y_1' coordinates, which requires a rotation of $+\theta$. Then $\mathbf{R}_b$ is given by

$$\mathbf{R}_b = \mathbf{R}(\theta)$$

and the new Sinclair matrix is

$$\mathbf{S}' = \mathbf{R}(\theta) \mathbf{S} \mathbf{R}^{-1}(\theta)$$

6.8. POLARIZATION FOR MAXIMUM AND MINIMUM POWER

In this section the transmitting and receiving antenna polarizations that give maximum and minimum received power for a target with symmetric Sinclair matrix are obtained, and the maximum power is found [3]. Co-polarization and cross-polarization null states are introduced, and the scattering matrix is given in terms of the null states. The radar cross section of a target is related to the target Sinclair matrix, and a form of the radar equation using the Sinclair matrix is developed.

Maximizing Backscattered Power

The open-circuit voltage induced in the receiving antenna of effective length $\mathbf{h}_r$ by a wave backscattered from a target is

$$V = \mathbf{h}_r^T \mathbf{E}^s = h_{rx} E_x^s + h_{ry} E_y^s \tag{6.111}$$

where both $\mathbf{h}_r$ and $\mathbf{E}^s$ are expressed in the $x_1 y_1 z_1$ coordinate system of Figure 6.9. The effective length is now treated as a 2×1 matrix, and superscript T signifies the transpose. Then the received power is

$$W = \frac{VV^*}{8R_a} = \frac{1}{8R_a} |h_{rx} E_x^s + h_{ry} E_y^s|^2 \tag{6.112}$$

Except for the multiplier, this equation is the same as (4.24), and W may be maximized just as it was in Section 4.5. The result is

$$W_m = \frac{1}{8R_a} |\mathbf{h}_r|^2 |\mathbf{E}^s|^2 \tag{6.113}$$

if we choose

$$\frac{h_{ry}}{h_{rx}} = \left(\frac{E_y^s}{E_x^s}\right)^* \tag{6.114}$$

subject to the constraint that the magnitude of $\mathbf{h}_r$ is constant. This equation is also obviously satisfied by the normalized effective length, or polarization state, denoted by $\hat{\mathbf{h}}$, so that

$$\mathbf{h}_r \cdot \mathbf{h}_r^* = C^2 \tag{6.115a}$$

$$\hat{\mathbf{h}}_r \cdot \hat{\mathbf{h}}_r^* = 1 \tag{6.115b}$$

We have thus determined the optimum polarization of the receiving antenna, given the scattered wave, to maximize received power. The polarization of the transmitting antenna for maximum received power must next be found. The received voltage, with transmitting antenna effective length $\mathbf{h}_t$ and receiver $\mathbf{h}_r$, is

$$V = \frac{jZ_0 I}{\sqrt{4\pi}(2\lambda r^2)} \mathbf{h}_r^T \mathbf{S} \mathbf{h}_t \tag{6.116}$$

We now define effective length vectors in a new polarization basis by the transformations

$$\mathbf{h}_t = \mathbf{R}\mathbf{H}_t \tag{6.117a}$$

$$\mathbf{h}_r = \mathbf{R}\mathbf{H}_r \tag{6.117b}$$

where $\mathbf{R}$ is a 2×2 unitary matrix. The received voltage is

$$V = \frac{jZ_0 I}{\sqrt{4\pi}(2\lambda r^2)} \mathbf{H}_r^T \mathbf{R}^T \mathbf{S} \mathbf{R} \mathbf{H}_t \tag{6.118}$$

It has been shown [4] that if **S** is a square, symmetric matrix (as it is for backscattering in a reciprocal medium, without Faraday rotation), then

$$\mathbf{D} = \mathbf{R}^T \mathbf{S} \mathbf{R} \tag{6.119}$$

is diagonal, with complex elements in general, so that

$$\mathbf{D} = \begin{bmatrix} D_{uu} & 0 \\ 0 & D_{vv} \end{bmatrix} \tag{6.120}$$

in the new basis system **u, v** reached by transformation **R**.

If the equation for received voltage is expanded, using the diagonal form for **D**, its magnitude becomes

$$|V| = \frac{Z_0 I}{\sqrt{4\pi(2\lambda r^2)}} |H_{tu} H_{ru} D_{uu} + H_{tv} H_{rv} D_{vv}| \tag{6.121}$$

Controlling the phase angle of the effective lengths maximizes the voltage magnitude. The requirement on angles is

$$\text{Ang}(H_{tu} H_{ru} D_{uu}) = \text{Ang}(H_{tv} H_{rv} D_{vv}) \tag{6.122}$$

which is readily met by setting

$$\text{Ang}(H_{tu}) = \text{Ang}(H_{ru}) = -\tfrac{1}{2}\text{Ang}(D_{uu}) \tag{6.123a}$$
$$\text{Ang}(H_{tv}) = \text{Ang}(H_{rv}) = -\tfrac{1}{2}\text{Ang}(D_{vv}) \tag{6.123b}$$

This gives a maximum voltage magnitude

$$|V|_m = |H_{tu}||H_{ru}||D_{uu}| + |H_{tv}||H_{rv}||D_{vv}| \tag{6.124}$$

Now it is easy to show that if

$$\mathbf{h} \cdot \mathbf{h}^* = C^2 \tag{6.125}$$

with C a real constant, then

$$\mathbf{H} \cdot \mathbf{H}^* = C^2 \tag{6.126}$$

with C the same constant, if **h** and **H** are related by the unitary transformation **R**. Then constraints on the transformed effective lengths are

$$|H_{tu}|^2 + |H_{tv}|^2 = C^2 \tag{6.127a}$$
$$|H_{ru}|^2 + |H_{rv}|^2 = C^2 \tag{6.127b}$$

POLARIZATION FOR MAXIMUM AND MINIMUM POWER   325

With these constraints the received voltage has its greatest magnitude when $|D_{uu}| > |D_{vv}|$ if

$$|H_{tu}| = |H_{ru}| = C \tag{6.128a}$$

$$|H_{tv}| = |H_{rv}| = 0 \tag{6.128b}$$

and when $|D_{uu}| < |D_{vv}|$ if

$$|H_{tu}| = |H_{ru}| = 0 \tag{6.129a}$$

$$|H_{tv}| = |H_{rv}| = C \tag{6.129b}$$

Both (6.128) and (6.129) are satisfied if

$$\mathbf{H}_t = \mathbf{H}_r \tag{6.130}$$

More generally, they are satisfied if

$$\mathbf{H}_t = \mathbf{H}_r e^{j\beta} \tag{6.131}$$

with β any angle, but the distinction is insignificant.

Returning to the effective lengths in xyz coordinates, (6.130) yields

$$\mathbf{h}_t = \mathbf{h}_r \tag{6.132}$$

The received power backscattered from a target with a symmetric Sinclair matrix reaches its maximum value if transmitting and receiving antennas have the same effective length. This is an important result; the same antenna can be used for transmitting and receiving, and if the effective length is properly chosen, no other antenna pair gives greater received power.

With this fact established we return to the condition on the receiving antenna for maximum received power, (6.114), which may be written as

$$\alpha \mathbf{h}_r^* = \mathbf{E}^s \tag{6.133}$$

where α is an arbitrary scalar, real or complex. The scattered wave from a target illuminated by an antenna of length $\mathbf{h}_t$,

$$\mathbf{E}^s = \frac{jZ_0 I}{\sqrt{4\pi}\,(2\lambda r^2)} \mathbf{S}\mathbf{h}_t \tag{6.134}$$

when combined with (6.133), leads to

$$\alpha \mathbf{h}_r^* = \frac{jZ_0 I}{\sqrt{4\pi}\,(2\lambda r^2)} \mathbf{S}\mathbf{h}_t \tag{6.135}$$

Since the transmitting and receiving antennas may be the same, without departing from the optimum for maximum received power, the relation becomes

$$\mathbf{Sh}_{opt} = \gamma \mathbf{h}^*_{opt} \tag{6.136}$$

where γ includes the constants of (6.135). This equation is also valid if the normalized form of the effective length, the polarization state, is used, so

$$\mathbf{S}\hat{\mathbf{h}}_{opt} = \gamma \hat{\mathbf{h}}^*_{opt} \tag{6.137}$$

with γ the same constant.

If this equation is combined with its conjugate, the result is

$$\frac{1}{\gamma}\mathbf{S}^*\mathbf{S}\hat{\mathbf{h}}_{opt} = \gamma^* \hat{\mathbf{h}}_{opt} \tag{6.138}$$

or

$$(\mathbf{S}^*\mathbf{S} - |\gamma|^2 \mathbf{I})\hat{\mathbf{h}}_{opt} = \mathbf{0} \tag{6.139}$$

where $\mathbf{I}$ is the identity matrix. This equation is valid only if the matrix multiplying $\mathbf{h}_{opt}$ is singular. Then the values of $|\gamma|^2$ are the eigenvalues of $\mathbf{S}^*\mathbf{S}$ found from

$$\|\mathbf{S}^*\mathbf{S} - |\gamma|^2 \mathbf{I}\| = 0 \tag{6.140}$$

Since $\mathbf{S}^*\mathbf{S}$ is Hermitian, the eigenvalues $|\gamma|^2$ are real. The $\hat{\mathbf{h}}_{opt}$ are normal eigenvectors.

Expanding the determinant leads to

$$|\gamma|^4 - B|\gamma|^2 + C = 0 \tag{6.141}$$

where

$$B = |S_{xx}|^2 + 2|S_{xy}|^2 + |S_{yy}|^2 \tag{6.142a}$$

$$C = |S_{xx}S_{yy}|^2 + |S_{xy}|^4 - 2\,\mathrm{Re}\!\left(S_{xx}S_{xy}^{*2}S_{yy}\right) \tag{6.142b}$$

If we are interested only in the scattering matrix with relative phase, S_{xy} may be taken as real, and C becomes

$$C = |S_{xx}S_{yy}|^2 + S_{xy}^4 - 2S_{xy}^2\,\mathrm{Re}(S_{xx}S_{yy}) \tag{6.143}$$

Now the received voltage with optimum polarization is

$$V = \mathbf{h}_{opt}^T \mathbf{E}^s = \frac{jZ_0 I}{\sqrt{4\pi}(2\lambda r^2)} \mathbf{h}_{opt}^T \mathbf{S}\mathbf{h}_{opt} \qquad (6.144)$$

where $\mathbf{h}_{opt}$ is the true effective length corresponding to the polarization state $\hat{\mathbf{h}}_{opt}$. Using (6.136), V becomes

$$V = \frac{jZ_0 I \gamma}{\sqrt{4\pi}(2\lambda r^2)} \mathbf{h}_{opt}^T \mathbf{h}_{opt}^* \qquad (6.145)$$

leading to received power

$$W = \frac{Z_0^2 I^2 |\gamma|^2 |\mathbf{h}_{opt}^T \mathbf{h}_{opt}^*|^2}{128\pi R_a \lambda^2 r^4} \qquad (6.146)$$

The maximum received power corresponds to the larger eigenvalue $|\gamma_1|^2$ given by solutions to (6.141),

$$|\gamma_1|^2 = \frac{B}{2} + \frac{1}{2}\sqrt{B^2 - 4C} \qquad (6.147a)$$

$$|\gamma_2|^2 = \frac{B}{2} - \frac{1}{2}\sqrt{B^2 - 4C} \qquad (6.147b)$$

and the maximum power is

$$W_m = \frac{Z_0^2 I^2 |\gamma_1|^2 |\mathbf{h}_{opt}^T \mathbf{h}_{opt}^*|^2}{128\pi R_a \lambda^2 r^4} \qquad (6.148)$$

A submaximum power corresponds to the smaller of the eigenvalues and is

$$W_{sm} = \frac{Z_0^2 I^2 |\gamma_2|^2 |\mathbf{h}_{opt}^T \mathbf{h}_{opt}^*|^2}{128\pi R_a \lambda^2 r^4} \qquad (6.149)$$

For some targets this submaximum power may in fact be as great as W_m, although that is not the general case. For convenience, both powers may be referred to later as "maximum." In Chapter 9 it is shown that it may also be useful to treat W_{sm} as a *minimum* power.

The optimum $\hat{\mathbf{h}}$ corresponding to maximum power is found from

$$\mathbf{S}^* \mathbf{S} \hat{\mathbf{h}} = |\gamma|^2 \hat{\mathbf{h}} \qquad (6.150)$$

This matrix form yields two scalar equations, from either of which the antenna polarization ratio may be found. They are

$$P = \frac{\hat{h}_y}{\hat{h}_x} = -\frac{S_{xx}S_{xy}^* + S_{xy}S_{yy}^*}{|S_{xy}|^2 + |S_{yy}|^2 - |\gamma|^2} \tag{6.151}$$

and

$$P = -\frac{|S_{xx}|^2 + |S_{xy}|^2 - |\gamma|^2}{S_{xx}^*S_{xy} + S_{xy}^*S_{yy}} \tag{6.152}$$

The eigenvalues of $\mathbf{S}^*\mathbf{S}$ are distinct unless

$$B^2 - 4C = 0 \tag{6.153}$$

which can be true only if

$$|S_{xx}| = |S_{yy}| \tag{6.154a}$$

$$S_{xy} = 0 \tag{6.154b}$$

For this case the eigenvalues reduce to

$$|\gamma_1|^2 = |\gamma_2|^2 = |S_{xx}|^2 = |S_{yy}|^2 \tag{6.155}$$

and the antenna polarization ratio, from either equation for it, is indeterminate.

Let us compare the antenna polarization ratios corresponding to the two eigenvalues $|\gamma_1|^2$ and $|\gamma_2|^2$, with $|\gamma_1|^2$ the larger. If we are interested only in the scattering matrix with relative phase, S_{xy} is real, and the polarization ratios are

$$P_1 = -S_{xy}\frac{S_{xx} + S_{yy}^*}{S_{xy}^2 + |S_{yy}|^2 - |\gamma_1|^2} = -\frac{1}{S_{xy}}\frac{|S_{xx}|^2 + S_{xy}^2 - |\gamma_1|^2}{S_{xx}^* + S_{yy}} \tag{6.156}$$

$$P_2 = -S_{xy}\frac{S_{xx} + S_{yy}^*}{S_{xy}^2 + |S_{yy}|^2 - |\gamma_2|^2} = -\frac{1}{S_{xy}}\frac{|S_{xx}|^2 + S_{xy}^2 - |\gamma_2|^2}{S_{xx}^* + S_{yy}} \tag{6.157}$$

From (6.142) and (6.147) it is readily seen that the denominator of the first equation, for P_1, is the negative of the numerator of the second equation, for P_2, and vice versa,

$$S_{xy}^2 + |S_{yy}|^2 - |\gamma_1|^2 = -\left(|S_{xx}|^2 + S_{xy}^2 - |\gamma_2|^2\right) \tag{6.158}$$

It follows that the two antenna polarization ratios are related by

$$P_2 = -\frac{1}{P_1^*} \tag{6.159}$$

This relationship is discussed in more detail later.

The normalized antenna effective lengths corresponding to the optimum polarization ratio P_1 and the suboptimum ratio P_2 are the eigenvectors of $\mathbf{S}^*\mathbf{S}$ and may be found from

$$\hat{\mathbf{h}}_1 = \begin{bmatrix} \hat{h}_{1x} \\ \hat{h}_{1y} \end{bmatrix} = \frac{1}{\sqrt{1 + |P_1|^2}} \begin{bmatrix} 1 \\ P_1 \end{bmatrix} \tag{6.160a}$$

$$\hat{\mathbf{h}}_2 = \frac{1}{\sqrt{1 + |P_2|^2}} \begin{bmatrix} 1 \\ P_2 \end{bmatrix} \tag{6.160b}$$

where the phase of $\hat{h}_x$ has been omitted. The polarization states are orthonormal, since

$$\hat{\mathbf{h}}_1^* \cdot \hat{\mathbf{h}}_1 = \hat{\mathbf{h}}_2^* \cdot \hat{\mathbf{h}}_2 = 1 \tag{6.161}$$

and, using (6.159),

$$\hat{\mathbf{h}}_1^* \cdot \hat{\mathbf{h}}_2 = \frac{1 + P_1^* P_2}{\sqrt{1 + |P_1|^2}\sqrt{1 + |P_2|^2}} = 0 \tag{6.162}$$

The concept of polarization efficiency for a two-antenna communication system is useful, and it is appropriate to consider the polarization efficiency of an antenna used both as transmitter and receiver in a monostatic radar. The received voltage is given by (6.116) with $\mathbf{h}_r = \mathbf{h}_t = \mathbf{h}$, and the maximum received voltage is given by (6.145) with $\gamma = \gamma_1$. The ratio of received powers obtained from these voltages then forms the definition of *backscatter polarization efficiency*,

$$\rho_s = \frac{|\hat{\mathbf{h}}^T \mathbf{S} \hat{\mathbf{h}}|^2}{|\gamma_1|^2} \tag{6.163}$$

where

$$\hat{\mathbf{h}}_{\text{opt}}^T \hat{\mathbf{h}}_{\text{opt}}^* = 1 \tag{6.164}$$

Minimum Backscattered Power

Consider next the antenna polarization for a monostatic radar, using one antenna for transmitting and receiving, that gives minimum backscattered power. From (6.116) the received voltage and power are zero if

$$\mathbf{h}^T \mathbf{Sh} = S_{xx} h_x^2 + 2S_{xy} h_x h_y + S_{yy} h_y^2 = 0 \tag{6.165}$$

If the antenna polarization ratio P is used, this equation becomes either

$$S_{yy} P^2 + 2 S_{xy} P + S_{xx} = 0 \tag{6.166}$$

or

$$S_{xx} \left(\frac{1}{P}\right)^2 + 2 S_{xy} \left(\frac{1}{P}\right) + S_{yy} = 0 \tag{6.167}$$

If $S_{yy} \neq 0$, P can be found from the first equation,

$$P = -\frac{S_{xy}}{S_{yy}} \pm \frac{1}{S_{yy}} \sqrt{S_{xy}^2 - S_{xx} S_{yy}} \tag{6.168}$$

and if $S_{xx} \neq 0$, from the second equation,

$$\frac{1}{P} = -\frac{S_{xy}}{S_{xx}} \pm \frac{1}{S_{xx}} \sqrt{S_{xy}^2 - S_{xx} S_{yy}} \tag{6.169}$$

The first form, (6.168), is used here, but the results that follow are essentially unaltered if the second form is used. The two roots of (6.168) are

$$P_3 = \frac{R - S_{xy}}{S_{yy}} \tag{6.170a}$$

$$P_4 = -\frac{R + S_{xy}}{S_{yy}} \tag{6.170b}$$

where

$$R = \sqrt{S_{xy}^2 - S_{xx} S_{yy}} \tag{6.171}$$

If the radar antenna used for transmitting and receiving has polarization ratio P_3 or P_4, the received power, backscattered from the target, is zero.

Copolarization and Cross-Polarization Nulls

The backscattered power from a target, when it is illuminated by a particular antenna, may be divided into a part that can be received by the antenna,

POLARIZATION FOR MAXIMUM AND MINIMUM POWER   331

called the *copolarized* signal, and a part that cannot be received, the *cross-polarized* signal. The cross-polarized signal can be received without polarization loss by an antenna orthogonal to the transmitting antenna. (By definition, orthogonal antennas radiate orthogonal waves. See Sections 3.11 and 4.8.) The voltage induced in a receiving antenna is

$$V = \frac{jZ_0 I}{\sqrt{4\pi}\,(2\lambda r^2)} \mathbf{h}_r^T \mathbf{S} \mathbf{h}_t \tag{6.116}$$

which may be written in terms of the antenna polarization ratios as

$$V = \frac{jZ_0 I}{\sqrt{4\pi}\,(2\lambda r^2)} h_{rx} h_{tx} \begin{bmatrix} 1 & P_r \end{bmatrix} \begin{bmatrix} S_{xx} & S_{xy} \\ S_{xy} & S_{yy} \end{bmatrix} \begin{bmatrix} 1 \\ P_t \end{bmatrix} \tag{6.172}$$

where the symmetry of the monostatic Sinclair matrix is used.

For orthogonal transmitting and receiving antennas, from Section 3.11,

$$P_r = -\frac{1}{P_t^*} \tag{6.173}$$

If this is substituted into the preceding equation, the cross-polarized received voltage becomes

$$V = \frac{jZ_0 I}{\sqrt{4\pi}\,(2\lambda r^2)} h_{rx} h_{tx} \left(S_{xx} + S_{xy} P_t - \frac{S_{xy}}{P_t^*} - \frac{S_{yy} P_t}{P_t^*} \right)$$

Now, polarization ratios P_1 and P_2 give a maximum copolarized backscattered power (a relative maximum in the case of P_2) and may be called *copolarization maxima* or, succinctly, *co-pol maxima*. Substitution of either P_1 or P_2, given by (6.156) and (6.157), into this equation for the cross-polarized voltage gives

$$V(P_1) = V(P_2) = 0$$

Thus at the co-pol maxima, none of the backscattered power is crosspolarized, and polarization ratios P_1 and P_2 may be called *cross-polarization null polarization ratios*, or, in a common phrase, *cross-pol nulls* or *X-pol nulls*. The co-pol maxima and the cross-pol nulls coincide.

Antenna polarization ratios P_3 and P_4 in a monostatic radar using one antenna for transmitting and receiving give zero copolarized power. They are therefore called *copolarization null polarization ratios* or *co-pol nulls*. The cross-polarized power to a colocated orthogonal receiving antenna when the transmitting antenna has polarizations P_3 or P_4 is of interest. From (6.116) this cross-polarized power is

$$W_x = \frac{Z_0^2 I^2}{128\pi R_a \lambda^2 r^4} |\mathbf{h}_r^T \mathbf{S} \mathbf{h}_t|^2 \tag{6.174}$$

where $\mathbf{h}_r$ and $\mathbf{h}_t$ are orthogonal. The ratio of this cross-polarized power to the maximum power (for polarization P_1) is

$$\frac{W_x}{W_m} = \frac{|\hat{\mathbf{h}}_r^T \mathbf{S} \hat{\mathbf{h}}_t|^2}{|\gamma_1|^2} \tag{6.175}$$

if the assumption is made that the magnitudes of the effective lengths of both transmitting and receiving antennas are equal to each other and to that used with polarization P_1 to obtain maximum power.

If this equation is expanded, using the orthogonality of the antennas,

$$\hat{\mathbf{h}}_r \cdot \hat{\mathbf{h}}_t^* = 0 \qquad P_r = -1/P_t^*$$

the cross-polarized power at transmitter polarizations P_3 or P_4 giving zero copolarized power is

$$\frac{W_x}{W_m} = \frac{|\gamma_2|}{|\gamma_1|} \tag{6.176}$$

This ratio can be quite large, but in general W_x is not the maximum cross-polarized power. Discussion of the transmitting polarization that gives maximum cross-polarized power is deferred to another section.

In the development leading to maximum copolarized power, the polarization ratio P_1 corresponded to the greater of the two eigenvalues, $|\gamma_1|^2$, and gave the greatest power. It is worth pointing out that in backscattering, substituting an orthogonally polarized antenna does not give minimum copolarized power. The antenna polarization ratio orthogonal to P_1 is P_2, which in general does not yield minimum copolarized power; in fact, for some targets it may give a power as great as that found by using P_1. Further, an antenna with polarization ratio P_3 or P_4 yields zero copolarized power but in general is not orthogonal to an antenna with ratio P_1.

Scattering Matrix from Polarization Nulls

The copolarization null ratios P_3 and P_4 may be used to obtain information about the target scattering matrix. From (6.170) and (6.171), we find

$$\frac{S_{xx}}{S_{xy}} = -\frac{2P_3 P_4}{P_3 + P_4} \tag{6.177a}$$

$$\frac{S_{yy}}{S_{xy}} = -\frac{2}{P_3 + P_4} \tag{6.177b}$$

Thus to within a multiplying factor, the relative target scattering matrix can be found from the co-pol nulls. The polarization ratios do not provide amplitude information, so S_{xy} cannot be found.

It has been shown that if the span of the relative scattering matrix

$$\text{Span}(\mathbf{S}) = |S_{xx}|^2 + 2|S_{xy}|^2 + |S_{yy}|^2$$

is known, then the relative scattering matrix, with amplitudes, can be obtained from a knowledge of the co-pol null pair, P_3 and P_4, or from a cross-pol null, P_1 or P_2, and a co-pol null, P_3 or P_4. The cross-pol null pair, P_1 and P_2, is not sufficient to find $\mathbf{S}$ [5].

The Radar Equation Reconsidered

In Chapter 1 the radar equation was introduced, but a discussion of how the polarization properties of antennas and target affect the power to the radar receiver was deferred. We are now in a position to consider these polarization effects.

In a form commonly used, the radar equation is

$$W_r = \frac{W_t G_t(\theta, \phi) A_{er}(\theta, \phi) \sigma_r}{(4\pi r_1 r_2)^2} \qquad (1.157)$$

where W_r is the received power, W_t the power accepted by the transmitting antenna, G_t the transmitting antenna gain, A_{er} the effective area of the receiving antenna, and r_1 and r_2 the transmitter–target and transmitter–receiver distances respectively. In this form, with no explicit factor to account for polarization effects, σ_r is the *radar cross section* of the target [6]. In determining it, the polarization states of both transmitting and receiving antennas must be specified, and it may be used only for cases involving those states. Typically, radar cross sections are specified as HH (horizontal transmitting, horizontal receiving antennas), HV (horizontal, vertical), LR (left, right circular), and so on.

Another target cross section, the *scattering cross section*, is often used [6]. It is defined, not in terms of received power, but by a power density. Specifically, it is the ratio of scattered power, assumed to be scattered isotropically, to the incident power flux density at the target, and appears in this form for power density at the receiver,

$$\mathscr{P} = \frac{W_t G_t \sigma_s}{(4\pi r_1 r_2)^2} \qquad (6.178)$$

For both incident and scattered waves, the total power density, without regard to polarization, is used.

334 POLARIZATION CHANGES BY REFLECTION AND TRANSMISSION

It must be emphasized that neither cross section can characterize the polarization behavior of a target. Since both are widely used, however, in this section they are related to each other and to the target scattering (Sinclair) matrix.

The scattered power density at the receiver is given by

$$\mathscr{P} = \frac{1}{2Z_0} \mathbf{E}^{sT} \mathbf{E}^{s*} = \frac{1}{8\pi Z_0 r_2^2} |\mathbf{SE}^i|^2 \qquad (6.179)$$

and, if the incident field $\mathbf{E}^i$ at the target is given in terms of the effective length of the transmitting antenna,

$$\mathscr{P} = \frac{Z_0 I^2}{32\pi\lambda^2 r_1^2 r_2^2} |\mathbf{Sh}_t|^2 \qquad (6.180)$$

If this is equated to the form for power density in terms of the scattering cross section, the result is

$$W_t G_t \sigma_s = \frac{\pi Z_0 I^2}{2\lambda^2} |\mathbf{Sh}_t|^2 \qquad (6.181)$$

Now the transmitter power is

$$W_t = \tfrac{1}{2} R_t I^2 \qquad (6.182)$$

where R_t is the transmitting antenna resistance. If this is used, and if it is noted from Section 4.3 that

$$G_t R_t = \frac{\pi Z_0}{\lambda^2} |\mathbf{h}_t|^2 \qquad (6.183)$$

then the scattering cross section becomes

$$\sigma_s = \frac{|\mathbf{Sh}_t|^2}{|\mathbf{h}_t|^2} \qquad (6.184)$$

where $\mathbf{h}_t$ may be taken as either the effective length of the transmitting antenna or its normalized value, the antenna polarization state. Note that σ_s depends on the transmitting antenna, but not on the receiving antenna.

To find a form for the radar cross section, an equation for received power using the scattering matrix must first be found. The open-circuit voltage induced in the receiving antenna is

$$V = \mathbf{h}_r^T \mathbf{E}^s = \frac{jZ_0 I}{(\sqrt{4\pi} r_2)(2\lambda r_1)} \mathbf{h}_r^T \mathbf{Sh}_t \qquad (6.185)$$

and the received power to an antenna with resistance R_a is

$$W_r = \frac{1}{8R_a}VV^* = \frac{Z_0^2 I^2}{128\pi \lambda^2 R_a r_1^2 r_2^2}|\mathbf{h}_r^T \mathbf{S}\mathbf{h}_t|^2 \tag{6.186}$$

Equating this to the form for received power in terms of the radar cross section leads to

$$W_t G_t A_{er} \sigma_r = \frac{\pi Z_0^2 I^2}{8\lambda^2 R_a}|\mathbf{h}_r^T \mathbf{S}\mathbf{h}_t|^2 \tag{6.187}$$

With the previously used substitutions for W_t and G_t this reduces to

$$A_{er}\sigma_r|\mathbf{h}_t|^2 = \frac{Z_0}{4R_a}|\mathbf{h}_r^T \mathbf{S}\mathbf{h}_t|^2 \tag{6.188}$$

It can also be seen from Section 4.3 that

$$A_{er} R_a = \frac{Z_0}{4}|\mathbf{h}_r|^2 \tag{6.189}$$

If this is used, the radar cross section becomes

$$\sigma_r = \frac{|\mathbf{h}_r^T \mathbf{S}\mathbf{h}_t|^2}{|\mathbf{h}_t|^2|\mathbf{h}_r|^2} \tag{6.190}$$

Note that, unlike the scattering cross section, the radar cross section of a target depends on both transmitting and receiving antennas.

The ratio of the two cross sections is the polarization efficiency of the receiving antenna, as may be seen here,

$$\frac{\sigma_r}{\sigma_s} = \frac{|\mathbf{h}_r^T \mathbf{S}\mathbf{h}_t|^2}{|\mathbf{S}\mathbf{h}_t|^2|\mathbf{h}_r|^2} = \frac{|\mathbf{h}_r^T \mathbf{E}^s|^2}{|\mathbf{E}^s|^2|\mathbf{h}_r|^2} = \rho \tag{6.191}$$

It is convenient to use the normalized polarization states for the antennas, and doing so causes the cross sections to take on the simpler forms

$$\sigma_r = |\hat{\mathbf{h}}_r^T \mathbf{S}\hat{\mathbf{h}}_t|^2 \tag{6.192}$$

$$\sigma_s = |\mathbf{S}\hat{\mathbf{h}}_t|^2 \tag{6.193}$$

One other form of the radar equation for backscattering should be noted. If (6.186) is compared to (6.148), it is seen that the maximum value of the radar cross section is

$$\sigma_{r\,max} = |\hat{\mathbf{h}}^T\mathbf{S}\hat{\mathbf{h}}|^2_{max} = |\gamma_1|^2 \tag{6.194}$$

where $|\gamma_1|^2$ is the largest eigenvalue of $\mathbf{S}^*\mathbf{S}$, given by (6.147a). Note also that for the optimum polarization that gives this maximum radar cross section the receiving antenna polarization efficiency is one, and $\sigma_{r\,max} = \sigma_{s\,max}$. A backscatter polarization efficiency was defined earlier as

$$\rho_s = \frac{|\hat{\mathbf{h}}^T\mathbf{S}\hat{\mathbf{h}}|^2}{|\gamma_1|^2} \tag{6.163}$$

and it may be seen now that the numerator is the radar cross section for backscattering and the denominator is its greatest value.

To summarize, perhaps the most useful form of the radar equation, neglecting propagation losses, impedance mismatch losses, and so on, is

$$W_r = \frac{W_t G_t(\theta,\phi) A_{er}(\theta,\phi)}{(4\pi r_1 r_2)^2} |\hat{\mathbf{h}}_r^T\mathbf{S}\hat{\mathbf{h}}_t|^2 \tag{6.195}$$

if the target scattering matrix is known. Equivalent forms use these equalities,

$$|\hat{\mathbf{h}}_r^T\mathbf{S}\hat{\mathbf{h}}_t|^2 = \sigma_r = \rho\sigma_s \tag{6.196}$$

for bistatic or monostatic radars, and

$$|\hat{\mathbf{h}}^T\mathbf{S}\hat{\mathbf{h}}|^2 = \rho_s|\gamma_1|^2 \tag{6.197}$$

for monostatic radars using the same antenna for transmitting and receiving, where ρ and ρ_s are given by (6.191) and (6.163) respectively.

6.9. THE SCATTERING MATRIX WITH GEOMETRIC VARIABLES

Characteristic radar polarization states that maximize and minimize received power from a target have been determined from its scattering matrix. In this section we shall see that the target's scattering matrix can be expressed in terms of geometric variables related to these characteristic polarization states.

The Diagonal Scattering Matrix

The optimum antenna effective length and the polarization state that give maximum received power in a monostatic radar both satisfy

$$\mathbf{S}\hat{\mathbf{h}}_{opt} = \gamma \hat{\mathbf{h}}^*_{opt} \tag{6.136}$$

where γ can take on two values whose magnitudes are given by (6.147). Then $\hat{\mathbf{h}}$ is one of two optimum values satisfying

$$\mathbf{S}\hat{\mathbf{h}}_1 = \gamma_1 \hat{\mathbf{h}}^*_1 \tag{6.198a}$$

$$\mathbf{S}\hat{\mathbf{h}}_2 = \gamma_2 \hat{\mathbf{h}}^*_2 \tag{6.198b}$$

In this section, the orthonormal polarization states $\hat{\mathbf{h}}_1$ and $\hat{\mathbf{h}}_2$ are used rather than the antenna effective lengths.

Form the matrix

$$\mathbf{U} = \begin{bmatrix} \hat{\mathbf{h}}_1 & \hat{\mathbf{h}}_2 \end{bmatrix} = \begin{bmatrix} \hat{h}_{1x} & \hat{h}_{2x} \\ \hat{h}_{1y} & \hat{h}_{2y} \end{bmatrix} = \begin{bmatrix} \mathbf{H}^T_x \\ \mathbf{H}^T_y \end{bmatrix} \tag{6.199}$$

noting that this also defines the vectors

$$\mathbf{H}_x = \begin{bmatrix} \hat{h}_{1x} \\ \hat{h}_{2x} \end{bmatrix} \tag{6.200a}$$

$$\mathbf{H}_y = \begin{bmatrix} \hat{h}_{1y} \\ \hat{h}_{2y} \end{bmatrix} \tag{6.200b}$$

It was shown earlier that $\hat{\mathbf{h}}_1$ and $\hat{\mathbf{h}}_2$ are orthonormal. Thus the columns of $\mathbf{U}$ considered as column vectors are orthonormal.

Consider $\mathbf{H}_x$ and $\mathbf{H}_y$ formed from the rows of $\mathbf{U}$. In terms of polarization ratios corresponding to $\hat{\mathbf{h}}_1$ and $\hat{\mathbf{h}}_2$, given by (6.160), we may write

$$\mathbf{H}_x \cdot \mathbf{H}^*_x = |\hat{h}_{1x}|^2 + |\hat{h}_{2x}|^2 = \frac{1}{1 + |P_1|^2} + \frac{1}{1 + |P_2|^2} \tag{6.201}$$

But

$$P_2 = -\frac{1}{P^*_1} \tag{6.159}$$

338 POLARIZATION CHANGES BY REFLECTION AND TRANSMISSION

so

$$\mathbf{H}_x \cdot \mathbf{H}_x^* = \frac{1}{1 + |P_1|^2} + \frac{|P_1|^2}{1 + |P_1|^2} = 1 \qquad (6.202)$$

Similarly,

$$\mathbf{H}_y \cdot \mathbf{H}_y^* = 1 \qquad (6.203)$$

Also

$$\mathbf{H}_x \cdot \mathbf{H}_y^* = \hat{h}_{1x}\hat{h}_{1y}^* + \hat{h}_{2x}\hat{h}_{2y}^* = \frac{P_1^*}{1 + |P_1|^2} + \frac{P_2^*}{1 + |P_2|^2} = 0 \qquad (6.204)$$

if the relation between P_1 and P_2 is used.

From this it is seen that the rows of **U** considered as row vectors are orthonormal. Taken with the orthonormality of the column vectors, this shows that **U** is unitary. This property may be used to diagonalize the scattering matrix, thus

$$\mathbf{U}^T \mathbf{S} \mathbf{U} = \mathbf{U}^T \mathbf{S} [\hat{\mathbf{h}}_1 \ \hat{\mathbf{h}}_2] = \mathbf{U}^T [\mathbf{S}\hat{\mathbf{h}}_1 \ \mathbf{S}\hat{\mathbf{h}}_2] \qquad (6.205)$$

This becomes, if (6.198) is used,

$$\mathbf{U}^T \mathbf{S} \mathbf{U} = \mathbf{U}^T [\gamma_1 \hat{\mathbf{h}}_1^* \ \gamma_2 \hat{\mathbf{h}}_2^*] = [\gamma_1 \mathbf{U}^T \hat{\mathbf{h}}_1^* \ \gamma_2 \mathbf{U}^T \hat{\mathbf{h}}_2^*] \qquad (6.206)$$

Now

$$\mathbf{U}^T \hat{\mathbf{h}}_1^* = \begin{bmatrix} \hat{\mathbf{h}}_1^T \\ \hat{\mathbf{h}}_2^T \end{bmatrix} \hat{\mathbf{h}}_1^* = \begin{bmatrix} \hat{\mathbf{h}}_1^T \hat{\mathbf{h}}_1^* \\ \hat{\mathbf{h}}_2^T \hat{\mathbf{h}}_1^* \end{bmatrix} = \begin{bmatrix} 1 \\ 0 \end{bmatrix} \qquad (6.207)$$

and

$$\mathbf{U}^T \hat{\mathbf{h}}_2^* = \begin{bmatrix} \hat{\mathbf{h}}_1^T \hat{\mathbf{h}}_2^* \\ \hat{\mathbf{h}}_2^T \hat{\mathbf{h}}_2^* \end{bmatrix} = \begin{bmatrix} 0 \\ 1 \end{bmatrix} \qquad (6.208)$$

Then

$$\mathbf{U}^T \mathbf{S} \mathbf{U} = \begin{bmatrix} \gamma_1 & 0 \\ 0 & \gamma_2 \end{bmatrix} \qquad (6.209)$$

This form will be referred to as the *diagonal scattering matrix*,

$$\mathbf{S}_d = \begin{bmatrix} \gamma_1 & 0 \\ 0 & \gamma_2 \end{bmatrix} = \mathbf{U}^T \mathbf{S} \mathbf{U} \qquad (6.210)$$

Note that this equation does not offer a means of obtaining the eigenvalues γ_1 and γ_2 from the scattering matrix of a target, since $\mathbf{U}$ is composed, not of the scattering matrix elements, but of the characteristic polarization states $\hat{\mathbf{h}}_1$ and $\hat{\mathbf{h}}_2$. These in turn are found after $|\gamma_1|^2$ and $|\gamma_2|^2$ are obtained from (6.147).

Before leaving this section note that the scattering matrix is readily found from the diagonal form by using the unitary property of $\mathbf{U}$ and multiplying (6.210) appropriately, thus,

$$\mathbf{U}^*\mathbf{U}^T\mathbf{S}\mathbf{U}\mathbf{U}^{T*} = \mathbf{U}^*\mathbf{S}_d\mathbf{U}^{T*}$$

$$\mathbf{S} = \mathbf{U}^*\mathbf{S}_d\mathbf{U}^{T*} \tag{6.211}$$

Geometric Form of the Characteristic Polarizations

In Chapter 3 we saw that an elliptically polarized wave can be written in rectangular coordinate form as

$$\mathbf{E} = \frac{m}{\cos\epsilon}\begin{bmatrix}\cos\tau & -\sin\tau \\ \sin\tau & \cos\tau\end{bmatrix}\begin{bmatrix}\cos\epsilon \\ j\sin\epsilon\end{bmatrix}e^{j\phi_0} \tag{3.54}$$

where τ and ϵ are the tilt and ellipticity angles, respectively, of the polarization ellipse, m is its semimajor axis, and ϕ_0 is a phase common to both horizontal and vertical components of the wave. It is then apparent that the characteristic polarization states for a target can be written

$$\hat{\mathbf{h}}_1 = \begin{bmatrix}\cos\tau_1 & -\sin\tau_1 \\ \sin\tau_1 & \cos\tau_1\end{bmatrix}\begin{bmatrix}\cos\epsilon_1 \\ j\sin\epsilon_1\end{bmatrix}e^{j\phi_1} \tag{6.212a}$$

$$\hat{\mathbf{h}}_2 = \begin{bmatrix}\cos\tau_2 & -\sin\tau_2 \\ \sin\tau_2 & \cos\tau_2\end{bmatrix}\begin{bmatrix}\cos\epsilon_2 \\ j\sin\epsilon_2\end{bmatrix}e^{j\phi_2} \tag{6.212b}$$

The orthonormality requirement

$$\hat{\mathbf{h}}_1 \cdot \hat{\mathbf{h}}_2^* = 0$$

can be satisfied if

$$\phi_1 = \phi_2 \tag{6.213a}$$

$$\tau_2 = \tau_1 + \pi/2 \tag{6.213b}$$

$$\epsilon_2 = -\epsilon_1 \tag{6.213c}$$

The common phase term ϕ_1 is of no significance and can be considered zero.

Then the polarization states are

$$\hat{\mathbf{h}}_1(\tau_1, \epsilon_1) = \begin{bmatrix} \cos \tau_1 & -\sin \tau_1 \\ \sin \tau_1 & \cos \tau_1 \end{bmatrix} \begin{bmatrix} \cos \epsilon_1 \\ j \sin \epsilon_1 \end{bmatrix} \qquad (6.214a)$$

$$\hat{\mathbf{h}}_2(\tau_2, \epsilon_2) = \hat{\mathbf{h}}_1(\tau_1 + \pi/2, -\epsilon_1) = \begin{bmatrix} -\sin \tau_1 & -\cos \tau_1 \\ \cos \tau_1 & -\sin \tau_1 \end{bmatrix} \begin{bmatrix} \cos \epsilon_1 \\ -j \sin \epsilon_1 \end{bmatrix}$$
$$(6.214b)$$

Geometric Interpretation of the Eigenvalues

The ability to write the diagonal scattering matrix with the eigenvalues γ_1 and γ_2 as elements,

$$\mathbf{S}_d = \begin{bmatrix} \gamma_1 & 0 \\ 0 & \gamma_2 \end{bmatrix} \qquad (6.215)$$

allows a geometrical interpretation of the eigenvalues. We may think of $\mathbf{S}_d$ as representing two targets, one backscattering only the portion of the incident wave with polarization state $\hat{\mathbf{h}}_1$ and polarization ratio P_1, and the other backscattering only the incident wave part with the orthogonal polarization $\hat{\mathbf{h}}_2$. As a simple, easily visualized case, a target for which $S_{xy} = 0$ in rectangular coordinates is already diagonal and has a diagonal scattering matrix

$$\mathbf{S}_d = \begin{bmatrix} S_{xx} & 0 \\ 0 & S_{yy} \end{bmatrix} \qquad (6.216)$$

It may be thought of as a horizontal wire, or grid, and a vertical wire, in general at a different distance from the radar. We may therefore think of γ_1 and γ_2 as representing orthogonal scatterers in the diagonalizing polarization base system. The two scatterers have a common phase determined by the coordinate value from which the phase is referenced and the common reflecting properties of the two scatterers (e.g., for the two perpendicular wires considered, the common phase is $-4\pi r/\lambda$ if the phase reference position is at the radar and r is the distance to one of the scatterers). The scatterers will also have a differential phase if they are not located at the same distance from the radar or have different reflecting properties. The diagonal scattering matrix may therefore be written as

$$\mathbf{S}_d = \begin{bmatrix} \gamma_1 & 0 \\ 0 & \gamma_2 \end{bmatrix} = \begin{bmatrix} |\gamma_1| & 0 \\ 0 & |\gamma_2|e^{j\eta} \end{bmatrix} e^{j\phi} \qquad (6.217)$$

Now the common phase is normally of little interest and may be omitted. In the next section we shall see that if the ratio of the eigenvalue magnitudes is written as $\tan^2 \beta$, then β has a geometric significance. When these changes are made, the diagonal scattering matrix can be written as

$$\mathbf{S}_d = |\gamma_1| \begin{bmatrix} 1 & 0 \\ 0 & \tan^2 \beta e^{j\eta} \end{bmatrix} \quad (6.218)$$

Phase angle η is the relative phase of the two orthogonal scatterers. Huynen has called $-\eta/4$ the *target skip angle* since its values are related to polarization changes dependent on the number of reflections (bounces) of the signal [2].

The Huynen Fork

The Stokes parameters corresponding to the polarization ratios P_1, P_2, P_3, and P_4, when plotted on the Poincaré sphere, form an interesting pattern that may be seen in Figure 6.12. It is somewhat like a fork with a handle and three tines and is known as the *polarization fork* or *Huynen fork* [2].

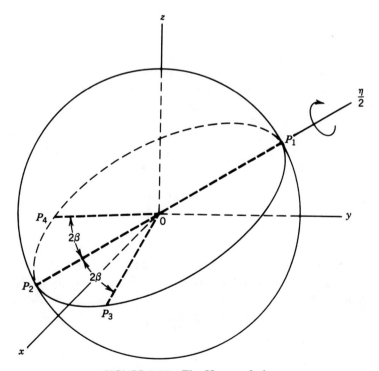

FIGURE 6.12 The Huynen fork.

342 POLARIZATION CHANGES BY REFLECTION AND TRANSMISSION

From (3.185) and (3.189) the Stokes parameters corresponding to polarization ratio P are

$$\frac{G_1}{G_0} = \frac{1 - |P|^2}{1 + |P|^2} \tag{6.219a}$$

$$\frac{G_2}{G_0} = \frac{2\,\text{Re}(P)}{1 + |P|^2} \tag{6.219b}$$

$$\frac{G_3}{G_0} = \frac{2\,\text{Im}(P)}{1 + |P|^2} \tag{6.219c}$$

It was noted earlier that polarization ratios P_1 and P_2 obtained by maximizing backscattered received power are related by

$$P_2 = -\frac{1}{P_1^*} \tag{6.159}$$

The Stokes parameters corresponding to these polarization ratios, found by substituting this equation into the preceding one, are related by

$$G_1^{(2)} = -G_1^{(1)} \tag{6.220a}$$
$$G_2^{(2)} = -G_2^{(1)} \tag{6.220b}$$
$$G_3^{(2)} = -G_3^{(1)} \tag{6.220c}$$

where the superscripts identify the polarization ratio to which the Stokes parameter corresponds. We see from this equation that the plotted polarization points corresponding to polarization ratios P_1 and P_2 lie at opposite ends of a diameter of the Poincaré sphere.

Next consider the location of points corresponding to P_3 and P_4, the polarization ratios giving minimum received backscattered power. If we think of a three-element vector composed of the three independent Stokes parameters normalized to G_0 (this is not the Stokes vector to be discussed later),

$$\mathbf{g} = \frac{1}{G_0} \begin{bmatrix} G_1 \\ G_2 \\ G_3 \end{bmatrix} \tag{6.221}$$

then the angle between two rays drawn from the sphere origin to points corresponding to polarization ratios P_a and P_b, on a Poincaré sphere of radius G_0, is given by

$$\cos \beta_{ab} = \mathbf{g}_a \cdot \mathbf{g}_b = \frac{G_1^a G_1^b}{G_0 G_0} + \frac{G_2^a G_2^b}{G_0 G_0} + \frac{G_3^a G_3^b}{G_0 G_0} \tag{6.222}$$

If a substitution of the Stokes parameters as functions of the polarization ratios is made, this equation becomes

$$\cos \beta_{ab} = \frac{1 - |P_a|^2 - |P_b|^2 + |P_a|^2|P_b|^2 + 2P_aP_b^* + 2P_a^*P_b}{(1 + |P_a|^2)(1 + |P_b|^2)} \quad (6.223)$$

If the polarization ratios P_1 and P_2, corresponding respectively to maximum received power and a submaximum power (which in some cases is equal to the maximum) and given by (6.156) and (6.157), are used in this equation for the central angle it is readily shown that

$$\cos \beta_{12} = -1 \quad (6.224)$$

This confirms what was already known, that points corresponding to P_1 and P_2 lie at opposite ends of a diameter of the Poincaré sphere.

If the pairs P_1, P_3 and P_1, P_4, where P_3 and P_4 are given by (6.170) and (6.171), are used, it can be shown by considerable manipulation that

$$\cos \beta_{13} = \cos \beta_{14} \quad (6.225)$$

Finally, if the appropriate polarization ratio pairs are used, it can be shown, again with considerable effort, that

$$\beta_{34} = 2\beta_{13} \quad (6.226)$$

The last two equations can be satisfied only if the points corresponding to P_1, P_2, P_3, and P_4 lie on a great circle of the Poincaré sphere, that is, all points lie in a plane bisecting the sphere. Further, the central angle formed by rays to the points P_1 and P_3 is equal to that between rays to P_1 and P_4; the two outer tines of the fork are symmetrically placed about the handle and center tine.

To plot the polarization fork (Huynen fork) on the Poincaré sphere, only the points corresponding to P_1 and P_3 need be determined, since P_2 is at the opposite end of a sphere diameter from P_1, and P_4 is the image of P_3 about the line $\overline{P_1P_2}$. The points may be located as Cartesian coordinates G_1, G_2, and G_3, with the Stokes parameters found from (6.219). (As a matter of notational convenience, references will generally be made to polarization points P on the sphere, although of course it is the Stokes parameters corresponding to P that are actually plotted.) An alternative procedure is to plot each point from its azimuth angle 2τ and elevation angle 2ϵ determined by combining (3.183), (3.185), and (3.189),

$$\tan 2\tau = \frac{G_2}{G_1} = \frac{2\,\mathrm{Re}(P)}{1 - |P|^2} \quad (6.227a)$$

$$\sin 2\epsilon = \frac{G_3}{G_0} = \frac{2\,\mathrm{Im}(P)}{1 + |P|^2} \quad (6.227b)$$

It should be noted that the location of P_1 may change if the coordinate reference changes, but the shape of the Huynen fork is dependent only on the target and is invariant to a rotation of either the radar or the target about the line of sight between radar and target.

The polarization fork is also determined by the two null polarizations P_3 and P_4. To find P_1 and P_2 from the known P_3 and P_4, a sphere diameter is drawn to bisect the central angle formed by P_3 and P_4. Then P_2 lies at the diameter end within the interior angle $\overline{P_3OP_4}$.

Other significant points characteristic of the target can be placed on the Poincaré sphere and related to the polarization fork. To avoid complexity, they are not discussed at this time.

It was stated earlier that if the ratio of eigenvalue magnitudes for maximum received power (a submaximum for γ_2) is taken as $\tan^2 \beta$, then β has a geometric significance. To see this, note that the central angle β_{23} of the Huynen fork is the angle between adjacent tines of the fork since the tines originate at the Poincaré sphere center. From (6.223), β may be shown to satisfy

$$\cos^2 \beta_{23} = \frac{B - 2RR^*}{B + 2RR^*} \tag{6.228}$$

where

$$B = |S_{xx}|^2 + 2|S_{xy}|^2 + |S_{yy}|^2 \tag{6.142a}$$

$$C = |S_{xx}S_{yy}|^2 + |S_{xy}|^4 - 2\,\text{Re}\!\left(S_{xx}S_{xy}^{*2}S_{yy}\right) \tag{6.142b}$$

and

$$R = \sqrt{S_{xy}^2 - S_{xx}S_{yy}} \tag{6.171}$$

If angle β is defined by the ratio of the eigenvalues giving maximum power (Note: This is a submaximum, a relative maximum, for γ_2, but for convenience, this is not pointed out each time this power is discussed.) according to

$$\tan^2 \beta = \frac{|\gamma_2|}{|\gamma_1|} = \frac{\left[B - (B^2 - 4C)^{1/2}\right]^{1/2}}{\left[B + (B^2 - 4C)^{1/2}\right]^{1/2}} \tag{6.229}$$

where the eigenvalues are taken from (6.147), it is not difficult to show that

$$\beta_{23} = 2\beta \tag{6.230}$$

The angle made by the tines of the Huynen fork is therefore related to the relative eigenvalue magnitudes. Angle β has been called the *characteristic angle* of a target [2] and also the *polarizability angle* [7]. Polarizability is the target characteristic that causes it to scatter an initially unpolarized wave (see Chapter 7) as one with a greater degree of polarization. As an example, an unpolarized wave has x and y components that are uncorrelated and have equal power densities. If it is incident on a target that backscatters the x but not the y component, then the backscattered wave is polarized. Targets with larger β values (more nearly equal eigenvalue magnitudes) are less able to polarize waves than those of smaller β values.

It is desirable to find out if the relative phase η of the scattering matrix eigenvalues appearing in the diagonal scattering matrix is a parameter of the Huynen fork. To do so, let the characteristic polarization ratio P_1 represent horizontal polarization and P_2 vertical. The shape of the polarization fork is not affected by this specialization. The diagonal scattering matrix may then be interpreted in Cartesian coordinates and written as

$$\mathbf{S}_d = \begin{bmatrix} S_{xx} & 0 \\ 0 & S_{yy} \end{bmatrix} = \begin{bmatrix} 1 & 0 \\ 0 & a^2 e^{j\eta} \end{bmatrix} \qquad 0 < a < 1 \qquad -\pi < \eta < \pi \quad (6.231)$$

The polarization fork for this case is completely determined by the location of the Poincaré sphere point corresponding to P'_3, where the prime denotes values obtained by treating the elements of the diagonal scattering matrix as rectangular elements in the equations for finding polarization ratios, tilt angles, and so on. Using (6.170) and (6.171) on the elements of $\mathbf{S}_d$ gives

$$P'_3 = \frac{j}{a} e^{-j\eta/2} \qquad (6.232)$$

The corresponding azimuth and elevation angles on the Poincaré sphere are, from (6.227),

$$\tan 2\tau'_3 = \frac{-2a \sin(\eta/2)}{1 - a^2} \qquad (6.233a)$$

$$\sin 2\epsilon'_3 = \frac{2a \cos(\eta/2)}{1 + a^2} \qquad (6.233b)$$

Figure 6.13 is a view of the Poincaré sphere from the vertical polarization position ($-x$ or $-G_1$ axis) with the point corresponding to P'_3 projected onto the yz (or $G_2 G_3$) plane. Using $2\tau'_3$ and $2\epsilon'_3$ in their geometrical significance as azimuth and elevation angles, we find lengths

$$\overline{AP'_3} = \sin 2\tau'_3 \cos 2\epsilon'_3 \qquad (6.234a)$$

$$\overline{BP'_3} = \sin 2\epsilon'_3 \qquad (6.234b)$$

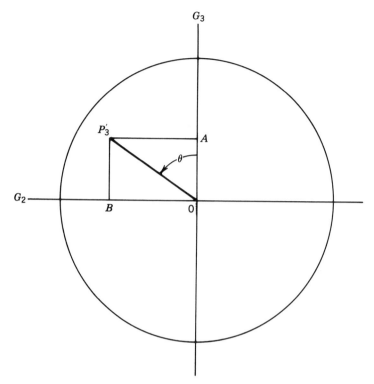

FIGURE 6.13 Rotation angle of the Huynen fork.

Then angle θ, which may be considered the angle by which the tines of the Huynen fork are rotated about the fork handle, is given by

$$\tan \theta = \frac{\sin 2\tau_3' \cos 2\epsilon_3'}{\sin 2\epsilon_3'} \qquad (6.235)$$

If the values found for $2\tau_3'$ and $2\epsilon_3'$ are used, it is readily seen that this rotation angle is

$$\theta = \frac{\eta}{2} \qquad (6.236)$$

The effect of the relative phase of the eigenvalues on the Huynen fork is now clear. For $\eta = 0$ the fork tines lie in a vertical $(G_1 G_3)$ plane (for this special case). As η increases, the tines rotate in a clockwise direction (looking from the G_1 axis) around the fork handle by angle $\eta/2$.

We noted earlier a means of constructing the polarization fork from the nondiagonalized target scattering matrix, determining first the characteristic polarization ratios P_1 and P_3 and plotting points corresponding to each by use of azimuth and elevation angles 2τ and 2ϵ on the Poincaré sphere. The geometric interpretation of β and η of the diagonal scattering matrix allows a different construction process: Determine the azimuth and elevation angles $2\tau_1$ and $2\epsilon_1$ corresponding to P_1 from the nondiagonalized scattering matrix. Locate P_1 by rotating $2\tau_1$ in the G_1G_2 plane of the Poincaré sphere and then $2\epsilon_1$ along a great circle passing through the sphere pole $(0,0,G_3)$. P_2 may then be located immediately. The great circle passing through P_1 and the pole determines a vertical plane and a new "horizontal" plane perpendicular to it and including the points P_1, P_2, and the sphere center. In the vertical plane, lay off an angle 2β from the line $\overline{P_1P_2}$ and construct a ray to the sphere surface. Rotating the Huynen fork around its handle (or the line $\overline{P_1P_2}$) in a clockwise direction by angle $\eta/2$ locates P_3 on the Poincaré sphere, and P_4 is its image around $\overline{P_1P_2}$. As an alternative to the use of angles β and η, the polarization ratio P_3' may be found from the diagonal scattering matrix. The point corresponding to P_3 is then found by rotating $2\tau_3'$ in the new "horizontal" plane from the point P_1 and then by $2\epsilon_3'$ in a new vertical plane perpendicular to this "horizontal" plane.

Some useful properties of the characteristic polarizations may be inferred from an examination of the Huynen fork:

1. The cross-polarization nulls P_1 and P_2 are distinct.
2. a. If one cross-polarization null represents linear polarization, so does the other.
   b. If one cross-polarization null represents circular polarization, so does the other.
3. If the copolarization nulls P_3 and P_4 are identical, they coincide with P_2, which is then the polarization for zero received power.
4. If the copolarization nulls P_3 and P_4 represent orthogonal waves, so that $P_3 = -1/P_4^*$, then $\beta = \pi/4$.
5. If one copolarization null, say P_3, is for left-circular polarization and the other, P_4, for right-circular, then $\beta = \pi/4$ and both cross-polarization nulls are linear.

The Scattering Matrix in Geometric Form

The stated goal of this section, expression of the target scattering matrix in terms of geometric variables of the characteristic polarization states, can now be realized.

The polarization states giving maximum received power from a target are

$$\hat{\mathbf{h}}_1(\tau_1, \epsilon_1) = \begin{bmatrix} \cos \tau_1 & -\sin \tau_1 \\ \sin \tau_1 & \cos \tau_1 \end{bmatrix} \begin{bmatrix} \cos \epsilon_1 \\ j \sin \epsilon_1 \end{bmatrix} \quad (6.214a)$$

$$\hat{\mathbf{h}}_2(\tau_2, \epsilon_2) = \hat{\mathbf{h}}_1(\tau_1 + \pi/2, -\epsilon_1) = \begin{bmatrix} -\sin \tau_1 & -\cos \tau_1 \\ \cos \tau_1 & -\sin \tau_1 \end{bmatrix} \begin{bmatrix} \cos \epsilon_1 \\ -j \sin \epsilon_1 \end{bmatrix}$$
$$(6.214b)$$

where τ_1 and ϵ_1 are the tilt and ellipticity angles of the polarization ellipse of the antenna of effective length $\mathbf{h}_1$ that maximizes received power. If matrices

$$\mathbf{V}(\tau_1) = \begin{bmatrix} \cos \tau_1 & -\sin \tau_1 \\ \sin \tau_1 & \cos \tau_1 \end{bmatrix} \tag{6.237}$$

$$\mathbf{W}(\epsilon_1) = \begin{bmatrix} \cos \epsilon_1 & j \sin \epsilon_1 \\ j \sin \epsilon_1 & \cos \epsilon_1 \end{bmatrix} \tag{6.238}$$

are defined, it is seen that the matrix $\mathbf{U}$ that diagonalizes the scattering matrix,

$$\mathbf{U} = \begin{bmatrix} \hat{\mathbf{h}}_1 & \hat{\mathbf{h}}_2 \end{bmatrix}$$
$$= \begin{bmatrix} \cos \tau_1 \cos \epsilon_1 - j \sin \tau_1 \sin \epsilon_1 & -\sin \tau_1 \cos \epsilon_1 + j \cos \tau_1 \sin \epsilon_1 \\ \sin \tau_1 \cos \epsilon_1 + j \cos \tau_1 \sin \epsilon_1 & \cos \tau_1 \cos \epsilon_1 + j \sin \tau_1 \sin \epsilon_1 \end{bmatrix},$$
$$\tag{6.239}$$

is the product

$$\mathbf{U}(\tau_1, \epsilon_1) = \mathbf{V}(\tau_1)\mathbf{W}(\epsilon_1) \tag{6.240}$$

The diagonal scattering matrix may also be written as a product. If matrices $\mathbf{L}$ and $\mathbf{N}$ are defined by

$$\mathbf{L}(\eta) = \begin{bmatrix} 1 & 0 \\ 0 & e^{j\eta/2} \end{bmatrix} \tag{6.241}$$

$$\mathbf{N}(\beta) = \begin{bmatrix} 1 & 0 \\ 0 & \tan^2 \beta \end{bmatrix} \tag{6.242}$$

then

$$\mathbf{S}_d(|\gamma_1|, \beta, \eta) = |\gamma_1|\mathbf{L}(\eta)\mathbf{N}(\beta)\mathbf{L}(\eta) \tag{6.243}$$

Finally, the target scattering matrix may be written as

$$\mathbf{S} = \mathbf{U}^*\mathbf{S}_d\mathbf{U}^{T*} = |\gamma_1|\mathbf{V}^*(\tau_1)\mathbf{W}^*(\epsilon_1)\mathbf{L}(\eta)\mathbf{N}(\beta)\mathbf{L}(\eta)\mathbf{W}^{T*}(\epsilon_1)\mathbf{V}^{T*}(\tau_1) \tag{6.244}$$

All the variables that appear in this equation, with the exception of $|\gamma_1|$, are those that determined the Huynen fork. It follows that if the Huynen fork is known, the scattering matrix of a target, to within a multiplicative constant, can be found.

6.10. SCATTERING MATRICES AND HUYNEN FORK PARAMETERS FOR SOME COMMON SCATTERERS

In this section the scattering matrices for several simple conducting scatterers will be developed, together with their characteristic polarizations and polarization fork parameters.

Flat Plate at Normal Incidence

For an infinite flat conducting plate at normal incidence, the incident and backscattered fields are related by

$$E_x^s = -E_x^i \qquad (6.245a)$$
$$E_y^s = -E_y^i \qquad (6.245b)$$

The scattering matrix is then

$$\mathbf{S} = C\begin{bmatrix} -1 & 0 \\ 0 & -1 \end{bmatrix} \qquad (6.246)$$

where C is some constant. For a disk of radius R, a physical optics approach gives a normal-incidence radar cross section for linear polarization [8]

$$\sigma_r = \frac{4\pi}{\lambda^2}(\pi R^2)^2 \qquad (6.247)$$

But from Section 6.8, the radar cross section of a target with scattering matrix $\mathbf{S}$ and incident wave polarization state $\hat{\mathbf{h}}$ (from an antenna with that normalized effective length) is

$$\sigma_r = |\hat{\mathbf{h}}^T \mathbf{S} \hat{\mathbf{h}}|^2 \qquad (6.248)$$

The disk scattering matrix is then

$$\mathbf{S} = \frac{2\sqrt{\pi}}{\lambda}(\pi R^2)\begin{bmatrix} -1 & 0 \\ 0 & -1 \end{bmatrix} \qquad (6.249)$$

since substitution of it into the preceding equation, using linear polarization, gives a radar cross section in agreement with the physical optics approximation.

Substitution of the scattering matrix into the equations for optimum polarization ratio and the Huynen fork parameters yields

$$|\gamma_1| = \frac{2\sqrt{\pi}}{\lambda}(\pi R^2) \quad \beta = \frac{\pi}{4} \quad \tau_1 \text{ (see below)} \qquad (6.250)$$
$$\epsilon_1 = 0 \qquad \qquad \eta = 0$$

350 POLARIZATION CHANGES BY REFLECTION AND TRANSMISSION

For a flat circular disk, it is apparent that rotation of the radar about its line of sight does not alter the received power, so tilt angle τ_1 is a matter of indifference. From the fact that the Poincaré sphere point corresponding to P_1 lies on the equator and from the values of β and η, it is apparent that P_3 and P_4 represent right- and left-circular polarization. Thus, either sense of circular polarization is a co-pol null for the flat circular disk at normal incidence.

For a noncircular plate, in general $S_{xx} \neq S_{yy}$ and the results given above do not hold.

Sphere

It was noted earlier that if a target is symmetric about a plane containing a ray from antenna to target in a monostatic radar system, the coordinate system can be chosen so that the off-diagonal elements in the rectangular scattering matrix are zero. For the sphere, any plane containing a ray from radar to sphere center is a symmetry plane. Further, the diagonal elements of the scattering matrix are equal. Then

$$\mathbf{S} = S_{xx} \begin{bmatrix} 1 & 0 \\ 0 & 1 \end{bmatrix} \tag{6.251}$$

The reader is referred to the literature to find cross sections of the sphere [8].

As for the flat circular disk, all linear polarizations are equivalent in received power. Then τ_1 is a matter of indifference, and the Huynen fork parameters of significance are

$$\epsilon_1 = 0 \quad \eta = 0 \quad \beta = \pi/4 \tag{6.252}$$

Like the flat circular disk, either sense of circular polarization is a co-pol null for the sphere. A spherical target is invisible to a circularly polarized radar. For this reason, it is common for radars to use circular polarization to suppress the return from raindrops, which to a first approximation are spherical. The return from a desired target may be reduced also with circular polarization, but generally the ratio of desired signal to rain clutter is improved [9].

If the raindrops are significantly deformed from the spherical so that, for example, $|S_{xx}| > |S_{yy}|$, then for maximum rejection of the rain signal it is appropriate to make $|h_x| < |h_y|$.

Dihedral Corner

Consider first a dihedral corner oriented so that its fold line (intersection line between the planes forming the dihedral) is parallel to the y axis. Figure 6.14 shows this corner with incident E_x and E_y fields (in the figure a dot

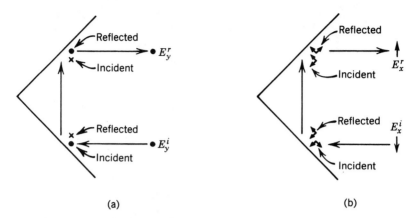

FIGURE 6.14 Electric field components for dihedral corner.

represents an electric field out of the paper and a cross one into the paper). It is easily seen from the figure that the E_y component is reflected unchanged (except for the phase change with distance, which is not shown), while the E_x component is reversed. No cross-polarized component exists. Then the scattering matrix for a corner constructed with semi-infinite plates is

$$\mathbf{S} = \begin{bmatrix} -1 & 0 \\ 0 & 1 \end{bmatrix} \quad (6.253)$$

Consider next a dihedral whose fold line is rotated in the xy plane by angle θ measured from the y axis toward the x axis (Figure 6.15). The dihedral plates are so large that edge effects are negligible. If the incident field is y directed, E_y^i, then components parallel and perpendicular to the dihedral fold line are

$$E_{x'} = -E_y^i \sin \theta \quad (6.254a)$$
$$E_{y'} = E_y^i \cos \theta \quad (6.254b)$$

After two reflections

$$E_{x'}^s = -E_{x'} = E_y^i \sin \theta \quad (6.255a)$$
$$E_{y'}^s = E_{y'} = E_y^i \cos \theta \quad (6.255b)$$

From the figure, the x and y components of the reflected field are

$$E_x^s = E_{x'}^s \cos \theta + E_{y'}^s \sin \theta = E_y^i \sin \theta \cos \theta + E_y^i \sin \theta \cos \theta \quad (6.256a)$$
$$E_y^s = E_{y'}^s \cos \theta - E_{x'}^s \sin \theta = E_y^i \cos^2 \theta - E_y^i \sin^2 \theta \quad (6.256b)$$

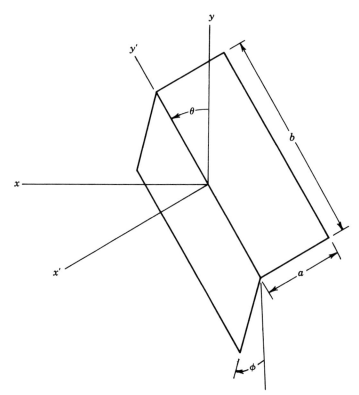

FIGURE 6.15 Tilted dihedral corner.

It follows that

$$S_{xy} = 2\sin\theta\cos\theta = \sin 2\theta \qquad (6.257a)$$

$$S_{yy} = \cos^2\theta - \sin^2\theta = \cos 2\theta \qquad (6.257b)$$

If the incident field is x directed,

$$E_{x'} = E_x^i \cos\theta \qquad (6.258a)$$

$$E_{y'} = E_x^i \sin\theta \qquad (6.258b)$$

After two reflections

$$E_{x'}^s = -E_x^i \cos\theta \qquad (6.259a)$$

$$E_{y'}^s = E_x^i \sin\theta \qquad (6.259b)$$

and

$$E_x^s = E_{x'}^s \cos\theta + E_{y'}^s \sin\theta = -E_x^i \cos^2\theta + E_x^i \sin^2\theta \quad (6.260a)$$
$$E_y^s = E_{y'}^s \cos\theta - E_{x'}^s \sin\theta = E_x^i \sin\theta\cos\theta + E_x^i \sin\theta\cos\theta \quad (6.260b)$$

It is then obvious that

$$S_{xx} = \sin^2\theta - \cos^2\theta = -\cos 2\theta \quad (6.261a)$$
$$S_{yx} = 2\sin\theta\cos\theta = \sin 2\theta \quad (6.261b)$$

For a dihedral corner with plate dimensions a and b (Fig. 6.15), the vertical polarization radar cross section with $\theta = 0$ is [8]

$$\sigma_r = \frac{16\pi a^2 b^2 \sin^2(\pi/4 + \phi)}{\lambda^2} \quad (6.262)$$

if the incident ray path is perpendicular to the dihedral fold line. The scattering matrix for a dihedral rotated by angle θ in the xy plane from the y axis is

$$\mathbf{S} = \frac{4\sqrt{\pi}\, ab\, \sin(\pi/4 + \phi)}{\lambda} \begin{bmatrix} -\cos 2\theta & \sin 2\theta \\ \sin 2\theta & \cos 2\theta \end{bmatrix} \quad (6.263)$$

since it yields the correct radar cross section when substituted in (6.248).

If (6.156) is used in an attempt to find the cross-polarization null P_1, it is indeterminate. However, the polarization ratio P_3, for minimum received power, is determinate and may be found from (6.170) and (6.171) to be

$$P_3 = \frac{1 - \sin 2\theta}{\cos 2\theta} \quad (6.264)$$

The diagonal form of the scattering matrix is readily found by setting $\theta = 0$ in (6.263). Neglecting the constant multiplier, it is

$$\mathbf{S}_d = -1\begin{bmatrix} 1 & 0 \\ 0 & -1 \end{bmatrix} \quad (6.265)$$

from which it is seen that $\beta = \pi/4$ and the central angle made by the Huynen fork tines is π. Then P_3 and P_4 are orthogonal. Further, since P_3 is real, P_3 and P_4 represent linear polarizations. The resultant plot of characteristic polarizations on the Poincaré sphere is shown in Figure 6.16. The copolarization nulls P_3 and P_4 lie on the sphere equator at opposite ends of a diameter. The cross-polarization nulls lie on the great circle defined by a plane through the sphere center perpendicular to the line $\overline{P_3 P_4}$.

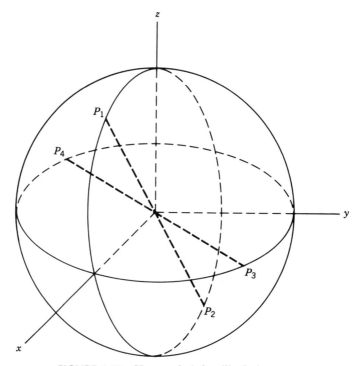

FIGURE 6.16 Huynen fork for dihedral corner.

The Huynen fork parameters of interest for the dihedral corner are then

$$|\gamma_1| = \frac{4\sqrt{\pi}\, ab \sin(\pi/4 + \phi)}{\lambda}$$

$$P_3 = \frac{1 - \sin 2\theta}{\cos 2\theta}$$

P_1—indeterminate subject to constraint (6.266)

$$\beta = \pi/4 \qquad 2\tau_3 = 2\theta + \pi/2$$

$$\eta = \pi \qquad 2\epsilon_3 = 0$$

Trihedral Corner

In Section 6.2 we saw that a trihedral corner reflector constructed of very large plates has the same polarization properties as a flat plate at normal incidence, namely $P^s = -P^i$. The scattering matrix of the trihedral corner of finite size can then be expected to be that for the flat plate, (6.246), modified to account for the finite cross section. Ruck et al. give the maximum cross

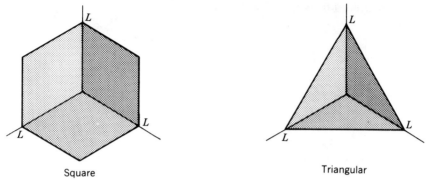

FIGURE 6.17 Trihedral corners.

sections of the square and triangular trihedral corners of Figure 6.17 [8]:

$$\text{Square} \quad \sigma_r = \frac{12\pi L^4}{\lambda^2} \qquad (6.267a)$$

$$\text{Triangular} \quad \sigma_r = \frac{4\pi L^4}{\lambda^2} \qquad (6.267b)$$

Then, if edge effects are neglected, the scattering matrices at angles giving maximum cross section are:

$$\text{Square} \quad \mathbf{S} = \frac{2\sqrt{3\pi}\,L^2}{\lambda}\begin{bmatrix} -1 & 0 \\ 0 & -1 \end{bmatrix} \qquad (6.268a)$$

$$\text{Triangular} \quad \mathbf{S} = \frac{2\sqrt{\pi}\,L^2}{\lambda}\begin{bmatrix} -1 & 0 \\ 0 & -1 \end{bmatrix} \qquad (6.268b)$$

From the scattering matrices, these parameters are found:

$$|\gamma_1| = \begin{cases} 2\sqrt{3\pi}\,L^2/\lambda & \text{Square} \\ 2\sqrt{\pi}\,L^2/\lambda & \text{Triangular} \end{cases} \qquad (6.269)$$

τ_1—arbitrary

$\beta = \pi/4 \quad \eta = 0$

Wire Grid or Single Long Wire

Consider a target of thin parallel wires or a single long wire in the xy plane, normal to the radar–target line of sight, making angle θ with the x axis. If the incident field is x directed, the induced current in each wire is

$$I = C_1 E_x^i \cos\theta \qquad (6.270)$$

and the scattered field components are

$$E_x^s = C_2 I \cos\theta = C_1 C_2 E_x^i \cos^2\theta \qquad (6.271a)$$
$$E_y^s = C_2 I \sin\theta = C_1 C_2 E_x^i \cos\theta \sin\theta \qquad (6.271b)$$

If the incident field is y directed, the wire current is

$$I' = C_1 E_y^i \sin\theta \qquad (6.272)$$

and the scattered fields are

$$E_x^s = C_1 C_2 E_y^i \cos\theta \sin\theta \qquad (6.273a)$$
$$E_y^s = C_1 C_2 E_y^i \sin^2\theta \qquad (6.273b)$$

The fields for the two cases lead to a scattering matrix

$$\mathbf{S} = C \begin{bmatrix} \cos^2\theta & \cos\theta \sin\theta \\ \cos\theta \sin\theta & \sin^2\theta \end{bmatrix} \qquad (6.274)$$

and Huynen fork parameters

$$\begin{aligned} P_1 &= \tan\theta & \beta &= 0 \\ \tau_1 &= 0 & \eta &\text{—indeterminate} \\ \epsilon_1 &= 0 \end{aligned} \qquad (6.275)$$

Two Diagonal Wires

A V-shaped structure of two wires as shown in Figure 6.18 has a scattering matrix that is the sum of scattering matrices of a wire at angle θ and another

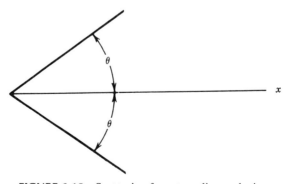

FIGURE 6.18 Scattering from two diagonal wires.

at $-\theta$. Adding the scattering matrices of two such wires gives

$$\mathbf{S} = 2C \begin{bmatrix} \cos^2 \theta & 0 \\ 0 & \sin^2 \theta \end{bmatrix} = 2C \cos^2 \theta \begin{bmatrix} 1 & 0 \\ 0 & \tan^2 \theta \end{bmatrix} \quad (6.276)$$

This is in a diagonal form since we chose the x axis to lie in the symmetry plane. The polarization fork parameters are quickly determined to be

$$\begin{array}{ccc} P_1 = 0 & \theta < \pi/4 & \beta = \theta \\ \tau_1 = 0 & \epsilon_1 = 0 & \eta = 0 \end{array} \quad (6.277)$$

Maximum power is received for linear horizontal polarization (and a submaximum for linear vertical). This is a rare target type for which the characteristic angle β corresponds to a physical angle of the target.

Symmetric Target

Target symmetry was discussed in this chapter when the scattering matrix was first introduced. A target is symmetric if a plane exists for which each point of the target has a corresponding point an equal distance on the opposite side of the plane and if a ray from the radar to the target lies in the plane. All targets that are symmetric about an axis, such as cylinders, ellipsoids, and cones, satisfy this definition of target symmetry at all aspect angles. Other targets may be symmetric at some aspect angles. If the coordinate system is chosen appropriately, perhaps by rotating the radar around the radar–target line of sight, the off-diagonal elements of the scattering matrix are zero. The diagonal terms of the scattering matrix may still be interpreted as rectangular elements. In other words, if the diagonal scattering matrix is of the form

$$\mathbf{S}_d = \begin{bmatrix} S_{xx} & 0 \\ 0 & S_{yy} \end{bmatrix} \quad (6.278)$$

in a properly chosen rectangular coordinate system, the target is symmetric. The Huynen fork parameters, determined with respect to the new coordinate system, are

$$\begin{array}{ccc} P_1 = 0 & & \tan^2 \beta = \left| \dfrac{S_{yy}}{S_{xx}} \right| \\ \tau_1 = 0 & \epsilon_1 = 0 & \eta = \mathrm{Ang}\left(\dfrac{S_{yy}}{S_{xx}} \right) \end{array} \quad (6.279)$$

We might consider extending this definition to define a "generally symmetric" target as one that has a diagonal scattering matrix in some orthogonal polarization base vector set. It is not a useful definition, however, since every target (with a scattering matrix) can be represented by a diagonal matrix in some base vector set. It is interesting, however, to consider a target that is "symmetric" in the circular-component sense.

"Symmetric" Target in Circular Components

Consider a target that, with an appropriate choice of coordinates, obtained perhaps by rotation of the radar about the radar–target line of sight, can be represented by a diagonal matrix whose elements satisfy

$$\begin{bmatrix} E_R^s \\ E_L^s \end{bmatrix} = \frac{1}{\sqrt{4\pi}\, r} \begin{bmatrix} S_{RR} & 0 \\ 0 & S_{LL} \end{bmatrix} \begin{bmatrix} E_R^i \\ E_L^i \end{bmatrix} \quad (6.280)$$

where the field vector elements are the right- and left-circular components of the incident and scattered waves. The diagonal scattering matrix is then

$$\mathbf{S}_d = \begin{bmatrix} S_{RR} & 0 \\ 0 & S_{LL} \end{bmatrix} \quad (6.281)$$

It is obvious that not all targets can be so represented. It should be noted, too, that if the diagonal matrix elements are in rectangular form, the corresponding target is symmetric in the common (or visual) sense. That should not be expected for this matrix, whose elements determine the circular components of the scattered wave.

To examine this "circularly symmetric" scattering matrix, use the relations of (6.94). Then in a rectangular polarization base the matrix (no longer diagonal) is

$$\mathbf{S} = \begin{bmatrix} S_{xx} & S_{xy} \\ S_{xy} & S_{yy} \end{bmatrix} = \frac{1}{2} \begin{bmatrix} S_{RR} + S_{LL} & j(S_{RR} - S_{LL}) \\ j(S_{RR} - S_{LL}) & -(S_{RR} + S_{LL}) \end{bmatrix} \quad (6.282)$$

which can be written as

$$\mathbf{S} = \begin{bmatrix} a & b \\ b & -a \end{bmatrix} \quad (6.283)$$

This target has been called an "N-target" or "nonsymmetric noise" target, and is associated with nonsymmetric target–noise components of the scattering from nonsymmetric rough surfaces [2]. Here we see that it is symmetric in a circular-component sense. It is not symmetric in the common usage of the word and cannot be made so by rotation of the radar.

The Huynen fork parameters are

$$P_1 = j \text{ if } |S_{LL}| > |S_{RR}| \quad \tan^2 \beta = \left|\frac{S_{LL}}{S_{RR}}\right| = \left|\frac{a + jb}{a - jb}\right|$$

$$\tau_1 \text{ arbitrary} \quad \epsilon_1 = \pi/4 \quad \eta = \text{Ang}\left(\frac{S_{LL}}{S_{RR}}\right) \quad (6.284)$$

Note that $(a + jb) \neq (a - jb)^*$ since a and b are complex.

"Helix" Target

If only S_{RR} of a target has value, the diagonal scattering matrix is

$$\mathbf{S}_d = \begin{bmatrix} S_{RR} & 0 \\ 0 & 0 \end{bmatrix} \quad (6.285)$$

and the target matrix in rectangular components is

$$\mathbf{S} = \begin{bmatrix} a & ja \\ ja & -a \end{bmatrix} = a \begin{bmatrix} 1 & j \\ j & -1 \end{bmatrix} \quad (6.286)$$

This matrix leads to

$$P_1 = -j \quad \beta = 0$$
$$\tau_1 \text{ arbitrary} \quad \epsilon_1 = -\pi/4 \quad \eta \text{ arbitrary} \quad (6.287)$$

The radar cross section found from

$$\sigma_r = |\hat{\mathbf{h}}^T \mathbf{S} \hat{\mathbf{h}}|^2 \quad (6.248)$$

if $\hat{\mathbf{h}}$ represents a linearly polarized antenna oriented at angle θ with respect to an axis transverse to the radar–target line of sight, is found to be independent of θ.

If only S_{LL} for the target has value,

$$\mathbf{S} = a \begin{bmatrix} 1 & -j \\ -j & -1 \end{bmatrix} \quad (6.288)$$

and

$$P_1 = j \quad \beta = 0$$
$$\tau_1 \text{ arbitrary} \quad \epsilon_1 = \pi/4 \quad \eta \text{ arbitrary} \quad (6.289)$$

These targets have been called "helix" targets [2] since they scatter right- and left-circular waves, and a helix antenna, in the axial mode, can radiate such waves. The helix targets here are right-handed (first one) and left-handed (second one). Note that an essential feature of the helix targets is that the ellipticity angle is $\pm\pi/4$. Any target with a non-zero ellipticity angle ϵ_1 is said to possess a property called *helicity* [2]. It was noted earlier that symmetric targets (in the common meaning) have zero ellipticity angles, and do not have this property. Nonsymmetric targets do have the helicity property.

6.11. ADDITIONAL CHARACTERISTIC POLARIZATIONS

In previous sections three characteristic polarization pairs were developed and plotted as the Huynen fork on the Poincaré sphere. They were the co-pol maxima P_1 and P_2, which coincide with the cross-pol nulls, and the co-pol nulls P_3 and P_4. It was pointed out that the co-pol nulls do not in general coincide with polarizations giving maximum cross-polarized power.

Boerner has derived the polarizations giving maximum cross-polarized power [10]. These are shown as polarizations P_5 and P_6 on the Poincaré sphere of Figure 6.19, on which the Huynen fork polarizations are also shown. Both P_5 and P_6 lie on the same great circle that includes the Huynen fork polarizations P_1, P_2, P_3, and P_4. They lie at opposite ends of a sphere

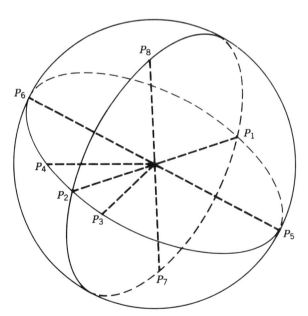

FIGURE 6.19 The Huynen–Boerner plot.

diameter that is perpendicular to the diameter $\overline{P_1 P_2}$. Following the method used to name the other characteristic polarizations, it is appropriate to call this polarization pair the *cross-pol maxima*.

If the transmitting antenna for a radar has either of these characteristic polarizations, the copolarized return power, normalized to the maximum copolarized received power, is

$$\frac{W_c}{W_{cm}} = \frac{\cos^2 2\beta}{4 \cos^4 \beta} \tag{6.290}$$

The cross-polarized power, which is a maximum, is given by

$$\frac{W_{xm}}{W_{cm}} = \frac{1}{4 \cos^4 \beta} \tag{6.291}$$

Note that polarizations P_3, P_4, P_5, and P_6 coincide for $\beta = \pi/4$.

Another characteristic polarization pair, P_7 and P_8, is shown in Figure 6.19. The points have been called "cross-pol saddlepoint extrema" [10]. According to Boerner, at one of these polarizations, the cross-polarization channel power return will increase in some directions symmetric to the point on the complex polarization plane. In some other orthogonal directions the power will decrease.

The copolarized channel power at these polarization states is

$$\frac{W_c}{W_{cm}} = \frac{1}{4 \cos^4 \beta} \tag{6.292}$$

and the cross-polarized power is

$$\frac{W_x}{W_{cm}} = \frac{\cos^2 2\beta}{4 \cos^4 \beta} \tag{6.293}$$

6.12. CHARACTERISTIC POLARIZATIONS WITH NONSYMMETRIC SINCLAIR MATRIX

The Sinclair matrix for bistatic radar is asymmetric, as is that for a target viewed by a monostatic radar if the region between radar and target exhibits Faraday rotation. In this chapter we determined the characteristic polarizations that lead to maximum and minimum received power for a target with a symmetric Sinclair matrix, which in effect limits the configuration to backscattering in a reciprocal medium. The extension to the nonsymmetric scattering matrix can be more readily understood after the Graves polarization power matrix is considered, and for that reason the asymmetric matrix case is deferred to Chapter 7.

REFERENCES

1. J. A. Stratton, *Electromagnetic Theory*, McGraw-Hill, New York, 1941.
2. J. R. Huynen, *Phenomenological Theory of Radar Targets*, Drukkerij Bronder-Offset, N. V., Rotterdam, 1970.
3. E. M. Kennaugh, "Effects of Type of Polarization on Echo Characteristics," Antenna Laboratory Report 389-4, Ohio State University, Columbus, 1950.
4. C. D. Graves, "Radar Polarization Power Scattering Matrix," *Proc. IRE*, **44**, 248–252, February 1956.
5. W-M. Boerner, M. B. El-Arini, C. Y. Chan, S. Saatchi, W. S. Ip, P. M. Mastoris, and B. Y. Foo, "Polarization Utilization in Radar Target Reconstruction," CL-EMID-NANRAR-81-01, University of Illinois, Chicago, IL, January 1981.
6. "IEEE Standard Definitions of Terms for Antennas," IEEE Std. 145-1983, The Institute of Electrical and Electronics Engineers, New York, 1983.
7. W A. Holm, "Polarimetric Fundamentals and Techniques," in *Principles of Modern Radar*, J. L. Eaves and E. K. Reedy, Eds., Van-Nostrand Reinhold, New York, 1987.
8. G. T. Ruck, D. E. Barrick, W. D. Stuart, and C. K. Krichbaum, *Radar Cross Section Handbook*, Plenum, New York, 1970.
9. P. Beckmann, "Optimum Polarization for Polarization Discrimination," *Proc. IEEE*, **56**(10), 1755–1756, October 1968.
10. W-M. Boerner and An-Qing Xi, "The Characteristic Radar Target Polarization State Theory for the Coherent Monostatic and Reciprocal Case Using the Generalized Polarization Transformation Ratio Formulation," *AEU*, **44**(6), June 1990.

PROBLEMS

6.1. For the dihedral corner reflector, let P_3 represent linear horizontal polarization. According to the text discussion, polarization P_1 for maximum received power lies anywhere on a great circle through the poles and passing through the Poincaré sphere point $G_2 = G_0$. Subject to this constraint, choose P_1 arbitrarily and show that the received power given by (6.195) is independent of the value of P_1.

6.2. A right-handed elliptical wave in air strikes the interface with a lossless dielectric material that has $\epsilon_R = 2$. The angle of incidence is 30°. The polarization ratio of the incident wave is $P^i = 2\exp(-j\pi/6)$. Find the polarization ratios of the reflected and transmitted waves.

6.3. A left-handed elliptical plane wave with power density 10 mW/m^2 strikes the interface with a lossless dielectric material that has $\epsilon_R = 2$. The incidence angle is $30°$. In a coordinate system chosen so that the x axis is perpendicular to the plane of incidence, the incident wave has an axial ratio of 2 and a tilt angle of $15°$. Find the axial ratio and tilt angle of the reflected wave with respect to a coordinate system whose x axis is perpendicular to the plane of reflection.

6.4. A left-circular wave from a radar is scattered from a target with Sinclair matrix proportional to

$$\begin{bmatrix} 2 & j1 \\ j1 & 4 \end{bmatrix}$$

Is the scattered wave left- or right-handed? Find its axial ratio.

6.5. Use the general polarization base transformation (3.93) to transform the incident wave on a target from rectangular- to circular-component form. Then use (6.110) to obtain the scattering matrix in circular-component form. Does the relationship thus obtained agree with (6.94)?

6.6. For a target with the scattering matrix of Problem 6.4 find the polarization ratios of a monostatic radar antenna to (a) maximize and (b) minimize copolarized received power.

Find the polarization states (normalized effective lengths) corresponding to the polarization ratios. Show that $\hat{\mathbf{h}}_1$ and $\hat{\mathbf{h}}_2$ are orthonormal.

6.7. Find the backscatter polarization efficiency for the target of Problem 6.4 if the radar has linear vertical polarization.

6.8. Substitute the orthogonal polarizations P_1 and P_2, given by (6.156) and (6.157), into (6.223) to show that P_1 and P_2 represent points at opposite ends of a diameter of the Poincaré sphere.

6.9. For a target with Sinclair matrix

$$\begin{bmatrix} 2 & e^{j\pi/6} \\ e^{j\pi/6} & j4 \end{bmatrix}$$

find the parameters necessary to define the Huynen fork.

6.10. The Huynen fork is completely defined by

$$P_1 = 2e^{j\pi/6} \qquad P_3 = -0.6e^{j\pi/12}$$

Sketch the Huynen fork. Find the scattering matrix (to within a constant multiplier) of the target.

6.11. Find the Sinclair matrix (to within a constant multiplier) of a target consisting of two equal-length wires. One is inclined at an angle of 30° from the x axis in a plane transverse to the radar–target line of sight. The other is inclined at a 45° angle in a plane one-eighth of a wavelength farther from the radar.

6.12. A monostatic radar having effective length $0.5\mathbf{u}_x$ and radiation resistance 10 Ω that is impedance-matched transmits 20 kW at wavelength 0.03 m. A target with Sinclair matrix

$$\begin{bmatrix} 2 & 1 \\ 1 & -2 \end{bmatrix}$$

is 10 km from the radar. Find the received power.

CHAPTER SEVEN

Partial Polarization

7.1. INTRODUCTION

We have considered so far only monochromatic waves. Such waves are *completely polarized*, with the end point of the field vector tracing an ellipse of constant axial ratio and tilt angle. Monochromatic waves are not observed in nature, however. Those arising from radiative transitions occupy a spectrum that is broadened by certain physical processes. Lower-frequency waves may be generated by resonant cavities that have a finite bandwidth because of loss. Moreover, the waves may be modulated to carry information, and the modulation process increases the bandwidth. The resulting waves are thus polychromatic. If the bandwidth is small, compared to the mean frequency, they are referred to as quasi-monochromatic. This chapter is concerned with quasi-monochromatic waves.

If a nonmonochromatic plane wave has only one linear field component, or if its two orthogonal components vary similarly, in a manner to be discussed later, the wave, like a monochromatic wave, is completely polarized. This situation is common for radio frequencies, although it may be altered by scattering from an object. As frequency increases, the likelihood of complete polarization decreases, and in the visible light region, most sources produce unpolarized waves.

Let a completely polarized wave be scattered by a *distributed target*, one that is extended in space and may be thought of as a collection of separated subscatterers. With respect to a fixed coordinate system, the scattered wave from each subtarget has two orthogonal components, generally with different amplitudes. If the relative positions of the subscatterers are fixed, the scattered wave components add as phasors, for example, the vertical components from each subscatterer add with phase angles dependent on their position. The scattered wave will not in general have the same polarization characteristics as the incident wave, but it will be completely polarized if the incident wave is completely polarized. If the relative scatterer positions vary, either deterministically with time or randomly, or if the aspect angle of the object varies, the phasor addition of the wave components will cause the two orthogonal components of the total scattered wave to vary. Further, the two components do not vary in the same manner. The two scattered field components in general are not constrained to obey the equation of an ellipse. Alternatively, they may be thought of as tracing an ellipse whose axial ratio and tilt angle (and amplitude) vary with time. Such a scattered wave is said to be *partially polarized*, or in the extreme case, *unpolarized*. We shall see that a receiving antenna cannot extract as much power from a partially polarized wave as it could from a complete polarized wave of equal power density.

A target whose scattered wave is partially polarized when the incident wave is completely polarized is said to *depolarize* the incident wave and is called a *depolarizing* target. Examples are a moving aircraft with various vibratory motions or a changing aspect angle, thunderstorms, the sea, and vegetation in wind-induced motion. Since the phase difference between two physically separated subscatterers is linear with frequency, the effect of target motion is more significant at higher frequencies. A target may, for example, not depolarize at one radar frequency but cause a significant depolarization at a higher frequency.

Normally a time-average power is measured rather than an instantaneous field intensity, both at radio and optical frequencies. This time-averaging period, or signal processing time, may be a determining factor in whether a wave is considered to be partially polarized or not. With a fast processor, compared, for example, to vibrating motion periods of a target, changes in the polarization ellipse of the scattered wave may be followed by the processor, and the wave can be considered completely polarized with slowly changing polarization characteristics. To a slow processor the same wave appears partially polarized. The concepts of partial polarization are closely related to those of coherence theory, but we shall not examine the relationship. The reader is referred to the excellent texts of Born and Wolf or Beran and Parrent [1, 2].

7.2. REPRESENTATIONS OF THE FIELDS

Coordinate Systems

Figure 7.1 shows the coordinate systems to be used to describe transmission between antennas, scattering by a target, and scattering (transmission) in a forward direction.

Think first of transmission between antennas. If antenna a is located at the origin of xyz coordinates and b is at the $\xi\eta\zeta$ origin, the fields may be converted from one system to the other if we, for example, need the

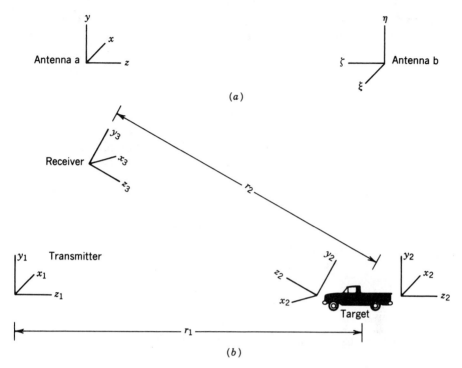

FIGURE 7.1 Coordinate systems for transmission and scattering: (a) transmission between antennas; (b) scattering.

transverse fields of b as x and y components. The subscripts of field quantities, such as x or ξ, identify the fields unambiguously.

In scattering problems the wave incident on the target is appropriately given in $x_1 y_1 z_1$ coordinates. In some radar problems it is convenient to express the scattered wave by its x_3 and y_3 components. This is particularly true for backscattering, where the $x_1 y_1 z_1$ and $x_3 y_3 z_3$ systems coincide. Then the scattered and incident fields for completely polarized waves are related by the Sinclair matrix $\mathbf{S}$ of Section 6.4, thus

$$\mathbf{E}^s = \frac{1}{\sqrt{4\pi}\, r} \mathbf{S} \mathbf{E}^i \tag{6.79}$$

where

$$\mathbf{S} = \begin{bmatrix} S_{x3x1} & S_{x3y1} \\ S_{y3x1} & S_{y3y1} \end{bmatrix} = \begin{bmatrix} S_{xx} & S_{xy} \\ S_{yx} & S_{yy} \end{bmatrix} \tag{7.1}$$

with the numerical subscripts omitted and the proper coordinate system remembered.

368 PARTIAL POLARIZATION

In this chapter we treat partially polarized waves, which are best described by time averages in a coherency matrix or Stokes vector formulation. It is conventional to define the coherency matrix and Stokes vector of a wave in a coordinate system with an axis pointing in the direction of wave propagation. For the scattered wave of Figure 7.1 this is the $x_2 y_2 z_2$ system. A vector equivalent to the Stokes vector, but using $x_3 y_3 z_3$ coordinates, is discussed later. At this time, we use principally the x_2 and y_2 components of the scattered fields, which are related to the incident wave components (for a completely polarized wave) by the Jones matrix of Section 6.4, thus

$$\mathbf{E}^s = \frac{1}{\sqrt{4\pi}\, r} \mathbf{T} \mathbf{E}^i \tag{6.82}$$

where

$$\mathbf{T} = \begin{bmatrix} T_{x2x1} & T_{x2y1} \\ T_{y2x1} & T_{y2y1} \end{bmatrix} = \begin{bmatrix} T_{xx} & T_{xy} \\ T_{yx} & T_{yy} \end{bmatrix} \tag{7.2}$$

Analytic Signals

It is customary to represent a real component of a monochromatic plane wave

$$\mathscr{E}^r(\mathbf{r}, t) = a(\mathbf{r})\cos[\omega t + \phi(\mathbf{r})] \tag{7.3}$$

by a complex time-invariant form

$$E(\mathbf{r}) = a(\mathbf{r}) e^{j\phi(\mathbf{r})} \tag{7.4}$$

obtained by adding to (7.3) the function

$$j\mathscr{E}^i(\mathbf{r}, t) = ja(\mathbf{r})\sin[\omega t + \phi(\mathbf{r})] \tag{7.5}$$

and suppressing the $\exp(j\omega t)$ multiplier. In this section we see that a similar representation can be used for polychromatic waves.

Let the real transverse components of a polychromatic wave traveling in the z direction be

$$\mathscr{E}_x^r(z, t) = a_x(t)\cos[\bar{\omega} t + \phi_x(z, t)] \tag{7.6a}$$
$$\mathscr{E}_y^r(z, t) = a_y(t)\cos[\bar{\omega} t + \phi_y(z, t)] \tag{7.6b}$$

We may think of one of these components as being made up of a large number of elementary sinusoidal waves with random amplitude, frequency, and phase. Then $\mathscr{E}_x^r$ and $\mathscr{E}_y^r$ may themselves be regarded as sample functions of a random process with mean frequency $\bar{\omega}$. The amplitudes a_x and a_y are positive and, in a lossless region, independent of z. The function form (7.6) is quite general, but it is particularly useful if the wave is *quasi-monochromatic*; that is, if it has a bandwidth small compared to the mean frequency. For such signals, $a(t)$ varies slowly when compared to $\cos \bar{\omega} t$, and $\phi(z, t)$ changes slowly compared to $\bar{\omega} t$. Then $a(t)$ is the envelope of a wave that approximates a cosine time function, and $\phi(z, t)$ is the associated phase. The methods discussed in this chapter require the quasi-monochromatic constraint.

The definition of narrow bandwidth depends on the application. An obvious restriction is that

$$\Delta \omega \ll \bar{\omega} \tag{7.7}$$

This allows frequency-dependent parameters to be evaluated at the mean frequency of the wave, which of course causes the methods of handling partially polarized waves to be approximate to a small degree. To develop another constraint, consider a target illuminated by a wave with extreme frequencies ω_1 and ω_2. If these two frequency components of the incident wave are in phase at the leading edge of the target, the phase difference of the reflected components is zero for the wave reflected from the target leading edge and

$$\Delta \phi = 4 \pi d \left(\frac{1}{\lambda_1} - \frac{1}{\lambda_2} \right) = 4 \pi d \left(\frac{f_1}{c} - \frac{f_2}{c} \right) \tag{7.8}$$

for the wave reflected from the trailing edge, if the target has length d. For a large phase difference, incident and reflected waves are substantially decorrelated. To keep $\Delta \phi$ small compared to 2π leads to

$$\Delta f \ll \frac{c}{2d} \tag{7.9}$$

This constraint may be less stringent than (7.7). A similar criterion may be used for an antenna or optical instrument.

Another criterion has been suggested for a linear antenna array [3]. For N elements spaced at half-wavelength intervals, the bandwidth should be no greater than that which gives

$$\frac{\sin\left[\frac{1}{2}(N-1)\Delta \omega T\right]}{\frac{1}{2}(N-1)\Delta \omega T} \approx 1 \tag{7.10}$$

where T is the propagation time between adjacent elements. This equation can be written as

$$\frac{\sin\left(\frac{1}{2} d \Delta \omega / c\right)}{\frac{1}{2} d \Delta \omega / c} \approx 1 \tag{7.11}$$

where d is the array length. The constraints (7.9) and (7.10) are not substantially different.

In (7.6), $a(t)$ and $\phi(z,t)$ are essentially arbitrary. The phase $\phi(z,t)$ may be chosen at will, subject only to the constraint that $a(t)$ be positive, and $a(t)$ assumes an appropriate form to yield the correct value for $\mathscr{E}^r(z,t)$. An aim of this section is to eliminate the arbitrariness in (7.6) and develop a unique representation for a polychromatic wave.

Assume that the field components of (7.6) possess Fourier and Hilbert transforms. A sufficient condition is that the component be square integrable, but if this condition is not met, we nevertheless assume the existence of the transforms. This assumption is common [2,4]. To obtain the transform, a truncated function equal to the field component in the interval $-T \le t \le T$, and zero elsewhere, is defined. The truncated function is transformed and the limit as $T \to \infty$ is taken to be the transform of the untruncated field component.

With the assumption of transformability, the field components may be written as Fourier integrals, and since they are real, as Fourier cosine integrals, thus

$$\mathscr{E}^r(z,t) = \int_0^\infty g(\omega)\cos[\omega t + \theta(z,\omega)]\, d\omega \qquad (7.12)$$

The relations (7.3)–(7.5) provide a suggestion on how to proceed. There we added to a real single-frequency wave another real function shifted in phase by $\pi/2$. A reasonable extension of this idea is that in (7.12) we shift the phase of each frequency component by $\pi/2$. We therefore define a second Fourier integral,

$$\mathscr{E}^i(z,t) = \int_0^\infty g(\omega)\sin[\omega t + \theta(z,\omega)]\, d\omega \qquad (7.13)$$

and add it to (7.12),

$$\mathscr{E}(z,t) = \mathscr{E}^r(z,t) + j\mathscr{E}^i(z,t) = \int_0^\infty g(\omega) e^{j[\omega t + \theta(z,\omega)]}\, d\omega \qquad (7.14)$$

This process is physically appealing because of its close relationship to the process used universally for monochromatic waves. It also has a firm mathematical foundation.

The integral (7.13) is known as the *allied integral* of the Fourier integral (7.12), and the functions $\mathscr{E}^r(z,t)$ and $\mathscr{E}^i(z,t)$ as *conjugate functions*. It is readily surmised that they are not independent, and it may be shown that they are related by the Hilbert transforms [5],

$$\mathscr{E}^i(z,t) = \frac{1}{\pi}\int_{-\infty}^\infty \frac{\mathscr{E}^r(z,t')}{t'-t}\, dt' \qquad (7.15a)$$

$$\mathscr{E}^r(z,t) = \frac{1}{\pi}\int_{-\infty}^\infty \frac{\mathscr{E}^i(z,t')}{t'-t}\, dt' \qquad (7.15b)$$

REPRESENTATIONS OF THE FIELDS   371

The bar across the integral symbols signifies the Cauchy principal value, that is,

$$\int_{-\infty}^{\infty} \frac{f(t')}{t'-t} dt' = \lim_{T \to t^-} \int_{-\infty}^{T} \frac{f(t')}{t'-t} dt' + \lim_{T \to t^+} \int_{T}^{\infty} \frac{f(t')}{t'-t} dt' \quad (7.16)$$

The function

$$\mathscr{E}(z,t) = \mathscr{E}^r(z,t) + j\mathscr{E}^i(z,t) \quad (7.17)$$

is called the *analytic signal* [6] associated with the real signal $\mathscr{E}^r(z,t)$ because $\mathscr{E}$, considered as a function of a complex variable, is analytic in a region of the complex plane. Equation (7.14) shows that it has no spectral components for negative frequencies.

Returning now to (7.6), the Hilbert transforms of the wave components are

$$\mathscr{E}_x^i(z,t) = a_x(t)\sin[\bar{\omega}t + \phi_x(z,t)] \quad (7.18a)$$
$$\mathscr{E}_y^i(z,t) = a_y(t)\sin[\bar{\omega}t + \phi_y(z,t)] \quad (7.18b)$$

Then the analytic signals for the transverse components of the quasi-monochromatic plane wave are

$$\mathscr{E}_x(z,t) = a_x(t)e^{j[\bar{\omega}t+\phi_x(z,t)]} \quad (7.19a)$$
$$\mathscr{E}_y(z,t) = a_y(t)e^{j[\bar{\omega}t+\phi_y(z,t)]} \quad (7.19b)$$

or, if the phase variation with z is made implicit,

$$\mathscr{E}_x(t) = a_x(t)e^{j[\bar{\omega}t+\phi_x(t)]} \quad (7.20a)$$
$$\mathscr{E}_y(t) = a_y(t)e^{j[\bar{\omega}t+\phi_y(t)]} \quad (7.20b)$$

We may now see that one aim of this section, to develop a unique representation of polychromatic fields, has been realized. From (7.17) and (7.19),

$$a_x(t) = \left\{[\mathscr{E}_x^r(z,t)]^2 + [\mathscr{E}_x^i(z,t)]^2\right\}^{1/2}$$
$$= [\mathscr{E}_x(z,t)\mathscr{E}_x^*(z,t)]^{1/2} = |\mathscr{E}_x(z,t)| \quad (7.21a)$$
$$a_y(t) = |\mathscr{E}_y(z,t)| \quad (7.21b)$$

Also

$$\phi_x(z,t) = \tan^{-1}\frac{\mathscr{E}_x^i(z,t)}{\mathscr{E}_x^r(z,t)} - \bar{\omega}t = -\tan^{-1}\left(j\frac{\mathscr{E}_x - \mathscr{E}_x^*}{\mathscr{E}_x + \mathscr{E}_x^*}\right) - \bar{\omega}t \quad (7.22a)$$

$$\phi_y(z,t) = -\tan^{-1}\left(j\frac{\mathscr{E}_y - \mathscr{E}_y^*}{\mathscr{E}_y + \mathscr{E}_y^*}\right) - \bar{\omega}t \quad (7.22b)$$

372 PARTIAL POLARIZATION

These equations show that the envelope amplitudes $a_x(t)$ and $a_y(t)$ are independent of the choice of phases, unlike the situation when the fields were represented by (7.6). Further, since $\bar{\omega}$ is the mean frequency of the random process of which $\mathscr{E}_x^r$ and $\mathscr{E}_y^r$ are sample functions, the phases ϕ_x and ϕ_y are completely specified by (7.22). These unique values of $a(t)$ and $\phi(t)$ may be used in (7.6) or (7.19) as desired. We have therefore succeeded, by use of the analytic signal, in obtaining a unique representation of a real quasi-monochromatic wave.

Finally, it is clear from (7.19) that $\exp(j\bar{\omega}t)$ may be omitted from the equations. Also, the phase terms may or may not include the distance variation, $-kz$, at our convenience, since the distance variation is commonly the same for both components. The wave components may then be written as

$$E_x(t) = a_x(t)e^{j\phi_x(t)} \tag{7.23a}$$

$$E_y(t) = a_y(t)e^{j\phi_y(t)} \tag{7.23b}$$

In this form, the phase variation with distance is implicit. It may be included or omitted according to need. The notation used here is a script letter for a field that includes an $\exp(j\bar{\omega}t)$ term and the corresponding italic or roman letter for a field that omits it. Strictly, (7.19) and (7.20) are the analytic signal representations of the wave, but more often than not we use the form (7.23) and refer to it as the analytic signal. No problems arise from this practice.

Time Averages of the Fields

In experimental work involving either antennas or optics it is common to measure a quantity proportional to the time average of the product of real field components. Consider as an illustration the time average

$$\langle \mathscr{E}_x^r(z,t)\mathscr{E}_y^r(z,t) \rangle = \langle a_x(t)\cos[\bar{\omega}t + \phi_x(z,t)]a_y(t)\cos[\bar{\omega}t + \phi_y(z,t)] \rangle \tag{7.24}$$

where the time average is defined as

$$\langle f(t) \rangle = \lim_{T \to \infty} \frac{1}{2T}\int_{-T}^{T} f(t)\,dt \tag{7.25}$$

Using this definition in (7.24) causes it to reduce to

$$\langle \mathscr{E}_x^r(z,t)\mathscr{E}_y^r(z,t) \rangle = \tfrac{1}{2}\langle a_x(t)a_y(t)\cos[\phi_y(z,t) - \phi_x(z,t)] \rangle \tag{7.26}$$

If we use either the strict analytic signal representation of the wave, (7.19), or the equivalent form (7.23), we find

$$\operatorname{Re}\langle \mathscr{E}_x^*(z,t)\mathscr{E}_y(z,t)\rangle = \langle a_x(t)a_y(t)\cos[\phi_y(z,t) - \phi_x(z,t)]\rangle \quad (7.27)$$

and see that

$$\langle \mathscr{E}_x^r(z,t)\mathscr{E}_y^r(z,t)\rangle = \tfrac{1}{2}\operatorname{Re}\langle \mathscr{E}_x^*(z,t)\mathscr{E}_y(z,t)\rangle = \tfrac{1}{2}\operatorname{Re}\langle E_x^*(t)E_y(t)\rangle \quad (7.28)$$

In the last term of (7.28), distance z is implied. The other product terms give similar results.

From time-average measurements the amplitudes and phases of the fields in the representation

$$E_x(t) = a_x(t)e^{j\phi_x(t)} \quad (7.23a)$$
$$E_y(t) = a_y(t)e^{j\phi_y(t)} \quad (7.23b)$$

cannot be determined. Since these are sample functions of a random process, the parameters will vary from one sample to the next. Neither can the mean values of the amplitudes and phases be found. The measurements are of

$$\langle \mathscr{E}_x^r \mathscr{E}_y^r \rangle = \tfrac{1}{2}\langle a_x(t)a_y(t)\cos\phi(z,t)\rangle \quad (7.29a)$$
$$\langle |\mathscr{E}_x^r|^2 \rangle = \tfrac{1}{2}\langle a_x^2(t)\rangle \quad (7.29b)$$
$$\langle |\mathscr{E}_y^r|^2 \rangle = \tfrac{1}{2}\langle a_y^2(t)\rangle \quad (7.29c)$$

where

$$\phi(z,t) = \phi_y(z,t) - \phi_x(z,t) \quad (7.30)$$

and it is obvious that the mean values of a_x, a_y, and ϕ cannot be determined.

It was pointed out earlier that the wave components $\mathscr{E}_x^r$ and $\mathscr{E}_y^r$ are sample functions of a random process. We assume that the process is ergodic so that the time average is equivalent to the statistical average across the ensemble. The time averages are auto- and cross-correlation functions, and we shall make use of their relationship to power spectral densities of the waves.

7.3. THE COHERENCY MATRIX

Coherency Matrix of a Quasi-Monochromatic Plane Wave

The concepts of a partially polarized plane wave come to our attention forcefully if we measure the intensity of such a wave. Let the orthogonal

374 PARTIAL POLARIZATION

components of the wave be

$$E_x(t) = a_x(t)e^{j\phi_x(t)} \tag{7.31a}$$
$$E_y(t) = a_y(t)e^{j\phi_y(t)} \tag{7.31b}$$

Although it is not essential to the development, we follow Born and Wolf [1] in subjecting the x component of the wave to a phase retardation (or equivalently advancing the y component) before measuring the intensity.* This provides valuable insight into the nature of the wave. Then the fields to be measured are

$$E'_x(t) = a_x(t)e^{j\phi_x(t)} = E_x(t) \tag{7.32a}$$
$$E'_y(t) = a_y(t)e^{j\phi_y(t)}e^{j\delta} = E_y(t)e^{j\delta} \tag{7.32b}$$

If the extreme frequencies of the wave are ω_1 and ω_2, we assume that

$$|\delta(\omega_1) - \delta(\omega_2)| \ll 2\pi \tag{7.33}$$

and δ in (7.32) is a mean value.

Let the measuring device (e.g., an antenna) respond to a field component at angle θ to the x axis. Then the induced voltage is proportional to the component

$$\begin{aligned} E(t,\theta,\delta) &= E'_x(t)\cos\theta + E'_y(t)\sin\theta \\ &= E_x(t)\cos\theta + E_y(t)e^{j\delta}\sin\theta \end{aligned} \tag{7.34}$$

In most cases, the measurement is that of a time-average power, which is proportional to the time average of the voltage magnitude squared, thus

$$\begin{aligned} I &= \langle E^*(t,\theta,\delta)E(t,\theta,\delta) \rangle \\ &= \langle E_x^*(t)E_x(t) \rangle \cos^2\theta + \langle E_y^*(t)E_y(t) \rangle \sin^2\theta \\ &\quad + \langle E_x^*(t)E_y(t) \rangle e^{j\delta}\cos\theta\sin\theta + \langle E_y^*(t)E_x(t) \rangle e^{-j\delta}\cos\theta\sin\theta \end{aligned} \tag{7.35}$$

The presence of the four time averages makes it desirable to order them in a matrix, called the *coherency matrix* of the wave [1]. We do so by

$$\mathbf{J} = \begin{bmatrix} J_{xx} & J_{xy} \\ J_{yx} & J_{yy} \end{bmatrix} = \begin{bmatrix} \langle E_x^*(t)E_x(t) \rangle & \langle E_x(t)E_y^*(t) \rangle \\ \langle E_x^*(t)E_y(t) \rangle & \langle E_y(t)E_y^*(t) \rangle \end{bmatrix} \tag{7.36}$$

---
*Born and Wolf retard the y component and arrive at the same form, but they use a time variation $\exp(-i\omega t)$.

which may also be written, using (7.23), as

$$\mathbf{J} = \begin{bmatrix} \langle a_x^2 \rangle & \langle a_x a_y e^{j(\phi_x - \phi_y)} \rangle \\ \langle a_x a_y e^{j(\phi_y - \phi_x)} \rangle & \langle a_y^2 \rangle \end{bmatrix} \qquad (7.37)$$

Note that **J** may be written concisely as

$$\mathbf{J} = \langle \mathbf{E}\mathbf{E}^{T*} \rangle \qquad (7.38)$$

The ordering of the off-diagonal terms is arbitrary.

If the amplitude and phase terms of (7.23) vary so slowly that the derivatives in the Maxwell equations, for example in

$$\nabla \times \mathcal{H}^r = \epsilon \frac{\partial \mathcal{E}^r}{\partial t} \qquad (7.39)$$

can be replaced by $j\bar{\omega}$, with $\bar{\omega}$ the mean frequency, the time-average Poynting vector of the field considered is proportional to the trace of the coherency matrix,

$$\mathcal{P} = \frac{1}{2Z_0}(\langle a_x^2 \rangle + \langle a_y^2 \rangle) = \frac{1}{2Z_0} \text{Tr}(\mathbf{J}) \qquad (7.40)$$

This requirement on the amplitudes and phases is another manifestation of the narrow bandwidth constraint.

The mixed terms of the coherency matrix may be normalized by setting

$$\mu_{xy} = \frac{J_{xy}}{\sqrt{J_{xx}} \sqrt{J_{yy}}} \qquad (7.41)$$

It may be shown by the Schwartz inequality that

$$|\mu_{xy}| \leq 1 \qquad (7.42)$$

The term μ_{xy} is a measure of the correlation between x and y components of the wave.

Equation (7.36) indicates that

$$J_{xy} = J_{yx}^* \qquad (7.43)$$

Then the coherency matrix determinant may be written as

$$\|\mathbf{J}\| = J_{xx} J_{yy} - |J_{xy}|^2 \qquad (7.44)$$

or as

$$\|\mathbf{J}\| = J_{xx}J_{yy}(1 - |\mu_{xy}|^2) \tag{7.45}$$

Since J_{xx} and J_{yy} are positive real and $|\mu_{xy}| \leq 1$,

$$\|\mathbf{J}\| \geq 0 \tag{7.46}$$

In terms of the coherency matrix elements, the time-average power in (7.35) is proportional to

$$I = J_{xx} \cos^2 \theta + J_{yy} \sin^2 \theta + \left(J_{xy}e^{-j\delta} + J_{yx}e^{j\delta}\right)\cos \theta \sin \theta \tag{7.47}$$

Unpolarized Waves

Waves that are unpolarized have the characteristic that the time-average power is independent of θ. In addition, the power is unaffected by the value of the phase retardation δ of a wave component. From (7.47) this clearly requires that

$$J_{xx} = J_{yy} \tag{7.48a}$$
$$J_{xy} = J_{yx} = 0 \tag{7.48b}$$

The wave components are uncorrelated, and the coherency matrix is

$$\mathbf{J} = \begin{bmatrix} J_{xx} & 0 \\ 0 & J_{yy} \end{bmatrix} = Z_0 \mathscr{P} \begin{bmatrix} 1 & 0 \\ 0 & 1 \end{bmatrix} \tag{7.49}$$

where $\mathscr{P}$ is the Poynting vector density of the wave.

Complete Polarization

First consider monochromatic waves. For these the terms a and ϕ of (7.37) are time-independent, and the coherency matrix is

$$\mathbf{J} = \begin{bmatrix} a_x^2 & a_x a_y e^{-j\phi} \\ a_x a_y e^{j\phi} & a_y^2 \end{bmatrix} \tag{7.50}$$

where

$$\phi = \phi_y - \phi_x \tag{7.51}$$

From the coherency matrix elements, it is seen that

$$\mu_{xy} = e^{-j\phi} \tag{7.52}$$

with ϕ a constant, and

$$|\mu_{xy}| = 1 \tag{7.53}$$

This monochromatic wave is of course completely polarized.

We may also have complete polarization for nonmonochromatic waves. Suppose that a_x, a_y, ϕ_x, and ϕ_y depend on time in such a way that the ratio of amplitudes and the difference in phase are independent of time; that is,

$$\frac{a_y(t)}{a_x(t)} = C_1 \tag{7.54a}$$

$$\phi = \phi_y(t) - \phi_x(t) = C_2 \tag{7.54b}$$

with C_1 and C_2 constant. It is clear that the tip of the electric vector smoothly traces an ellipse as time varies. The ellipse size varies slowly as the amplitudes change slowly, but no significant change occurs during one trace of the ellipse. The polarization ellipse, considered distinct from the path of the electric vector tip, is constant. The angular position of the electric vector is deterministic, and its rotation rate is given by (3.60) with appropriate notation changes. The polarization ratio is constant, and from (3.73) is

$$P = C_1 e^{jC_2} \tag{7.55}$$

The wave is then completely polarized. Its coherency matrix is

$$\mathbf{J} = \begin{bmatrix} \langle a_x^2 \rangle & C_1 \langle a_x^2 \rangle e^{-jC_2} \\ C_1 \langle a_x^2 \rangle e^{jC_2} & C_1^2 \langle a_x^2 \rangle \end{bmatrix} \tag{7.56}$$

and

$$|\mu_{xy}| = |e^{-jC_2}| = 1 \tag{7.57}$$

just as for the monochromatic case. The coherency matrix is the same as for a monochromatic wave with components

$$E_x = \sqrt{\langle a_x^2 \rangle}\, e^{j\phi_x} \tag{7.58a}$$

$$E_y = C_1 \sqrt{\langle a_x^2 \rangle}\, e^{j(\phi_x + C_2)} \tag{7.58b}$$

The monochromatic wave and the quasi-monochromatic wave obeying (7.54), both completely polarized, lead to the relation $|\mu_{xy}| = 1$, and no other wave does so. We may therefore consider (7.53) to be the defining relationship for complete polarization. From (7.45) and (7.53) it is also clear that for a completely polarized wave, and only for such a wave,

$$\|\mathbf{J}\| = 0 \qquad (7.59)$$

Linear Polarization

For linear polarization the wave must satisfy the requirements for complete polarization, and in addition

$$\phi = 0, \pm \pi, \pm 2\pi, \ldots \qquad (7.60)$$

Then the coherency matrices for monochromatic linear and completely polarized quasi-monochromatic linear waves are, respectively,

$$\mathbf{J} = \begin{bmatrix} a_x^2 & (-1)^m a_x a_y \\ (-1)^m a_x a_y & a_y^2 \end{bmatrix} \quad m = 0, 1, 2, \ldots \qquad (7.61)$$

and

$$\mathbf{J} = \begin{bmatrix} \langle a_x^2 \rangle & (-1)^m C_1 \langle a_x^2 \rangle \\ (-1)^m C_1 \langle a_x^2 \rangle & C_1^2 \langle a_x^2 \rangle \end{bmatrix} \quad m = 0, 1, 2, \ldots \qquad (7.62)$$

More particularly, the coherency matrices

$$\mathbf{J} = 2Z_0 \mathscr{P} \begin{bmatrix} 1 & 0 \\ 0 & 0 \end{bmatrix}, 2Z_0 \mathscr{P} \begin{bmatrix} 0 & 0 \\ 0 & 1 \end{bmatrix}, Z_0 \mathscr{P} \begin{bmatrix} 1 & 1 \\ 1 & 1 \end{bmatrix}, Z_0 \mathscr{P} \begin{bmatrix} 1 & -1 \\ -1 & 1 \end{bmatrix} \qquad (7.63)$$

represent linear polarizations that are, respectively, x directed, y directed, 45° from the x axis, and 135° from the x axis.

Circular Polarization

For circular polarization the rectangular component amplitudes are equal and

$$\phi = \pm \tfrac{1}{2} \pi \qquad (7.64)$$

for left- (upper sign) and right-circular waves. The coherency matrices for left- and right-circular waves, respectively, are

$$\mathbf{J} = Z_0 \mathscr{P} \begin{bmatrix} 1 & -j \\ j & 1 \end{bmatrix}, \quad Z_0 \mathscr{P} \begin{bmatrix} 1 & j \\ -j & 1 \end{bmatrix} \qquad (7.65)$$

Degree of Polarization

Suppose that N independent electromagnetic waves propagate in the same direction. The electric field intensities are additive, thus

$$E_x = \sum_{n=1}^{N} E_x^{(n)} \tag{7.66a}$$

$$E_y = \sum_{n=1}^{N} E_y^{(n)} \tag{7.66b}$$

The coherency matrix elements of the total wave are

$$J_{kl} = \langle E_k E_l^* \rangle = \sum_n \sum_m \langle E_k^{(n)} E_l^{(m)*} \rangle \tag{7.67}$$

where k and l independently represent x and y. This may be written as

$$J_{kl} = \sum_n \langle E_k^{(n)} E_l^{(n)*} \rangle + \sum_{n \neq m} \langle E_k^{(n)} E_l^{(m)*} \rangle \tag{7.68}$$

Since the waves are independent, each term in the last summation is zero and

$$J_{kl} = \sum_{n=1}^{N} J_{kl}^{(n)} \tag{7.69}$$

where $J_{kl}^{(n)}$ are the coherency matrix elements of the nth wave. Therefore, for independent waves propagating in the same direction, the coherency matrix elements of the individual waves may be added to obtain the overall coherency matrix. It is clear that we may consider the inverse of this result and treat any wave as the sum of independent waves chosen at will.

In particular, consider a quasi-monochromatic plane wave to be the sum of a completely polarized wave and an unpolarized wave. We can show that this representation is unique by showing that any coherency matrix can be uniquely expressed in the form

$$\mathbf{J} = \mathbf{J}^{(1)} + \mathbf{J}^{(2)} \tag{7.70}$$

where

$$\mathbf{J}^{(1)} = \begin{bmatrix} A & 0 \\ 0 & A \end{bmatrix} \tag{7.71a}$$

$$\mathbf{J}^{(2)} = \begin{bmatrix} B & D \\ D^* & C \end{bmatrix} \tag{7.71b}$$

with

$$A \geq 0 \quad B \geq 0 \quad C \geq 0 \quad BC - DD^* = 0 \quad (7.72)$$

If (7.71) is compared to the special case (7.49), it is seen that $\mathbf{J}^{(1)}$ is the coherency matrix for an unpolarized wave. If we use $\|\mathbf{J}\| = 0$ (which gives $|\mu_{xy}| = 1$) as the defining relation for complete polarization, then $\mathbf{J}^{(2)}$ is the coherency matrix of a completely polarized wave.

We must next show that the decomposition into unpolarized and completely polarized waves is unique. This we do by obtaining the elements of $\mathbf{J}^{(1)}$ and $\mathbf{J}^{(2)}$ from the known elements of $\mathbf{J}$. From (7.70) and (7.71) we may write

$$A + B = J_{xx} \quad (7.73a)$$
$$D = J_{xy} \quad (7.73b)$$
$$D^* = J_{yx} \quad (7.73c)$$
$$A + C = J_{yy} \quad (7.73d)$$

Substitution into the last equation of (7.72) gives

$$(J_{xx} - A)(J_{yy} - A) - J_{xy}J_{yx} = 0 \quad (7.74)$$

which is a quadratic in A with solution

$$A = \tfrac{1}{2}(J_{xx} + J_{yy}) \pm \tfrac{1}{2}\left[(J_{xx} + J_{yy})^2 - 4\|\mathbf{J}\|\right]^{1/2} \quad (7.75)$$

Substituting this result into (7.73a) gives

$$B = \tfrac{1}{2}(J_{xx} - J_{yy}) \mp \tfrac{1}{2}\left[(J_{xx} + J_{yy})^2 - 4\|\mathbf{J}\|\right]^{1/2} \quad (7.76)$$

The negative sign for the last term is not allowed since it would make B negative, contrary to our hypothesis. Then the values of A, B, C, D are found uniquely from

$$A = \tfrac{1}{2}(J_{xx} + J_{yy}) - \tfrac{1}{2}\left[(J_{xx} + J_{yy})^2 - 4\|\mathbf{J}\|\right]^{1/2} \quad (7.77a)$$

$$B = \tfrac{1}{2}(J_{xx} - J_{yy}) + \tfrac{1}{2}\left[(J_{xx} + J_{yy})^2 - 4\|\mathbf{J}\|\right]^{1/2} \quad (7.77b)$$

$$C = \tfrac{1}{2}(J_{yy} - J_{xx}) + \tfrac{1}{2}\left[(J_{xx} + J_{yy})^2 - 4\|\mathbf{J}\|\right]^{1/2} \quad (7.77c)$$

$$D = J_{xy} \quad (7.77d)$$

$$D^* = J_{yx} \quad (7.77e)$$

The Poynting vector magnitude of the total wave is

$$\mathscr{P}_t = \frac{1}{2Z_0} \text{Tr}(\mathbf{J}) = \frac{1}{2Z_0}(J_{xx} + J_{yy}) \tag{7.78}$$

and that of the polarized part of the wave is

$$\mathscr{P}_p = \frac{1}{2Z_0} \text{Tr}[\mathbf{J}^{(2)}] = \frac{1}{2Z_0}(B + C) = \frac{1}{2Z_0}\left[(J_{xx} + J_{yy})^2 - 4\|\mathbf{J}\|\right]^{1/2} \tag{7.79}$$

The ratio of the power densities of the polarized part and the total wave is called the *degree of polarization* of the wave. It is given by

$$R = \frac{\mathscr{P}_p}{\mathscr{P}_t} = \left[1 - \frac{4\|\mathbf{J}\|}{(J_{xx} + J_{yy})^2}\right]^{1/2} \tag{7.80}$$

Now

$$\|\mathbf{J}\| \le J_{xx} J_{yy} \le \tfrac{1}{4}(J_{xx} + J_{yy})^2$$

and, therefore,

$$0 \le R \le 1 \tag{7.81}$$

Consider the two extreme values of R. For $R = 1$ (7.80) requires that $\|\mathbf{J}\| = 0$, which is the condition for complete polarization. Then $|\mu_{xy}| = 1$, and the x and y wave components are completely correlated (Born and Wolf use the term "mutually coherent" [1]). For $R = 0$ (7.80) requires that $(J_{xx} - J_{yy})^2 + 4|J_{xy}|^2 = 0$, which can be satisfied only by

$$J_{xx} = J_{yy} \quad J_{xy} = J_{yx} = 0$$

It follows that $|\mu_{xy}| = 0$, and $E_x(t)$ and $E_y(t)$ are completely decorrelated (mutually incoherent).

Now $R = 0$ requires that $E_x(t)$ and $E_y(t)$ be uncorrelated. The converse is not true. For complete decorrelation, $J_{xy} = J_{yx} = 0$ and $|\mu_{xy}| = 0$. Then

$$R = \left[1 - \frac{4 J_{xx} J_{yy}}{(J_{xx} + J_{yy})^2}\right]^{1/2} = \frac{|J_{xx} - J_{yy}|}{J_{xx} + J_{yy}}$$

This shows that $|\mu_{xy}| = 0$ is not sufficient to give an unpolarized wave. To make it completely unpolarized, we must also have $J_{xx} = J_{yy}$. It is easy to see why this is so. A wave with a large monochromatic x component, for example, cannot be substantially depolarized by the addition of a small independent y component.

A partially polarized wave may also be treated as though it were the sum of two independent completely polarized orthogonal waves. To show this and to find the component waves, we write the coherency matrix as the sum of coherency matrices, thus

$$\mathbf{J} = \begin{bmatrix} J_{xx} & J_{xy} \\ J_{yx} & J_{yy} \end{bmatrix} = \mathbf{J}^{(3)} + \mathbf{J}^{(4)} = \begin{bmatrix} A & B \\ B^* & C \end{bmatrix} + \begin{bmatrix} D & E \\ E^* & F \end{bmatrix} \quad (7.82)$$

Since the component matrices $\mathbf{J}^{(3)}$ and $\mathbf{J}^{(4)}$ represent completely polarized waves, their elements obey the relations

$$\begin{array}{ll} A \geq 0 & D \geq 0 \\ C \geq 0 & F \geq 0 \quad A, C, D, F \text{ real} \\ AC = |B|^2 & DF = |E|^2 \end{array} \quad (7.83)$$

In another section we show that the linear polarization ratio of a completely polarized wave is

$$P = \frac{J_{yx}}{J_{xx}} \quad (7.84)$$

Then the linear polarization ratios corresponding to $\mathbf{J}^{(3)}$ and $\mathbf{J}^{(4)}$ are

$$P_3 = \frac{B^*}{A} \quad (7.85a)$$

$$P_4 = \frac{E^*}{D} \quad (7.85b)$$

If these are substituted into the coherency matrices and the relationships of (7.83) are used, the component matrices of the partially polarized wave become

$$\mathbf{J}^{(3)} = A \begin{bmatrix} 1 & P_3^* \\ P_3 & |P_3|^2 \end{bmatrix} \quad (7.86a)$$

$$\mathbf{J}^{(4)} = D \begin{bmatrix} 1 & P_4^* \\ P_4 & |P_4|^2 \end{bmatrix} \quad (7.86b)$$

THE COHERENCY MATRIX

If the two component waves are orthogonal, their linear polarization ratios are related by

$$P_4 = -\frac{1}{P_3^*} \tag{7.87}$$

Then the coherency matrix of the total wave is

$$\mathbf{J} = A \begin{bmatrix} 1 & P_3^* \\ P_3 & |P_3|^2 \end{bmatrix} + D \begin{bmatrix} 1 & -\dfrac{1}{P_3} \\ -\dfrac{1}{P_3^*} & \dfrac{1}{|P_3|^2} \end{bmatrix} \tag{7.88}$$

The unknown parameters can be found from the three relations

$$J_{xx} = J_{xx}^{(3)} + J_{xx}^{(4)} \tag{7.89a}$$

$$J_{xy} = J_{xy}^{(3)} + J_{xy}^{(4)} \tag{7.89b}$$

$$J_{yy} = J_{yy}^{(3)} + J_{yy}^{(4)} \tag{7.89c}$$

The first is satisfied if

$$J_{xx}^{(3)} = A = \frac{J_{xx}}{2} - \delta \tag{7.90a}$$

$$J_{xx}^{(4)} = D = \frac{J_{xx}}{2} + \delta \tag{7.90b}$$

The second gives

$$J_{xy} = \left(\frac{J_{xx}}{2} - \delta\right) P_3^* - \frac{J_{xx}/2 + \delta}{P_3} \tag{7.91}$$

which may be solved to give

$$\delta = \frac{(J_{xx}/2)(|P_3|^2 - 1) - J_{xy} P_3}{|P_3|^2 + 1} \tag{7.92}$$

The third relationship of (7.89) gives

$$J_{yy} = \left(\frac{J_{xx}}{2} - \delta\right) |P_3|^2 + \frac{J_{xx}/2 + \delta}{|P_3|^2} \tag{7.93}$$

which leads to

$$\delta = \frac{(J_{xx}/2)(|P_3|^4 + 1) - J_{yy}|P_3|^2}{|P_3|^4 - 1} \tag{7.94}$$

384 PARTIAL POLARIZATION

If the two forms for δ are equated, the result,

$$\frac{|P_3|^2 - 1}{P_3^*} = \frac{J_{yy} - J_{xx}}{J_{xy}} \qquad (7.95)$$

can be solved for P_3. We are free to require either that $|P_3| > 1$ or $|P_3| < 1$ since two orthogonal polarizations are used. Then we choose

$$|P_3| > 1 \quad \text{if} \quad J_{yy} > J_{xx}$$
$$|P_3| < 1 \quad \text{if} \quad J_{yy} < J_{xx}$$

The solution for P_3 is then, in either case,

$$|P_3| = \frac{1}{2|J_{xy}|}\left(J_{yy} - J_{xx} + \sqrt{(J_{yy} - J_{xx})^2 + 4|J_{xy}|^2}\right) \qquad (7.96a)$$

$$\text{Ang}(P_3) = -\text{Ang}(J_{xy}) = \text{Ang}(J_{yx}) \qquad (7.96b)$$

The value of δ is then found from (7.92) which becomes, if the phase angle of P_3 is used,

$$\delta = \frac{(J_{xx}/2)(|P_3|^2 - 1) - |J_{xy}||P_3|}{|P_3|^2 + 1} \qquad (7.97)$$

Finally, from P_3 and δ the elements of the component matrices of the partially polarized wave can be found.

The coherency matrix can be transformed to a new polarization basis by the same unitary matrix used in previous chapters to transform the fields and scattering matrix. The coherency matrix is

$$\mathbf{J} = \langle \mathbf{E}\mathbf{E}^{T*} \rangle \qquad (7.38)$$

where the angle brackets indicate a time average, and the field vector is in xy coordinates. Vector $\mathbf{E}$ can be expressed in polarization basis set $\mathbf{u}_1, \mathbf{u}_2$ by a product, thus

$$\mathbf{E}' = \mathbf{E}_{\mathbf{u}_1\mathbf{u}_2} = \mathbf{R}\mathbf{E} \qquad (7.98)$$

where $\mathbf{R}$ is a unitary matrix given by

$$\mathbf{R} = \begin{bmatrix} \mathbf{u}_1^* \cdot \mathbf{u}_x & \mathbf{u}_1^* \cdot \mathbf{u}_y \\ \mathbf{u}_2^* \cdot \mathbf{u}_x & \mathbf{u}_2^* \cdot \mathbf{u}_y \end{bmatrix} \qquad (7.99)$$

The coherency matrix in the new basis set may be defined as

$$\mathbf{J}' = \langle \mathbf{E}'\mathbf{E}'^{T*} \rangle \qquad (7.100)$$

and altered to

$$\mathbf{J}' = \langle \mathbf{REE}^{T*}\mathbf{R}^{T*}\rangle = \mathbf{R}\langle \mathbf{EE}^{T*}\rangle \mathbf{R}^{T*} = \mathbf{RJR}^{-1} \quad (7.101)$$

This transformation is the same one that transformed the Jones matrix in Chapter 6.

Let the transformation be restricted to a rotation of the coordinate axes so that

$$\mathbf{R} = \begin{bmatrix} \cos\theta & \sin\theta \\ -\sin\theta & \cos\theta \end{bmatrix} \quad (7.102)$$

Then

$$\mathbf{J}' = \begin{bmatrix} \cos\theta & \sin\theta \\ -\sin\theta & \cos\theta \end{bmatrix} \begin{bmatrix} J_{xx} & J_{xy} \\ J_{xy}^* & J_{yy} \end{bmatrix} \begin{bmatrix} \cos\theta & -\sin\theta \\ \sin\theta & \cos\theta \end{bmatrix} \quad (7.103)$$

Straightforward multiplication shows that

$$\mathrm{Tr}(\mathbf{J}') = \mathrm{Tr}(\mathbf{J}) \quad (7.104a)$$
$$\|\mathbf{J}'\| = \|\mathbf{J}\| \quad (7.104b)$$

Thus the trace and determinant of the coherency matrix are invariant with a coordinate rotation. It follows that the Poynting vector magnitude, (7.78), and the degree of polarization, given by (7.80), are also unchanged by a rotation of the coordinate axes, results that were expected.

Let us apply a coordinate rotation to a coherency matrix that is the sum of an unpolarized wave and a completely polarized wave,

$$\mathbf{J} = \mathbf{J}^{(1)} + \mathbf{J}^{(2)} \quad (7.70)$$

where the matrix elements are given by (7.71). Then

$$\mathbf{J}' = \mathbf{J}'^{(1)} + \mathbf{J}'^{(2)} = \mathbf{RJ}^{(1)}\mathbf{R}^{-1} + \mathbf{RJ}^{(2)}\mathbf{R}^{-1} \quad (7.105)$$

The first product is

$$\mathbf{J}'^{(1)} = \mathbf{RJ}^{(1)}\mathbf{R}^{-1} = \begin{bmatrix} A & 0 \\ 0 & A \end{bmatrix} = \mathbf{J}^{(1)} \quad (7.106)$$

and it can thus be seen that the coherency matrix of an unpolarized wave is unaltered by a coordinate rotation. The coherency matrix of the polarized part of the wave is altered by the rotation, as expected. However, the new coherency matrix $\mathbf{J}'^{(2)}$ still satisfies the conditions to represent a completely polarized wave, and

$$\mathrm{Tr}[\mathbf{J}'^{(2)}] = B + C = \mathrm{Tr}[\mathbf{J}^{(2)}] \quad (7.107)$$

Since the trace of the coherency matrix is proportional to the power density of the wave, the transformed completely polarized part of the wave has the same power density as the original completely polarized part. In summary, a partially polarized wave written as the sum of an unpolarized wave and a completely polarized wave may be transformed by a coordinate rotation. The wave after rotation is still the sum of an unpolarized and a completely polarized wave, and their power densities are the same as those of the wave in the original coordinate system.

If the rotation transformation is applied to the coherency matrix of a wave expressed as the sum of two independent completely polarized waves,

$$\mathbf{J} = \mathbf{J}^{(3)} + \mathbf{J}^{(4)} \tag{7.108}$$

the result

$$\mathbf{J}' = \mathbf{R}\mathbf{J}^{(3)}\mathbf{R}^{-1} + \mathbf{R}\mathbf{J}^{(4)}\mathbf{R}^{-1} \tag{7.109}$$

remains the sum of two independent completely polarized waves. The power density in each component is the same as in each component of the wave in unrotated coordinates, and the two components are still orthogonal.

Finally, an unpolarized wave can be thought of as being composed of two independent linearly polarized waves orthogonal to each other, each with equal power density. To see this, note that (7.49), the coherency matrix of an unpolarized wave, can be split, thus

$$\mathbf{J} = Z_0 \mathscr{P} \begin{bmatrix} 1 & 0 \\ 0 & 1 \end{bmatrix} = Z_0 \mathscr{P} \begin{bmatrix} 1 & 0 \\ 0 & 0 \end{bmatrix} + Z_0 \mathscr{P} \begin{bmatrix} 0 & 0 \\ 0 & 1 \end{bmatrix} \tag{7.110}$$

and each resulting matrix represents a linearly polarized wave.

Just as readily, we could have written

$$\mathbf{J} = Z_0 \frac{\mathscr{P}}{2} \begin{bmatrix} 1 & j \\ -j & 1 \end{bmatrix} + Z_0 \frac{\mathscr{P}}{2} \begin{bmatrix} 1 & -j \\ j & 1 \end{bmatrix} \tag{7.111}$$

showing that with equal validity an unpolarized wave can be considered the sum of two independent circular waves of opposite rotation sense and equal power density.

Received Power

A wave with field intensity $\mathbf{E}$ falling on a receiving antenna with effective length $\mathbf{h}$ produces an open-circuit voltage at the antenna terminals,

$$V = \mathbf{E} \cdot \mathbf{h} \tag{7.112}$$

This holds if $\mathbf{E}$ is either completely or partially polarized, but we are concerned here primarily with partial polarization and shall accordingly consider the power supplied to a matched load on the antenna to be [7]

$$W = \frac{\langle VV^* \rangle}{8R_a} = \frac{\langle |\mathbf{E} \cdot \mathbf{h}|^2 \rangle}{8R_a} \qquad (7.113)$$

where R_a is the antenna resistance (radiation resistance plus loss resistance).

If the wave incident on a receiving antenna is given by its coherency matrix, (7.113) cannot be used directly to find received power. We must then find an alternative way to describe the antenna, preferably one that includes $\mathbf{h}$. Now it was noted earlier that the polarization ratio of an antenna is by definition the polarization ratio of the wave it transmits, so in a similar manner we consider defining a *coherency matrix of an antenna* as that of the wave it transmits when fed by a monochromatic signal. The wave coherency matrix from an antenna supplied by a single frequency is

$$\mathbf{J} = \begin{bmatrix} |E_x|^2 & E_x E_y^* \\ E_x^* E_y & |E_y|^2 \end{bmatrix} \qquad (7.114)$$

and using (4.2),

$$\mathbf{J} = \frac{Z_0^2 I^2}{4\lambda^2 r^2} \begin{bmatrix} |h_x|^2 & h_x h_y^* \\ h_x^* h_y & |h_y|^2 \end{bmatrix} \qquad (7.115)$$

The multiplier is inappropriate for an antenna description, so we define an *antenna coherency matrix* as [7]

$$\mathbf{J}_A = \begin{bmatrix} |h_x|^2 & h_x h_y^* \\ h_x^* h_y & |h_y|^2 \end{bmatrix} \qquad (7.116)$$

Note that it is proper to consider a monochromatic signal in this definition. If the antenna is supplied with a polychromatic signal, the radiated wave is polychromatic, but this feature should not be assigned to the antenna. On the other hand, (7.116) allows an antenna fed by a single frequency to radiate a partially polarized wave.

Consider two antennas with antenna a receiving and antenna b transmitting. All polarization descriptors for antenna a, including the coherency matrix, are given in a coordinate system appropriate to a as a transmitter, for example, the xyz system of Figure 7.1, even though a is receiving. Polarization descriptors for b, including the coherency matrix of its transmitted wave, are given in coordinates appropriate to b as a transmitter, for example, the $\xi\eta\zeta$ system of Figure 7.1.

The coherency matrix of the wave incident on the receiving antenna is

$$\mathbf{J}^i = \begin{bmatrix} \langle |E_\xi|^2 \rangle & \langle E_\xi E_\eta^* \rangle \\ \langle E_\xi^* E_\eta \rangle & \langle |E_\eta|^2 \rangle \end{bmatrix} \quad (7.117)$$

It would be convenient if we could carry out a simple operation with the coherency matrices of the incident wave and antenna to find the received power, although there is no a priori reason to believe the operation possible. To see how to proceed, expand (7.113), noting that **E** and **h** must be expressed in the same coordinates,

$$W = \frac{1}{8R_a} \langle |E_\xi|^2 |h_\xi|^2 + E_\xi E_\eta^* h_\xi h_\eta^* + E_\xi^* E_\eta h_\xi^* h_\eta + |E_\eta|^2 |h_\eta|^2 \rangle \quad (7.118)$$

The operation we consider on $\mathbf{J}_A$ and $\mathbf{J}^i$ should give this power. A straightforward product gives a 2×2 matrix, so a further operation, such as obtaining the trace of the product, is necessary. Even with this, neither the product $\mathbf{J}^i \mathbf{J}_A$ nor $\mathbf{J}_A \mathbf{J}^i$ yields the correct power. Therefore we define a new matrix, constrained by the requirement

$$W = \frac{1}{8R_a} \text{Tr}(\mathbf{J}^i \mathbf{J}_R) \quad (7.119)$$

It is easy to see from (7.117) and (7.118) that the power constraint is satisfied if $\mathbf{J}_R$ is defined by

$$\mathbf{J}_R = \begin{bmatrix} |h_\xi|^2 & h_\xi^* h_\eta \\ h_\xi h_\eta^* & |h_\eta|^2 \end{bmatrix} = \begin{bmatrix} |h_x|^2 & -h_x^* h_y \\ -h_x h_y^* & |h_y|^2 \end{bmatrix} \quad (7.120)$$

$\mathbf{J}_R$ may be considered the *receiving coherency matrix* of the antenna, which has an antenna coherency matrix $\mathbf{J}_A$. It is formed by first taking the conjugate of $\mathbf{J}_A$ and then replacing h_x and h_y by h_ξ and h_η. (Note: This is not the normal coordinate change that replaces h_ξ by $-h_x$.) In the multiplication of (7.119), both $\mathbf{J}^i$ and $\mathbf{J}_R$ must be in the same coordinates.

The linear polarization ratio of a completely polarized wave is related to the coherency matrix elements by

$$P = \frac{J_{yx}}{J_{xx}} \quad (7.84)$$

Then if P_A is the antenna polarization ratio corresponding to $\mathbf{J}_A$ of (7.116),

$$P_A = \frac{h_y}{h_x} \tag{7.121}$$

the polarization ratio corresponding to $\mathbf{J}_R$ is

$$P_R = \frac{-h_x h_y^*}{|h_x|^2} = -\left(\frac{h_y}{h_x}\right)^* = -P_A^* \tag{7.122}$$

Then P_R is the receiving polarization ratio of the antenna as defined in Chapter 4.

In general, the effective length $\mathbf{h}$ and the coherency matrix $\mathbf{J}_A$ of the antenna are frequency dependent. Then $\mathbf{h}$ is determined at the mean frequency of the incident wave, and (7.119) is to a small degree approximate. It is clear from

$$\mathbf{J} = Z_0 \mathscr{P} \begin{bmatrix} 1 & 0 \\ 0 & 1 \end{bmatrix} \tag{7.49}$$

which is the coherency matrix of an unpolarized wave, taken here as an example, that two waves with different bandwidths and the same power density, $\mathscr{P}$, can have equal coherency matrices. An antenna with a frequency-dependent effective length will not receive the waves equally well, however, even if they have the same mean frequencies. If the incident wave has zero bandwidth, or if $\mathbf{h}$ is independent of frequency over the bandwidth of the incident wave, then the approximate relationship between (7.119) and (7.113) is exact. This point will become clearer when power spectral densities of the fields are considered.

Scattering by a Target

Consider scattering by a target illuminated by a quasi-monochromatic wave, or, in the case of an optical wave, transmission through some optical instrument. Figure 7.1b is applicable to both problems. In Section 6.4 it was noted that the $x_1 y_1 z_1$ and $x_3 y_3 z_3$ coordinate systems are commonly used for bistatic scattering of rectangular field components and the $x_1 y_1 z_1$ system alone for backscattering (the two systems coincide for backscattering). Following the convention that we have used thus far, all polarization descriptors other than the rectangular field components are given in $x_1 y_1 z_1$ coordinates for the incident wave and $x_2 y_2 z_2$ coordinates for the reflected or transmitted wave. Then the scattering form normal for use with coherency matrices is the Jones matrix of Chapter 6,

$$\mathbf{T} = \begin{bmatrix} T_{xx} & T_{xy} \\ T_{yx} & T_{yy} \end{bmatrix} = \begin{bmatrix} T_{x2x1} & T_{x2y1} \\ T_{y2x1} & T_{y2y1} \end{bmatrix} \tag{7.123}$$

The scattered field is

$$\mathbf{E}^s = \frac{1}{\sqrt{4\pi}\, r} \mathbf{T} \mathbf{E}^i \tag{6.82}$$

with $\mathbf{E}^s$ in $x_2 y_2 z_2$ coordinates.

The coherency matrix of the wave incident on the target is

$$\mathbf{J}^i = \langle \mathbf{E}^i \mathbf{E}^{iT*} \rangle \tag{7.124}$$

where the superscripts T and * identify the conjugate transpose matrix. The coherency matrix of the scattered (or transmitted) wave is

$$\mathbf{J}^s = \langle \mathbf{E}^s \mathbf{E}^{sT*} \rangle = \frac{1}{4\pi r^2} \langle \mathbf{T} \mathbf{E}^i \mathbf{E}^{iT*} \mathbf{T}^{T*} \rangle \tag{7.125}$$

If the Jones matrix $\mathbf{T}$ can be removed from the time average, a condition that requires $\mathbf{T}$ to be frequency-invariant over the bandwidth of the incident wave and to be invariant to time during the period in which the time average is taken, then the coherency matrix of the scattered wave is

$$\mathbf{J}^s = \frac{1}{4\pi r^2} \mathbf{T} \langle \mathbf{E}^i \mathbf{E}^{iT*} \rangle \mathbf{T}^{T*} = \frac{1}{4\pi r^2} \mathbf{T} \mathbf{J}^i \mathbf{T}^{T*} \tag{7.126}$$

If a transmitting antenna a is located at the origin of $x_1 y_1 z_1$ coordinates in Figure 7.1, the coherency matrix of the wave incident on the target is

$$\mathbf{J}^i = \frac{Z_0^2 I_a^2}{4\lambda^2 r_1^2} \mathbf{J}_A^{(a)} \tag{7.127}$$

where λ is the mean wavelength of the incident wave, I_a is a current in the transmitting antenna, and $\mathbf{J}_A^{(a)}$ is the coherency matrix of the transmitting antenna. Note that, in accordance with common radar practice, the feed current to the transmitting antenna is monochromatic.

The coherency matrix of the wave reflected toward the receiving antenna b at the origin of $x_3 y_3 z_3$ coordinates is

$$\mathbf{J}^s = \frac{1}{4\pi r_2^2} \mathbf{T} \mathbf{J}^i \mathbf{T}^{T*} = \frac{Z_0^2 I_a^2}{16\pi \lambda^2 r_1^2 r_2^2} \mathbf{T} \mathbf{J}_A^{(a)} \mathbf{T}^{T*} \quad (7.128)$$

If the receiving antenna has coherency matrix $\mathbf{J}_A^{(b)}$, defined in transmitting coordinates $x_3 y_3 z_3$, we define a matrix $\mathbf{J}_R^{(b)}$ in the manner prescribed earlier: the conjugate matrix of $\mathbf{J}_A^{(b)}$ is found, and h_{x_3} is replaced by $-h_{x_3}$ $(= h_{x_2})$ and h_{y_3} by h_{y_2}. The received power is then found by

$$W = \frac{1}{8R_a} \operatorname{Tr}(\mathbf{J}^s \mathbf{J}_R^{(b)}) = \frac{Z_0^2 I_a^2}{128\pi R_a \lambda^2 r_1^2 r_2^2} \operatorname{Tr}(\mathbf{T} \mathbf{J}_A^{(a)} \mathbf{T}^{T*} \mathbf{J}_R^{(b)}) \quad (7.129)$$

This simplifies somewhat for backscattering, with the same antenna transmitting and receiving. The distances are the same and the antenna identifiers may be omitted. The backscattered power is

$$W = \frac{Z_0^2 I^2}{128\pi R_a \lambda^2 r^4} \operatorname{Tr}(\mathbf{T} \mathbf{J}_A \mathbf{T}^{T*} \mathbf{J}_R) \quad (7.130)$$

This form, applicable to partially polarized waves, may be compared to the single-frequency power scattered by a target with the same antenna transmitting and receiving,

$$W = \frac{Z_0^2 I^2}{128\pi R_a \lambda^2 r^4} \mathbf{h}^T \mathbf{S} \mathbf{h} \mathbf{h}^{T*} \mathbf{S}^* \mathbf{h}^* \quad (7.131)$$

7.4. STOKES VECTOR OF PARTIALLY POLARIZED WAVES

The Stokes Vector

The use of the Stokes vector to describe a quasi-monochromatic wave allows a more concise formulation of reception and scattering problems than can be achieved with the coherency matrix. Four parameters were introduced by Stokes in his studies of partially polarized light. They were given earlier for a monochromatic wave,

$$G_0 = |E_x|^2 + |E_y|^2 \qquad (3.173a)$$
$$G_1 = |E_x|^2 - |E_y|^2 \qquad (3.173b)$$
$$G_2 = 2|E_x||E_y|\cos\phi \qquad (3.173c)$$
$$G_3 = 2|E_x||E_y|\sin\phi \qquad (3.173d)$$

For quasi-monochromatic waves a more general definition, which reduces to (3.173) for time-independent amplitude and phase of the wave components (a monochromatic wave), is

$$G_0 = \langle |E_x(t)|^2 + |E_y(t)|^2 \rangle = \langle |a_x|^2 + |a_y|^2 \rangle \qquad (7.132a)$$
$$G_1 = \langle |E_x(t)|^2 - |E_y(t)|^2 \rangle = \langle |a_x|^2 - |a_y|^2 \rangle \qquad (7.132b)$$
$$G_2 = \langle 2\operatorname{Re}[E_x^*(t)E_y(t)] \rangle = 2\langle a_x a_y \cos\phi \rangle \qquad (7.132c)$$
$$G_3 = \langle 2\operatorname{Im}[E_x^*(t)E_y(t)] \rangle = 2\langle a_x a_y \sin\phi \rangle \qquad (7.132d)$$

where

$$\phi = \phi(t) = \phi_y(t) - \phi_x(t) \qquad (7.133)$$

Comparison of these parameters to the elements of the coherency matrix, (7.36), shows that

$$G_0 = (J_{xx} + J_{yy}) \qquad (7.134a)$$
$$G_1 = (J_{xx} - J_{yy}) \qquad (7.134b)$$
$$G_2 = (J_{xy} + J_{yx}) \qquad (7.134c)$$
$$G_3 = j(J_{xy} - J_{yx}) \qquad (7.134d)$$

Inverting allows the coherency matrix to be found from the Stokes parameters, and it is seen that the Stokes parameters are sufficient to characterize our knowledge of the wave.

It is desirable to order the Stokes parameters, and we therefore introduce the *Stokes vector*.

$$\mathbf{G} = \begin{bmatrix} G_0 \\ G_1 \\ G_2 \\ G_3 \end{bmatrix} = \begin{bmatrix} \langle |E_x|^2 + |E_y|^2 \rangle \\ \langle |E_x|^2 - |E_y|^2 \rangle \\ 2\operatorname{Re}\langle E_x^* E_y \rangle \\ 2\operatorname{Im}\langle E_x^* E_y \rangle \end{bmatrix} \qquad (7.135)$$

Unlike the coherency matrix, all elements of the Stokes vector are real. We saw earlier that two quasi-monochromatic waves with different bandwidths and the same power density can have equal coherency matrices. It follows from the relationship between the Stokes parameters and coherency matrix elements that two waves with different bandwidths can have the same Stokes vector. We must therefore recognize that some equations utilizing the Stokes vector for reception and scattering require the quasi-monochromatic constraint and are—to a slight degree—approximate.

The Stokes vector elements are not independent for monochromatic waves since

$$G_0^2 = G_1^2 + G_2^2 + G_3^2 \quad \text{(monochromatic waves)} \quad (3.174)$$

This relationship does not hold for partially polarized waves. The coherency matrix elements in terms of the Stokes parameters are

$$J_{xx} = \tfrac{1}{2}(G_0 + G_1) \quad (7.136a)$$
$$J_{yy} = \tfrac{1}{2}(G_0 - G_1) \quad (7.136b)$$
$$J_{xy} = \tfrac{1}{2}(G_2 - jG_3) \quad (7.136c)$$
$$J_{yx} = \tfrac{1}{2}(G_2 + jG_3) \quad (7.136d)$$

From these

$$\|\mathbf{J}\| = \tfrac{1}{4}\left(G_0^2 - G_1^2 - G_2^2 - G_3^2\right) \quad (7.137)$$

But we know that

$$\|\mathbf{J}\| \geq 0 \quad (7.46)$$

for partially polarized waves, and it follows that

$$G_0^2 \geq G_1^2 + G_2^2 + G_3^2 \quad (7.138)$$

Unpolarized Waves

For a completely unpolarized wave, we found earlier that

$$J_{xx} = J_{yy} \quad (7.48a)$$
$$J_{xy} = J_{yx} = 0 \quad (7.48b)$$

where J_{xx} is Z_0 times the power density, $\mathscr{P}$, of the wave. It follows that for such a wave,

$$G_0 = 2Z_0\mathscr{P} \quad (7.139a)$$
$$G_1 = G_2 = G_3 = 0 \quad (7.139b)$$

Complete Polarization

A monochromatic wave and a quasi-monochromatic wave that obeys the condition (7.54) both satisfy

$$\|\mathbf{J}\| = 0 \qquad (7.59)$$

and it follows from (7.137) that

$$G_0^2 = G_1^2 + G_2^2 + G_3^2 \qquad (7.140)$$

Degree of Polarization

We saw earlier that the coherency matrix elements of independent waves propagating in the same direction are additive. It follows that the Stokes vector elements of such waves can also be added. We saw also that the coherency matrix of any wave can be uniquely expressed as the sum of coherency matrices for unpolarized and completely polarized waves. The Stokes vector of any wave may then be written as the Stokes vector of an unpolarized wave plus the Stokes vector of a completely polarized wave, and this representation is unique. We then write

$$\mathbf{G} = \mathbf{G}^{(1)} + \mathbf{G}^{(2)} \qquad (7.141)$$

where $\mathbf{G}^{(1)}$ represents an unpolarized wave and obeys (7.139), and $\mathbf{G}^{(2)}$ represents a completely polarized wave satisfying (7.140). It follows that

$$G_0 = G_0^{(1)} + G_0^{(2)} \qquad (7.142a)$$
$$G_1 = G_1^{(2)} \qquad (7.142b)$$
$$G_2 = G_2^{(2)} \qquad (7.142c)$$
$$G_3 = G_3^{(2)} \qquad (7.142d)$$

Three elements of $\mathbf{G}^{(1)}$ are zero, while three elements of $\mathbf{G}^{(2)}$ may be found from this equation and a knowledge of $\mathbf{G}$. The remaining elements can be found from (7.142) and the fact that $\mathbf{G}^{(2)}$ must satisfy (7.140). This gives

$$G_0^{(2)} = \sqrt{G_1^2 + G_2^2 + G_3^2} \qquad (7.143)$$

and

$$G_0^{(1)} = G_0 - \sqrt{G_1^2 + G_2^2 + G_3^2} \qquad (7.144)$$

The degree of polarization was defined in a previous section as the ratio of power densities of the polarized part and the total wave. But $G_0^{(2)}$ is proportional to the density of the polarized part, and G_0 is proportional, with the same proportionality constant, to the density of the total wave. Then the degree of polarization in terms of the Stokes parameters is

$$R = \frac{G_0^{(2)}}{G_0} = \frac{\sqrt{G_1^2 + G_2^2 + G_3^2}}{G_0} \quad (7.145)$$

This form is consistent with (7.80) if the relationship between **J** and **G** is used. Equation (7.138) shows that, as expected, R ranges from zero to one.

We saw earlier that a partially polarized wave may be treated as the sum of two independent completely polarized orthogonal waves, and we found the coherency matrices of the two components. The Stokes vector of the partially polarized wave may also be written as a sum,

$$\mathbf{G} = \mathbf{G}^{(3)} + \mathbf{G}^{(4)} \quad (7.146)$$

where $\mathbf{G}^{(3)}$ and $\mathbf{G}^{(4)}$ correspond to the coherency matrices $\mathbf{J}^{(3)}$ and $\mathbf{J}^{(4)}$ of (7.82) and represent completely polarized independent orthogonal waves.

If the elements found for the decomposition $\mathbf{J}^{(3)}$ and $\mathbf{J}^{(4)}$ are used in the relationships of Stokes vector and coherency matrix elements, the Stokes vector written as the sum of two completely polarized Stokes vectors is found to be

$$\mathbf{G} = \left[\tfrac{1}{4}(G_0 + G_1) + \delta\right] \begin{bmatrix} 1 + |P_3|^2 \\ 1 - |P_3|^2 \\ P_3^* + P_3 \\ j(P_3^* - P_3) \end{bmatrix}$$

$$+ \left[\tfrac{1}{4}(G_0 + G_1) + \delta\right] \begin{bmatrix} 1 + 1/|P_3|^2 \\ 1 - 1/|P_3|^2 \\ -(1/P_3^* + 1/P_3) \\ j(1/P_3^* - 1/P_3) \end{bmatrix} \quad (7.147)$$

where

$$|P_3| = \frac{RG_0 - G_1}{\sqrt{G_2^2 + G_3^2}} \quad (7.148a)$$

$$\text{Ang}(P_3) = \tan^{-1} \frac{G_3}{G_2} \quad (7.148b)$$

$$\delta = \frac{1}{4} \frac{(G_0 + G_1)(|P_3|^2 - 1) - 2\sqrt{G_2^2 + G_3^2}\,|P_3|}{|P_3|^2 + 1} \quad (7.149)$$

with R the degree of polarization, given by (7.145).

The Poincaré Sphere

Equation (7.145) may be rewritten as

$$G_1^2 + G_2^2 + G_3^2 = (RG_0)^2 \tag{7.150}$$

and from this it is clear that the Stokes parameters may still be regarded as the rectangular coordinates of a point on the Poincaré sphere. The sphere radius for partially polarized waves is RG_0, and for unpolarized waves becomes zero. A polarization ellipse is not defined for the total wave, so the interpretation of latitude and azimuth angles of points on the sphere in terms of axial ratio and tilt angle must be limited to the polarized part of the wave.

If the wave is separated into unpolarized and completely polarized components according to (7.141), the Poincaré sphere for the total wave is also that for the completely polarized part, since

$$G_0^{(2)} = RG_0 \tag{7.151}$$

If this is used in (7.150) and the relationships in (7.142) noted, it follows that

$$\left[G_1^{(2)}\right]^2 + \left[G_2^{(2)}\right]^2 + \left[G_3^{(2)}\right]^2 = \left[G_0^{(2)}\right]^2 \tag{7.152}$$

Then, on the Poincaré sphere for the total wave, we may freely interpret points in terms of axial ratio and tilt angle of the completely polarized part of the wave.

Measurement of the Stokes Vector

The concepts of a partially polarized plane wave were introduced earlier by considering the measurement of the power of a wave component at angle θ to the x axis after the x component of the wave is delayed by mean phase angle δ. The resulting time-average power is proportional to

$$I = J_{xx} \cos^2 \theta + J_{yy} \sin^2 \theta + \left(J_{xy} e^{-j\delta} + J_{yx} e^{j\delta}\right) \cos \theta \sin \theta \tag{7.47}$$

The constant of proportionality is readily found from (7.119) and (7.120), and the power measured by an antenna of the wave component at angle θ is

$$W(\theta, \delta) = \frac{|\mathbf{h}|^2}{8R_a} \left[J_{xx} \cos^2 \theta + J_{yy} \sin^2 \theta + \left(J_{xy} e^{-j\delta} + J_{yx} e^{j\delta}\right) \cos \theta \sin \theta \right] \tag{7.153}$$

where $\mathbf{h}$ is the antenna effective length and R_a its resistance.

If the Stokes vector elements are substituted into the equation for power, it becomes

$$W(\theta, \delta) = \frac{|\mathbf{h}|^2}{8R_a}\left[\tfrac{1}{2}(G_0 + G_1 \cos 2\theta) + (G_2 \cos \delta - G_3 \sin \delta)\cos \theta \sin \theta\right]$$
(7.154)

Certain choices of θ and δ for the power measurements lead quickly to the Stokes parameters, which in terms of the measured power are

$$G_0 = \frac{8R_a}{|\mathbf{h}|^2}[W(0°, 0) + W(90°, 0)] \quad (7.155\text{a})$$

$$G_1 = \frac{8R_a}{|\mathbf{h}|^2}[W(0°, 0) - W(90°, 0)] \quad (7.155\text{b})$$

$$G_2 = \frac{8R_a}{|\mathbf{h}|^2}[W(45°, 0) - W(135°, 0)] \quad (7.155\text{c})$$

$$G_3 = \frac{8R_a}{|\mathbf{h}|^2}[W(135°, \pi/2) - W(45°, \pi/2)] \quad (7.155\text{d})$$

Note that the only phase shift necessary is $\pi/2$, which is readily obtained at both optical and lower frequencies.

The measurements that give the Stokes vector can be used for the coherency matrix elements also, since the coherency matrix can be found from the Stokes vector.

Received Power

If the same rationale that led us to define a coherency matrix for an antenna is applied again, a Stokes vector for an antenna can be defined,

$$\mathbf{G}_A = \begin{bmatrix} |h_x|^2 + |h_y|^2 \\ |h_x|^2 - |h_y|^2 \\ 2\,\text{Re}\langle h_x^* h_y \rangle \\ 2\,\text{Im}\langle h_x^* h_y \rangle \end{bmatrix} \quad (7.156)$$

If the antenna is fed with a polychromatic signal it will radiate a polychromatic wave, but the definition is unaffected. On the other hand, if the feeding signal is a single frequency, the radiated wave may be partially polarized, and this definition of an antenna Stokes vector allows such a wave to be treated.

398 PARTIAL POLARIZATION

The antenna Stokes vector is related to the coherency matrix elements by

$$\mathbf{G}_A = \begin{bmatrix} 1 & 0 & 0 & 1 \\ 1 & 0 & 0 & -1 \\ 0 & 1 & 1 & 0 \\ 0 & j & -j & 0 \end{bmatrix} \begin{bmatrix} J_{Axx} \\ J_{Axy} \\ J_{Ayx} \\ J_{Ayy} \end{bmatrix} \quad (7.157)$$

The same relation applies to the wave radiated by the antenna.

Consider again the two-antenna configuration, with a receiving and b transmitting, that was previously treated with coherency matrices. The Stokes vector of the incident wave, in coordinates appropriate to b as a transmitter, is

$$\mathbf{G}^i = \begin{bmatrix} \langle |E_\xi|^2 + |E_\eta|^2 \rangle \\ \langle |E_\xi|^2 - |E_\eta|^2 \rangle \\ 2\,\mathrm{Re}\langle E_\xi^* E_\eta \rangle \\ 2\,\mathrm{Im}\langle E_\xi^* E_\eta \rangle \end{bmatrix} \quad (7.158)$$

It was desirable in a previous section to define a receiving coherency matrix for an antenna in order to determine the received power. It is also useful to define in an analogous fashion a *receiving Stokes vector* for an antenna with Stokes vector $\mathbf{G}_A$. If (7.134), relating the Stokes parameters to the coherency matrix elements, is applied to the receiving coherency matrix of the antenna, (7.120), the receiving Stokes vector is found to be

$$\mathbf{G}_R = \begin{bmatrix} |h_\xi|^2 + |h_\eta|^2 \\ |h_\xi|^2 - |h_\eta|^2 \\ 2\,\mathrm{Re}(h_\xi^* h_\eta) \\ -2\,\mathrm{Im}(h_\xi^* h_\eta) \end{bmatrix} = \begin{bmatrix} |h_x|^2 + |h_y|^2 \\ |h_x|^2 - |h_y|^2 \\ -2\,\mathrm{Re}(h_x^* h_y) \\ 2\,\mathrm{Im}(h_x^* h_y) \end{bmatrix} \quad (7.159)$$

If the transpose of the incident Stokes vector is multiplied with this receiving Stokes vector, the product is (using both vectors in the same coordinates)

$$\mathbf{G}^{iT}\mathbf{G}_R = 2\big(\langle |E_\xi|^2 \rangle |h_\xi|^2 + \langle E_\xi E_\eta^* \rangle h_\xi h_\eta^* + \langle E_\xi^* E_\eta \rangle h_\xi^* h_\eta + \langle |E_\eta|^2 \rangle |h_\eta|^2 \big) \quad (7.160)$$

If $\mathbf{h}$ is frequency-invariant over the bandwidth of the incident wave and time-invariant on the scale of time in which averaging takes place, conditions that were discussed earlier, the components of **h** can be taken inside the time averages and the product written as

$$\mathbf{G}^{iT}\mathbf{G}_R = 2\langle |\mathbf{E} \cdot \mathbf{h}|^2 \rangle \tag{7.161}$$

It follows that the received power to an impedance-matched load is

$$W = \frac{1}{16R_a} \mathbf{G}^{iT}\mathbf{G}_R \tag{7.162}$$

7.5. POLARIZATION RATIO OF PARTIALLY POLARIZED WAVES

The polarization ratio, linear or circular, and the polarization ellipse characteristics of the polarized part of a partially polarized wave can be obtained just as for the completely polarized wave. We treat the wave as the sum of an unpolarized wave (1) and a completely polarized wave (2), just as we did previously. From (3.181a) the linear polarization ratio of the polarized part of the wave is

$$P = \frac{G_2^{(2)} + jG_3^{(2)}}{G_0^{(2)} + G_1^{(2)}} \tag{7.163}$$

Equations (7.142) and (7.143) can be used to find P in terms of the parameters of the total wave, giving

$$P = \frac{G_2 + jG_3}{G_1 + \sqrt{G_1^2 + G_2^2 + G_3^2}} \tag{7.164}$$

The inverse circular polarization ratio is found in the same way, using (3.181c), to be

$$q = \frac{G_1^{(2)} - jG_2^{(2)}}{G_0^{(2)} - G_3^{(2)}} = \frac{G_1 - jG_2}{\sqrt{G_1^2 + G_2^2 + G_3^2} - G_3} \tag{7.165}$$

In terms of the coherency matrix elements for the partially polarized wave, the linear polarization ratio becomes, on using the relationships between Stokes vector and coherency matrix elements,

$$P = \frac{2J_{yx}}{(R+1)J_{xx} + (R-1)J_{yy}} \tag{7.166}$$

where R is the degree of polarization of the wave.

400 PARTIAL POLARIZATION

For complete polarization $R = 1$ and

$$P = \frac{J_{yx}}{J_{xx}} \quad (7.167)$$

which becomes, in terms of the analytic signal representations for the fields, using (7.19), (7.54), and (7.55),

$$P = \frac{\mathscr{E}_y(z,t)}{\mathscr{E}_x(z,t)} \quad (7.168)$$

7.6. RECEPTION OF PARTIALLY POLARIZED WAVES

We have formally considered the reception of partially polarized waves by representing the antenna with a receiving coherency matrix or a receiving Stokes vector. Such a representation is convenient and valuable, but does not provide the physical insight that might be gained by treating the antenna in terms of its effective length. In this section we once again find the received power from a wave incident on a receiving antenna, this time using the antenna effective length and restricting the antenna to be time-independent. The effective length that maximizes the received power will be found.

A wave with field intensity $\mathbf{E}$ falling on a receiving antenna with effective length $\mathbf{h}$ produces an open-circuit voltage at the antenna terminals,

$$V = \mathbf{E} \cdot \mathbf{h} \quad (7.112)$$

This holds whether $\mathbf{E}$ is coherent or not, but we are concerned here with partially polarized waves and will accordingly consider the power supplied to a matched load on the antenna to be

$$W = \frac{\langle VV^* \rangle}{8R_a} \quad (7.169)$$

where R_a is the antenna resistance (radiation resistance plus loss resistance). By using (4.8) and (4.11), R_a may be put into the form

$$R_a = \frac{Z_0 \mathbf{h} \cdot \mathbf{h}^*}{4A_e} \quad (7.170)$$

where A_e is the effective area of the antenna. If this and (7.112) are used in the power equation it becomes

$$W = \frac{A_e}{2Z_0 \mathbf{h} \cdot \mathbf{h}^*} \langle (\mathbf{E} \cdot \mathbf{h})(\mathbf{E} \cdot \mathbf{h})^* \rangle \quad (7.171)$$

If the receiving antenna is time-independent, time averaging for it is unnecessary, and the received power is

$$W = \frac{A_e}{2Z_0 \mathbf{h} \cdot \mathbf{h}^*} \left(|h_x|^2 \langle E_x E_x^* \rangle + h_x h_y^* \langle E_x E_y^* \rangle + h_x^* h_y \langle E_x^* E_y \rangle \right.$$
$$\left. + |h_y|^2 \langle E_y E_y^* \rangle \right) \quad (7.172)$$

This becomes, on using the coherency matrix elements of the incident wave,

$$W = \frac{A_e}{2Z_0 \mathbf{h} \cdot \mathbf{h}^*} \left(|h_x|^2 J_{xx} + h_x h_y^* J_{xy} + h_x^* h_y J_{yx} + |h_y|^2 J_{yy} \right) \quad (7.173)$$

Consider the partially polarized plane wave to be the sum of a completely polarized wave and an unpolarized wave. The coherency matrix elements of the component waves are given by (7.71). If they are substituted into the equation for received power it becomes

$$W = \frac{A_e}{2Z_0 \mathbf{h} \cdot \mathbf{h}^*} \left[|h_x|^2 (A + B) + h_x h_y^* D + h_x^* h_y D^* + |h_y|^2 (A + C) \right]$$
$$(7.174)$$

This form may be separated to give

$$W = W' + W'' = \frac{A_e A}{2Z_0} + \frac{A_e}{2Z_0 \mathbf{h} \cdot \mathbf{h}^*} \left(|h_x|^2 B + h_x h_y^* D + h_x^* h_y D^* + |h_y|^2 C \right)$$
$$(7.175)$$

where the first term,

$$W' = \frac{A_e A}{2Z_0} \quad (7.176)$$

which represents the power received from the unpolarized portion of the wave, is independent of the polarization characteristics of the receiving antenna. It is informative to express this power in terms of the degree of polarization of the wave. From (7.77a) and (7.80),

$$A = \tfrac{1}{2}(J_{xx} + J_{yy})(1 - R) \quad (7.177)$$

Using the relation between the sum of the coherency matrix elements and the power density, the received power W' becomes on substitution

$$W' = A_e \frac{\mathscr{P}_t}{2} (1 - R) \quad (7.178)$$

402 PARTIAL POLARIZATION

Note that if the wave is unpolarized ($R = 0$), the maximum power that can be extracted from the wave is one-half the power that could be utilized from a completely polarized wave matched in polarization to the receiving antenna.

We need not be concerned further with W' since nothing we can do with the polarization of the receiving antenna will either increase or decrease it. We therefore turn our attention to the power received from the completely polarized part of the wave and attempt to maximize it. Now B and C in

$$W'' = \frac{A_e}{2Z_0 \mathbf{h} \cdot \mathbf{h}^*}\left(|h_x|^2 B + h_x h_y^* D + h_x^* h_y D^* + |h_y|^2 C\right) \quad (7.179)$$

are by definition positive real. We therefore first maximize the sum of the two middle terms of W'' by setting

$$h_x = |h_x| e^{j\beta_x} \quad (7.180a)$$
$$h_y = |h_y| e^{j\beta_y} \quad (7.180b)$$
$$D = |D| e^{j\delta} \quad (7.180c)$$

It is at once obvious that the sum $h_x h_y^* D + h_x^* h_y D^*$ is maximum if we choose

$$\beta_y - \beta_x = \delta \quad (7.181)$$

Then W'' becomes

$$W''_m = \frac{A_e}{2Z_0 \mathbf{h} \cdot \mathbf{h}^*}\left(|h_x|^2 B + 2|h_x||h_y||D| + |h_y|^2 C\right) \quad (7.182)$$

This "maximum" power can be further maximized by varying h_x and h_y while holding $\mathbf{h} \cdot \mathbf{h}^*$ constant. This is an appropriate constraint and was discussed in Section 4.5. Differentiating W''_m with respect to $|h_x|$, given by

$$|h_x| = \left(\mathbf{h} \cdot \mathbf{h}^* - |h_y|^2\right)^{1/2} \quad (7.183)$$

and setting the derivative to zero gives

$$\frac{|h_y|^2 - |h_x|^2}{|h_x||h_y|} = \frac{C - B}{|D|}$$

with solution

$$\frac{|h_y|}{|h_x|} = \frac{C}{|D|} \quad (7.184a)$$

$$\frac{|h_x|}{|h_y|} = \frac{B}{|D|} \quad (7.184b)$$

Note that $|D| \neq 0$ unless the wave is completely unpolarized. One of these forms is the inverse of the other, and this leads to the requirement

$$BC - |D|^2 = 0$$

which agrees with (7.72). Since $|D| \neq 0$, then both B and C are nonzero.

Combining (7.184) and (7.181) leads to the relations

$$\frac{h_y}{h_x} = \frac{C}{D^*} \quad (7.185a)$$

$$\frac{h_x}{h_y} = \frac{B}{D} \quad (7.185b)$$

If these values are substituted into the equation for W''_m, the power received from the polarized part of the wave becomes

$$W''_{mm} = \frac{A_e}{2Z_0}(B + C) \quad (7.186)$$

which is obviously maximum, rather than minimum.

By using (7.77)–(7.79) and the definition for degree of polarization, R, the true maximum power that can be received from the polarized part of the wave is

$$W''_{mm} = A_e \mathscr{P}_p = A_e \mathscr{P}_t R \quad (7.187)$$

where $\mathscr{P}_p$ is the power density of the polarized part of the wave and $\mathscr{P}_t$ is that of the total wave.

It can be shown that if the wave is completely polarized the choices made for the receiving antenna effective lengths in (7.185) are the same as those made in (4.44). This is left as an exercise.

It was noted earlier that for the unpolarized part of the wave the maximum power that can be received is one-half the power that could be received from a polarized wave of the same power density using a polarization-matched receiver. The received power from the unpolarized part of the wave is independent of the receiver polarization. Then in order to maximize total received power, we need only to match the receiver to the completely polarized portion of the wave, using (7.185). The total received power is then the sum

$$W_m = W' + W''_{mm} = \tfrac{1}{2} A_e \mathscr{P}_t (1 + R) \quad (7.188)$$

7.7. SCATTERING BY A TARGET: THE MUELLER MATRIX

The scattering of a quasi-monochromatic plane wave by a target and transmission through an optical instrument are readily treated by the use of

PARTIAL POLARIZATION

Stokes vectors. To assist in developing the scattering equations, it is useful to reorder the coherency matrix elements into four-vector form and develop a formal relation between that vector and the Stokes vector.

The *coherency vector* $\mathscr{J}$ of a plane wave is defined as the *direct or Kronecker product* of the fields by [8, 9]

$$\mathscr{J} = \langle \mathbf{E} \times \mathbf{E}^* \rangle = \left\langle \begin{bmatrix} E_x \mathbf{E}^* \\ E_y \mathbf{E}^* \end{bmatrix} \right\rangle = \begin{bmatrix} \langle E_x E_x^* \rangle \\ \langle E_x E_y^* \rangle \\ \langle E_y E_x^* \rangle \\ \langle E_y E_y^* \rangle \end{bmatrix} = \begin{bmatrix} J_{xx} \\ J_{xy} \\ J_{yx} \\ J_{yy} \end{bmatrix} \quad (7.189)$$

If **Q** is defined by

$$\mathbf{Q} = \begin{bmatrix} 1 & 0 & 0 & 1 \\ 1 & 0 & 0 & -1 \\ 0 & 1 & 1 & 0 \\ 0 & j & -j & 0 \end{bmatrix} \quad (7.190)$$

the Stokes vector is related to $\mathscr{J}$ by

$$\mathbf{G} = \mathbf{Q}\mathscr{J} \quad (7.191)$$

A wave scattered by a target has coherency vector

$$\mathscr{J}^s = \langle \mathbf{E}^s \times \mathbf{E}^{s*} \rangle \quad (7.192)$$

If the scattered and incident field vectors are related by the Jones matrix, the scattered coherency vector is

$$\mathscr{J}^s = \frac{1}{4\pi r^2} \langle (\mathbf{T}\mathbf{E}^i) \times (\mathbf{T}^* \mathbf{E}^{i*}) \rangle \quad (7.193)$$

In this form, the target may be time-varying, but shortly **T** will be removed from the time-averaging process.

Now we use an identity relating the direct product and matrix product [9]

$$(\mathbf{A} \times \mathbf{B})(\mathbf{C} \times \mathbf{D}) = (\mathbf{AC}) \times (\mathbf{BD}) \quad (7.194)$$

In the form needed, **A** and **B** are 2×2, and **C** and **D** are 2×1. The direct product of **A** and **B** is defined as [9, 10]

$$\mathbf{A} \times \mathbf{B} = \begin{bmatrix} A_{11} & A_{12} \\ A_{21} & A_{22} \end{bmatrix} \times \begin{bmatrix} B_{11} & B_{12} \\ B_{21} & B_{22} \end{bmatrix} = \begin{bmatrix} A_{11}\mathbf{B} & A_{12}\mathbf{B} \\ A_{21}\mathbf{B} & A_{22}\mathbf{B} \end{bmatrix}$$

$$= \begin{bmatrix} A_{11}B_{11} & A_{11}B_{12} & A_{12}B_{11} & A_{12}B_{12} \\ A_{11}B_{21} & A_{11}B_{22} & A_{12}B_{21} & A_{12}B_{22} \\ A_{21}B_{11} & A_{21}B_{12} & A_{22}B_{11} & A_{22}B_{12} \\ A_{21}B_{21} & A_{21}B_{22} & A_{22}B_{21} & A_{22}B_{22} \end{bmatrix} \quad (7.195)$$

With the identity (7.194) we can write

$$\mathscr{I}^s = \frac{1}{4\pi r^2}\langle(\mathbf{T} \times \mathbf{T}^*)(\mathbf{E}^i \times \mathbf{E}^{i*})\rangle$$
$$= \frac{1}{4\pi r^2}(\mathbf{T} \times \mathbf{T}^*)\langle(\mathbf{E}^i \times \mathbf{E}^{i*})\rangle = \frac{1}{4\pi r^2}(\mathbf{T} \times \mathbf{T}^*)\mathscr{I}^i \quad (7.196)$$

The Stokes vector of the scattered wave then is

$$\mathbf{G}^s = \mathbf{Q}\mathscr{I}^s = \frac{1}{4\pi r^2}\mathbf{Q}(\mathbf{T} \times \mathbf{T}^*)\mathbf{Q}^{-1}\mathbf{G}^i \quad (7.197)$$

Note that $\mathbf{Q}$ as defined is not quite unitary. Instead

$$\mathbf{Q}^{-1} = \tfrac{1}{2}\mathbf{Q}^{T*} \quad (7.198)$$

The matrix

$$\mathbf{M} = \mathbf{Q}(\mathbf{T} \times \mathbf{T}^*)\mathbf{Q}^{-1} \quad (7.199)$$

which relates the incident and scattered wave Stokes vectors by

$$\mathbf{G}^s = \frac{1}{4\pi r^2}\mathbf{M}\mathbf{G}^i \quad (7.200)$$

has been called by various names, but the common designation is that of *Mueller matrix* [11–13]. It is sometimes appropriate to omit the $4\pi r^2$ variation.

In these equations, the incident field quantities $\mathbf{E}^i$, $\mathscr{I}^i$, and $\mathbf{G}^i$ are in $x_1 y_1 z_1$ coordinates of Figure 7.1b, and the scattered field terms are in $x_2 y_2 z_2$ coordinates. The Mueller matrix then naturally involves both coordinate sets. For this reason, the Jones matrix, which relates the x_2 and y_2 components of the scattered field to the x_1 and y_1 components of the incident field, rather than the Sinclair matrix, was used in the development of the Mueller matrix. In radar, particularly monostatic radar, it is often desirable to use only one coordinate system, the $x_1 y_1 z_1$ system. In a later section a matrix similar to the Mueller matrix but involving only one set of coordinates is developed.

Since a product of $\mathbf{T}$ was removed from the time average, it is clear that $\mathbf{T}$ must be time-invariant on the time scale of the averages in $\mathscr{I}$. It will be clear later that it must be frequency-invariant over the bandwidth of the incident wave.

The multiplication of (7.199) has been carried out to give the Mueller matrix elements in terms of the elements of the Jones matrix, and the results are presented in Appendix A. The relationship may be written in a concise form if the subscripts x and y of the Jones matrix elements are replaced respectively by 1 and 2 [8]. The Mueller matrix is

$$M = \begin{bmatrix} \frac{1}{2}(E_1 + E_2 + E_3 + E_4) & \frac{1}{2}(E_1 - E_2 - E_3 + E_4) & F_{13} + F_{42} & -G_{13} - G_{42} \\ \frac{1}{2}(E_1 - E_2 + E_3 - E_4) & \frac{1}{2}(E_1 + E_2 - E_3 - E_4) & F_{13} - F_{42} & -G_{13} + G_{42} \\ F_{14} + F_{32} & F_{14} - F_{32} & F_{12} + F_{34} & -G_{12} + G_{34} \\ G_{14} + G_{32} & G_{14} - G_{32} & G_{12} + G_{34} & F_{12} - F_{34} \end{bmatrix}$$

(7.201)

where

$$E_i = T_i T_i^* = |T_i|^2 \quad i = 1, 2, 3, 4$$

$$F_{ij} = F_{ji} = \operatorname{Re}(T_i T_j^*) = \operatorname{Re}(T_j T_i^*) \quad i, j = 1, 2, 3, 4 \quad (7.202)$$

$$G_{ij} = -G_{ji} = \operatorname{Im}(T_i^* T_j) = -\operatorname{Im}(T_j^* T_i) \quad i, j = 1, 2, 3, 4$$

In the last equation, the notation

$$T_1 = T_{11} \qquad T_2 = T_{22} \qquad T_3 = T_{12} \qquad T_4 = T_{21} \quad (7.203)$$

has been used for convenience.

The Mueller matrix may be found in terms of the Sinclair matrix **S** in a straightforward manner by replacing the Jones matrix elements according to

$$\begin{bmatrix} T_{xx} & T_{xy} \\ T_{yx} & T_{yy} \end{bmatrix} = \begin{bmatrix} -S_{xx} & -S_{xy} \\ S_{yx} & S_{yy} \end{bmatrix} \quad (6.87)$$

It must be noted, however, that when this is done the Stokes vector of the scattered wave is still in $x_2 y_2 z_2$ coordinates, that is,

$$G_0^s = \langle |E_{x2}|^2 + |E_{y2}|^2 \rangle$$

and so on. In the alternative formulation of a later section, using only one coordinate system, a four-vector equivalent to the Stokes vector and a 4×4 matrix equivalent to the Mueller matrix allow concerns about appropriate coordinates for the reflected wave to be dismissed.

It was noted earlier that the Stokes vectors for incident and reflected waves are real. Then all elements of the Mueller matrix should be real, and (7.201) and (7.202) show that this is so. Mueller matrices for some common reflectors and optical elements are given in Appendix A.

Now let the wave incident on the target come from a transmitting antenna a at the origin of coordinates $x_1 y_1 z_1$ in Figure 7.1. The Stokes vector of the incident wave is

$$\mathbf{G}^i = \frac{Z_0^2 I_a^2}{4\lambda^2 r_1^2} \mathbf{G}_A^{(a)} \tag{7.204}$$

where λ is the mean wavelength of the incident wave, I_a is a current in the transmitting antenna, and $\mathbf{G}_A^{(a)}$ is the Stokes vector of the transmitting antenna. In this form, it is implied that the feed current to the transmitting antenna is single-frequency and that the radiated wave is polychromatic and partially polarized because of the characteristics of the transmitting antenna. The radiated wave can also be partially polarized if the feed current has a finite bandwidth and if the antenna has a different frequency response for its x and y components. If this is the case, I_a is an equivalent current giving the correct power, and $\mathbf{G}_A^{(a)}$ can no longer be considered strictly an attribute of the antenna but must depend to some degree on the feed current.

The reflected wave has a Stokes vector at the receiving antenna b,

$$\mathbf{G}^s = \frac{1}{4\pi r_2^2} \mathbf{M} \mathbf{G}^i = \frac{Z_0^2 I_a^2}{16\pi \lambda^2 r_1^2 r_2^2} \mathbf{M} \mathbf{G}_A^{(a)} \tag{7.205}$$

The scattered wave incident on the receiving antenna gives a received power, from (7.162),

$$W = \frac{1}{16 R_a} \mathbf{G}^{sT} \mathbf{G}_R^{(b)} = \frac{Z_0^2 I_a^2}{256 \pi R_a \lambda^2 r_1^2 r_2^2} \mathbf{G}_A^{(a)T} \mathbf{M}^T \mathbf{G}_R^{(b)} \tag{7.206}$$

If the same antenna is used for transmitting and receiving, the backscattered power simplifies to

$$W = \frac{Z_0^2 I_a^2}{256 \pi R_a \lambda^2 r^4} \mathbf{G}_A^T \mathbf{M}^T \mathbf{G}_R \tag{7.207}$$

Target Polarizability

The wave scattered by certain targets may be more highly polarized than the incident wave. For example, if an unpolarized wave strikes a dielectric interface at the Brewster angle, the reflected wave is completely polarized. This property is described by the *polarizability* of the target. Note that the scattered polarized wave has a smaller power density than the incident wave.

Let a wave incident on a target be unpolarized, with Stokes vector

$$\mathbf{G}^i = \begin{bmatrix} G_0^i & 0 & 0 & 0 \end{bmatrix}^T \tag{7.208}$$

408 PARTIAL POLARIZATION

From (7.200), the Stokes vector of the scattered wave is

$$\mathbf{G}^s = \frac{G_0^i}{4\pi r^2}[M_{11} \ M_{21} \ M_{31} \ M_{41}]^T \tag{7.209}$$

A reasonable definition of target polarizability is the degree of polarization of the scattered wave divided by the maximum possible degree of polarization (which we have seen to be one) when the incident wave is unpolarized. Then the target polarizability is, using (7.145),

$$\Pi = \frac{\left[(G_1^s)^2 + (G_2^s)^2 + (G_3^s)^2\right]^{1/2}}{G_0^s} = \frac{\left[(M_{21})^2 + (M_{31})^2 + (M_{41})^2\right]^{1/2}}{M_{11}} \tag{7.210}$$

It is interesting to see what types of target give the extreme values, 0 and 1, of polarizability. For $\Pi = 1$,

$$(M_{21})^2 + (M_{31})^2 + (M_{41})^2 = (M_{11})^2 \tag{7.211}$$

A polarizing target can be expected to have a scattering matrix, and if the relation between **M** and **S** from Appendix A is used in (7.211) it becomes, for the backscattering case, using the relative scattering matrix,

$$|S_{xx}|^2|S_{yy}|^2 = S_{xy}^2(S_{xx}S_{yy} + S_{xx}^*S_{yy}^*) - S_{xy}^4 \tag{7.212}$$

A target constructed of parallel thin wires oriented at angle α with the x axis has scattering matrix

$$\mathbf{S} = C\begin{bmatrix} \cos^2\alpha & \cos\alpha\sin\alpha \\ \cos\alpha\sin\alpha & \sin^2\alpha \end{bmatrix}$$

It is readily seen that (7.212) is satisfied by this target, and its polarizability is one. We recognize on physical grounds that the wire grid is a polarizing target and its unity polarizability is expected.

If the scattered wave is unpolarized,

$$M_{21} = M_{31} = M_{41} = 0$$

For a target with a scattering matrix, this is equivalent to

$$|S_{xx}| = |S_{yy}| \quad S_{xy} = 0$$

For such a target the equal-power x and y components of the incident unpolarized wave are reflected as equal-power x and y components of an unpolarized wave.

Change of Base: Stokes Vector and Mueller Matrix

It was noted earlier that if a four-vector of coherency matrix elements is defined by the direct product

$$\mathscr{J} = \langle \mathbf{E} \times \mathbf{E}^* \rangle = \left\langle \begin{bmatrix} E_x \mathbf{E}^* \\ E_y \mathbf{E}^* \end{bmatrix} \right\rangle = \begin{bmatrix} \langle E_x E_x^* \rangle \\ \langle E_x E_y^* \rangle \\ \langle E_y E_x^* \rangle \\ \langle E_y E_y^* \rangle \end{bmatrix} = \begin{bmatrix} J_{xx} \\ J_{xy} \\ J_{yx} \\ J_{yy} \end{bmatrix} \quad (7.189)$$

the Stokes vector can be written as

$$\mathbf{G} = \mathbf{Q}\mathscr{J} \quad (7.191)$$

where

$$\mathbf{Q} = \begin{bmatrix} 1 & 0 & 0 & 1 \\ 1 & 0 & 0 & -1 \\ 0 & 1 & 1 & 0 \\ 0 & j & -j & 0 \end{bmatrix} \quad (7.190)$$

Now $\mathscr{J}$, in xy coordinates, can be transformed to a new set of polarization basis vectors $\mathbf{u}_1$ and $\mathbf{u}_2$ by the same matrix

$$\mathbf{R} = \begin{bmatrix} \mathbf{u}_1^* \cdot \mathbf{u}_x & \mathbf{u}_1^* \cdot \mathbf{u}_y \\ \mathbf{u}_2^* \cdot \mathbf{u}_x & \mathbf{u}_2^* \cdot \mathbf{u}_y \end{bmatrix} \quad (7.99)$$

used earlier to transform the fields and coherency matrix. In the new basis, the fields $\mathbf{E}'$ and four-vector $\mathscr{J}'$ are related by

$$\mathscr{J}' = \langle \mathbf{E}' \times \mathbf{E}'^* \rangle = \langle (\mathbf{R}\mathbf{E}) \times (\mathbf{R}^* \mathbf{E}^*) \rangle \quad (7.213)$$

If the identity (7.194) is used, then

$$\mathscr{J}' = \langle (\mathbf{R} \times \mathbf{R}^*)(\mathbf{E} \times \mathbf{E}^*) \rangle = (\mathbf{R} \times \mathbf{R}^*)\langle \mathbf{E} \times \mathbf{E}^* \rangle = (\mathbf{R} \times \mathbf{R}^*)\mathscr{J} \quad (7.214)$$

The Stokes vector in the new basis is

$$\mathbf{G}' = \mathbf{Q}\mathscr{J}' \quad (7.215)$$

and may be related to the Stokes vector in the old base by

$$\mathbf{G}' = \mathbf{Q}(\mathbf{R} \times \mathbf{R}^*)\mathscr{J} = \mathbf{Q}(\mathbf{R} \times \mathbf{R}^*)\mathbf{Q}^{-1}\mathbf{G} = \mathbf{L}\mathbf{G} \quad (7.216)$$

where

$$\mathbf{L} = \mathbf{Q}(\mathbf{R} \times \mathbf{R}^*)\mathbf{Q}^{-1} \quad (7.217)$$

410 PARTIAL POLARIZATION

The Mueller matrix relates the incident and scattered Stokes vectors by

$$\mathbf{G}^s = \frac{1}{4\pi r^2} \mathbf{M} \mathbf{G}^i \qquad (7.200)$$

in the old basis and by

$$\mathbf{G}'^s = \frac{1}{4\pi r^2} \mathbf{M}' \mathbf{G}'^i \qquad (7.218)$$

in the new. If the same polarization basis transform is applied to incident and scattered waves, then

$$\mathbf{G}'^s = \mathbf{L}\mathbf{G}^s = \frac{1}{4\pi r^2} \mathbf{L}\mathbf{M}\mathbf{G}^i = \frac{1}{4\pi r^2} \mathbf{L}\mathbf{M}\mathbf{L}^{-1} \mathbf{G}'^i \qquad (7.219)$$

In the transformed basis system the Mueller matrix is therefore

$$\mathbf{M}' = \mathbf{L}\mathbf{M}\mathbf{L}^{-1} = [\mathbf{Q}(\mathbf{R} \times \mathbf{R}^*)\mathbf{Q}^{-1}]\mathbf{M}[\mathbf{Q}(\mathbf{R} \times \mathbf{R}^*)\mathbf{Q}^{-1}]^{-1} \qquad (7.220)$$

For a general transformation matrix $\mathbf{R}$, the direct product in this equation for the new basis Mueller matrix is cumbersome. If $\mathbf{R}$ is the rotation matrix,

$$\mathbf{R} = \begin{bmatrix} \cos\theta & \sin\theta \\ -\sin\theta & \cos\theta \end{bmatrix} \qquad (7.102)$$

it simplifies to

$$\mathbf{R} \times \mathbf{R}^* = \begin{bmatrix} \cos^2\theta & \cos\theta\sin\theta & \cos\theta\sin\theta & \sin^2\theta \\ -\cos\theta\sin\theta & \cos^2\theta & -\sin^2\theta & \cos\theta\sin\theta \\ -\cos\theta\sin\theta & -\sin^2\theta & \cos^2\theta & \cos\theta\sin\theta \\ \sin^2\theta & -\cos\theta\sin\theta & -\cos\theta\sin\theta & \cos^2\theta \end{bmatrix} \qquad (7.221)$$

and $\mathbf{L}$ becomes

$$\mathbf{L} = \begin{bmatrix} 1 & 0 & 0 & 0 \\ 0 & \cos 2\theta & \sin 2\theta & 0 \\ 0 & -\sin 2\theta & \cos 2\theta & 0 \\ 0 & 0 & 0 & 1 \end{bmatrix} \qquad (7.222)$$

Mueller Matrix Relationships

We remarked previously, "If the scattered and incident field vectors are related by the Jones matrix...," implying that for some targets they are not.

The Mueller matrix was introduced here in terms of the Jones matrix for convenience, but fundamentally it is a ratio of the Stokes vectors of scattered and incident waves, according to (7.200). Since the Stokes vector exists for all waves, the Mueller matrix is defined for all targets. If the target has a Jones matrix, the Mueller matrix can be found in terms of the Jones matrix elements by means of the equations presented here. Interesting questions are: If the Mueller matrix is known, can a corresponding Jones matrix be found? More fundamentally, does a Jones (or Sinclair) matrix exist for all targets? We shall see that the answer to the first question is "not always" and to the second, "no." A thorough discussion of the conditions that must be met if there is to be a Jones matrix corresponding to a given Mueller matrix for a target will not be given here, but one of the references has a short bibliography on the subject [13]. The radar observation time is a significant factor, and this is discussed further in the introduction to Chapter 9, where it is pointed out that *all* targets have Jones and Sinclair matrices if the observation time is infinitesimal.

The Mueller matrix elements are given in Appendix A in terms of the elements of the Jones matrix **T**. There are sixteen elements for the 4 × 4 Mueller matrix. Now the Jones and Sinclair matrices

$$\mathbf{S} = \begin{bmatrix} S_{xx} & S_{xy} \\ S_{yx} & S_{yy} \end{bmatrix} \qquad \mathbf{T} = \begin{bmatrix} T_{xx} & T_{xy} \\ T_{yx} & T_{yy} \end{bmatrix}$$

have eight independent parameters for the bistatic case. It is customary, however, to choose the phase of one element as a reference. The resulting matrix, with seven independent parameters, four amplitudes and three phases, is called the relative scattering matrix. It follows that the sixteen Mueller elements are not independent for a target with a scattering matrix. There must be nine relationships among the elements of (A.4), and such a set, while not unique, is given here in the notation used in this text [14],

$$(M_{11} - M_{22})^2 - (M_{12} - M_{21})^2 = (M_{33} - M_{44})^2 + (M_{34} + M_{43})^2 \quad (7.223\text{a})$$

$$M_{13}M_{23} + M_{14}M_{24} = M_{11}M_{21} - M_{12}M_{22} \quad (7.223\text{b})$$

$$M_{31}M_{32} + M_{41}M_{42} = M_{11}M_{12} - M_{21}M_{22} \quad (7.223\text{c})$$

$$M_{13}M_{14} - M_{23}M_{24} = M_{33}M_{34} + M_{43}M_{44} \quad (7.223\text{d})$$

$$M_{31}M_{41} - M_{32}M_{42} = M_{33}M_{43} + M_{34}M_{44} \quad (7.223\text{e})$$

$$M_{13}^2 + M_{23}^2 + M_{14}^2 + M_{24}^2 = M_{11}^2 - M_{12}^2 + M_{21}^2 - M_{22}^2 \quad (7.223\text{f})$$

$$M_{31}^2 + M_{32}^2 + M_{41}^2 + M_{42}^2 = M_{11}^2 + M_{12}^2 - M_{21}^2 - M_{22}^2 \quad (7.223\text{g})$$

$$M_{13}^2 - M_{23}^2 - M_{14}^2 + M_{24}^2 = M_{33}^2 - M_{34}^2 + M_{43}^2 - M_{44}^2 \quad (7.223\text{h})$$

$$M_{31}^2 - M_{32}^2 - M_{41}^2 + M_{42}^2 = M_{33}^2 + M_{34}^2 - M_{43}^2 - M_{44}^2 \quad (7.223\text{i})$$

412 PARTIAL POLARIZATION

If these relationships are satisfied by the Mueller matrix elements of a target, then the target scattering (Sinclair or Jones) matrix exists and can be found.

Consider a target of nonzero size, made up of N subtargets that may be point scatterers or may be space extended. Each subscatterer, if space extended, is assumed to have fixed dimensions and to be representable by a scattering matrix and a Mueller matrix. Further, the position of each subscatterer is assumed to be random and independent of the location of all other subscatterers. If the orientation of a subtarget is random it may not have a constant scattering matrix, but it can be further subdivided until it does. The random positions of the subscatterers cause the phases of the waves reflected by them to be random and independent.

We noted earlier that the Stokes vector elements of independent waves propagating in the same direction may be added. If all subscatterers are illuminated by the same incident wave (having the same Stokes vector, not necessarily the same phase), it is clear from (7.200) that the Stokes parameters of the reflected waves can be added only if the Mueller matrices of the N subscatterers are additive. Therefore, the Mueller matrix of the target is the sum of the Mueller matrices of the subtargets, or

$$\mathbf{M} = \sum_{}^{N} \mathbf{M}^{(n)} \qquad (7.224)$$

Consider a target with two subscatterers and examine equation (7.223b) for example. The additivity of the Mueller elements gives

$$\left[M_{13}^{(1)} + M_{13}^{(2)}\right]\left[M_{23}^{(1)} + M_{23}^{(2)}\right] + \left[M_{14}^{(1)} + M_{14}^{(2)}\right]\left[M_{24}^{(1)} + M_{24}^{(2)}\right]$$
$$= \left[M_{11}^{(1)} + M_{11}^{(2)}\right]\left[M_{21}^{(1)} + M_{21}^{(2)}\right] - \left[M_{12}^{(1)} + M_{12}^{(2)}\right]\left[M_{22}^{(1)} + M_{22}^{(2)}\right]$$
$$(7.225)$$

Each submatrix individually must satisfy the same equation, so that

$$M_{13}^{(1)}M_{23}^{(1)} + M_{14}^{(1)}M_{24}^{(1)} = M_{11}^{(1)}M_{21}^{(1)} - M_{12}^{(1)}M_{22}^{(1)} \qquad (7.226a)$$

$$M_{13}^{(2)}M_{23}^{(2)} + M_{14}^{(2)}M_{24}^{(2)} = M_{11}^{(2)}M_{21}^{(2)} - M_{12}^{(2)}M_{22}^{(2)} \qquad (7.226b)$$

If (7.225) is multiplied and (7.226) is used to cancel terms, the result is

$$M_{13}^{(1)}M_{23}^{(2)} + M_{13}^{(2)}M_{23}^{(1)} + M_{14}^{(1)}M_{24}^{(2)} + M_{14}^{(2)}M_{24}^{(1)}$$
$$= M_{11}^{(1)}M_{21}^{(2)} + M_{11}^{(2)}M_{21}^{(1)} - M_{12}^{(1)}M_{22}^{(2)} - M_{12}^{(2)}M_{22}^{(1)} \qquad (7.227)$$

This equation cannot in general be satisfied, and it follows that (7.223b)—and by extension all of (7.223)—is not satisfied by the Mueller matrix of the overall target. The Mueller matrix of the overall target must exist, however. It follows that no scattering matrix exists for targets extended in space with scattering centers randomly and independently located. The Mueller matrix exists, however, and has sixteen independent elements that may be determined by measurement of the Stokes vectors of the scattered wave when appropriate incident waves are utilized. This matrix may be used to characterize targets that cause the polarization of the scattered wave to fluctuate with a random component and for which no scattering matrix exists. It is commonly called the *average Mueller matrix* [15–17] and is discussed further in the following section.

If backscattering is considered for a target, the scattering matrix elements S_{xy} and S_{yx}, if they exist, must be equal, and the scattering matrix has five independent parameters. There must then be eleven equations relating the Mueller matrix elements to each other. Six come from specializing (A.4), and five come from (7.223), which reduces to that number if the first six relationships are used therein. Then for backscattering, the Mueller elements must obey the following relationships if the target is representable by a scattering matrix,

$$M_{21} = M_{12} \qquad (7.228a)$$
$$M_{31} = -M_{13} \qquad (7.228b)$$
$$M_{32} = -M_{23} \qquad (7.228c)$$
$$M_{41} = M_{14} \qquad (7.228d)$$
$$M_{42} = M_{24} \qquad (7.228e)$$
$$M_{43} = -M_{34} \qquad (7.228f)$$
$$(M_{11} - M_{22})^2 = (M_{33} - M_{44})^2 \qquad (7.228g)$$
$$M_{13}M_{23} + M_{14}M_{24} = M_{12}(M_{11} - M_{22}) \qquad (7.228h)$$
$$M_{13}M_{14} - M_{23}M_{24} = M_{34}(M_{33} - M_{44}) \qquad (7.228i)$$
$$M_{13}^2 + M_{23}^2 + M_{14}^2 + M_{24}^2 = M_{11}^2 - M_{22}^2 \qquad (7.228j)$$
$$M_{13}^2 - M_{23}^2 - M_{14}^2 + M_{24}^2 = M_{33}^2 - M_{44}^2 \qquad (7.228k)$$

Mueller backscattering matrices for some common targets that have scattering matrices, and for transmission through some common optical elements, are given in Appendix A.

The Mueller matrix of a target that completely depolarizes all incident waves is of interest. If the reflected wave is unpolarized, then (7.145) requires that $(G_1^s)^2 + (G_2^s)^2 + (G_3^s)^2 = 0$. If this requirement is imposed on

$$\mathbf{G}^s = \frac{1}{4\pi r^2} \mathbf{M} \mathbf{G}^i \qquad (7.200)$$

for completely polarized incident waves representing, successively, waves that are linear horizontal, linear vertical, linear at $\pm \pi/4$, and right- and left-circular, only the first row of the Mueller matrix can be nonzero. If it is noted further that G_0^s is independent of the polarization of the incident wave, only M_{11} has value. For a target that completely depolarizes all waves, with power independent of incident polarization,

$$\mathbf{M} = M_{11} \begin{bmatrix} 1 & 0 & 0 & 0 \\ 0 & 0 & 0 & 0 \\ 0 & 0 & 0 & 0 \\ 0 & 0 & 0 & 0 \end{bmatrix} \quad (7.229)$$

The Average Mueller Matrix

The Mueller matrix was first developed in this section for a target with a scattering matrix, but we saw that it exists even for a target without a scattering matrix as a ratio of Stokes vectors of reflected and incident waves.

Consider a distributed target of nonzero size. At any instant of time, its subscatterers have fixed spatial relationships to each other, and an instantaneous scattering matrix exists. During the time required to process the scattered signal, however, the relative positions of the subscatterers change, and the scattering matrix at any one instant does not give a correct scattered wave. For this case, let us see if a Mueller matrix can be found in terms of time averages of the scattering matrix.

The coherency vector of the scattered wave is

$$\mathcal{J}^s = \frac{1}{4\pi r^2} \langle (\mathbf{TE})^i \times (\mathbf{T}^* \mathbf{E}^{i*}) \rangle \quad (7.193)$$

where $\mathbf{T}$ is the instantaneous target Jones matrix. By use of an identity used previously, this becomes

$$\mathcal{J}^s = \frac{1}{4\pi r^2} \langle (\mathbf{T} \times \mathbf{T}^*)(\mathbf{E}^i \times \mathbf{E}^{i*}) \rangle \quad (7.230)$$

If the target is time-varying and the incident wave is completely polarized, the direct product involving the incident fields can be removed from the time average, and $\mathcal{J}^s$ becomes

$$\mathcal{J}^s = \frac{1}{4\pi r^2} \langle (\mathbf{T} \times \mathbf{T}^*) \rangle (\mathbf{E}^i \times \mathbf{E}^{i*}) = \frac{1}{4\pi r^2} \langle (\mathbf{T} \times \mathbf{T}^*) \rangle \mathcal{J}^i \quad (7.231)$$

The Stokes vector of the scattered wave then is

$$\mathbf{G}^s = \mathbf{Q} \mathcal{J}^s = \frac{1}{4\pi r^2} \mathbf{Q} \langle \mathbf{T} \times \mathbf{T}^* \rangle \mathbf{Q}^{-1} \mathbf{G}^i \quad (7.232)$$

where the incident Stokes vector is for a completely polarized wave. The product

$$\mathbf{M}_{av} = \mathbf{Q}\langle \mathbf{T} \times \mathbf{T}^* \rangle \mathbf{Q}^{-1} \qquad (7.233)$$

is a Mueller matrix applicable to the situation of a time-varying target and a completely polarized incident wave. It is commonly called the *average Mueller matrix*.

A target represented by an average Mueller matrix and not possessing a scattering matrix may appropriately be called a *depolarizing target*, since an incident completely polarized wave is scattered by it with a degree of polarization less than unity. For backscattering, the average Mueller matrix has elements with the same symmetries as for the target with a scattering matrix, since the symmetries are a consequence of reciprocity and not of the particular target. It then has nine independent parameters [11, 14, 15]. Of the sixteen elements, six are eliminated by symmetry (or antisymmetry) and one by the relation [15]*

$$M_{11} = M_{22} - M_{33} + M_{44} \qquad (7.234)$$

In its most general form, then, the Mueller matrix relates the incident and scattered Stokes vectors of a target, is independent of whether the target has a scattering matrix or not, and does not require a completely polarized incident wave. If the target has a scattering matrix, the Mueller matrix can be found from its elements; again the incident wave need not be completely polarized. In a third case, if the incident wave is completely polarized, an average Mueller matrix, related to time averages of scattering matrix products, can be found.

7.8. THE KENNAUGH MATRIX

An important alternative to the Mueller matrix formulation is widely used in radar scattering problems. In the equations used, all incident field parameters are in $x_1 y_1 z_1$ coordinates of Figure 7.1 and the scattered field quantities are in $x_3 y_3 z_3$ coordinates. For backscattering the coordinate systems coincide.

Let the field incident on a target with Sinclair matrix $\mathbf{S}$ be $\mathbf{E}^i$. The open-circuit voltage at the terminals of a receiving antenna with effective length $\mathbf{h}$, at distance r_2, is

$$V = \frac{1}{\sqrt{4\pi}\, r_2} \mathbf{h}^T \mathbf{S} \mathbf{E}^i \qquad (7.235)$$

---

*The reference uses all positive signs in the equation because of the coordinate system choice.

416 PARTIAL POLARIZATION

and the received power is

$$W = \frac{1}{8R_a}\langle VV^*\rangle = \frac{1}{32\pi R_a r_2^2}\langle (\mathbf{h}^T\mathbf{SE}^i)(\mathbf{h}^{T*}\mathbf{S}^*\mathbf{E}^{i*})\rangle \quad (7.236)$$

Carrying out the matrix multiplication gives

$$W = \frac{1}{32\pi R_a r_2^2}\langle |h_x S_{xx} E_x^i + h_x S_{xy} E_y^i + h_y S_{yx} E_x^i + h_y S_{yy} E_y^i|^2\rangle \quad (7.237)$$

The sixteen terms obtained when the square of the magnitude is found by multiplication can be put into a matrix product form in a natural manner, thus

$$W = \frac{1}{32\pi R_a r_2^2}$$

$$\times \left\langle [h_x h_x^* \quad h_x h_y^* \quad h_y h_x^* \quad h_y h_y^*] \begin{bmatrix} S_{xx}S_{xx}^* & S_{xx}S_{xy}^* & S_{xy}S_{xx}^* & S_{xy}S_{xy}^* \\ S_{xx}S_{yx}^* & S_{xx}S_{yy}^* & S_{xy}S_{yx}^* & S_{xy}S_{yy}^* \\ S_{yx}S_{xx}^* & S_{yx}S_{xy}^* & S_{yy}S_{xx}^* & S_{yy}S_{xy}^* \\ S_{yx}S_{yx}^* & S_{yx}S_{yy}^* & S_{yy}S_{yx}^* & S_{yy}S_{yy}^* \end{bmatrix} \begin{bmatrix} E_x^i E_x^{i*} \\ E_x^i E_y^{i*} \\ E_y^i E_x^{i*} \\ E_y^i E_y^{i*} \end{bmatrix} \right\rangle$$

(7.238)

The matrices in this equation may be recognized as Kronecker products, thus*

$$W = \frac{1}{32\pi R_a r_2^2}\langle (\mathbf{h}\times\mathbf{h}^*)^T(\mathbf{S}\times\mathbf{S}^*)(\mathbf{E}^i\times\mathbf{E}^{i*})\rangle \quad (7.239)$$

The first vector can be removed from the time average, and so can the last if the incident wave is completely polarized, giving

$$W = \frac{1}{32\pi R_a r_2^2}(\mathbf{h}\times\mathbf{h}^*)^T\langle\mathbf{S}\times\mathbf{S}^*\rangle(\mathbf{E}^i\times\mathbf{E}^{i*}) \quad (7.240)$$

The last four-vector is recognizable as the coherency vector $\mathcal{J}^i$ of a completely polarized incident wave. The first is similar in form and may be considered the coherency four-vector of the receiving antenna (the normal coherency vector of the antenna as a transmitter). Then

$$W = \frac{1}{32\pi R_a r_2^2}(\mathcal{J}_A^r)^T\langle\mathbf{S}\times\mathbf{S}^*\rangle\mathcal{J}^i \quad (7.241)$$

---

*The order of terms in (7.238) differs from that of Kennaugh [15], who used this formulation, so that it can be written as a Kronecker product.

Note that for the bistatic case $\mathcal{S}^i$ is in $x_1y_1z_1$ coordinates and $\mathcal{S}_A^r$ in $x_3y_3z_3$, while for monostatic radar both vectors are in the same coordinate system.

The Stokes vector of the incident wave is related to $\mathcal{S}^i$ by

$$\mathbf{G}^i = \mathbf{Q}\mathcal{S}^i = \mathbf{Q}(\mathbf{E}^i \times \mathbf{E}^{i*}) \tag{7.242}$$

where $\mathbf{Q}$ is

$$\mathbf{Q} = \begin{bmatrix} 1 & 0 & 0 & 1 \\ 1 & 0 & 0 & -1 \\ 0 & 1 & 1 & 0 \\ 0 & j & -j & 0 \end{bmatrix} \tag{7.190}$$

The coherency vector of the receiving antenna transformed by the same operator is the Stokes vector [not the receiving Stokes vector of (7.159)] of the receiving antenna, thus

$$\mathbf{G}_A^r = \mathbf{Q}\mathcal{S}_A^r = \mathbf{Q}(\mathbf{h} \times \mathbf{h}^*) \tag{7.243}$$

The equation for received power becomes, with these changes,

$$W = \frac{1}{32\pi R_a r_2^2} \mathbf{G}_A^{rT}(\mathbf{Q}^{-1})^T \langle \mathbf{S} \times \mathbf{S}^* \rangle \mathbf{Q}^{-1} \mathbf{G}^i \tag{7.244}$$

In this equation, a 4×4 matrix

$$\mathbf{K} = 2(\mathbf{Q}^{-1})^T \langle \mathbf{S} \times \mathbf{S}^* \rangle \mathbf{Q}^{-1} \tag{7.245}$$

plays a part like that of the Mueller matrix of previous work. The factor 2 is included to make the terms of $\mathbf{K}$ comparable to the Mueller matrix elements, since

$$\mathbf{Q}^{-1} = \tfrac{1}{2}\mathbf{Q}^{T*} \tag{7.246}$$

Then $\mathbf{K}$ may be found from

$$\mathbf{K} = \tfrac{1}{2}\mathbf{Q}^* \langle \mathbf{S} \times \mathbf{S}^* \rangle \mathbf{Q}^{T*} \tag{7.247}$$

and the received power is

$$W = \frac{1}{64\pi R_a r_2^2} \mathbf{G}_A^{rT} \mathbf{K} \mathbf{G}^i \tag{7.248}$$

If it is desired to use the Stokes vector of the transmitting antenna $\mathbf{G}_A^t$, rather than that of the wave incident on the target, they are related by

$$\mathbf{G}^i = \frac{Z_0^2 I^2}{4\lambda^2 r_1^2} \mathbf{G}_A^t \tag{7.249}$$

The received power for the bistatic case is then

$$W = \frac{Z_0^2 I^2}{256\pi R_a \lambda^2 r_1^2 r_2^2} \mathbf{G}_A^{rT} \mathbf{K} \mathbf{G}_A^t \tag{7.250}$$

and for the monostatic configuration

$$W = \frac{Z_0^2 I^2}{256\pi R_a \lambda^2 r^4} \mathbf{G}_A^T \mathbf{K} \mathbf{G}_A \tag{7.251}$$

where the superscripts r and t denoting receiving and transmitting antennas have been dropped (they can be retained for colocated separate antennas).

The 4×4 matrix $\mathbf{K}$ plays the same role in these equations as the Mueller matrix in equations using $x_2 y_2 z_2$ coordinates for the scattered field. It will be called here the *Kennaugh scattering matrix* (or simply the Kennaugh matrix). The scattered wave is characterized by the four-vector

$$\mathbf{G}_K = \frac{1}{4\pi r_2^2} \mathbf{K} \mathbf{G}^i \tag{7.252}$$

which will be called the *Kennaugh vector*. It is important that the Kennaugh matrix and Kennaugh vector be distinguished from the Mueller matrix and Stokes vector that they resemble (although a Kennaugh vector for the incident wave would be the same as the Stokes vector). Although the differences arise because of the choice of coordinate systems, conversion from the Mueller matrix to the Kennaugh matrix is not a trivial problem. Failure to maintain the distinction between Kennaugh matrix and vector and the Mueller matrix—Stokes vector formulation can lead to considerable confusion, both in communication between workers in disparate interest areas and in determining such quantities as received voltage, where the incident field and receiving antenna effective length must be in the same coordinate system.

This formulation for received power in terms of the Kennaugh vector and Kennaugh matrix is important for radar applications, particularly for monostatic radar, which comprises most of the cases of interest. It is the equivalent, in the domain of partial polarization, of using the Sinclair matrix rather than the Jones matrix for a target in the completely polarized case. For a monostatic radar, it allows one to bypass many of the problems arising from the use of two coordinate systems.

The elements of **K**, in terms of the Sinclair matrix, are given in Appendix B.

7.9. POWER SPECTRAL DENSITIES OF THE FIELDS

In this section we explicitly consider what has only been alluded to previously, the frequency dependence of antennas and targets. It will be seen that appropriate terms for the equations of transmission, reception, and scattering are power spectral densities. These spectral densities can be arranged in coherency matrix and Stokes vector forms. Equations for scattering in terms of the common coherency matrix, Stokes vector, scattering matrix, and Mueller matrix are specializations of these more general equations.

Coherency Matrix Elements as Correlation Functions

It may have been noted that the coherency matrix and Stokes vector elements are specializations of the autocorrelation and crosscorrelation functions of analytic signals representing the orthogonal field components of the wave. The auto- and crosscorrelation functions of two complex time functions may be defined as

Autocorrelation

$$R_{uu}(\tau) = \lim_{T \to \infty} \frac{1}{2T} \int_{-T}^{T} u^*(t) u(t + \tau) \, dt = \langle u^*(t) u(t + \tau) \rangle \quad (7.253)$$

Crosscorrelation

$$R_{uv}(\tau) = \lim_{T \to \infty} \frac{1}{2T} \int_{-T}^{T} u^*(t) v(t + \tau) \, dt = \langle u^*(t) v(t + \tau) \rangle \quad (7.254)$$

With these definitions, the correlation functions for the field components of a plane wave can be written as

$$R_{xx}(\tau) = \langle E_x^*(t) E_x(t + \tau) \rangle \quad (7.255a)$$
$$R_{xy}(\tau) = \langle E_x^*(t) E_y(t + \tau) \rangle \quad (7.255b)$$
$$R_{yx}(\tau) = \langle E_y^*(t) E_x(t + \tau) \rangle \quad (7.255c)$$
$$R_{yy}(\tau) = \langle E_y^*(t) E_y(t + \tau) \rangle \quad (7.255d)$$

It is clear that the coherency matrix elements are proportional to these correlation functions with time difference $\tau = 0$.*

---

*The choice of R_{xy} in these equations causes $J_{xy} = R_{yx}(0)$, but this does not lead to any difficulty.

420 PARTIAL POLARIZATION

Relationship to Spectral Densities

The power spectral density of an analytic signal is related to the autocorrelation function by the equation [18]*

$$S_{uu}(\omega) = \int_{-\infty}^{\infty} R_{uu}(\tau) e^{-j\omega\tau} d\tau \qquad (7.256)$$

That is, the power spectral density is the Fourier transform of the signal autocorrelation function. Likewise, the cross spectral density of two functions is the Fourier transform of their crosscorrelation function,

$$S_{uv}(\omega) = \int_{-\infty}^{\infty} R_{uv}(\tau) e^{-j\omega\tau} d\tau \qquad (7.257)$$

The correlation functions are the inverse Fourier transforms

$$R_{uu}(\tau) = \frac{1}{2\pi} \int_{-\infty}^{\infty} S_{uu}(\omega) e^{j\omega\tau} d\omega \qquad (7.258a)$$

$$R_{uv}(\tau) = \frac{1}{2\pi} \int_{-\infty}^{\infty} S_{uv}(\omega) e^{j\omega\tau} d\omega \qquad (7.258b)$$

Measurement of the Correlation Functions

If we are to utilize the correlation functions in a study of partially polarized waves, a way to measure them must be found. No difficulty exists at lower frequencies. Two linearly polarized antennas, x- and y-directed, frequency-independent over the bandwidth of interest, receive voltages proportional to the fields E_x and E_y. The auto- and crosscorrelation functions of these signals may then be found with standard instruments.

At optical frequencies, measurement is more difficult. We assume here that we can measure not merely received power but the power spectral densities of x- and y-directed wave components, $S_{xx}(\omega)$ and $S_{yy}(\omega)$. From these measurements, the autocorrelation functions $R_{xx}(\tau)$ and $R_{yy}(\tau)$ may be found from (7.258a).

Consider one of the crosscorrelation terms, say

$$R_{xy}(\tau) = \langle E_x^*(t) E_y(t+\tau) \rangle \qquad (7.255b)$$

In this form, the fields do not include the $\exp(j\bar{\omega}t)$ term, but the autocorrelation would be unchanged if the form

$$\mathscr{E}_y(t) = a_y(t) e^{j[\bar{\omega}t + \phi_y(t)]} \qquad (7.20b)$$

---

*In this section S is used as a symbol for power spectral density and not for the Sinclair matrix. The Sinclair matrix is not considered in this section.

were used. At the advanced time $t + \tau$, the field is

$$\mathscr{E}_y(t + \tau) = a_y(t + \tau)e^{j[\bar{\omega}(t+\tau)+\phi_y(t+\tau)]} \qquad (7.259)$$

Now the change in $\mathscr{E}_y$ due to $\exp(j\bar{\omega}t)$ is much greater than that caused by the changed argument in either the amplitude a_y or phase ϕ_y. Then to a very good approximation

$$\mathscr{E}_y(t + \tau) = a_y(t)e^{j[\bar{\omega}t+\phi_y(t)]}e^{j\bar{\omega}\tau} = \mathscr{E}_y(t)e^{j\bar{\omega}\tau} \qquad (7.260)$$

Thus the time advance τ is equivalent to a phase advance $\bar{\omega}\tau$. We may use in (7.255b)

$$E_y(t + \tau) = E_y(t)e^{j\bar{\omega}\tau} \qquad (7.261)$$

To measure the crosscorrelation functions of a wave, let the y component of the wave be advanced by phase β, which corresponds to time advance

$$\tau = \frac{\beta}{\omega} \qquad (7.262)$$

Then the quantities to be measured are

$$\langle E_x^*(t)E_x(t) \rangle \qquad \langle E_x^*(t)E_y(t + \tau) \rangle$$
$$\langle E_y^*(t + \tau)E_x(t) \rangle \qquad \langle E_y^*(t + \tau)E_y(t + \tau) \rangle$$

which become, assuming stationarity,

$$\langle E_x^*(t)E_x(t) \rangle \qquad \langle E_x^*(t)E_y(t + \tau) \rangle$$
$$\langle E_y^*(t)E_x(t - \tau) \rangle \qquad \langle E_y^*(t)E_y(t) \rangle$$

These time averages may be measured as specified in an earlier section on measurement of the Stokes vector. Thus if one wave component is delayed by a known time τ, by means of an optical compensator, the correlation functions of the wave can be measured.

Radiation

Let an analytic signal $v(t)$ be applied to a terminal pair of a transmitting antenna, as in Figure 7.2. The antenna radiates a wave with x and y components. We may think of the effective lengths h_x and h_y as transducers converting $v(t)$ to the wave components $E_x(t)$ and $E_y(t)$, and we note that the effective lengths are functions of frequency,

$$\mathbf{h}(\omega) = \mathbf{u}_x h_x(\omega) + \mathbf{u}_y h_y(\omega) \qquad (7.263)$$

422   PARTIAL POLARIZATION

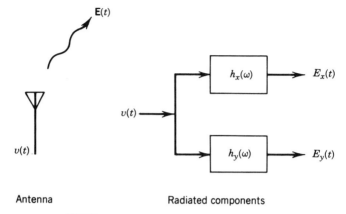

FIGURE 7.2   Radiation of analytic signal.

The effective lengths are of course dependent on angles θ and ϕ, and the dependence is implicit here. It must be noted that lumped parameter theory is used here, so the antenna elements are considered to be small. Time delays between radiating elements can be accounted for. Such a process is standard [3].

If the signal $v(t)$ has a nonzero bandwidth, no direct relationship between $v(t)$ and the radiated field components can be found unless **h** is independent of frequency. Instead, the spectral densities are related by the transfer functions. Let $v(t)$ have autocorrelation $R_{vv}(\tau)$ and power spectral density $S_{vv}(\omega)$. It is appropriate to define the autocorrelation of $v(t)$ in terms of the antenna current I used in the definition (4.2) of effective length,

$$R_{vv}(\tau) = \langle I^*(t)I(t+\tau)\rangle \tag{7.264}$$

Then I^2 in any form coming from (4.2) can be replaced by the inverse transform of the more general spectral density [compare (7.206) and (7.322)]. The signal power spectral densities and cross spectral densities in terms of the antenna effective length components are*

$$S_{xx}(\omega) = \frac{Z_0^2}{4\lambda^2 r^2}|h_x(\omega)|^2 S_{vv}(\omega) \tag{7.265a}$$

$$S_{yy}(\omega) = \frac{Z_0^2}{4\lambda^2 r^2}|h_y(\omega)|^2 S_{vv}(\omega) \tag{7.265b}$$

$$S_{xy}(\omega) = \frac{Z_0^2}{4\lambda^2 r^2} h_x^*(\omega) h_y(\omega) S_{vv}(\omega) \tag{7.265c}$$

$$S_{yx}(\omega) = \frac{Z_0^2}{4\lambda^2 r^2} h_y^*(\omega) h_x(\omega) S_{vv}(\omega) \tag{7.265d}$$

---
*See Compton [3] for a similar treatment of received signals in an array.

POWER SPECTRAL DENSITIES OF THE FIELDS

The correlation functions are the inverse transforms, and specifically the elements of the coherency matrix are

$$J_{xx} = R_{xx}(0) = \frac{Z_0^2}{8\pi\lambda^2 r^2}\int_{-\infty}^{\infty}|h_x(\omega)|^2 S_{vv}(\omega)\,d\omega \qquad (7.266a)$$

$$J_{yy} = R_{yy}(0) = \frac{Z_0^2}{8\pi\lambda^2 r^2}\int_{-\infty}^{\infty}|h_y(\omega)|^2 S_{vv}(\omega)\,d\omega \qquad (7.266b)$$

$$J_{xy} = R_{yx}(0) = \frac{Z_0^2}{8\pi\lambda^2 r^2}\int_{-\infty}^{\infty}h_y^*(\omega)h_x(\omega) S_{vv}(\omega)\,d\omega \qquad (7.266c)$$

$$J_{yx} = R_{xy}(0) = \frac{Z_0^2}{8\pi\lambda^2 r^2}\int_{-\infty}^{\infty}h_x^*(\omega)h_y(\omega) S_{vv}(\omega)\,d\omega \qquad (7.266d)$$

These equations seem to imply that the coherency matrix of the radiated wave is independent of frequency, but that is not the case. While the integral limits are theoretically infinite, in reality the integrations extend only over the signal bandwidth. If the center frequency of the signal changes, the integrations are performed over a different frequency range, and the different values of the effective length components give a different coherency matrix.

Consider now the special case where **h** is constant over all frequencies for which $S_{vv}(\omega)$ has value. Then

$$J_{xx} = |h_x|^2 C \qquad (7.267a)$$
$$J_{xy} = h_y^* h_x C \qquad (7.267b)$$
$$J_{yx} = h_x^* h_y C \qquad (7.267c)$$
$$J_{yy} = |h_y|^2 C \qquad (7.267d)$$

where

$$C = \frac{Z_0^2}{8\pi\lambda^2 r^2}\int_{-\infty}^{\infty} S_{vv}(\omega)\,d\omega \qquad (7.268)$$

From (7.41), we find that

$$|\mu_{xy}| = \left|\frac{J_{xy}}{\sqrt{J_{xx}}\sqrt{J_{yy}}}\right| = \left|\frac{h_x h_y^*}{|h_x||h_y|}\right| = 1 \qquad (7.269)$$

This is the condition for complete polarization. It is seen then that if the effective length components of an antenna are frequency independent over the bandwidth of the applied signal, the radiated wave is completely polarized.

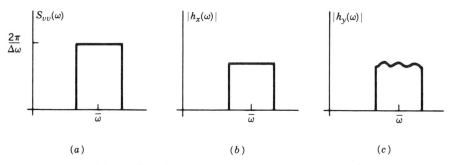

FIGURE 7.3 Signal spectrum and effective length components.

The radiated wave is also completely polarized if the effective length components vary in the same manner with frequency. If the antenna polarization ratio

$$P = \frac{h_y(\omega)}{h_x(\omega)} \tag{7.270}$$

is independent of ω, then (7.266) and (7.269) show the wave to be completely polarized.

To determine how effective lengths that do not change in the same manner with frequency affect the radiated wave, it is necessary to assume a power spectral density for the signal $v(t)$. We take it as a flat bandlimited spectrum with center frequency $\bar{\omega}$ and bandwidth $\Delta\omega$, as shown in Figure 7.3a. The amplitude is chosen to make the total power unity. Any depolarization (decrease in the degree of polarization) of the radiated wave is related to the difference between $h_x(\omega)$ and $h_y(\omega)$. We therefore choose a simple form, $h_x(\omega) = 1$, over the bandwidth $\Delta\omega$ of the signal.

Now $h_y(\omega)$ will, in general, differ from h_x in both amplitude and phase. The choice

$$h_y(\omega) = 1 + \alpha e^{j\gamma(\omega - \bar{\omega})}$$

can approximate a wide variety of differences, with a large or small number of ripples across the passband and an arbitrary ripple amplitude, as shown in Figure 7.3c.

The coherency matrix elements can be found from (7.266). Straightforward integration, noting that $v(t)$ has no negative frequencies, leads to

$$J_{xx} = \frac{Z_0^2}{4\lambda^2 r^2}$$

$$J_{xy} = \frac{Z_0^2}{4\lambda^2 r^2}\left[1 + \alpha \operatorname{sinc}\left(\frac{\gamma \Delta\omega}{2}\right)\right]$$

$$J_{yy} = \frac{Z_0^2}{4\lambda^2 r^2}\left[1 + 2\alpha \operatorname{sinc}\left(\frac{\gamma \Delta\omega}{2}\right) + \alpha^2\right]$$

where

$$\text{sinc}(x) = \frac{\sin x}{x} \qquad (7.271)$$

and $\Delta\omega$ is the signal bandwidth. From (7.80), the degree of polarization is

$$R = \left\{ \frac{[2\alpha \,\text{sinc}(\gamma\Delta\omega/2) + \alpha^2]^2 + 4[1 + \alpha \,\text{sinc}(\gamma\Delta\omega/2)]^2}{[2 + 2\alpha \,\text{sinc}(\gamma\Delta\omega/2) + \alpha^2]^2} \right\}^{1/2}$$

Examination of this equation shows R to be unity for either $\Delta\omega = 0$ (single frequency) or $\alpha = 0$ ($h_y = h_x$). Further, R has smaller values for negative values of the sinc function. Since we are interested in seeing how much the radiated wave can be depolarized by the difference in h_x and h_y, consider a case leading to considerable depolarization. Let

$$\frac{\gamma\Delta\omega}{2} = 4.493 \qquad \text{sinc}\left(\frac{\gamma\Delta\omega}{2}\right) = -0.2172$$

which is the first, and greatest magnitude, minimum of the sinc function. Figure 7.4 shows R as a function of α for this choice. It can be seen from the figure that substantial depolarization can occur, with the loss in the polarized part of the power somewhat greater than 1 dB for $\alpha = 0.5$. Figure 7.4 also shows R for

$$\frac{\gamma\Delta\omega}{2} = 1.895 \qquad \text{sinc}\left(\frac{\gamma\Delta\omega}{2}\right) = 0.5$$

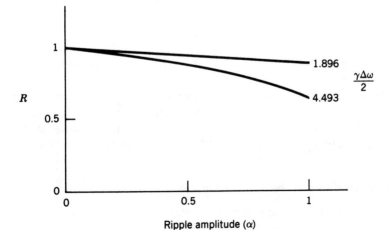

FIGURE 7.4 Degree of polarization of radiated wave.

426 PARTIAL POLARIZATION

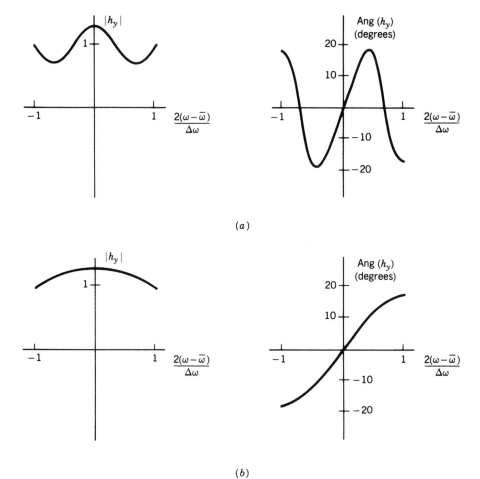

FIGURE 7.5 Magnitude and phase of h_y: (a) $\gamma\Delta\omega/2 = 4.493$, $\alpha = 0.3$; (b) $\gamma\Delta\omega/2 = 1.896$, $\alpha = 0.3$.

which represents a smaller value of γ, corresponding to a better phase matching of h_x and h_y.

Figure 7.5 shows the magnitude and phase of h_y for the chosen values of $\gamma\Delta\omega/2$, and $\alpha = 0.3$. The slower phase variation across the frequency band gives a higher degree of polarization of the radiated wave in this case, although a general conclusion that this is always true cannot be drawn.

Characterizations of the Radiated Wave

We have seen that a quasi-monochromatic plane wave can be characterized by the power spectral densities and cross-spectral densities of its orthogonal

POWER SPECTRAL DENSITIES OF THE FIELDS   427

field components. It is reasonable to order these spectral densities, just as we did for the corresponding power terms, in the form of a coherency matrix and a Stokes vector. To accomplish this, we first define a *generalized coherency matrix* or *correlation matrix* and a *Stokes correlation vector* as

$$\mathbf{J}(\tau) = \begin{bmatrix} R_{xx}(\tau) & R_{yx}(\tau) \\ R_{xy}(\tau) & R_{yy}(\tau) \end{bmatrix} = \begin{bmatrix} \langle E_x^*(t)E_x(t+\tau) \rangle & \langle E_y^*(t)E_x(t+\tau) \rangle \\ \langle E_x^*(t)E_y(t+\tau) \rangle & \langle E_y^*(t)E_y(t+\tau) \rangle \end{bmatrix}$$

(7.272)

and

$$\mathbf{G}(\tau) = \begin{bmatrix} R_{xx}(\tau) + R_{yy}(\tau) \\ R_{xx}(\tau) - R_{yy}(\tau) \\ R_{xy}(\tau) + R_{yx}(\tau) \\ -j[R_{xy}(\tau) - R_{yx}(\tau)] \end{bmatrix}$$

$$= \begin{bmatrix} \langle E_x^*(t)E_x(t+\tau) + E_y^*(t)E_y(t+\tau) \rangle \\ \langle E_x^*(t)E_x(t+\tau) - E_y^*(t)E_y(t+\tau) \rangle \\ \langle E_x^*(t)E_y(t+\tau) + E_y^*(t)E_x(t+\tau) \rangle \\ -j\langle E_x^*(t)E_y(t+\tau) - E_y^*(t)E_x(t+\tau) \rangle \end{bmatrix}$$

(7.273)

For $\tau = 0$, these reduce to the common forms (7.36) and (7.135). Note that in general $R_{xy}(\tau) \neq R_{yx}^*(\tau)$, so that $\mathbf{J}(\tau)$ is not Hermitian, and $\mathbf{G}(\tau)$ is not real. We next define a *spectral coherency matrix* as the transform

$$\mathscr{J}(\omega) = \int_{-\infty}^{\infty} \mathbf{J}(\tau) e^{-j\omega\tau} \, d\tau = \begin{bmatrix} S_{xx}(\omega) & S_{yx}(\omega) \\ S_{xy}(\omega) & S_{yy}(\omega) \end{bmatrix}$$

(7.274)

The same symbol is used for this spectral coherency matrix and the coherency vector used to introduce the Stokes vector in Section 7.7, but the coherency vector is little used and use of the same symbol should cause no difficulty. A *spectral Stokes vector* is also defined as

$$\mathscr{G}(\omega) = \int_{-\infty}^{\infty} \mathbf{G}(\tau) e^{-j\omega\tau} \, d\tau = \begin{bmatrix} S_{xx}(\omega) + S_{yy}(\omega) \\ S_{xx}(\omega) - S_{yy}(\omega) \\ S_{xy}(\omega) + S_{yx}(\omega) \\ -j[S_{xy}(\omega) - S_{yx}(\omega)] \end{bmatrix}$$

(7.275)

where

$$S_{xx}(\omega) = \int_{-\infty}^{\infty} R_{xx}(\tau) e^{-j\omega\tau} \, d\tau = \int_{-\infty}^{\infty} \langle E_x^*(t) E_x(t+\tau) \rangle e^{-j\omega\tau} \, d\tau \quad (7.276a)$$

$$S_{xy}(\omega) = \int_{-\infty}^{\infty} R_{xy}(\tau) e^{-j\omega\tau} \, d\tau = \int_{-\infty}^{\infty} \langle E_x^*(t) E_y(t+\tau) \rangle e^{-j\omega\tau} \, d\tau \quad (7.276b)$$

$$S_{yx}(\omega) = \int_{-\infty}^{\infty} R_{yx}(\tau) e^{-j\omega\tau} \, d\tau = \int_{-\infty}^{\infty} \langle E_y^*(t) E_x(t+\tau) \rangle e^{-j\omega\tau} \, d\tau \quad (7.276c)$$

$$S_{yy}(\omega) = \int_{-\infty}^{\infty} R_{yy}(\tau) e^{-j\omega\tau} \, d\tau = \int_{-\infty}^{\infty} \langle E_y^*(t) E_y(t+\tau) \rangle e^{-j\omega\tau} \, d\tau \quad (7.276d)$$

It is not difficult to show from this equation set that

$$S_{yx}(\omega) = S_{xy}^*(\omega) \quad (7.277)$$

Then the spectral coherency matrix $\mathscr{J}$ is Hermitian, and the spectral Stokes vector $\mathscr{G}$ is real.

The coherency matrix and Stokes vector of an antenna were defined previously. The definitions are rewritten here to show explicitly the element dependence on frequency,

$$\mathbf{J}_A = \begin{bmatrix} |h_x(\omega)|^2 & h_x(\omega) h_y^*(\omega) \\ h_x^*(\omega) h_y(\omega) & |h_y(\omega)|^2 \end{bmatrix} \quad (7.278)$$

$$\mathbf{G}_A = \begin{bmatrix} |h_x(\omega)|^2 + |h_y(\omega)|^2 \\ |h_x(\omega)|^2 - |h_y(\omega)|^2 \\ 2\operatorname{Re}[h_x^*(\omega) h_y(\omega)] \\ 2\operatorname{Im}[h_x^*(\omega) h_y(\omega)] \end{bmatrix} = \begin{bmatrix} |h_x(\omega)|^2 + |h_y(\omega)|^2 \\ |h_x(\omega)|^2 - |h_y(\omega)|^2 \\ h_x^*(\omega) h_y(\omega) + h_x(\omega) h_y^*(\omega) \\ -j[h_x^*(\omega) h_y(\omega) - h_x(\omega) h_y^*(\omega)] \end{bmatrix} \quad (7.279)$$

If the antenna is fed by a source $v(t)$ having a power spectral density S_{vv}, the product

$$\begin{bmatrix} S_{xx}(\omega) & S_{yx}(\omega) \\ S_{xy}(\omega) & S_{yy}(\omega) \end{bmatrix} = \frac{Z_0^2}{4\lambda^2 r^2} \begin{bmatrix} |h_x(\omega)|^2 & h_y^*(\omega) h_x(\omega) \\ h_x^*(\omega) h_y(\omega) & |h_y(\omega)|^2 \end{bmatrix} S_{vv} \quad (7.280)$$

clearly satisfies (7.265). Then the spectral coherency matrix of the radiated wave, from (7.274), is

$$\mathscr{J}(\omega) = \frac{Z_0^2}{4\lambda^2 r^2} \mathbf{J}_A S_{vv} \quad (7.281)$$

The spectral Stokes vector of the wave in terms of the antenna Stokes vector is found in a similar manner to be

$$\mathscr{G}(\omega) = \frac{Z_0^2}{4\lambda^2 r^2} G_A S_{vv} \quad (7.282)$$

For a single-frequency signal, these equations reduce to (7.127) and (7.204) respectively.

Received Power Using the Spectral Coherency Matrix

Consider once more a transmitting–receiving antenna configuration with a receiving and b transmitting. The wave incident on a has a spectral coherency matrix

$$\mathscr{J}^i(\omega) = \begin{bmatrix} S_{\xi\xi}(\omega) & S_{\eta\xi}(\omega) \\ S_{\xi\eta}(\omega) & S_{\eta\eta}(\omega) \end{bmatrix} \quad (7.283)$$

We may think of the receiving antenna as two frequency-dependent transducers that take us from two incident field components to a received signal, as in Figure 7.6. The antenna effective lengths are given as h_ξ and h_η because for reception the effective length components need to be in the same coordinates as the incident wave.

From Figure 7.6, the output signal is

$$w(t) = u(t) + v(t) \quad (7.284)$$

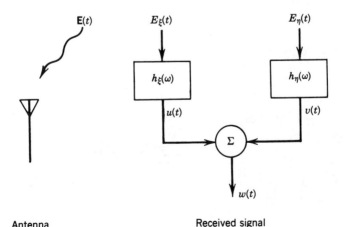

FIGURE 7.6 Equivalent circuit of antenna.

with autocorrelation function

$$\begin{aligned}
R_{ww}(\tau) &= \langle w^*(t)w(t+\tau)\rangle = \langle [u^*(t)+v^*(t)][u(t+\tau)+v(t+\tau)]\rangle \\
&= \langle u^*(t)u(t+\tau)\rangle + \langle u^*(t)v(t+\tau)\rangle \\
&\quad + \langle v^*(t)u(t+\tau)\rangle + \langle v^*(t)v(t+\tau)\rangle \\
&= R_{uu}(\tau) + R_{uv}(\tau) + R_{vu}(\tau) + R_{vv}(\tau)
\end{aligned} \quad (7.285)$$

The spectral densities are related in the same way,

$$S_{ww}(\omega) = S_{uu}(\omega) + S_{uv}(\omega) + S_{vu}(\omega) + S_{vv}(\omega) \quad (7.286)$$

By the use of conventional network theory, the spectral densities can be found from the spectral densities of the incident fields. Doing so gives

$$\begin{aligned}
S_{ww}(\omega) &= |h_\xi(\omega)|^2 S_{\xi\xi}(\omega) + h_\xi^*(\omega)h_\eta(\omega)S_{\xi\eta}(\omega) \\
&\quad + h_\eta^*(\omega)h_\xi(\omega)S_{\eta\xi}(\omega) + |h_\eta(\omega)|^2 S_{\eta\eta}(\omega)
\end{aligned} \quad (7.287)$$

In a previous section devoted to reception it was necessary to define a receiving coherency matrix in receiving coordinates for the antenna by taking the conjugate and replacing x and y by ξ and η. We do that here also. Then, altering (7.278) according to this rule, the receiving coherency matrix of the antenna in receiving coordinates, is defined as

$$\mathbf{J}_R(\omega) = \begin{bmatrix} |h_\xi(\omega)|^2 & h_\xi^*(\omega)h_\eta(\omega) \\ h_\xi(\omega)h_\eta^*(\omega) & |h_\eta(\omega)|^2 \end{bmatrix} \quad (7.288)$$

The product of the spectral coherency matrix of the incident wave and the antenna receiving coherency matrix is

$$\mathscr{J}^i(\omega)\mathbf{J}_R(\omega) = \begin{bmatrix} S_{\xi\xi}(\omega) & S_{\eta\xi}(\omega) \\ S_{\xi\eta}(\omega) & S_{\eta\eta}(\omega) \end{bmatrix} \begin{bmatrix} |h_\xi(\omega)|^2 & h_\xi^*(\omega)h_\eta(\omega) \\ h_\xi(\omega)h_\eta^*(\omega) & |h_\eta(\omega)|^2 \end{bmatrix} \quad (7.289)$$

Comparison to (7.287) shows that

$$S_{ww}(\omega) = \text{Tr}[\mathscr{J}^i(\omega)\mathbf{J}_R(\omega)] \quad (7.290)$$

The inverse Fourier transform of S_{ww} is the autocorrelation of the received voltage,

$$\langle w^*(t)w(t+\tau)\rangle = \mathscr{F}^{-1}[S_{ww}(\omega)] \quad (7.291)$$

and the received power is

$$W = \frac{1}{8R_a}\langle w^*(t)w(t)\rangle = \frac{1}{8R_a}\frac{1}{2\pi}\int_{-\infty}^{\infty} S_{ww}(\omega)\,d\omega$$

$$= \frac{1}{8R_a}\frac{1}{2\pi}\int_{-\infty}^{\infty} \text{Tr}\big[\mathscr{J}^i(\omega)\mathbf{J}_R(\omega)\big]\,d\omega \tag{7.292}$$

If the effective length components of the antenna are independent of frequency over the bandwidth of the incident wave, the received power is

$$W = \frac{1}{8R_a}\text{Tr}\bigg[\mathbf{J}_R\frac{1}{2\pi}\int_{-\infty}^{\infty}\mathscr{J}^i(\omega)\,d\omega\bigg] \tag{7.293}$$

If (7.117) and (7.283) are compared, and if it is recognized that the elements of (7.117) are the inverse Fourier transforms (for $\tau = 0$) of those in (7.283), it is seen that

$$W = \frac{1}{8R_a}\text{Tr}(\mathbf{J}_R\mathbf{J}^i) = \frac{1}{8R_a}\text{Tr}(\mathbf{J}^i\mathbf{J}_R) \tag{7.294}$$

which agrees with (7.119).

It was said following (7.122) that (7.119) only approximates the received power if **h** depends on frequency. It was not obvious then that the statement was correct, since frequency had not been explicitly considered. This development of received power in terms of spectral densities and the reduction to (7.294) for frequency-independent $\mathbf{J}_R$ make it clear that the constraint on **h** is valid.

We saw earlier that if $h_x(\omega)$ and $h_y(\omega)$ do not vary with frequency in the same way across the bandwidth of the signal, the wave radiated by a transmitting antenna is depolarized. It is clear that the same phenomenon is encountered for a receiving antenna. Even if the incident wave is completely polarized (but not monochromatic), frequency-dependent differences between the effective length components will decorrelate the received signal, and power is lost. The effect is of the same order as the polarized power lost on transmission.

Received Power Using the Spectral Stokes Vector

Consider, with the use of the Stokes vector, the power received by antenna *a* of the transmitting–receiving configuration of the previous section. The spectral Stokes vector of the incident wave is

$$\mathscr{S}^i = \begin{bmatrix} S_{\xi\xi}(\omega) + S_{\eta\eta}(\omega) \\ S_{\xi\xi}(\omega) - S_{\eta\eta}(\omega) \\ S_{\xi\eta}(\omega) + S_{\eta\xi}(\omega) \\ -j[S_{\xi\eta}(\omega) - S_{\eta\xi}(\omega)] \end{bmatrix} \quad (7.295)$$

It was desirable in a previous section to define a receiving Stokes vector for an antenna. By comparing the antenna Stokes vector of (7.156) with the receiving Stokes vector of (7.159), it is seen that the receiving Stokes vector is formed by changing from transmitting coordinates xy to receiving coordinates $\xi\eta$ and reversing the sign of G_3. Carrying out the same procedure on (7.279), or merely rewriting (7.159) to show explicitly the dependence on frequency, gives the receiving Stokes vector of the antenna in receiving coordinates:

$$\mathbf{G}_R = \begin{bmatrix} |h_\xi(\omega)|^2 + |h_\eta(\omega)|^2 \\ |h_\xi(\omega)|^2 - |h_\eta(\omega)|^2 \\ h_\xi^*(\omega)h_\eta(\omega) + h_\xi(\omega)h_\eta^*(\omega) \\ j[h_\xi^*(\omega)h_\eta(\omega) - h_\xi(\omega)h_\eta^*(\omega)] \end{bmatrix} \quad (7.296)$$

It may be written in transmitting coordinates by replacing h_ξ with $-h_x$ and h_η with h_y, thus

$$\mathbf{G}_R = \begin{bmatrix} |h_x(\omega)|^2 + |h_y(\omega)|^2 \\ |h_x(\omega)|^2 - |h_y(\omega)|^2 \\ -2\,\mathrm{Re}[h_x^*(\omega)h_y(\omega)] \\ 2\,\mathrm{Im}[h_x^*(\omega)h_y(\omega)] \end{bmatrix} \quad (7.297)$$

The product of the spectral Stokes vector transpose of the incident wave with $\mathbf{G}_R$ (in receiving coordinates) is

$$\mathscr{S}^{iT}\mathbf{G}_R = 2\Big(|h_\xi(\omega)|^2 S_{\xi\xi}(\omega) + h_\xi^*(\omega)h_\eta(\omega)S_{\xi\eta}(\omega) \\ + h_\xi(\omega)h_\eta^*(\omega)S_{\eta\xi}(\omega) + |h_\eta(\omega)|^2 S_{\eta\eta}(\omega)\Big) \quad (7.298)$$

and a comparison with (7.287) shows the power spectral density of the received signal to be

$$S_{ww}(\omega) = \tfrac{1}{2}\mathscr{S}^{iT}\mathbf{G}_R \quad (7.299)$$

The received power may be found from (7.292), leading to

$$W = \frac{1}{32\pi R_a} \int_{-\infty}^{\infty} \mathscr{G}^{iT} \mathbf{G}_R \, d\omega \tag{7.300}$$

If the receiving Stokes vector $\mathbf{G}_R$ is frequency-independent over the bandwidth, this equation reduces to (7.162).

In (4.133) the polarization efficiency for two antennas was given with a negative sign for the product $G_2^a G_2^b$. That is consistent with the sign reversal of G_2 in (7.297) when going from the antenna Stokes vector to its receiving Stokes vector.

Scattering by a Target

We have treated the antenna in this section as two frequency-dependent transducers converting a narrow-band feed signal to radiated field components, or incident field components to a received signal. A radar target, or optical transmission instrument, can be dealt with in the same manner, since

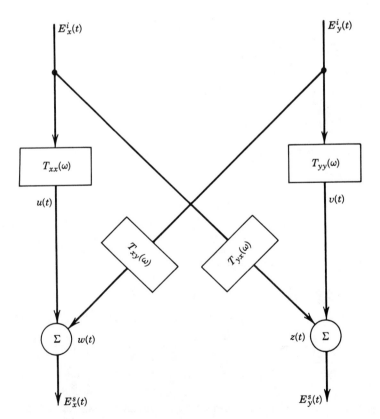

FIGURE 7.7 Equivalent circuit for scattering.

434 PARTIAL POLARIZATION

it converts an incident wave component into two scattered or transmitted components. An equivalent circuit, applicable to this situation, is shown in Figure 7.7. The incident wave is in $x_1 y_1 z_1$ coordinates of Figure 7.1 and the scattered wave in $x_2 y_2 z_2$, so it is appropriate to use the Jones matrix **T** in the development. Later, the Sinclair matrix with incident and scattered fields in the same coordinates (for backscattering) will be considered. The elements of **T** are treated here as frequency-dependent transfer functions. The known quantities are the elements of $\mathbf{T}(\omega)$, the input spectral densities S^i_{xx}, S^i_{xy}, S^i_{yx}, and S^i_{yy}, and the corresponding correlations R^i_{xx}, R^i_{xy}, R^i_{yx}, and R^i_{yy}. The field components and intermediate signals u, v, w, and z are analytic signals. The $\sqrt{4\pi} r$ variation in the output fields is neglected at this time.

From Figure 7.7,

$$E^s_x(t) = u(t) + w(t) \tag{7.301a}$$

$$E^s_y(t) = v(t) + z(t) \tag{7.301b}$$

Then the auto- and crosscorrelations of the outputs are

$$R^s_{xx}(\tau) = \langle E^{s*}_x(t) E^s_x(t+\tau) \rangle = \langle [u^*(t) + w^*(t)][u(t+\tau) + w(t+\tau)] \rangle$$
$$= \langle u^*(t)u(t+\tau) \rangle + \langle u^*(t)w(t+\tau) \rangle + \langle w^*(t)u(t+\tau) \rangle$$
$$+ \langle w^*(t)w(t+\tau) \rangle$$
$$= R_{uu}(\tau) + R_{uw}(\tau) + R_{wu}(\tau) + R_{ww}(\tau) \tag{7.302a}$$

$$R^s_{xy}(\tau) = \langle E^{s*}_x(t) E^s_y(t+\tau) \rangle$$
$$= R_{uv}(\tau) + R_{uz}(\tau) + R_{wv}(\tau) + R_{wz}(\tau) \tag{7.302b}$$

$$R^s_{yx}(\tau) = \langle E^{s*}_y(t) E^s_x(t+\tau) \rangle$$
$$= R_{vu}(\tau) + R_{vw}(\tau) + R_{zu}(\tau) + R_{zw}(\tau) \tag{7.302c}$$

$$R^s_{yy}(\tau) = R_{vv}(\tau) + R_{vz}(\tau) + R_{zv}(\tau) + R_{zz}(\tau) \tag{7.302d}$$

The spectral densities are linear transforms, so

$$S^s_{xx}(\omega) = S_{uu}(\omega) + S_{uw}(\omega) + S_{wu}(\omega) + S_{ww}(\omega) \tag{7.303a}$$

$$S^s_{xy}(\omega) = S_{uv}(\omega) + S_{uz}(\omega) + S_{wv}(\omega) + S_{wz}(\omega) \tag{7.303b}$$

$$S^s_{yx}(\omega) = S_{vu}(\omega) + S_{vw}(\omega) + S_{zu}(\omega) + S_{zw}(\omega) \tag{7.303c}$$

$$S^s_{yy}(\omega) = S_{vv}(\omega) + S_{vz}(\omega) + S_{zv}(\omega) + S_{zz}(\omega) \tag{7.303d}$$

The spectral densities of the intermediate signals can all be obtained from the input spectral densities and the transfer functions of $\mathbf{T}(\omega)$. Therefore, omitting the functional dependence on ω for conciseness,

$$S_{xx}^s = |T_{xx}|^2 S_{xx}^i + T_{xx}^* T_{xy} S_{xy}^i + T_{xy}^* T_{xx} S_{yx}^i + |T_{xy}|^2 S_{yy}^i \quad (7.304\text{a})$$

$$S_{xy}^s = T_{xx}^* T_{yy} S_{xy}^i + T_{xx}^* T_{yx} S_{xx}^i + T_{xy}^* T_{yy} S_{yy}^i + T_{xy}^* T_{yx} S_{yx}^i \quad (7.304\text{b})$$

$$S_{yx}^s = T_{yy}^* T_{xx} S_{yx}^i + T_{yy}^* T_{xy} S_{yy}^i + T_{yx}^* T_{xx} S_{xx}^i + T_{yx}^* T_{xy} S_{xy}^i \quad (7.304\text{c})$$

$$S_{yy}^s = |T_{yy}|^2 S_{yy}^i + T_{yy}^* T_{yx} S_{yx}^i + T_{yx}^* T_{yy} S_{xy}^i + |T_{yx}|^2 S_{xx}^i \quad (7.304\text{d})$$

The output auto- and crosscorrelations are the inverse Fourier transforms. Normally, it is the correlations for $\tau = 0$ that are of interest; they are

$$R_{\alpha\beta}(0) = \frac{1}{2\pi} \int_{-\infty}^{\infty} S_{\alpha\beta}(\omega)\, d\omega \quad (7.305)$$

where α and β independently take on the values x and y. If the scattering matrix is independent of frequency over the signal bandwidth, the scattering matrix elements may be removed from the integrals, giving

$$R_{xx}^s(0) = |T_{xx}|^2 R_{xx}^i(0) + T_{xx}^* T_{xy} R_{xy}^i(0) + T_{xy}^* T_{xx} R_{yx}^i(0) + |T_{xy}|^2 R_{yy}^i(0)$$
$$(7.306\text{a})$$

$$R_{xy}^s(0) = T_{xx}^* T_{yy} R_{xy}^i(0) + T_{xx}^* T_{yx} R_{xx}^i(0) + T_{xy}^* T_{yy} R_{yy}^i(0) + T_{xy}^* T_{yx} R_{yx}^i(0)$$
$$(7.306\text{b})$$

$$R_{yx}^s(0) = T_{yy}^* T_{xx} R_{yx}^i(0) + T_{yy}^* T_{xy} R_{yy}^i(0) + T_{yx}^* T_{xx} R_{xx}^i(0) + T_{yx}^* T_{xy} R_{xy}^i(0)$$
$$(7.306\text{c})$$

$$R_{yy}^s(0) = |T_{yy}|^2 R_{yy}^i(0) + T_{yy}^* T_{yx} R_{yx}^i(0) + T_{yx}^* T_{yy} R_{xy}^i(0) + |T_{yx}|^2 R_{xx}^i(0)$$
$$(7.306\text{d})$$

If

$$\mathbf{E}^s = \mathbf{T}\mathbf{E}^i \quad (7.307)$$

is expanded and the correlations of $\mathbf{E}^s$ found at $\tau = 0$, they will agree with (7.306).

It is desirable to cast the spectral densities of (7.304) into the spectral Stokes vector form. At this time, the distance term $4\pi r^2$ is reinserted. Then (7.304) becomes

$$\begin{bmatrix} S_{xx}^s \\ S_{xy}^s \\ S_{yx}^s \\ S_{yy}^s \end{bmatrix} = \frac{1}{4\pi r^2} \begin{bmatrix} |T_{xx}|^2 & T_{xx}^* T_{xy} & T_{xy}^* T_{xx} & |T_{xy}|^2 \\ T_{xx}^* T_{yx} & T_{xx}^* T_{yy} & T_{xy}^* T_{yx} & T_{xy}^* T_{yy} \\ T_{yx}^* T_{xx} & T_{yx}^* T_{xy} & T_{yy}^* T_{xx} & T_{yy}^* T_{xy} \\ |T_{yx}|^2 & T_{yx}^* T_{yy} & T_{yy}^* T_{yx} & |T_{yy}|^2 \end{bmatrix} \begin{bmatrix} S_{xx}^i \\ S_{xy}^i \\ S_{yx}^i \\ S_{yy}^i \end{bmatrix} \quad (7.308)$$

which may be written as

$$\mathbf{V}^s = \frac{1}{4\pi r^2}(\mathbf{T}^* \times \mathbf{T})\mathbf{V}^i \tag{7.309}$$

where $\mathbf{V}$ is the four-vector of the spectral densities. It is readily seen that the spectral Stokes vector is related to $\mathbf{V}$ by

$$\mathscr{G} = \mathbf{Q}^*\mathbf{V} = \begin{bmatrix} 1 & 0 & 0 & 1 \\ 1 & 0 & 0 & -1 \\ 0 & 1 & 1 & 0 \\ 0 & -j & j & 0 \end{bmatrix}\mathbf{V} \tag{7.310}$$

This can be combined with the preceding equation to give

$$\mathscr{G}^s = \mathbf{Q}^*\mathbf{V}^s = \frac{1}{4\pi r^2}\mathbf{Q}^*(\mathbf{T}^* \times \mathbf{T})(\mathbf{Q}^*)^{-1}\mathscr{G}^i \tag{7.311}$$

which relates the spectral Stokes vectors of incident and scattered waves. Let

$$\mathscr{M} = \mathbf{Q}^*(\mathbf{T}^* \times \mathbf{T})(\mathbf{Q}^*)^{-1} \tag{7.312}$$

Then the scattering equation is

$$\mathscr{G}^s(\omega) = \frac{1}{4\pi r^2}\mathscr{M}(\omega)\mathscr{G}^i(\omega) \tag{7.313}$$

The elements of $\mathscr{M}$ may be determined by multiplication but are found more easily by noting that

$$(\mathbf{T}^* \times \mathbf{T}) = (\mathbf{T} \times \mathbf{T}^*)^* \tag{7.314}$$

and $\mathbf{Q}$ is given by (7.190). Note that now $\mathbf{T}$ is to be considered a function of frequency. It follows that

$$\mathscr{M}(\omega) = \mathbf{M}^*$$

where $\mathbf{M}$ is the Mueller matrix. Since the Mueller elements are real,

$$\mathscr{M}(\omega) = \mathbf{M} \tag{7.315}$$

The spectral Stokes vectors of incident and scattered waves are therefore related by the Mueller matrix, with elements considered as functions of frequency, thus,

$$\mathscr{G}^s(\omega) = \frac{1}{4\pi r^2}\mathbf{M}(\omega)\mathscr{G}^i(\omega) \tag{7.316}$$

According to the definitions of this section, the Stokes vector is the inverse transform, at $\tau = 0$, of the spectral Stokes vector. Then for the scattering problem considered,

$$\mathbf{G}^s = \{\mathscr{F}^{-1}[\mathscr{G}^s(\omega)]\}_{\tau=0} = \frac{1}{4\pi r^2}\{\mathscr{F}^{-1}[\mathbf{M}(\omega)\mathscr{G}^i(\omega)]\}_{\tau=0} \quad (7.317)$$

If **M** is independent of frequency, this reduces to

$$\mathbf{G}^s = \frac{1}{4\pi r^2}\mathbf{M}\mathbf{G}^i \quad (7.200)$$

More generally, the ordinary Stokes vector of the scattered wave is

$$\mathbf{G}^s = \frac{1}{4\pi r^2}\frac{1}{2\pi}\int_{-\infty}^{\infty}\mathbf{M}(\omega)\mathscr{G}^i(\omega)\,d\omega \quad (7.318)$$

Now let the wave incident on the target originate at transmitting antenna a at the origin of coordinates $x_1 y_1 z_1$ of Figure 7.1. The spectral Stokes vector of the incident wave is

$$\mathscr{G}^i = \frac{Z_0^2}{4\lambda^2 r_1^2}\mathbf{G}_A^{(a)}S_{vv}^{(a)} \quad (7.319)$$

where $S_{vv}^{(a)}$ is the power spectral density of the signal transmitted from a. The scattered wave, in $x_2 y_2 z_2$ coordinates at the receiving antenna b, is

$$\mathscr{G}^s = \frac{1}{4\pi r_2^2}\mathbf{M}(\omega)\mathscr{G}^i \quad (7.320)$$

and the power spectral density of the received signal following antenna b is

$$S_{ww}^{(b)} = \tfrac{1}{2}\mathscr{G}^{sT}\mathbf{G}_R^{(b)} = \frac{Z_0^2 S_{vv}^{(a)}}{32\pi\lambda^2 r_1^2 r_2^2}\mathbf{G}_A^{(a)T}\mathbf{M}^T\mathbf{G}_R^{(b)} \quad (7.321)$$

Finally, the received power is proportional to the inverse transform,

$$W = \frac{1}{8R_a}\{\mathscr{F}^{-1}[S_{ww}^{(b)}]\}_{\tau=0} = \frac{Z_0^2}{512\pi^2 R_a \lambda^2 r_1^2 r_2^2}\int_{-\infty}^{\infty}S_{vv}^{(a)}\mathbf{G}_A^{(a)T}\mathbf{M}^T\mathbf{G}_R^{(b)}\,d\omega$$

$$(7.322)$$

If the Mueller matrix and the two antenna matrices are frequency-independent, this equation reduces to (7.206).

We saw earlier that all scatterers can be represented by Mueller matrices, but not all have Sinclair or Jones matrices. This situation is not changed by the introduction of the frequency-dependent scattering and Mueller matrices. The elements of $\mathbf{M}(\omega)$ must still satisfy an equation set like (7.223) if a scattering matrix is to exist, and they cannot do so for some targets, for example a distributed target with elementary scatterers whose positions are random.

The Frequency-Dependent Kennaugh Matrix

The development leading to received power for a narrow-band signal and frequency-dependent target and antennas was in terms of a frequency-dependent Mueller matrix. It is clear, however, that the coordinate system used for the scattered wave has no effect on the signals. The Jones matrix elements were treated as transducers changing an incident signal to a scattered signal, but the elements of the Sinclair matrix $\mathbf{S}$ could have been used instead. If this development were repeated, using the Kennaugh formulation, it would be appropriate to define a *spectral Kennaugh vector* of the scattered wave, related to the spectral Kennaugh (or Stokes) vector of the incident wave by

$$\mathscr{G}_K(\omega) = \frac{1}{4\pi r_2^2} \mathbf{K}(\omega) \mathscr{G}^i(\omega) \tag{7.323}$$

where $\mathbf{K}(\omega)$ is a *frequency-dependent Kennaugh matrix* of the target.

The received power, corresponding to (7.322), is, for the Kennaugh form,

$$W = \frac{Z_0^2}{512\pi^2 R_a \lambda^2 r_1^2 r_2^2} \int_{-\infty}^{\infty} S_{vv} \mathbf{G}_A^{rT} \mathbf{K}^T \mathbf{G}_A^t \, d\omega \tag{7.324}$$

where S_{vv} is the power spectral density of the signal fed to the transmitting antenna. If the Kennaugh matrix and the two antenna matrices are frequency-independent, this equation reduces to (7.250).

Limitations and Example

In this section, conventional network theory has been used. Since it is valid for small element dimensions, it is necessary to examine the maximum size of targets. We do this by example. One way to study the problem would be to obtain the degree of coherence of the reflected wave [1], but we have equations for the degree of polarization, so we shall use them.

Consider a simple target that consists of two wires transverse to the propagation axis and separated by distance d. The Jones matrix elements, referenced to the front surface of the target, are

$$T_{xx} = 1 + e^{-j2kd} = 1 + e^{-j2\omega d/c}$$
$$T_{xy} = -T_{yx} = e^{-j2\omega d/c}$$
$$T_{yy} = e^{-j2\omega d/c}$$

The incident wave has a flat spectrum with power spectral density like that of Figure 7.3a,

$$S^i_{xx} = \frac{2\pi}{\Delta\omega} \qquad \bar{\omega} - \frac{\Delta\omega}{2} < \omega < \bar{\omega} + \frac{\Delta\omega}{2}$$

Power spectral densities of the reflected wave, neglecting a constant multiplier, are, from (7.304),

$$S^s_{xx} = |T_{xx}|^2 S^i_{xx} = 2[1 + \cos(2\omega d/c)] S^i_{xx}$$
$$S^s_{xy} = T^*_{xx} T_{yx} S^i_{xx} = -(1 + e^{-j2\omega d/c}) S^i_{xx}$$
$$S^s_{yx} = T^*_{yx} T_{xx} S^i_{xx} = -(1 + e^{j2\omega d/c}) S^i_{xx}$$
$$S^s_{yy} = |T_{yx}|^2 S^i_{xx} = S^i_{xx}$$

The coherency matrix elements are proportional to the correlation functions at $\tau = 0$. Again neglecting the multiplier, they are

$$J_{xx} = \frac{2}{\Delta\omega} \int_{\bar{\omega}-\Delta\omega/2}^{\bar{\omega}+\Delta\omega/2} \left[1 + \cos\left(2\omega\frac{d}{c}\right)\right] d\omega = 2 + 2\cos\left(2\bar{\omega}\frac{d}{c}\right) \operatorname{sinc}\left(\Delta\omega\frac{d}{c}\right)$$

$$J_{xy} = -\frac{1}{\Delta\omega} \int_{\bar{\omega}-\Delta\omega/2}^{\bar{\omega}+\Delta\omega/2} (1 + e^{j2\omega d/c}) d\omega = -\left[1 + e^{j2\bar{\omega}d/c} \operatorname{sinc}\left(\Delta\omega\frac{d}{c}\right)\right]$$

$$J_{yy} = \frac{1}{\Delta\omega} \int_{\bar{\omega}-\Delta\omega/2}^{\bar{\omega}+\Delta\omega/2} d\omega = 1$$

The determinant of **J** is simple,

$$\|\mathbf{J}\| = 1 - \operatorname{sinc}^2(\Delta\omega d/c)$$

It may be seen from this that the scattered wave is completely polarized for $d = 0$, as expected, and completely depolarized at the roots of the sinc function. The first root occurs for the sinc function argument equal to π, or at $d/\lambda = 1/2B$, where $B = \Delta\omega/\bar{\omega}$ is the fractional bandwidth.

It is evident that the depolarization depends, not merely on the target dimension d, but on the manner in which it scatters and on the incident wave

440 PARTIAL POLARIZATION

itself. This example led to a greater depolarization, and hence a smaller allowable target length, than certain other targets. Some extended targets do not depolarize the reflected wave, and the allowable target length must be determined using other criteria. In this example, the incident wave is completely polarized but is depolarized by reflection, whereas a monochromatic incident wave would not be depolarized.

7.10. THE POWER SCATTERING MATRIX

The Graves Matrix

An important class of scattering problems is that for which the incident wave is completely polarized and the scatterer cannot be represented by a scattering matrix. Typical problems are radar returns from vegetation in wind-induced motion and scattering by meteorological phenomena such as thunderstorms. It is relatively easy at radar frequencies to generate a completely polarized incident wave, but the scatterer may significantly reduce the degree of polarization of the reflected wave. These problems may be treated by use of the Mueller or Kennaugh matrices, but a simpler treatment that takes advantage of the complete polarization of the incident wave uses the *polarization power scattering matrix* of Graves [19]. The Graves matrix is also applicable to configurations with nonsymmetric Sinclair matrices, such as bistatic scattering or scattering in a nonreciprocal medium with Faraday rotation.

It is convenient to introduce the power scattering matrix, or Graves matrix, by relating it to the scattering matrix, although its existence does not depend on that of the scattering matrix. Accordingly, we write the scattered field, with receiver at distance r, as

$$\mathbf{E}^s = \frac{1}{\sqrt{4\pi}\,r}\mathbf{S}\mathbf{E}^i \qquad (6.79)$$

where $\mathbf{E}^i$ and $\mathbf{E}^s$ are in $x_1 y_1 z_1$ and $x_3 y_3 z_3$ coordinates respectively. For backscattering the two coordinate systems coincide. The power density of the wave scattered to the receiver is

$$\mathscr{P}^s = \frac{1}{2Z_0}\mathbf{E}^{sT*}\mathbf{E}^s = \frac{1}{8\pi Z_0 r^2}\mathbf{E}^{iT*}\mathbf{S}^{T*}\mathbf{S}\mathbf{E}^i = \frac{1}{8\pi Z_0 r^2}\mathbf{E}^{iT*}\boldsymbol{\sigma}\mathbf{E}^i \quad (7.325)$$

where

$$\boldsymbol{\sigma} = \mathbf{S}^{T*}\mathbf{S} \qquad (7.326)$$

The matrix $\boldsymbol{\sigma}$ is the *Graves matrix* or the *polarization power scattering matrix*.

Multiplication gives the matrix elements

$$\sigma = \begin{bmatrix} |S_{xx}|^2 + |S_{yx}|^2 & S_{xx}^* S_{xy} + S_{yx}^* S_{yy} \\ S_{xx} S_{xy}^* + S_{yx} S_{yy}^* & |S_{xy}|^2 + |S_{yy}|^2 \end{bmatrix} \quad (7.327)$$

In this form the matrix is valid for bistatic scattering. For backscattering, **S** is symmetric, and, if the relative scattering matrix with S_{xy} real is used, σ becomes

$$\sigma = \begin{bmatrix} |S_{xx}|^2 + S_{xy}^2 & S_{xy}(S_{xx}^* + S_{yy}) \\ S_{xy}(S_{xx} + S_{yy}^*) & S_{xy}^2 + |S_{yy}|^2 \end{bmatrix} \quad (7.328)$$

Decomposition of the Power Scattering Matrix

Equation (7.325) gives no information about the polarization of the scattered wave and is of limited utility. The power scattering matrix can, however, be decomposed into two parts, σ_H, which leads to the power that can be received by a horizontally (x) polarized receiving antenna, and σ_V, which specifies the power that can be received by a vertically (y) polarized antenna. To effect the decomposition, consider a horizontally polarized receiving antenna. The open-circuit voltage induced in it by the scattered field $\mathbf{E}^s$ is

$$V_H = \mathbf{h}^T \mathbf{E}^s = \frac{1}{\sqrt{4\pi}\, r}[h_x \ 0]\mathbf{S}\mathbf{E}^i = \frac{h_x}{\sqrt{4\pi}\, r}\left(S_{xx} E_x^i + S_{xy} E_y^i \right) \quad (7.329)$$

The received power is proportional to

$$V_H V_H^* = \frac{|h_x|^2}{4\pi r^2}\left[|S_{xx}|^2 |E_x^i|^2 + S_{xx} S_{xy}^* E_x^i E_y^{i*} + S_{xx}^* S_{xy} E_x^{i*} E_y^i + |S_{xy}|^2 |E_y^i|^2 \right] \quad (7.330)$$

which may be written as

$$V_H V_H^* = \frac{|h_x|^2}{4\pi r^2} \mathbf{E}^{iT*} \begin{bmatrix} |S_{xx}|^2 & S_{xx}^* S_{xy} \\ S_{xx} S_{xy}^* & |S_{xy}|^2 \end{bmatrix} \mathbf{E}^i \quad (7.331)$$

The voltage in a vertically (y) polarized antenna, found in a similar manner, is

$$V_V = \frac{h_y}{\sqrt{4\pi}\, r}\left(S_{yx} E_x^i + S_{yy} E_y^i \right) \quad (7.332)$$

which leads to

$$V_V V_V^* = \frac{|h_y|^2}{4\pi r^2} \mathbf{E}^{iT*} \begin{bmatrix} |S_{yx}|^2 & S_{yx}^* S_{yy} \\ S_{yx} S_{yy}^* & |S_{yy}|^2 \end{bmatrix} \mathbf{E}^i \quad (7.333)$$

Now define the 2×2 matrices appearing in these power equations as

$$\boldsymbol{\sigma}_H = \begin{bmatrix} |S_{xx}|^2 & S_{xx}^* S_{xy} \\ S_{xx} S_{xy}^* & |S_{xy}|^2 \end{bmatrix} \quad (7.334a)$$

$$\boldsymbol{\sigma}_V = \begin{bmatrix} |S_{yx}|^2 & S_{yx}^* S_{yy} \\ S_{yx} S_{yy}^* & |S_{yy}|^2 \end{bmatrix} \quad (7.334b)$$

Comparison with (7.327) shows the power-scattering matrix $\boldsymbol{\sigma}$ to be the sum

$$\boldsymbol{\sigma} = \boldsymbol{\sigma}_H + \boldsymbol{\sigma}_V \quad (7.335)$$

Further, the total scattered power density of (7.325) can be written as the sum

$$\mathscr{P}^s = \mathscr{P}^s_H + \mathscr{P}^s_V = \frac{1}{8\pi Z_0 r^2} \mathbf{E}^{iT*} (\boldsymbol{\sigma}_H + \boldsymbol{\sigma}_V) \mathbf{E}^i \quad (7.336)$$

where

$$\mathscr{P}^s_H = \frac{1}{8\pi Z_0 r^2} \mathbf{E}^{iT*} \boldsymbol{\sigma}_H \mathbf{E}^i \quad (7.337)$$

represents the scattered power density that is horizontally (x) polarized, and

$$\mathscr{P}^s_V = \frac{1}{8\pi Z_0 r^2} \mathbf{E}^{iT*} \boldsymbol{\sigma}_V \mathbf{E}^i \quad (7.338)$$

is the scattered power density that is vertically (y) polarized.

The power received by a horizontally polarized receiving antenna with effective length h_x and antenna resistance R_a is, from (7.331),

$$W_H = \frac{V_H V_H^*}{8R_a} = \frac{|h_x|^2}{32\pi R_a r^2} \mathbf{E}^{iT*} \boldsymbol{\sigma}_H \mathbf{E}^i \quad (7.339)$$

and that to the load of a vertically polarized receiver is

$$W_V = \frac{|h_y|^2}{32\pi R_a r^2} \mathbf{E}^{iT*} \boldsymbol{\sigma}_V \mathbf{E}^i \quad (7.340)$$

THE POWER SCATTERING MATRIX   443

The elements of $\boldsymbol{\sigma}_H$ and $\boldsymbol{\sigma}_V$ can be measured, and it is apparent from (7.334) that knowledge of the measured elements is sufficient to allow the elements of **S**, the scattering matrix with relative phase, to be found. On the other hand, **S** cannot be found from measurements of the total power scattering matrix $\boldsymbol{\sigma}$. The relative phase information necessary to find the scattering matrix is contained in off-diagonal terms that appear as sums in $\boldsymbol{\sigma}$ [16]. The decomposition of the power scattering matrix into horizontal and vertical matrices is essential to obtaining the target scattering matrix from measurements of the power scattering matrix elements.

Target Without a Scattering Matrix

The preceding section was based on the target scattering matrix, but all targets of interest (depolarizing targets) do not possess such matrices. It is therefore important to inquire if a target possesses a power scattering matrix if it does not have a scattering matrix.

The power density of the scattered wave is given by (7.325) and also in terms of the Stokes vector of the incident wave and Mueller matrix of the target,

$$\mathbf{G}^s = \frac{1}{4\pi r^2} \mathbf{M} \mathbf{G}^i \qquad (7.200)$$

The scattered power density in terms of the scattered Stokes vector is

$$\mathscr{P}^s = \frac{G_0^s}{2Z_0} = \frac{1}{8\pi Z_0 r^2}\left(M_{11}G_0^i + M_{12}G_1^i + M_{13}G_2^i + M_{14}G_3^i\right) \quad (7.341)$$

Multiplying (7.325) and equating it to (7.341) specialized to the monochromatic equivalent of a completely polarized wave leads to

$$\sigma_{xx}|E_x^i|^2 + \sigma_{xy}E_x^{i*}E_y^i + \sigma_{yx}E_x^iE_y^{i*} + \sigma_{yy}|E_y^i|^2$$
$$= M_{11}\left(|E_x^i|^2 + |E_y^i|^2\right) + M_{12}\left(|E_x^i|^2 - |E_y^i|^2\right)$$
$$+ M_{13}\left(E_x^{i*}E_y^i + E_x^iE_y^{i*}\right) - jM_{14}\left(E_x^{i*}E_y^i - E_x^iE_y^{i*}\right) \quad (7.342)$$

Examination of this equation shows the power scattering matrix in terms of the Mueller matrix elements to be

$$\boldsymbol{\sigma} = \begin{bmatrix} M_{11} + M_{12} & M_{13} - jM_{14} \\ M_{13} + jM_{14} & M_{11} - M_{12} \end{bmatrix} \qquad (7.343)$$

In essence, this is a definition of the polarization power scattering matrix for

444 PARTIAL POLARIZATION

targets that do not have a scattering matrix. It may be used, however, only if the incident wave is completely polarized and if the matrix elements are such that the reflected wave is quasi-monochromatic.

Power Scattering Matrix as a Time Average

Consider a radar target composed of subscatterers whose relative positions vary with time. If the scatterer positions vary slowly relative to a period of the wave, as is often the case, we may consider it to have a scattering matrix that is a function of time, $\mathbf{S}(t)$. If the target is illuminated by a monochromatic wave, the scattered field is

$$\mathbf{E}^s(t) = \frac{1}{\sqrt{4\pi}\, r} \mathbf{S}(t) \mathbf{E}^i \tag{7.344}$$

The Poynting vector of the reflected wave is

$$\mathscr{P}(t) = \frac{1}{2Z_0} \mathbf{E}^{sT*}(t) \mathbf{E}^s(t) \tag{7.345}$$

Now in this form of the Poynting vector, a time average has already been taken, but we recognize that it has been taken over a time interval comparable to a period of the illuminating wave, which is very short compared to the time intervals we are considering here. Then the "time-average" Poynting vector is properly considered to be a function of time.

Combining the last two equations yields

$$\mathscr{P}(t) = \frac{1}{8\pi Z_0 r^2} \mathbf{E}^{iT*} \mathbf{S}^{T*}(t) \mathbf{S}(t) \mathbf{E}^i \tag{7.346}$$

If the measuring instrument responds slowly on a time scale set by the target motion, it is appropriate to obtain a time average of this Poynting vector. In doing so, note that the incident field (in its analytic signal representation) is independent of time, and therefore

$$\mathscr{P}_{av} = \langle \mathscr{P}(t) \rangle = \frac{1}{8\pi Z_0 r^2} \mathbf{E}^{iT*} \langle \mathbf{S}^{T*}(t) \mathbf{S}(t) \rangle \mathbf{E}^i \tag{7.347}$$

If this is compared to (7.325) it is seen that the power scattering matrix for a target whose scatterers move slowly compared to a wave period but rapidly compared to the time-averaging period of the receiver is

$$\boldsymbol{\sigma} = \langle \mathbf{S}^{T*}(t) \mathbf{S}(t) \rangle \tag{7.348}$$

Polarization for Maximum Power Density

In Section 6.8 the polarization of an antenna of a monostatic radar that maximizes the received power was found. Let us now determine the polarization of the transmitting antenna that maximizes the scattered power *density* from a target that is described by its power-scattering matrix $\boldsymbol{\sigma}$.

It is seen from (7.325) that the scattered power density is maximum if

$$F = \hat{\mathbf{h}}^{T*} \boldsymbol{\sigma} \hat{\mathbf{h}} \tag{7.349}$$

is maximized by choosing the normalized effective length $\hat{\mathbf{h}}$ of the transmitting antenna properly, subject to the constraint

$$\hat{\mathbf{h}}^{T*} \hat{\mathbf{h}} = 1 \tag{7.350}$$

To assist in maximizing F, define a new effective length $\mathbf{H}$ by

$$\hat{\mathbf{h}} = \mathbf{R}\mathbf{H} \tag{7.351}$$

where $\mathbf{R}$ is 2×2 with properties to be discussed later. Then

$$F = \mathbf{H}^{T*} \mathbf{R}^{T*} \boldsymbol{\sigma} \mathbf{R} \mathbf{H} \tag{7.352}$$

Now (7.343) shows $\boldsymbol{\sigma}$ to be Hermitian. For a Hermitian matrix a unitary matrix exists that diagonalizes the matrix [10]. Let $\mathbf{R}$ be such a matrix. Then

$$\mathbf{R}^{T*} \boldsymbol{\sigma} \mathbf{R} = \text{diag}(\lambda_1, \lambda_2, \ldots, \lambda_N) \tag{7.353}$$

where the λ_i are eigenvalues of $\boldsymbol{\sigma}$. Then for the 2×2 polarization power scattering matrix,

$$\mathbf{R}^{T*} \boldsymbol{\sigma} \mathbf{R} = \begin{bmatrix} \lambda_1 & 0 \\ 0 & \lambda_2 \end{bmatrix} \tag{7.354}$$

The columns of $\mathbf{R}$ are the eigenvectors of $\boldsymbol{\sigma}$.

The eigenvalues of $\boldsymbol{\sigma}$ are found from

$$\|\boldsymbol{\sigma} - \lambda \mathbf{I}\| = \begin{vmatrix} \sigma_{xx} - \lambda & \sigma_{xy} \\ \sigma_{xy}^* & \sigma_{yy} - \lambda \end{vmatrix} = 0 \tag{7.355}$$

which is a quadratic equation with solutions

$$\lambda_1 = \tfrac{1}{2}(\sigma_{xx} + \sigma_{yy}) + \tfrac{1}{2}\sqrt{(\sigma_{xx} - \sigma_{yy})^2 + 4|\sigma_{xy}|^2} \tag{7.356a}$$

$$\lambda_2 = \tfrac{1}{2}(\sigma_{xx} + \sigma_{yy}) - \tfrac{1}{2}\sqrt{(\sigma_{xx} - \sigma_{yy})^2 + 4|\sigma_{xy}|^2} \tag{7.356b}$$

Note that λ_1 has been chosen to be the larger eigenvalue.

The eigenvectors of $\boldsymbol{\sigma}$ are found from

$$(\boldsymbol{\sigma} - \lambda_1 \mathbf{I})\mathbf{e}_1 = \mathbf{0} \tag{7.357a}$$
$$(\boldsymbol{\sigma} - \lambda_2 \mathbf{I})\mathbf{e}_2 = \mathbf{0} \tag{7.357b}$$

which lead to the equations

$$e_{1y} = -\frac{\sigma_{xx} - \lambda_1}{\sigma_{xy}} e_{1x} \tag{7.358a}$$

$$e_{2y} = -\frac{\sigma_{xx} - \lambda_2}{\sigma_{xy}} e_{2x} \tag{7.358b}$$

The unitary matrix is formed from the eigenvectors as

$$\mathbf{R} = [\mathbf{e}_1 \quad \mathbf{e}_2] = \begin{bmatrix} e_{1x} & e_{2x} \\ e_{1y} & e_{2y} \end{bmatrix} \tag{7.359}$$

Next, F is maximized, using

$$F = \mathbf{H}^{T*}\mathbf{R}^{T*}\boldsymbol{\sigma}\mathbf{R}\mathbf{H} = [H_\alpha^* \quad H_\beta^*] \begin{bmatrix} \lambda_1 & 0 \\ 0 & \lambda_2 \end{bmatrix} \begin{bmatrix} H_\alpha \\ H_\beta \end{bmatrix} = |H_\alpha|^2 \lambda_1 + |H_\beta|^2 \lambda_2 \tag{7.360}$$

where α and β are the basis vectors of $\mathbf{H}$. Now if

$$\hat{\mathbf{h}}^{T*}\hat{\mathbf{h}} = 1 \tag{7.350}$$

and $\mathbf{R}$ is unitary, it is easily shown that $\mathbf{H}$ in

$$\hat{\mathbf{h}} = \mathbf{RH} \tag{7.351}$$

also satisfies

$$\mathbf{H}^{T*}\mathbf{H} = |H_\alpha|^2 + |H_\beta|^2 = 1 \tag{7.361}$$

Since in (7.360), λ_1 is the larger eigenvalue, F is maximum if

$$|H_\alpha|^2 = 1 \quad |H_\beta|^2 = 0$$

giving

$$F_{\max} = \lambda_1 \tag{7.362}$$

If (7.349) is substituted into (7.325), using the maximum value of F, and if the relationship between the incident field and the effective length of a transmitting antenna at distance r_1 from the target is used, then the maximum power density at distance r_2 from the target is

$$\mathscr{P}^s = \frac{Z_0 \lambda_1 |\mathbf{h}|^2 I^2}{32\pi \lambda^2 r_1^2 r_2^2} \tag{7.363}$$

With the choice of **H** made to maximize F, the polarization state corresponding to maximum power density is

$$\hat{\mathbf{h}}_{opt} = \mathbf{RH} = \begin{bmatrix} e_{1x} & e_{2x} \\ e_{1y} & e_{2y} \end{bmatrix} \begin{bmatrix} H_\alpha \\ 0 \end{bmatrix} = e_{1x} H_\alpha \begin{bmatrix} 1 \\ -\dfrac{\sigma_{xx} - \lambda_1}{\sigma_{xy}} \end{bmatrix} \tag{7.364}$$

and the transmitting antenna polarization ratio that maximizes the scattered power density is

$$P_{opt} = -\frac{\sigma_{xx} - \lambda_1}{\sigma_{xy}} \tag{7.365}$$

A polarization efficiency can be defined in terms of power densities. For a transmitting antenna of length **h**, the ratio of the scattered power density at any point in the field to the maximum density is

$$\rho_s = \frac{|\mathbf{h}^{T*} \sigma \mathbf{h}|}{|\mathbf{h}|^2 \lambda_1} \tag{7.366}$$

In terms of the polarization ratio of the transmitting antenna, the efficiency is

$$\rho_s = \frac{|\sigma_{xx} + 2\,\mathrm{Re}(\sigma_{xy} P) + \sigma_{yy}|P|^2|}{\lambda_1 (1 + |P|^2)} \tag{7.367}$$

If the target has scattering matrix **S**, it is of interest to see if the optimum polarization for maximum power density is related to the optimum polarization found in Chapter 6 for maximum received power to a monostatic radar using the same antenna for transmitting and receiving. If the elements of σ from (7.327) are substituted into the eigenvalue equations (7.356) they become

$$\lambda_1, \lambda_2 = |\gamma_1|^2, |\gamma_2|^2 \tag{7.368}$$

where $|\gamma_1|^2$ and $|\gamma_2|^2$ are eigenvalues found in maximizing the received power (copolarized power) for a radar using the same antenna for transmitting and receiving. They are given by

$$|\gamma_1|^2 = \frac{B}{2} + \frac{1}{2}\sqrt{B^2 - 4C} \qquad (6.147a)$$

$$|\gamma_2|^2 = \frac{B}{2} - \frac{1}{2}\sqrt{B^2 - 4C} \qquad (6.147b)$$

with

$$B = |S_{xx}|^2 + 2|S_{xy}|^2 + |S_{yy}|^2 \qquad (6.142a)$$

$$C = |S_{xx}S_{yy}|^2 + |S_{xy}|^4 - 2\operatorname{Re}(S_{xx}S_{xy}^{*2}S_{yy}) \qquad (6.142b)$$

Corresponding to $|\gamma_1|^2$ and $|\gamma_2|^2$ are antenna polarization ratios P_1 and P_2 that yield maximum copolarized backscattered power proportional to $|\gamma_1|^2$ (for P_1) and a submaximum power proportional to $|\gamma_2|^2$ (for P_2). Substitution shows that the optimum polarization for maximum power density, (7.365), becomes that for maximum monostatic received power, (6.156), for a target with Sinclair matrix **S**, thus

$$P_{\text{opt}} = -\frac{\sigma_{xx} - \lambda_1}{\sigma_{xy}} = P_1 = -\frac{1}{S_{xy}} \frac{|S_{xx}|^2 + S_{xy}^2 - |\gamma_1|^2}{S_{xx}^* + S_{yy}} \qquad (7.369)$$

where the relative scattering matrix, with S_{xy} real, is used. The polarization ratio giving a suboptimum power density likewise is the same as that giving a submaximum monostatic radar received power, or

$$-\frac{\sigma_{xx} - \lambda_2}{\sigma_{xy}} = P_2 = -\frac{1}{S_{xy}} \frac{|S_{xx}|^2 + S_{xy}^2 - |\gamma_2|^2}{S_{xx}^* + S_{yy}} \qquad (7.370)$$

The antenna polarization that gives maximum copolarized power in a monostatic radar is the same as that which produces maximum power density in the scattered field. It will be recalled from Chapter 6 that the monostatic radar antenna polarization that gives maximum copolarized power simultaneously gives zero crosspolarized power. The co-pol maxima coincide with the cross-pol nulls. It is then not surprising that the polarization ratio P_{opt} giving maximum scattered power density also gives maximum copolarized power.

Maximum Horizontal and Vertical Power Densities

The total scattered power density could be maximized because σ is Hermitian. Now σ_H and σ_V are Hermitian also. Then a transmitting antenna polarization can be selected to maximize either

$$\mathcal{P}_H^s = \frac{1}{8\pi Z_0 r^2} \mathbf{E}^{iT*} \boldsymbol{\sigma}_H \mathbf{E}^i \qquad (7.337)$$

or

$$\mathcal{P}_V^s = \frac{1}{8\pi Z_0 r^2} \mathbf{E}^{iT*} \boldsymbol{\sigma}_V \mathbf{E}^i \qquad (7.338)$$

the scattered power densities that are horizontally and vertically polarized, respectively. The selection process parallels that used to maximize the total power density. Equations (7.356) for the eigenvalues and (7.365) for the optimum antenna polarization ratio are valid if the matrix elements used in them are taken from $\boldsymbol{\sigma}_H$ and $\boldsymbol{\sigma}_V$ as desired. In general, the polarizations to maximize horizontal and vertical scattered power densities are not the same, nor is one of them equal to the polarization to maximize the total scattered power density.

Polarization for Minimum Power Density

The scattered power density polarization efficiency has been found to be, in terms of the transmitting antenna polarization ratio and the Graves matrix elements of the target,

$$\rho_s = \frac{|\sigma_{xx} + 2\operatorname{Re}(\sigma_{xy} P) + \sigma_{yy}|P|^2|}{\lambda_1 (1 + |P|^2)} \qquad (7.367)$$

On physical grounds we recognize that it may not be possible to select P to give a null in ρ_s in a specified direction. For example, a flat plate illuminated from a direction perpendicular to the plate has a nonzero backscattered power density regardless of the incident wave's polarization. The scattered power density can, however, be minimized for many targets. Consider a target for which no null of ρ_s exists in a specified direction.

In (7.367), σ_{xx} and σ_{yy} are positive real, and ρ_s is therefore minimized by choosing the phase angle of P so that

$$\operatorname{Re}(\sigma_{xy} P) = -|\sigma_{xy}||P| \qquad (7.371)$$

giving a minimized efficiency

$$\rho_{sm} = \frac{\sigma_{xx} - 2|\sigma_{xy}||P| + \sigma_{yy}|P|^2}{\lambda_1 (1 + |P|^2)} \qquad (7.372)$$

where by hypothesis the numerator is nonnegative.

If this minimized efficiency is differentiated with respect to $|P|$ and the derivative set to zero, the resulting equation is

$$|\sigma_{xy}||P|^2 + (\sigma_{yy} - \sigma_{xx})|P| - |\sigma_{xy}| = 0 \qquad (7.373)$$

It has one positive real root. Further, the slope of ρ_{sm} is negative at $|P| = 0$, so the root corresponds to a minimum of ρ_{sm}. The root, together with the phase angle found from (7.371), determines the value of P, the transmitting antenna polarization ratio, that minimizes the scattered power density from a target.

In general, the polarization for minimum scattered power density does not coincide with the polarization for minimum copolarized received power from the same target (the co-pol nulls). The phase angle of P, from (7.371), is clearly not the same as the angle of the co-pol null polarizations, P_3 and P_4, found from (6.170), so P is not equal to either of those polarizations.

In certain directions from the target the power matrix elements may be such that, for some values of P, the scattered power density is zero. An example is a flat plate illuminated from broadside, with scattering in other directions considered (thus requiring consideration of the bistatic power scattering matrix). If ρ_s in (7.367) is set to zero, a unique solution for P may not exist. For a solution to exist, $\text{Re}(\sigma_{xy}P)$ must be negative and with a magnitude large enough to make the sum of the numerator terms nonpositive. For sufficiently large $|\sigma_{xy}|$ a range of angles of P will satisfy this constraint. Then a value of $|P|$ can be chosen for a specified angle of P to make ρ_s zero.

In summary, a null polarization for the scattered power density may not exist. If it does not, a polarization for minimum scattered power density may exist and be found. If a null polarization exists, it may not be unique.

Polarization Designations

The polarization state $\hat{\mathbf{h}}_1$ corresponding to the eigenvalue λ_1 of the Graves matrix $\boldsymbol{\sigma}$ is an eigenvector of $\boldsymbol{\sigma}$ and can appropriately be called the *maximum-power-density eigenvector*. The eigenvector $\hat{\mathbf{h}}_2$, corresponding to λ_2, in general leads to a lesser power density than $\hat{\mathbf{h}}_1$ since λ_2 was chosen to be smaller than λ_1 (they are equal only for specialized targets). The scattered power density corresponding to $\hat{\mathbf{h}}_2$ is a suboptimum or submaximum value, and other values of $\hat{\mathbf{h}}$ may yield smaller power densities, as discussed in the immediately preceding subsection. The values of $\hat{\mathbf{h}}$ giving smaller power density are not readily found by methods directly applicable to all targets, however. For this reason, while $\hat{\mathbf{h}}_2$ is properly referred to as a *suboptimum power-density eigenvector*, it can also be called the *minimum-power-density eigenvector*.

7.11. TARGET COHERENCY AND COVARIANCE MATRICES

The Mueller and Kennaugh matrices are not the only matrices able to characterize a target whose polarization varies because of its motion. In this section, we shall examine two other matrices that contain the same information about a time-varying target as the average Mueller matrix. Both are developed from the conventional Sinclair matrix S and as a consequence are expressed in terms of the $x_1 y_1 z_1$ and $x_3 y_3 z_3$ coordinates used in this text with simplification for backscattering.

The Coherency Matrix

The target scattering matrix

$$\mathbf{S} = \begin{bmatrix} S_{xx} & S_{xy} \\ S_{yx} & S_{yy} \end{bmatrix} \tag{7.1}$$

can be arranged in a convenient four-vector form by the use of the 2×2 identity matrix and the Pauli spin matrices [20],

$$\begin{aligned}\boldsymbol{\sigma}_0 &= \begin{bmatrix} 1 & 0 \\ 0 & 1 \end{bmatrix} & \boldsymbol{\sigma}_1 &= \begin{bmatrix} 1 & 0 \\ 0 & -1 \end{bmatrix} \\ \boldsymbol{\sigma}_2 &= \begin{bmatrix} 0 & 1 \\ 1 & 0 \end{bmatrix} & \boldsymbol{\sigma}_3 &= \begin{bmatrix} 0 & -j \\ j & 0 \end{bmatrix}\end{aligned} \tag{7.374}$$

Next a *target vector* **k** may be defined as

$$\mathbf{k} = \begin{bmatrix} k_0 & k_1 & k_2 & k_3 \end{bmatrix}^\mathrm{T} \tag{7.375}$$

where the elements are

$$k_i = \mathrm{Tr}(\mathbf{S}\boldsymbol{\sigma}_i) \quad i = 0, 1, 2, 3 \tag{7.376}$$

If this trace operation is carried out, the relation between the target vector and the Sinclair matrix is found to be

$$\mathbf{k} = \frac{1}{2} \begin{bmatrix} S_{xx} + S_{yy} \\ S_{xx} - S_{yy} \\ S_{xy} + S_{yx} \\ j(S_{xy} - S_{yx}) \end{bmatrix} \tag{7.377}$$

$$\mathbf{S} = \begin{bmatrix} k_0 + k_1 & k_2 - jk_3 \\ k_2 + jk_3 & k_0 - k_1 \end{bmatrix} \tag{7.378}$$

Note that while **S** appears to be Hermitian, it is not.

452 PARTIAL POLARIZATION

The *target coherency matrix** is defined as [21]

$$\mathbf{L} = \langle \mathbf{k}\mathbf{k}^{T*} \rangle \qquad (7.379)$$

where $\langle \ldots \rangle$ denotes the time average. Performing the matrix multiplication indicated leads to

$$\mathbf{L} = \left\langle \begin{bmatrix} |k_0|^2 & k_0 k_1^* & k_0 k_2^* & k_0 k_3^* \\ k_1 k_0^* & |k_1|^2 & k_1 k_2^* & k_1 k_3^* \\ k_2 k_0^* & k_2 k_1^* & |k_2|^2 & k_2 k_3^* \\ k_3 k_0^* & k_3 k_1^* & k_3 k_2^* & |k_3|^2 \end{bmatrix} \right\rangle \qquad (7.380)$$

If the k_i in terms of the scattering matrix elements are used in this equation for **L** it is apparent that the elements of the average Mueller matrix or the Kennaugh matrix may be found from the elements of **L**, and vice versa. The matrices then contain equivalent information about the target. The coherency matrix has been used in studies of target decomposition into the sum of simpler targets [21].

The Covariance Matrix

It is obvious that the Pauli matrices are not the only ones that can be used to construct a target vector from the scattering matrix. If the simple set of basis matrices

$$\boldsymbol{\gamma}_0 = \begin{bmatrix} 1 & 0 \\ 0 & 0 \end{bmatrix} \quad \boldsymbol{\gamma}_1 = \begin{bmatrix} 0 & 1 \\ 0 & 0 \end{bmatrix}$$

$$\boldsymbol{\gamma}_2 = \begin{bmatrix} 0 & 0 \\ 1 & 0 \end{bmatrix} \quad \boldsymbol{\gamma}_3 = \begin{bmatrix} 0 & 0 \\ 0 & 1 \end{bmatrix} \qquad (7.381)$$

is used, a target vector **X** can be formed as

$$\mathbf{X} = \begin{bmatrix} X_0 & X_1 & X_2 & X_3 \end{bmatrix}^T \qquad (7.382)$$

where

$$X_i = \mathrm{Tr}(\mathbf{S}\boldsymbol{\gamma}_i) \qquad i = 1, 2, 3, 4 \qquad (7.383)$$

---

*There should be no confusion between this matrix for a target and the coherency matrix or coherency vector for a wave or an antenna.

Then **X** and **S** are related by

$$\mathbf{X} = \begin{bmatrix} S_{xx} & S_{yx} & S_{xy} & S_{yy} \end{bmatrix}^T \qquad (7.384)$$

$$\mathbf{S} = \begin{bmatrix} X_0 & X_2 \\ X_1 & X_3 \end{bmatrix} \qquad (7.385)$$

The *target covariance matrix* is defined by [21]

$$\mathbf{\Sigma} = \langle \mathbf{X}\mathbf{X}^{T*} \rangle \qquad (7.386)$$

and it may be written in terms of the scattering matrix elements as

$$\mathbf{\Sigma} = \left\langle \begin{bmatrix} |S_{xx}|^2 & S_{xx}S_{yx}^* & S_{xx}S_{xy}^* & S_{xx}S_{yy}^* \\ S_{yx}S_{xx}^* & |S_{yx}|^2 & S_{yx}S_{xy}^* & S_{yx}S_{yy}^* \\ S_{xy}S_{xx}^* & S_{xy}S_{yx}^* & |S_{xy}|^2 & S_{xy}S_{yy}^* \\ S_{yy}S_{xx}^* & S_{yy}S_{yx}^* & S_{yy}S_{xy}^* & |S_{yy}|^2 \end{bmatrix} \right\rangle \qquad (7.387)$$

This target covariance matrix is equivalent to the Kennaugh matrix or the average Mueller matrix in that they contain the same target information.

Covariance Matrix for Backscattering

The target scattering matrix is symmetric for backscattering, and a target three-vector can be formed using only

$$\mathbf{X} = \begin{bmatrix} X_0 & X_2 & X_3 \end{bmatrix}^T = \begin{bmatrix} S_{xx} & S_{xy} & S_{yy} \end{bmatrix}^T \qquad (7.388)$$

The backscattering covariance matrix can then be formed as

$$\mathbf{\Sigma} = \langle \mathbf{X}\mathbf{X}^{T*} \rangle = \left\langle \begin{bmatrix} |S_{xx}|^2 & S_{xx}S_{xy}^* & S_{xx}S_{yy}^* \\ S_{xy}S_{xx}^* & |S_{xy}|^2 & S_{xy}S_{yy}^* \\ S_{yy}S_{xx}^* & S_{yy}S_{xy}^* & |S_{yy}|^2 \end{bmatrix} \right\rangle \qquad (7.389)$$

Received Power

Consider the power received after scattering by a target with known covariance matrix [21]. Assume the incident wave is completely polarized and consider only backscattering. The open-circuit received voltage in a receiving

454 PARTIAL POLARIZATION

antenna of effective length $\mathbf{h}^r$ is

$$V = \frac{jZ_0 I}{\sqrt{4\pi(2\lambda r^2)}} \mathbf{h}^{rT} \mathbf{S} \mathbf{h}^t \qquad (7.390)$$

where $\mathbf{h}^t$ is the effective length of the transmitting antenna and I is its feed current.

Assume next that this open-circuit voltage can also be expressed in terms of the target vector $\mathbf{X}$ by

$$V = \frac{jZ_0 I}{\sqrt{4\pi(2\lambda r^2)}} \mathbf{N}^{T*} \mathbf{X} \qquad (7.391)$$

where $\mathbf{N}$ is a three-vector

$$\mathbf{N} = \begin{bmatrix} N_{xx} & N_{xy} & N_{yy} \end{bmatrix}^T \qquad (7.392)$$

If the two forms for V are equated and the matrix multiplication is performed, the result is

$$h_x^r S_{xx} h_x^t + h_x^r S_{xy} h_y^t + h_y^r S_{xy} h_x^t + h_y^r S_{yy} h_y^t = N_{xx}^* S_{xx} + N_{xy}^* S_{xy} + N_{yy}^* S_{yy} \qquad (7.393)$$

It follows from this equation that

$$N_{xx}^* = h_x^r h_x^t \qquad (7.394a)$$
$$N_{xy}^* = h_x^r h_y^t + h_y^r h_x^t \qquad (7.394b)$$
$$N_{yy}^* = h_y^r h_y^t \qquad (7.394c)$$

The received power may be found from

$$W = \frac{1}{8R_a} \langle VV^* \rangle = \frac{Z_0^2 I^2}{64\pi R_a \lambda^2 r^4} \langle \mathbf{N}^{T*} \mathbf{X} \mathbf{N}^T \mathbf{X}^* \rangle = \frac{Z_0^2 I^2}{64\pi R_a \lambda^2 r^4} \langle \mathbf{N}^{T*} \mathbf{X} \mathbf{X}^{T*} \mathbf{N} \rangle \qquad (7.395)$$

where R_a is the resistance of the receiving antenna. Since $\mathbf{N}$ is time-invariant,

$$W = \frac{Z_0^2 I^2}{64\pi R_a \lambda^2 r^4} \mathbf{N}^{T*} \langle \mathbf{X} \mathbf{X}^{T*} \rangle \mathbf{N} = \frac{Z_0^2 I^2}{64\pi R_a \lambda^2 r^4} \mathbf{N}^{T*} \mathbf{\Sigma} \mathbf{N} \qquad (7.396)$$

This is a relatively easy form to use in finding the received power backscattered by a fluctuating target. Note that it is restricted to an incident wave that is completely polarized.

7.12. CHARACTERISTIC POLARIZATIONS WITH NONSYMMETRIC SCATTERING MATRIX*

In Section 6.8 the antenna polarizations were determined that give maximum and minimum received powers for a radar operating in a reciprocal medium (without Faraday rotation) for which the target scattering matrix is symmetric. It was shown there that, with a symmetric Sinclair matrix, the use of one antenna for transmitting and receiving is sufficient to give maximum received power, and no greater power can be received with separate antennas for transmitting and receiving. In Section 7.10, the concern was not for received power but for power density at the receiver, and a transmitter polarization was found to maximize that power density. It was also shown in that section that the polarizations for maximum power density and maximum received power were the same for a target with a symmetric matrix.

Consider now a target with a nonsymmetric Sinclair matrix. No matter what the polarization of the wave scattered from it, a receiving antenna can always be found, using equations of Chapter 4, to maximize the received power from the wave. To put this another way, the received power, with a properly chosen receiving antenna, depends *only* on the power density of the scattered wave (assuming a receiving antenna of unit length). Then, in order to maximize the received power, a transmitting polarization should be chosen to maximize the scattered power density and a receiving antenna selected to extract maximum power from the scattered wave. No other antenna, or pair of antennas, will result in greater received power. The first part of that prescription, maximizing power density at the receiver, was considered in Section 7.10 using the Graves matrix. It is summarized here.

The Graves polarization power scattering matrix is

$$\boldsymbol{\sigma} = \mathbf{S}^{T*}\mathbf{S} \qquad (7.326)$$

where $\mathbf{S}$ is the target Sinclair matrix. If $\boldsymbol{\sigma}$ is expanded as

$$\boldsymbol{\sigma} = \begin{bmatrix} |S_{xx}|^2 + |S_{yx}|^2 & S_{xx}^* S_{xy} + S_{yx}^* S_{yy} \\ S_{xx} S_{xy}^* + S_{yx} S_{yy}^* & |S_{xy}|^2 + |S_{yy}|^2 \end{bmatrix} \qquad (7.327)$$

it is seen to be Hermitian, whether $\mathbf{S}$ is symmetric or not. The eigenvalues found in Section 7.10 are therefore unchanged,

$$\lambda_1 = \tfrac{1}{2}(\sigma_{xx} + \sigma_{yy}) + \tfrac{1}{2}\sqrt{(\sigma_{xx} - \sigma_{yy})^2 + 4|\sigma_{xy}|^2} \qquad (7.356a)$$

$$\lambda_2 = \tfrac{1}{2}(\sigma_{xx} + \sigma_{yy}) - \tfrac{1}{2}\sqrt{(\sigma_{xx} - \sigma_{yy})^2 + 4|\sigma_{xy}|^2} \qquad (7.356b)$$

---

*This section has been placed here rather than with other discussions of characteristic polarizations in Chapter 6 in order to use the developments of the Graves polarization power-scattering matrix of Section 7.10.

456 PARTIAL POLARIZATION

and the transmitting antenna polarization ratio for maximum power density at the receiver, which need not be colocated with the transmitter, adapted from (7.365), is

$$P_{t1} = -\frac{\sigma_{xx} - \lambda_1}{\sigma_{xy}} \tag{7.397}$$

A suboptimum polarization ratio corresponds to λ_2 and is

$$P_{t2} = -\frac{\sigma_{xx} - \lambda_2}{\sigma_{xy}} \tag{7.398}$$

Substitution shows that P_{t1} and P_{t2} describe orthogonal polarizations even if the Sinclair matrix **S** is not symmetric.

If **S** is symmetric, it was shown in Section 7.10 that the eigenvalues λ_1 and λ_2 reduce to the eigenvalues that give maximum received power for one antenna transmitting and receiving; that is,

$$\lambda_1, \lambda_2 = |\gamma_1|^2, |\gamma_2|^2 \tag{7.368}$$

$$P_{t1}, P_{t2} = P_1, P_2 \tag{7.399}$$

where P_1 and P_2 are polarization ratios developed in Section 6.8 that give maximum received backscattered power with symmetric Sinclair matrix and the same antenna (or identical antennas) transmitting and receiving.

After the transmitting antenna polarization to yield maximum power density at a receiver is found, the receiving antenna polarization to extract maximum power from the scattered wave can be determined by using a development of Chapter 4. The incident field at the target, when a transmitting antenna of effective length

$$\mathbf{h}_{t1} = \frac{|\mathbf{h}_{t1}|}{\sqrt{1 + |P_{t1}|^2}} \begin{bmatrix} 1 \\ P_{t1} \end{bmatrix} \tag{7.400}$$

is used, is

$$\mathbf{E}_1^i = \frac{jZ_0 I}{2\lambda r_1} \mathbf{h}_{t1} \tag{7.401}$$

where I is the current in the transmitting antenna, λ the wavelength, and r_1 the distance from transmitter to target. Then the scattered field at the receiver, which may not be colocated with the transmitter, is

$$\mathbf{E}_1^s = \frac{1}{\sqrt{4\pi} r_2} \mathbf{S} \mathbf{E}_1^i \tag{7.402}$$

with r_2 the target–receiver distance.

CHARACTERISTIC POLARIZATIONS WITH NONSYMMETRIC SCATTERING MATRIX   457

In Chapter 4, the polarization state of a receiving antenna to maximize received power with a given incident wave was developed and expressed in (4.43). That equation is satisfied by an effective length of the receiving antenna

$$\mathbf{h}_{r1} = C(\mathbf{E}_1^s)^* = C_1(\mathbf{Sh}_{t1})^*$$

where C and C_1 are constants. The effective length to receive maximum power may then be written as

$$\mathbf{h}_{r1} = |\mathbf{h}_{r1}| \frac{\mathbf{S}^* \mathbf{h}_{t1}^*}{|\mathbf{Sh}_{t1}|} \qquad (7.403)$$

If the suboptimum polarization ratio is used instead of P_{t1}, the effective length of the optimum receiving antenna is

$$\mathbf{h}_{r2} = |\mathbf{h}_{r2}| \frac{\mathbf{S}^* \mathbf{h}_{t2}^*}{|\mathbf{Sh}_{t2}|} \qquad (7.404)$$

Recall that $\mathbf{h}_{t2}$ corresponds to the eigenvalue of the Graves matrix that yields a smaller power density than that obtained with $\mathbf{h}_{t1}$. It is sometimes convenient, therefore, to refer to it as the minimum-power-density effective length in contrast to the maximum-power-density effective length $\mathbf{h}_{t1}$. It must be remembered that $\mathbf{h}_{t2}$ may not lead to the smallest power density, but it is the smallest readily determined in all cases.

It was noted earlier that the transmitting polarizations are orthogonal, that is,

$$\hat{\mathbf{h}}_{t1}^T \hat{\mathbf{h}}_{t2}^* = 0 \qquad (7.405a)$$

$$P_{t2} = -\frac{1}{P_{t1}^*} \qquad (7.405b)$$

To see if $\hat{\mathbf{h}}_{r1}$ and $\hat{\mathbf{h}}_{r2}$ are also orthogonal, form the product

$$\hat{\mathbf{h}}_{r1}^T \hat{\mathbf{h}}_{r2}^* = C \hat{\mathbf{h}}_{t1}^{T*} \mathbf{S}^{T*} \mathbf{S} \hat{\mathbf{h}}_{t2} = C' \begin{bmatrix} 1 & P_{t1}^* \end{bmatrix} \boldsymbol{\sigma} \begin{bmatrix} 1 & P_{t2} \end{bmatrix}^T$$

where C and C' are constants. Multiplying and making use of the Hermitian property of $\boldsymbol{\sigma}$ and the orthogonality relationship of P_{t1} and P_{t2} gives the product

$$\hat{\mathbf{h}}_{r1}^T \hat{\mathbf{h}}_{r2}^* = 0 \qquad (7.406)$$

Since the transmitting polarizations leading to a maximum power density, $\hat{\mathbf{h}}_{t1}$, and the suboptimum, $\hat{\mathbf{h}}_{t2}$, are orthogonal, the receiving antenna polarizations for extracting maximum power are also orthogonal.

The receiving antenna polarizations $\hat{\mathbf{h}}_{r1}$ and $\hat{\mathbf{h}}_{r2}$ are analogous to the polarization states $\hat{\mathbf{h}}_1$ and $\hat{\mathbf{h}}_2$ discussed in Section 6.8 that maximize received power when only one antenna is used in a configuration with a symmetric Sinclair matrix. Those maximum power polarization states also make zero the cross-polarized power and are called cross-polarization nulls. Then $\hat{\mathbf{h}}_{r1}$ and $\hat{\mathbf{h}}_{r2}$ for the nonsymmetric Sinclair matrix case can properly be called cross-polarization receiving nulls. It must be kept in mind that their value depends on the transmitted polarizations, which are selected to maximize power density at the receiver, in the case of $\hat{\mathbf{h}}_{t1}$, or to "minimize" it (in the sense of selecting in a straightforward manner a transmitter polarization to give a smaller power than the maximum), in the case of $\hat{\mathbf{h}}_{t2}$.

With the transmitting antenna polarization selected to maximize received power density, with polarization ratio P_{t1}, or its suboptimum value P_{t2}, a receiving antenna polarization can be chosen to make received power zero. To do so, set the received voltage to zero,

$$\mathbf{h}_r^T \mathbf{S} \mathbf{h}_t = 0$$

where the effective length $\mathbf{h}_t$ can take on subscripts 1 and 2. The corresponding values of $\mathbf{h}_r$ are $\mathbf{h}_{r3}$ and $\mathbf{h}_{r4}$. This equation can be written in terms of polarization ratios as

$$\begin{bmatrix} 1 & P_r \end{bmatrix} \mathbf{S} \begin{bmatrix} 1 & P_t \end{bmatrix}^T = 0$$

and if it is multiplied and solved for P_r, the two zero-power polarizations for the receiving antenna are

$$P_{r3} = -\frac{S_{xx} + S_{xy} P_{t1}}{S_{yx} + S_{yy} P_{t1}} \tag{7.407a}$$

$$P_{r4} = -\frac{S_{xx} + S_{xy} P_{t2}}{S_{yx} + S_{yy} P_{t2}} \tag{7.407b}$$

Note that while these are polarization ratios for a receiving antenna, they are defined in accordance with the conventions of this book as if the antenna were transmitting.

Relatively easy multiplication shows that

$$P_{r3} P_{r4}^* = -1$$

proving that the null polarization states of (7.407) are orthogonal. These receiving antenna polarizations are analogous to the copolarization nulls obtained for a symmetric scattering matrix with a single transmitting–receiving antenna.

In summary, if the target Sinclair matrix is asymmetric, as it is for the bistatic case or a target in a nonreciprocal medium, the use of one antenna for transmitting and receiving neither maximizes nor minimizes the received power. The received power can be maximized by choosing the transmitting antenna effective length $\mathbf{h}_{t1}$ to maximize the power density at the receiver, using the equations developed for the Graves polarization power scattering matrix. The received power is then maximized by choosing receiver effective length $\mathbf{h}_{r1}$ according to procedures developed in Chapter 3 or minimized (forced to zero) by selecting receiver polarization P_{r3} of (7.407). Alternatively, the power density is "minimized" (given its suboptimum value) by choosing transmitter effective length $\mathbf{h}_{t2}$. From this lesser power density the received power is "maximized" by selecting receiver polarization $\mathbf{h}_{r2}$ (although this power is less than that obtained by the antenna pair $\mathbf{h}_{t1}$ and $\mathbf{h}_{r1}$) or again set to zero by selecting receiver length $\mathbf{h}_{r4}$ corresponding to the polarization ratio P_{r4} of (7.407). The procedure has been called the "three-stage procedure" [22]. In the first stage, the transmitting antenna polarizations are found, in the second the scattered fields determined, and in the third the receiving antenna polarizations obtained. The process can also be used if the target matrix is symmetric (in which case the work is simplified) and may well be less tedious than the equations of Section 6.8 for the symmetric matrix.

REFERENCES

1. M. Born and E. Wolf, *Principles of Optics*, Pergamon, New York, 1965.
2. M. J. Beran and G. B. Parrent, Jr., *Theory of Partial Coherence*, Prentice-Hall, Englewood Cliffs, NJ, 1964.
3. R. T. Compton, Jr., *Adaptive Antennas*, Prentice-Hall, Englewood Cliffs, NJ, 1988.
4. E. Wolf, "Coherence Properties of Partially Polarized Electromagnetic Radiation," *Nuovo Cimento*, **XIII**(6), 1165–1181, 1959.
5. E. C. Titchmarsh, *Theory of Fourier Integrals*, 2nd ed., Oxford University Press, London, 1948.
6. D. Gabor, "Theory of Communication," *J. Inst. Elec. Eng.*, **93**, Part III, 429–457, 1946.
7. H. C. Ko, "The Interaction of Radio Antennas with Statistical Radiation," Notes for Short Course, Ohio State University, Columbus OH, 1965.
8. R. M. A. Azzam and N. M. Bashara, *Ellipsometry and Polarized Light*, North-Holland, New York, 1986.
9. R. Bellman, *Introduction to Matrix Analysis*, McGraw-Hill, New York, 1960.
10. M. C. Pease, III, *Methods of Matrix Algebra*, Academic, New York, 1965.
11. J. R. Huynen, "Phenomenological Theory of Radar Targets," Ph. D. Dissertation, Drukkerij Bronder-Offset, N. V., Rotterdam, 1970.

12. E. L. O'Neill, *Introduction to Statistical Optics*, Addison-Wesley, Reading, MA, 1963.
13. K. Kim, L. Mandel, and E. Wolf, "Relationship Between Jones and Mueller Matrices for Random Media," *J. Opt. Soc. Am.*, Part A, **4**, 433–437, March 1987.
14. J. J. van Zyl, "On the Importance of Polarization in Radar Scattering Problems," Ph.D. Dissertation, California Institute of Technology, Pasadena, CA, 1986.
15. E. M. Kennaugh, "Effects of the Type of Polarization on Echo Characteristics," Antenna Laboratory, Ohio State University, Report 389-9, 1951.
16. C-Y Chan, "Studies on the Power Scattering Matrix of Radar Targets," M. S. Thesis, University of Illinois, Chicago, IL, 1981.
17. S. R. Cloude, "Polarization Techniques in Radar Signal Processing," *Microwave J.*, 119–127, July 1983.
18. R. E. Ziemer and W. H. Tranter, *Principles of Communications*, Houghton Mifflin, Boston, MA, 1976.
19. C. D. Graves, "Radar Polarization Power Scattering Matrix," *Proc. Inst. Radio Eng.*, **44**(2), 248–252, February 1956.
20. L. I. Schiff, *Quantum Mechanics*, McGraw-Hill, New York, 1955.
21. S. R. Cloude, private communication.
22. A. B. Kostinski and W-M. Boerner, "On Foundations of Radar Polarimetry," *IEEE Trans. Antennas Prop.* **AP-34**(12), 1395–1404, December 1986.

PROBLEMS

7.1. For the example following (7.324), show that the value of d/λ that completely depolarizes a quasi-monochromatic incident wave does not meet the criterion (7.9) for correlation between incident and scattered waves.

7.2. Obtain the antenna effective length components analogous to those of (7.185) if it is desired to minimize the power received from the polarized part of a partially polarized wave.

7.3. Derive (7.186). Begin with (7.182).

7.4. Show that for a monochromatic wave the choice (7.185) for effective length that maximizes received power reduces to the choice made for the monochromatic wave in (4.44).

7.5. A wave with mean frequency 10 GHz has components

$$\mathscr{E}_x = [1 + 0.2\cos(20000\pi t)]e^{j(\pi/6 + \bar{\omega}t - kz)}$$
$$\mathscr{E}_y = [1 + 0.3\sin(20000\pi t)]e^{j(\bar{\omega}t - kz)}$$

The receiver response time is such that the wave can be considered partially polarized.
(a) Find the coherency matrix of the wave.
(b) Find its degree of polarization.

7.6. Find the polarization ratio of an antenna to maximize the received power from the wave of Problem 7.5. If the optimum receiving antenna is used, find the ratio of received power to the power that could be received from a completely polarized wave with the same power density.

7.7. A target has a Sinclair matrix

$$\mathbf{S} = \begin{bmatrix} 2 & e^{j\pi/4} \\ e^{j\pi/4} & 3 \end{bmatrix}$$

Find the Graves matrices σ_H, σ_V, and σ.

7.8. A target has Graves matrices

$$\sigma_H = \begin{bmatrix} 4 & 2e^{j\pi/6} \\ 2e^{-j\pi/6} & 1 \end{bmatrix} \quad \sigma_V = \begin{bmatrix} 1 & 4e^{-j\pi/6} \\ 4e^{j\pi/6} & 16 \end{bmatrix}$$

Find the relative scattering (Sinclair) matrix.

7.9. Find the Kennaugh matrix of the target of Problem 7.7.

7.10. An optical device has a Mueller matrix,

$$\mathbf{M} = \begin{bmatrix} 1 & 0 & 0 & 0 \\ 0 & 0.866 & -0.500 & 0 \\ 0 & 0.500 & 0.866 & 0 \\ 0 & 0 & 0 & 1 \end{bmatrix}$$

Does it have a Jones matrix?

7.11. A modulating voltage of frequency ω is used to modulate a carrier of frequency $\bar{\omega}$ according to

$$\mathscr{E}^r = (1 + a \cos \omega t) \cos \bar{\omega} t$$

The modulated signal is radiated in the z direction with propagation constant k. Write the analytic signal for the radiated wave in the strict form (7.19). Write it in the looser form (7.23).

7.12. The coherency matrix of a wave is

$$\mathbf{J} = \begin{bmatrix} J_{xx} & J_{xy} \\ J_{yx} & J_{yy} \end{bmatrix} = \begin{bmatrix} 4 & 3e^{-j\pi/6} \\ 3e^{j\pi/6} & 3 \end{bmatrix}$$

Find the degree of polarization. Write $\mathbf{J}$ as the sum of an unpolarized wave $\mathbf{J}^{(1)}$ and a completely polarized wave $\mathbf{J}^{(2)}$. Find the elements of the two matrices.

7.13. A target has Mueller matrix

$$\mathbf{M} = \begin{bmatrix} 9 & 0 & 0 & 0 \\ 0 & 3 & 4 & 0 \\ 0 & -4 & 3 & 0 \\ 0 & 0 & 0 & 5 \end{bmatrix}$$

A wave incident on the target from a monostatic radar is completely polarized and horizontal with

$$\mathbf{E}^i = \mathbf{u}_x E_x^i$$

Find the Stokes vector of the scattered wave. Find the degree of polarization. Why is it less than one? If $E_x^i = 200 \ \mu V/m$, find the power to the radar receiver with effective length $\mathbf{h} = 1\mathbf{u}_x$ and matched load of resistance 20 Ω located 10 km from the target.

7.14. Find the Graves matrix of the target of Problem 7.13.

7.15. A bistatic radar has a Sinclair matrix proportional to

$$\begin{bmatrix} 2 & e^{j\pi/6} \\ e^{-j\pi/6} & 4 \end{bmatrix}$$

Find the polarization ratio of the transmitting antenna to maximize the scattered power density.

7.16. Find the polarization ratio of the receiving antenna of Problem 7.15 to maximize the received power.

CHAPTER EIGHT

Polarization Measurements

8.1. INTRODUCTION

In this chapter we consider measurement of the polarization parameters of an electromagnetic wave, an antenna, and a target. For waves and antennas these parameters include the polarization ratio, polarization ellipse parameters, Stokes parameters, and elements of the coherency matrix. For targets they include the absolute and relative scattering matrices, the Graves polarization power scattering matrix, and the Mueller and Kennaugh matrices. Discussions of instrumentation and range requirements for the measurements are beyond the scope of this book and may be found elsewhere [1–4]; therefore, we consider primarily the parameters to be measured. No attempt is made to minimize the number of measurements to obtain a desired parameter; rather the emphasis is on practical measurements with available standard antennas.

8.2. MONOCHROMATIC POLARIZATION PARAMETERS

We need not consider as separate problems measurement of the polarization state of a wave and an antenna, since by definition the polarization state of a wave is that of the antenna that transmits it. We therefore consider the polarization state of a transmitting antenna and use certain standard antennas, or an antenna whose orientation is varied, as receivers. It is of course equally correct for a reciprocal antenna [5] to measure the response of the antenna under test while transmitting toward it waves of known polarization.

The Linear-Component Method

Apart from an absolute phase term that depends on distance from the transmitting antenna and is normally of little interest, the polarization of a wave is completely defined by the linear polarization ratio

$$P = \frac{E_y}{E_x} = \frac{|E_y|}{|E_x|} e^{j\phi} \tag{8.1}$$

Measurement of the amplitudes of the x- and y-directed field components and the phase difference between them using two linearly polarized receiving antennas is an effective method for determining the polarization ratio. It is desirable to use one receiver and switch it from one antenna to the other in measuring amplitudes to insure the same gain in both antenna channels. The phase difference ϕ can be measured if the two receiving antenna signals are applied simultaneously to a phase comparator. A slotted line method has been suggested for this purpose [6].

At lower frequencies, linear dipoles are satisfactory linearly polarized standards for use as the orthogonal receiving antennas in the measurement system. At higher frequencies, standard gain horns can be used. Typically, on axis, their axial ratios are on the order of 40 dB, which is satisfactory for most measurements [5]. Gains of the standard antennas can be measured at the desired frequencies prior to their use in the polarization measurement system, or, if the antennas are carefully constructed, the gains may be taken as equal. Placement of the two receiving standards is critical when measuring the phase difference between the wave components, and for this reason the method is more susceptible to error at higher frequencies.

The Circular-Component Method

The polarization state of a wave is specified completely by its inverse circular polarization ratio

$$q = \frac{E_L}{E_R} = \frac{|E_L|}{|E_R|} e^{-j\theta} \tag{8.2}$$

It follows that with two antennas having equal gains and impedances, one left-circularly polarized and the other right-circular, we can use the procedures outlined for the linear-component method to measure polarization states. Kraus [6] suggests the use of helices for the standard antennas, but Rubin [7] points out the difficulty of constructing identical (except for rotation sense) antennas, particularly when it is necessary to cover a wide frequency range. Another problem also exists. Since the axial ratio of an N-turn helix is not 1 but is given by [6],

$$\text{AR} = \frac{2N + 1}{2N} \tag{8.3}$$

if we require that the helix polarization approach circular as closely as the standard gain horn approaches linear polarization (AR → 40 dB), the helix must have a large number of turns (see Problem 8.1). This may make the method impractical for very precise measurements. An alternative to the use of helices is to synthesize each circularly polarized antenna by two linearly polarized antennas with outputs combined in quadrature, or a horn with orthogonal outputs combined in quadrature.

An alternative to the measurement of θ is the use of a linearly polarized receiving antenna to measure the tilt angle of the polarization ellipse [7]. Then θ is found from the tilt angle by (3.112). The problem of precise antenna placement, necessary to measure a phase difference, is eliminated. The tilt angle, together with the rotation sense, obtainable from $|q|$, and the axial ratio, found from

$$\text{AR} = \left| \frac{1 + |q|}{1 - |q|} \right| \tag{3.110}$$

define the polarization state completely.

The Polarization Pattern

Equation (4.96) for the polarization efficiency of two antennas with the same rotation sense and (4.100) for antennas of opposite sense both reduce to

$$\rho = \frac{\text{AR}_1^2 \cos^2(\tau_1 + \tau_2) + \sin^2(\tau_1 + \tau_2)}{\text{AR}_1^2 + 1} \tag{8.4}$$

if antenna 2 is linearly polarized, with $\text{AR}_2 \to \infty$. This equation leads to a widely used method for obtaining the polarization ellipse of an antenna experimentally.

Antenna 2 is rotated around a line drawn between the two antennas, say the z axis of Figure 4.3. Further, antenna 2 is so oriented that it cannot receive any z-directed wave components as it is rotated. For a dipole the rotation axis is perpendicular to the dipole.

At $\tau_2 = -\tau_1$, which corresponds to coincidence between the major axes of the ellipses for the two antennas,

$$\rho = \frac{\text{AR}_1^2}{\text{AR}_1^2 + 1} \tag{8.5}$$

which is a maximum. At $\tau_2 = -\tau_1 \pm \pi/2$, which corresponds to the major axis of the linearly polarized antenna coinciding with the minor axis of the antenna being tested,

$$\rho = \frac{1}{\text{AR}_1^2 + 1} \tag{8.6}$$

which is a minimum.

The open-circuit voltage is proportional to the square root of ρ, so the ratio of maximum to minimum open-circuit voltage, in magnitude, is

$$\frac{|V_{max}|}{|V_{min}|} = AR_1 \tag{8.7}$$

We thus have the axial ratio of the antenna undergoing test, and of course we have its tilt angle from the known rotation angle of the linear antenna when maximum power is received (or better, from the angle for minimum power plus 90°, since the minimum-power angle is more sharply defined than the maximum-power angle).

A plot of the square root of ρ from (8.4) is called the *polarization pattern* of the antenna whose polarization is being measured. Figure 8.1 shows the polar form of the pattern for (a) an antenna with an axial ratio of 2 and (b) a linearly polarized antenna. Since the ratio of maximum to minimum values of the polarization pattern is the axial ratio of the antenna under test, the polarization ellipse can be inscribed in the polarization pattern, as shown in Figure 8.1.

The polarization pattern method lends itself well to the rapid testing of an antenna's polarization properties as a function of angle from beam maximum. The linear sampling antenna is rotated rapidly while the antenna under test is scanned slowly. A recording of the received voltage shows the

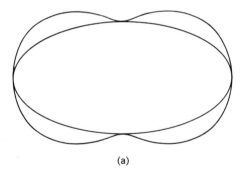

(a)

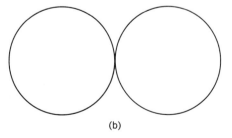

(b)

FIGURE 8.1 Polarization patterns for measuring axial ratio and tilt angle of the polarization ellipse: (*a*) AR = 2, with inscribed polarization ellipse; (*b*) AR → ∞.

antenna pattern with a rapid cyclic variation on it caused by the spinning of the sampling antenna. The ratio of amplitudes of adjacent maxima and minima will yield the axial ratio of the antenna being tested if the antenna pattern does not change significantly while the sampling antenna rotates through one-half revolution [5]. This automated process will clearly be more effective for antennas almost circularly polarized than for linearly polarized antennas.

An obvious deficiency of the polarization pattern method is its failure to give the rotation sense of the antenna under test. This information sometimes may be inferred from the antenna construction. It may also be obtained by making additional measurements with two equal-gain, opposite-sense, circularly polarized antennas.

Since the sampling antenna is mechanically rotated, care must be taken that the received power is not affected by the motion. In particular, rotary joints must have a constant output or be calibrated.

As a final remark about the polarization-pattern method, if the antenna undergoing test is nominally linearly polarized, a measurement of its axial ratio will be inaccurate unless the axial ratio of the sampling antenna is much greater than that of the antenna being tested.

Polarization Properties from Amplitude Measurements

The polarization properties of a wave can be determined from amplitude measurements alone, using appropriate standard receiving antennas. Thus the need for a phase measurement and the attendant requirement for precise antenna placement are eliminated. We found in Section 4.11 that if the point corresponding to the polarization ratio of a transmitting antenna and that for the receiving polarization ratio of a receiving antenna (or vice versa) are plotted on a Poincaré sphere by means of the Stokes parameters, using (3.185) and (3.189), the polarization match factor between the antennas is given by

$$\rho = \cos^2(\tfrac{1}{2}\beta) \qquad (4.136)$$

where β is the angle between the rays from the sphere center to the two plotted points. For a transmitting antenna with polarization unknown and a receiving antenna with known polarization, a circle drawn on the Poincaré sphere, with radius compatible with (4.136) and center at the receiver polarization point, will pass through the sphere point defining the transmitter polarization.* If we take a second receiving antenna, a circle with its receiving polarization as center will also pass through the transmitter polarization point. In general, the two circles will intersect at two points on the Poincaré sphere. A third receiving antenna can be used to remove the

---
*We do not consider the case of transmitter and receiver orthogonal.

ambiguity. As a general rule, the three circles generated by using three receiving antennas paired with the transmitting antenna will intersect at one point on the Poincaré sphere, thus uniquely defining the the polarization of the transmitting antenna. Note that amplitude measurements only are needed [8].

If the circles on the Poincaré sphere intersect at small angles, small errors in the amplitude measurements can lead to significant uncertainty in the polarization. Prior knowledge of the antenna under test can be used to assist in the selection of the sampling antennas, and in fact, if the rotation sense of the antenna being tested is known, it may be possible to eliminate one measurement.

The polarization match factor ρ is not measured directly; rather power to a receiver load is measured, and this is determined by polarization, antenna gains, transmitted power, and so on. It is then clear that additional measurements are needed to determine an antenna's polarization properties.

A convenient method of handling the requirement for additional information is by using pairs of receiving antennas that have the same gains but are orthogonally polarized, such as left- and right-circular antennas. Power ratios are then used to determine the polarization of the antenna being tested. We illustrate the method by using three pairs of receiving antennas, linear horizontal (x-directed) and vertical, linear at 45° and 135° from the x axis, and left and right circular.

Linear Vertical and Horizontal In terms of the linear polarization ratio, the polarization match factor between a transmitting antenna (1) and a receiving antenna (2) is

$$\rho = \frac{(1 - P_1 P_2)(1 - P_1^* P_2^*)}{(1 + P_1 P_1^*)(1 + P_2 P_2^*)} \quad (4.57)$$

If a linear vertical receiving antenna is used, with $P_2 \to \infty$, the power received is

$$W_V = C_1 \rho_V = C_1 \frac{|P_1|^2}{1 + |P_1|^2}$$

where C_1 is a constant that includes the antenna gains, power transmitted, receiver gain, impedance match, and antenna separation, but not polarization.

If a linear horizontal receiving antenna is used, with $P_2 = 0$, keeping all other factors the same, the power received is

$$W_H = C_1 \rho_H = \frac{C_1}{1 + |P_1|^2}$$

and the ratio of the two received powers is

$$\frac{W_V}{W_H} = \frac{\rho_V}{\rho_H} = |P_1|^2 \tag{8.8}$$

Linear 45° and 135° If a linearly polarized antenna tilted at 45° ($P_2 = 1$) and one at 135° ($P_2 = -1$) are used successively with the antenna under test, the ratio of powers received is

$$\frac{W_{45}}{W_{135}} = \frac{1 + |P_1|^2 - 2\operatorname{Re}(P_1)}{1 + |P_1|^2 + 2\operatorname{Re}(P_1)} \tag{8.9}$$

if the gains of the two receiving antennas are equal.

Left and Right Circular Using left-circular ($P_2 = j$) and right-circular ($P_2 = -j$) antennas leads to a power ratio

$$\frac{W_R}{W_L} = \frac{1 + |P_1|^2 + 2\operatorname{Im}(P_1)}{1 + |P_1|^2 - 2\operatorname{Im}(P_1)} \tag{8.10}$$

These last three equations can be solved to give

$$\operatorname{Re}(P_1) = \frac{\left(1 + \dfrac{W_V}{W_H}\right)\left(1 - \dfrac{W_{45}}{W_{135}}\right)}{2\left(1 + \dfrac{W_{45}}{W_{135}}\right)} \tag{8.11a}$$

$$\operatorname{Im}(P_1) = \frac{\left(1 + \dfrac{W_V}{W_H}\right)\left(\dfrac{W_R}{W_L} - 1\right)}{2\left(\dfrac{W_R}{W_L} + 1\right)} \tag{8.11b}$$

These equations show that six amplitude measurements lead to the polarization of a general antenna. In the equations for P_1 the term W_V/W_H can be replaced by $|P_1|^2$, and since this is equal to the sums of the squares of the two equations of (8.11), it appears that P_1 can be determined from the two remaining power ratios of (8.11). This leads to ambiguities in P_1, however.

The antenna pairs utilized to obtain (8.11) may not be optimum for the measurement of a general antenna, but the chosen antennas are easily obtained. One linearly polarized antenna may be used for four of the measurements. If helices are used for the circularly polarized receiving

470 POLARIZATION MEASUREMENTS

antennas, their small departure from the circular will not affect polarization measurements substantially. The problem remains of constructing equal-gain helices of opposite rotation. Nevertheless, the freedom from measuring phase makes this method attractive.

8.3. THE STOKES VECTOR

In Section 7.4 a preliminary discussion of the measurement of the Stokes parameters was given, and it is continued here. Equation (7.155) suggests that appropriate receiving antennas to measure the Stokes vector elements of an incident wave are linear (horizontal, vertical, and tilted) and circular. Since the Stokes vector is valid for nonmonochromatic as well as monochromatic waves, the standard antennas must maintain their polarization properties over the bandwidth of the received signal. Consider the same antenna pairs used to find the polarization ratio of an incident monochromatic wave by amplitude measurements alone.

The received power from an incident wave with Stokes vector $\mathbf{G}^i$ is, from (7.159) and (7.162),

$$W = \frac{1}{16R_a} [G_0^i \ G_1^i \ G_2^i \ G_3^i] \begin{bmatrix} |h_\xi|^2 + |h_\eta|^2 \\ |h_\xi|^2 - |h_\eta|^2 \\ 2\,\mathrm{Re}(h_\xi^* h_\eta) \\ -2\,\mathrm{Im}(h_\xi^* h_\eta) \end{bmatrix} \quad (8.12)$$

using the coordinate systems of Figure 7.1.

Linear Horizontal and Vertical

If linear horizontal and vertical receiving antennas are used sequentially, the received powers are

$$W_H = \frac{|h_\xi|^2}{16R_a}(G_0^i + G_1^i) \quad (8.13a)$$

$$W_V = \frac{|h_\eta|^2}{16R_a}(G_0^i - G_1^i) \quad (8.13b)$$

If the antennas have equal impedances and effective lengths, the ratio of received powers is

$$\frac{W_V}{W_H} = \frac{G_0^i - G_1^i}{G_0^i + G_1^i} \quad (8.14)$$

Linear 45° and 135°

For a linear antenna tilted at 45° in transmitting coordinates, $h_\eta = -h_\xi$, and for a 135° tilt $h_\eta = h_\xi$. Then, for identical antennas,

$$\frac{W_{45}}{W_{135}} = \frac{G_0^i - G_2^i}{G_0^i + G_2^i} \tag{8.15}$$

Left and Right Circular

For circular antennas, $h_\eta = h_\xi \exp(\pm j\pi/2)$, with the positive sign corresponding to right circular. Then for antennas identical except for rotation sense,

$$\frac{W_R}{W_L} = \frac{G_0^i - G_3^i}{G_0^i + G_3^i} \tag{8.16}$$

These equations are readily solved to give

$$\frac{G_1^i}{G_0^i} = \frac{W_H - W_V}{W_H + W_V} \tag{8.17a}$$

$$\frac{G_2^i}{G_0^i} = \frac{W_{135} - W_{45}}{W_{135} + W_{45}} \tag{8.17b}$$

$$\frac{G_3^i}{G_0^i} = \frac{W_L - W_R}{W_L + W_R} \tag{8.17c}$$

For monochromatic waves, the Stokes vector is completely determined by these equations, since G_0^i is fixed by the remaining three parameters. For partially polarized waves, the system must be calibrated to establish

$$\frac{|h_\xi|^2}{R_a} \left(= \frac{|h_\eta|^2}{R_a} \right)$$

and then G_0^i may be found from (8.13),

$$G_0^i = \frac{8R_a}{|h_\xi|^2}(W_H + W_V) \tag{8.18}$$

This additional information may also be used to eliminate one of the tilted linear measurements if the same antenna is used for the tilted measurement as for the horizontal and vertical.

The same measurement set can be used for the Stokes vector and for the polarization ratio, P, with an additional calibration required for the Stokes vector of a partially polarized wave. The same measurements also suffice to determine the coherency matrix since its elements can be obtained from the Stokes vector.

8.4. THE SCATTERING MATRIX

In this section we consider primarily the backscattering Sinclair matrix, although the procedures are adaptable to the bistatic situation and to the Jones matrix. The absolute and relative scattering matrices require different measurement techniques and are discussed separately.

The Absolute Scattering Matrix

The open-circuit voltage induced in a receiving antenna is

$$V = \mathbf{h}_r^T \mathbf{E}^s \tag{8.19}$$

where $\mathbf{h}_r$ is its effective length and $\mathbf{E}^s$ the wave scattered from the target, given by

$$\mathbf{E}^s = \frac{1}{\sqrt{4\pi}\,r}\mathbf{S}\mathbf{E}^i = \frac{1}{\sqrt{4\pi}\,r}\begin{bmatrix} S_{xx} & S_{xy} \\ S_{yx} & S_{yy} \end{bmatrix}\begin{bmatrix} E_x^i \\ E_y^i \end{bmatrix} \tag{8.20}$$

If the wave incident on the target has only a horizontal (x) component of the electric field, the open-circuit voltages induced in x- and y-polarized receiving antennas are

$$V_{xx} = \frac{1}{\sqrt{4\pi}\,r} h_x S_{xx} E_x^i \tag{8.21a}$$

$$V_{yx} = \frac{1}{\sqrt{4\pi}\,r} h_y S_{yx} E_x^i \tag{8.21b}$$

Similarly, if the transmitted wave is y-directed, the received voltages are

$$V_{xy} = \frac{1}{\sqrt{4\pi}\,r} h_x S_{xy} E_y^i \tag{8.22a}$$

$$V_{yy} = \frac{1}{\sqrt{4\pi}\,r} h_y S_{yy} E_y^i \tag{8.22b}$$

A radar system that can measure the amplitude and phase of the received signal can, with the use of (8.21) and (8.22), determine the elements of the scattering matrix. It has been suggested that an appropriate system transmits two orthogonal signals (in this case x- and y-linear) sequentially and receives the orthogonal components simultaneously, using two receiver channels [1]. The phase measurements require that the transmitted signal be up-converted from the output of a coherent oscillator, which is also mixed with the received signals and applied to phase detectors [1, 9]. The system may be calibrated with a standard target.

For backscattering, the matrix is symmetric and the measurement to find either S_{xy} or S_{yx} is not needed.

The Relative Scattering Matrix

The need for a coherent processor is removed if the relative scattering matrix, rather than the absolute, is desired. For backscattering, the scattering matrix is symmetric and the off-diagonal element S_{xy} may be used as a phase reference. The received powers associated with (8.21) and (8.22) are

$$W_{xx} = \frac{1}{32 R_a r^2} |h_x|^2 |S_{xx}|^2 |E_x^i|^2 \qquad (8.23a)$$

$$W_{yx} = \frac{1}{32 R_a r^2} |h_y|^2 |S_{yx}|^2 |E_x^i|^2 \qquad (8.23b)$$

for an incident horizontal field and

$$W_{xy} = \frac{1}{32 R_a r^2} |h_x|^2 |S_{xy}|^2 |E_y^i|^2 \qquad (8.24a)$$

$$W_{yy} = \frac{1}{32 R_a r^2} |h_y|^2 |S_{yy}|^2 |E_y^i|^2 \qquad (8.24b)$$

for an incident vertically polarized wave. The matrix element magnitudes can be found from these powers if the system has been calibrated with a standard target. With an incident x-polarized wave, a phase detector can be used to find the angle of S_{xx} by the relation

$$\text{ang}(S_{xx}) = \text{ang}(V_{xx}) - \text{ang}(V_{yx}) \qquad (8.25)$$

If the incident wave is y-polarized,

$$\text{ang}(S_{yy}) = \text{ang}(V_{yy}) - \text{ang}(V_{xy}) \qquad (8.26)$$

If $S_{xy} = 0$, this method fails, since the phase angles of the diagonal elements are compared to that of S_{xy}. For this, the radar or the target may be rotated around the line of sight until a suitable off-diagonal element is measured. The scattering matrix in the rotated coordinates is then transformed to the desired coordinate system. Alternatively, the linear transmitting antenna may be rotated 45° around the line of sight [10]. The simultaneously received voltages on linear x- and y-directed antennas are then proportional to

$$V_x = h_x S_{xx} E_x^i \tag{8.27a}$$
$$V_y = h_y S_{yy} E_y^i \tag{8.27b}$$

(with the incident fields decreased by $\sqrt{2}$ from their previous value). The amplitudes of the scattering coefficients are found from the voltage magnitudes, and the phase difference between S_{xx} and S_{yy} is found by phase-comparing the voltages.

Relative Scattering Matrix from Amplitudes

It was seen earlier that the polarization properties of a wave, including phase, can be found from amplitude measurements only. This allows the relative scattering matrix to be found from amplitude measurements. If a horizontally polarized transmitter is used, the scattered field from a target is

$$\begin{bmatrix} E_x^s \\ E_y^s \end{bmatrix} = \frac{1}{\sqrt{4\pi}\,r} \begin{bmatrix} S_{xx} E_x^i \\ S_{yx} E_x^i \end{bmatrix} \tag{8.28}$$

The amplitude-only technique allows the measurement of the magnitudes and the relative phase of the scattered field components in this equation, and from them the scattering matrix elements in the equation can be found. Thus $|S_{xx}|$ and $|S_{yx}|$ and their relative phase can be found from six amplitude measurements if the transmitted wave is x-polarized. Similarly, $|S_{xy}|$ and $|S_{yy}|$ and their relative phase may be found from six amplitude measurements if the wave incident on the target is y-polarized. It has been pointed out that seven measurements are the minimum necessary to find the relative scattering matrix [10], so this set of twelve is clearly not optimum. It is a convenient set, however, with readily obtainable receiving and transmitting antennas. It was noted earlier, also, that one measurement can be eliminated with a calibrated system.

8.5. POLARIZATION POWER SCATTERING MATRIX

The power density of a wave backscattered from a target is

$$\mathscr{P} = \frac{1}{8\pi Z_0 r^2} \mathbf{E}^{iT*} \boldsymbol{\sigma} \mathbf{E}^i \tag{7.325}$$

where σ is the Graves polarization power scattering matrix. If the target possesses a scattering matrix, σ can be decomposed into parts giving the horizontal and vertical component power densities of the scattered wave,

$$\sigma = \sigma_H + \sigma_V \qquad (7.335)$$

The powers received by horizontally (x) and vertically (y) polarized antennas are

$$W_H = \frac{|h_x|^2}{32\pi R_a r^2} \mathbf{E}^{iT*} \sigma_H \mathbf{E}^i \qquad (7.339)$$

$$W_V = \frac{|h_y|^2}{32\pi R_a r^2} \mathbf{E}^{iT*} \sigma_V \mathbf{E}^i \qquad (7.340)$$

Measurement of σ

The sum of the powers received simultaneously on horizontally and vertically polarized receiving antennas of equal effective length $|h_x| = |h_y| = |h|$ and equal impedance is

$$W = W_H + W_V = \frac{|h|^2}{32\pi R_a r^2} \mathbf{E}^{iT*} \sigma \mathbf{E}^i = C \mathbf{E}^{iT*} \begin{bmatrix} \sigma_{xx} & \sigma_{xy} \\ \sigma_{xy}^* & \sigma_{yy} \end{bmatrix} \mathbf{E}^i \qquad (8.29)$$

where C is a constant and σ is Hermitian.

If the wave directed at the target is x-polarized, with field E_0, the measured power is

$$W = C\sigma_{xx}|E_0|^2 \qquad (8.30)$$

and if it is y-polarized

$$W = C\sigma_{yy}|E_0|^2 \qquad (8.31)$$

To find the off-diagonal element, an incident wave at 45°,

$$\mathbf{E}^i = E_0 \begin{bmatrix} 1 \\ 1 \end{bmatrix}$$

can be used, giving

$$W = C|E_0|^2 \left[\sigma_{xx} + 2\operatorname{Re}(\sigma_{xy}) + \sigma_{yy}\right] \qquad (8.32)$$

Since this provides only the real part of σ_{xy}, an additional measurement is needed. This can be done with a circularly polarized transmitter, for which

$$\mathbf{E}^i = E_0 \begin{bmatrix} 1 \\ \pm j \end{bmatrix}$$

yielding a sum of powers

$$W = C|E_0|^2 [\sigma_{xx} \mp 2\,\text{Im}(\sigma_{xy}) + \sigma_{yy}] \quad (8.33)$$

The power scattering matrix can be determined from these four measurements after system calibration [10].

Measurement of σ_H and σ_V

Equation (7.334) shows that if the target possesses a scattering matrix both σ_H and σ_V are Hermitian. Then the same method used to measure σ can be used to measure σ_H and σ_V if horizontally and vertically polarized receivers are used separately, rather than taking a sum of powers as was done to measure σ. To measure σ_H a horizontally polarized receiver is used with the results shown here for four incident waves:

$$W_H = \begin{cases} C|E_0|^2 \sigma_{Hxx} & x\text{-directed} \\ C|E_0|^2 \sigma_{Hyy} & y\text{-directed} \\ C|E_0|^2 [\sigma_{Hxx} + 2\,\text{Re}(\sigma_{Hxy}) + \sigma_{Hyy}] & 45° \text{ tilt} \\ C|E_0|^2 [\sigma_{Hxx} \mp 2\,\text{Im}(\sigma_{Hxy}) + \sigma_{Hyy}] & \text{circular} \end{cases} \quad \text{Incident Wave} \quad (8.34)$$

The σ_V matrix can be found using a vertically polarized receiving antenna and the same measurement set and equations.

8.6. MUELLER AND KENNAUGH MATICES

If a target possesses a scattering matrix, it is convenient to measure it and determine the Mueller and Kennaugh matrices from it. We therefore consider targets without a scattering matrix, and for convenience only backscattering.

The symmetries that arise in the Mueller and Kennaugh matrices are a consequence of reciprocity and exist whether the target has a scattering matrix or not. Then, we specifically consider the Kennaugh formulation of the backscattered wave,

$$\mathbf{G}_K^s = C \begin{bmatrix} K_{11} & K_{12} & K_{13} & K_{14} \\ K_{12} & K_{22} & K_{23} & K_{24} \\ K_{13} & K_{23} & K_{33} & K_{34} \\ K_{14} & K_{24} & K_{34} & K_{44} \end{bmatrix} \mathbf{G}_K^i \qquad (8.35)$$

The diagonal elements are related by [11]

$$K_{11} = K_{22} + K_{33} + K_{44} \qquad (8.36)$$

so that only nine elements of the Kennaugh (or Mueller) matrix are independent. Nine measurements suffice to determine the matrix, and the following set has been suggested [10]. If it is used, the received power, from (7.251), will be proportional to the listed sums.

Transmitting Antenna	Receiving Antenna	Received Power
Linear horizontal (x)	Linear horizontal (x)	$K_{11} + 2K_{12} + K_{22}$
Linear vertical (y)	Linear vertical (y)	$K_{11} - 2K_{12} + K_{22}$
Linear, 45° to x	Linear, 45° to x	$K_{11} + 2K_{13} + K_{33}$
Linear, 135° to x	Linear, 135° to x	$K_{11} - 2K_{13} + K_{33}$
Left circular	Left circular	$K_{11} + 2K_{14} + K_{44}$
Right circular	Right circular	$K_{11} - 2K_{14} + K_{44}$
Linear horizontal (x)	Left circular	$K_{11} + K_{12} + K_{14} + K_{24}$
Linear, 45° to x	Linear horizontal (x)	$K_{11} + K_{12} + K_{13} + K_{23}$
Linear, 45° to x	Left circular	$K_{11} + K_{13} + K_{14} + K_{34}$

The matrix elements can be determined to within a constant from the nine measured powers.

REFERENCES

1. J. R. Huynen, "Measurement of the Target Scattering Matrix," *Proc. IEEE*, **53**(8), 936–946, August 1965.
2. P. Blacksmith, Jr., R. E. Hiatt, and R. B. Mack, "Introduction to Radar Cross-Section Measurements," *Proc. IEEE*, **53**(8), 901–920, August 1965.
3. R. G. Kouyoumjian and L. Peters, Jr., "Range Requirements in Radar Cross-Section Measurements," *Proc. IEEE*, **53**(8), 920–928, August 1965.
4. H. C. Marlow, D. C. Watson, C. H. van Hoozer, and C. C. Freeny, "The RAT SCAT Cross-Section Facility," *Proc. IEEE*, **53**(8), 946–954, August 1965.

478 POLARIZATION MEASUREMENTS

5. T. G. Hickman, J. S. Hollis, and L. Clayton, Jr., "Polarization Measurements," Chapter 10 in J. S. Hollis, T. J. Lyon, and L. Clayton, Jr., *Microwave Antenna Measurements*, Scientific-Atlanta, Atlanta, GA, 1970.
6. J. D. Kraus, *Antennas*, 2nd ed., McGraw-Hill, New York, 1988.
7. R. Rubin, "Antenna Measurements," Chapter 34 in H. Jasik, Ed., *Antenna Engineering Handbook*, McGraw-Hill, New York, 1961.
8. G. H. Knittel, "The Polarization Sphere as a Graphical Aid in Determining the Polarization of an Antenna by Amplitude Measurements Only," *IEEE Trans. Antennas Prop.*, **AP-15**(2), 217–221, March 1967.
9. M. I. Skolnik, *Introduction to Radar Systems*, 2nd Ed., McGraw-Hill, New York, 1980.
10. C.-Y. Chan, "Studies on the Power Scattering Matrix of Radar Targets," M.S. Thesis, University of Illinois, Chicago, IL, 1981.
11. E. M. Kennaugh, "Effects of Type of Polarization on Echo Characteristics," Report No. 389-9, Antenna Laboratory, Ohio State University, June 16, 1951.

PROBLEMS

8.1. A helix is desired for polarization measurements with a quality as high as that of a good linearly polarized antenna (AR = 40 dB). Define equal quality to mean that $|E_R|/|E_L|$ for the helix is the same as $|E_y|/|E_x|$ for the linear antenna. Find the number of turns needed for the helix.

8.2. A dipole, a left-circular, and a right-circular antenna are used to determine the polarization ratio of a transmitting antenna. The two circularly polarized antennas have the same gain and impedance. With respect to a coordinate system at the receiving antenna having its z axis directed toward the transmitter, the dipole is successively oriented among the y axis (vertical), along the x axis, and at 45° and 135° from the x axis. The same receiver load impedance is used for all antennas. The received powers are (in milliwatts)

Vertical:	3.82	135°:	4.04
Horizontal:	0.95	Right Circular:	7.80
45°:	0.73	Left Circular:	3.34

Find the polarization ratio of the transmitting antenna.

CHAPTER NINE

Target Detection

9.1. INTRODUCTION

Target detection, discrimination, and recognition are names applied to the process by which a radar separates and classifies the return from a resolution cell (range, angle, velocity) as coming from a target (an object of interest), clutter (objects of lesser interest), or noise. The definitions are radar dependent since a target of interest to one radar may be clutter to another. Although the concept is not universal, the processes may be thought of as occurring in a sequence [1]. The first step is detection, in which some technique such as amplitude thresholding or exploitation of Doppler frequency is used to effect an initial separation of target and clutter returns. In the second step, discrimination, somewhat more subtle differences in target and clutter returns are utilized to refine the separation, with frequency agility or polarimetric processing as potential tools [1]. Target recognition involves a further refinement in separation and classification. An example of recognition is the decision that a target is a truck and not an automobile. The boundaries between the processes are imprecisely defined, but no attempt will be made here to make them more exact. Clearly, all three steps are not taken by some radars. Early pulse radars, for example, used only amplitude thresholding, which falls into the category of detection as defined here, and did not take the discrimination and recognition steps.

In this chapter, techniques for utilizing differences between targets and clutter, or between two targets, that affect the measurable polarization properties of their scattered signals are discussed. Some of the techniques may be used for any one of the processes of detection, discrimination, or recognition, and in this discussion detection or discrimination is used to mean any or all of the processes. The emphasis is on targets and clutter that are difficult to separate by simpler means, such as amplitude thresholding and Doppler processing.

The time behavior of targets and clutter makes it necessary to consider the length of time they are observed by the radar before detection decisions are made. In a pulse radar, this may be as short as the duration of one pulse, on the order of nanoseconds or microseconds. In a radar that uses pulse integration in its processing algorithm the observation time may require many interpulse periods, each on the order of hundreds of microseconds, or milliseconds. In a scanning radar the upper time limit for observation is the time the target remains in the scanning beam. This dwell time may be milliseconds or tens of milliseconds. In a nonscanning radar the observation time may range from nanoseconds to any desired value.

The volume of the radar cell, determined by its down-range and cross-range dimensions, is also important. Polarization behavior can be different if the resolution cell includes the full target or only a portion of it.

Targets

An extended radar target, such as an aircraft, has a large number of regions, or scattering centers, from which the incident wave is reflected. The wave scattered to the radar receiver is the coherent sum of the contributions from the scattering centers. A constituent wave from a scattering center generally does not have the same polarization as the incident wave. In the terminology of this book, it is repolarized but not depolarized, that is, the degree of polarization is unchanged. The scattering may be, for example, from a curved conducting surface. If local radii of surface curvature differ in orthogonal directions, as for an ellipsoid or cylinder, the tilt angle of an incident wave may be changed by reflection. The phasor addition of the constituent waves depends on the relative positions of the scattering centers, which in turn are affected by the aspect angle of the target and relative motions of the scattering centers caused, for example, by vibration. It follows that the overall wave scattered by the target has a different polarization from that of the incident wave, and the scattered wave polarization in general is a function of time. Then the wave scattered from the target is depolarized, that is, the degree of wave polarization is decreased by scattering.

It is clear that, if an infinitesimal observation time is considered, a scattering matrix can be determined for any target; it is also clear from the preceding discussion that the matrix is time-dependent. It is then important to decide if a target with a time-varying scattering matrix, observed by a radar for a finite time period, can be described by a constant scattering matrix (constant during the observation period), or if it must be represented by a Mueller or Kennaugh matrix, suitable for partially polarized waves.* Clearly the decision must be made from experience. Some information is available on amplitude changes of the backscattered wave from aircraft because of aspect angle changes. Skolnik has suggested modulation frequencies up to 15 Hz for

---

*In keeping with common radar practice, the illuminating wave is completely polarized.

a turning target. He also reports on observations of an aircraft flying in a straight line. The periods of amplitude modulation for linear polarization were found to be a few tenths of a second [2]. If the assumption is made that polarization changes occur at a rate on the order of amplitude modulation rates, because their origin is also primarily aspect angle changes, then a polarization change rate of 15 Hz or less appears reasonable. This indicates that a significant polarization change in the scattered wave will not occur unless the radar observation time is tens of milliseconds or longer.

Measurements that have an important influence on the scattering matrix–Kennaugh matrix decision have been reported by Giuli [3]. An aircraft taking off, gaining altitude, and turning was observed from approximately 30 km away by a monostatic S-band radar with pulse length 1 μs, pulse repetition frequency 1 kHz, and dwell time on target about 20 ms. The polarization of each received pulse was plotted on an orthographic polarization chart. Right-circular polarization was transmitted and both right- and left-circular polarizations were received. Two polarization charts (not shown here) present Giuli's data. The first shows the points representing the polarization of each received pulse during one dwell time (20 pulses in 20 ms) in which the aircraft was moving approximately transverse to the radar line of sight. These points are tightly grouped, with a tilt angle extreme difference of about 10° and an ellipticity angle extreme difference of about 5° or less. The second chart shows the polarization points in one dwell time about 20 s later, with the plane turning away from its transverse direction. The extreme differences in this group are greater for both tilt angle and ellipticity angle than for the first group, being about 35° and 20° respectively, indicating a smaller degree of polarization. Standard deviations appear from Giuli's chart to be about 11° and 6° for tilt and ellipticity angles respectively, so even this divergent group does not exhibit a large change in polarization during a dwell time. Overall, the polarization point means for the two scans, separated in time by 20 s, differ greatly in tilt angle, about 60°, but not greatly in ellipticity angle, less than 10°. It then appears reasonable to consider the scattered signal from a target of this type to be completely polarized and to represent the target by its scattering matrix if the radar observation time is of the order of milliseconds or a few tens of milliseconds.*

Clutter

Much of the clutter surrounding a target comes from objects that move with respect to each other and have velocities that vary with time (waves on the sea, rainfall, or movement of vegetation in the wind). The scattering matrix for clutter, like that for the target, is time-varying, and we must also decide

---

*Helicopter rotors, airplane propellers, and jet engine fans modulate the radar return at a much higher frequency than the frequencies due to aspect angle changes. They are excluded from our discussion.

482   TARGET DETECTION

for it if a constant scattering matrix for the observation period can be used or if a Mueller or Kennaugh matrix is necessary. Again the answer must come from experience.

The clutter power spectrum for wooded terrain has been reported to be down 10 dB at about 15 Hz for a linearly polarized incident wave at 9.2 cm, with a wind speed of 50 mph [4]. For a moderately heavy rain at X-band with linear vertical polarization, the 10-dB frequency has been measured to be about 80 Hz [5]. For sea clutter, a half-power spectral width of about 35 Hz has been given [6]. It appears then that we are justified in taking clutter power as constant during an observation time of a few milliseconds, and if we assume amplitudes and phases of all field components to vary no more rapidly than power, clutter can be represented by a constant scattering matrix for observation times of a few milliseconds. Another common experience validates this conclusion for rain. In the next section, it is noted that the use of circular polarization can reduce rain clutter by as much as 30 dB. Such a cancellation would not be possible if rain significantly depolarizes (decreases the degree of polarization) the incident wave on a time scale comparable to the observation time. A similar remark about sea clutter and horizontal polarization can be made.

Giuli also has reported the polarization of land clutter returns using the radar described previously for the aircraft target [3]. He found the polarization points for the pulses in one scan for strong point clutter in an urban area to be tightly grouped, with a tilt angle variation in one group appearing to be about 5° and the ellipticity angle variation about 16°. A second scan, 8 s later, of the same small area showed a somewhat greater tilt angle variation and smaller ellipticity angle variation. The group means for the two scans were quite close together. For distributed land clutter in an urban area, the degree of polarization of the clutter returns was found to be less, with the tilt angle variation among all the pulse returns in one scan about 45° and the ellipticity angle variation about 35°. The polarization point grouping was noted to be quite similar from one scan to another 8 s later.

It appears appropriate then, for typical radar observation times, to treat clutter by means of its scattering matrix, using a constant matrix during the observation time. The treatment becomes more exact as the observation time and the resolution cell size become smaller.

9.2. RAIN CLUTTER

To a meteorologist rainfall may be the desired radar target, but to others it is clutter, with the target an aircraft or ground vehicle. Rainfall surrounding a target for a monostatic radar attenuates the radar signal and backscatters to the receiver clutter returns in the target resolution cell (assuming a resolution cell larger than the target) and in other resolution cells, leading to possible missed detections and false alarms. Both the attenuation and the backscattering cross section increase with rainfall rate and, in the range of radar frequencies, strongly with frequency [7, 8]. The backscattering cross section is proportional to the resolution cell volume and for a rainfall rate of 10 mm/hr and wavelength 3 cm has been found to be approximately 3×10^{-2} cm^2/m^3 with linear polarization [7].

In theory, if a circularly polarized wave is incident on either a conducting or dielectric sphere, the backscattered wave has the opposite rotation sense, and this may be used to reduce backscattering from raindrops. In rainfall, drop sizes are distributed over a range of values. In very light rain (0.25 mm/hr), most of the drops are small (< 1 mm) in diameter but become larger for increasing rainfall rates [8]. The smallest drops (< 0.028 cm) are spherical, and those larger are oblate spheroids with the minor axis vertical in the absence of wind forces. The ratio of minor to major axis decreases with increasing drop size. Drops with an equivalent diameter (diameter of sphere with equal volume) greater than 3 mm may assume complex shapes because of hydrodynamic instability [9].

The ratio of the rain radar cross section using circular polarization to the cross section using linear vertical polarization can be taken as a measure of the clutter cancellation. The nonspherical form of the raindrops just noted will diminish the cancellation obtainable with circular polarization. Consider a spheroidal raindrop with minor axis vertical. Its scattering matrix, neglecting an absolute value, is

$$\mathbf{S} = \begin{bmatrix} S_{xx} & S_{xy} \\ S_{yx} & S_{yy} \end{bmatrix} = \begin{bmatrix} 1 & 0 \\ 0 & a \end{bmatrix} \qquad 0 \leq a \leq 1 \qquad (9.1)$$

where the radar x axis is horizontal. The radar cross section, if the same antenna is used for transmitting and receiving, can be obtained from

$$\sigma_r = |\hat{\mathbf{h}}^T \mathbf{S} \hat{\mathbf{h}}|^2 \qquad (6.192)$$

where the magnitude of the normalized antenna effective length is one. If linear vertical and circular polarizations are considered, the ratio of circular and linear cross sections is

$$\frac{\sigma_{r\,\text{circ}}}{\sigma_{r\,\text{lin}}} = \frac{1}{4}\left(\frac{1-a}{a}\right)^2 \qquad (9.2)$$

The clutter cancellation of this ideal spheroid ranges from 6 dB (for $a = 0.5$) to 25 dB (for $a = 0.9$), and this provides us with a rough guide about cancellation with spheroidal raindrops.

In light rainfall, it is evident from the spherical form of the majority of raindrops that rain clutter will be much less for a circular transmitted wave and the same sense reception than for a linear transmitted wave with the same received polarization. In practice, clutter cancellation of 30 dB has been reported [7]. In heavier rain the nonspherical raindrops cause the clutter cancellation with circular polarization to be less effective, and a cancellation of 5 dB has been noted [7].

Larger raindrops have a smaller vertical (in the absence of wind) dimension than horizontal and consequently a smaller vertical–vertical cross section than for horizontal transmitting and receiving polarizations. The rain clutter for a monostatic radar can then be decreased by making the transmit and receive polarization elliptical with the same rotation sense. The major axis of the polarization ellipse should be vertical and the phase angle between vertical and horizontal components 90°. An improvement in clutter cancellation in heavy rain with elliptical polarization (compared to circular) of 12 dB has been reported [7].

Several factors detract from the effectiveness of this clutter cancellation method. First, the desired polarization depends on the rainfall rate, and for optimum performance, a variable-polarization radar is required. Second, the spheroidal nature of the larger raindrops causes the vertical and horizontal wave components to be attenuated differently, making the polarization nonoptimum over part of the wave path. Third, the cross section of the target may be different for circular and linear polarizations. Measurements on aircraft have shown a cross section decrease of as much as 2.5 dB for circular polarization as compared to linear, although that does not always occur [7]. For this reason, radars for use with aircraft targets may use linear polarization in the absence of rain and switch to circular during rainfall. Finally, canting of the raindrops by wind forces affects the optimum polarization. These factors are not great enough to negate the benefits, however, and the use of circular polarization to reduce rain clutter is widespread.

9.3. CHAFF

Chaff consists usually of aluminum or aluminum-coated mylar, glass, or nylon strips or cylinders, dispensed in such a manner that they form an airborne cloud. The strip length is designed to be a half wavelength at the radar frequency of interest, or in some cases a range of lengths is used to cover a radar band. After dispersal a chaff cloud can present a very large cross section to confuse enemy radars. In fact, it has been calculated that a 1-lb chaff package can yield a radar cross section of 1000 m^2 [10]. After dispersal, chaff quickly loses its initial velocity and falls slowly while drifting in the wind. The spectra for chaff and rain are similar except that chaff shows less wind shear effect and a different fall velocity profile with altitude than rain [10].

The chaff dipoles are not randomly oriented in a chaff cloud. Their aerodynamic properties are not isotropic, and their orientation is therefore influenced by their fall and by wind. They are also not packaged with random orientation. It has been noted that the dipoles can retain a common orientation, to a significant degree, during their fall. The radar return may then be sensitive to polarization. With a polarization-adaptive radar, a significant suppression of chaff clutter can be achieved [3].

9.4. SEA CLUTTER

A radar located above the sea surface and illuminating the surface may have a significant amount of power backscattering to it from the surface. In many cases, this sea clutter limits the ability of the radar to detect a target [11]. Measurements and theory indicate that the clutter depends on radar frequency and polarization, the grazing angle of the incident wave, and the state of the sea and wind.

Clutter characteristics of the sea are described by a clutter cross section per unit area, $\sigma°$, of the illuminated sea surface. As a function of grazing angle, measurements indicate that this cross section varies somewhat like that of Figure 9.1. In the interference region, the direct wave and that scattered from the surface are out of phase and destructively interfere. The waves that exceed the average height are the scatterers [6, 7]. In the plateau region, the main scatterers are waves comparable in dimension to the radar wavelength [7], and in the quasi-specular region, the scattering is considered to be specular from faceted surfaces perpendicular to the radar line of sight [7]. The boundary angles between the interference and plateau regions and plateau and quasi-specular regions are called the critical and transition

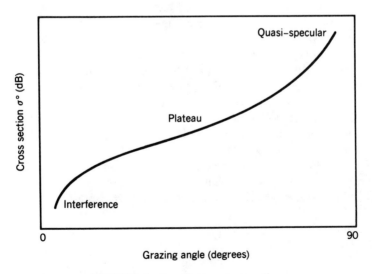

FIGURE 9.1 Sea clutter cross section.

angles respectively. The critical angle depends on frequency, polarization, and sea state [11].

As frequency increases, the sea clutter cross section of Figure 9.1 increases significantly in the interference region and by a lesser amount in the plateau region. It is common to treat $\sigma°$ as varying as f^2 in the interference region, linearly in the plateau region, and little or not at all in the quasi-specular region, but it is difficult to verify these relationships either theoretically or experimentally [11]. Experimental data indicate a critical angle that decreases with frequency.

It is difficult to separate the effects of wind and sea state on the sea-clutter cross section. Wind, which can quickly excite short, so-called capillary waves on the ocean surface, appears to be important at higher microwave frequencies (X-band), while at lower frequencies wave characteristics play a more important role. In general, $\sigma°$ increases with increasing wave height except in the quasi-specular region. There $\sigma°$ may actually decrease [11].

The behavior of the sea clutter cross section with radar polarization is our primary concern, and its behavior with other variables was considered in order to clarify the clutter–polarization relationship. Measurements indicate that $\sigma°$ varies little with polarization (linear vertical and horizontal) except in the interference region, with low grazing angles. There the cross section is considerably greater for vertical polarization than for horizontal with low to moderate sea states [5, 11]. In rough seas the difference between vertical and horizontal polarization is less, or even nonexistent. The difference is also smaller for the higher microwave frequencies [11]. It should be noted that most of the measurements reported have not considered polarization other than linear vertical and horizontal, nor in general were the cross-polarized components measured. As a final remark, measurements of sea clutter cannot be carried out under controlled conditions and are subject to considerable variation in measured cross sections.

Nevertheless, in spite of these caveats, linear horizontal polarization does show an advantage over linear vertical in reducing sea clutter at low grazing angles and at the lower microwave frequencies. The low grazing angle requirement suggests use of linear horizontal polarization for ship-mounted radars that have other surface vessels or coastlines as targets. The majority of radars used for that purpose are horizontally polarized, although some are vertical and some have circular polarization [12]. Most of these radars operate above 9 GHz, where the advantage of horizontal polarization in reducing sea clutter is not as great as at lower frequencies, although a lower frequency (about 3 GHz) is available. The use of circular polarization to reduce rain clutter might be a better choice for this application.

9.5. TARGET GLINT

The use of a variable-polarization radar to reduce glint in an angle-tracking radar is not exactly an application of a polarization-based technique to target

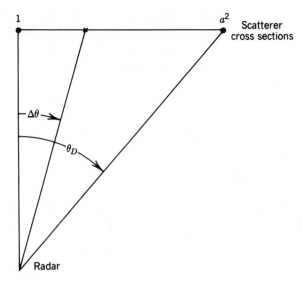

FIGURE 9.2 Apparent radar center.

discrimination, but it does involve the use of polarization in a relatively simple way to ameliorate a real radar problem.

When an angle-tracking radar, such as a phase-monopulse radar, has an angle-extended target, it points to some equivalent position, the apparent target angular position at which the error signal is zero. It is well known that this apparent target center of reflection may not be at the target centroid, and in fact the radar may point to some spot not lying within the angular boundaries of the target. As the target aspect angle changes, the apparent radar center of the target may wander [7]. The phenomenon is called target *glint* and presents serious problems at short ranges to tracking radars.

The apparent radar center of a target is determined by the phasor addition of signal returns from numerous scattering centers of a complex target. For a simplified target model consisting of two scatterers only, as shown in Figure 9.2, the apparent angular target position, which is the angle at which the angle-tracking radar tries to point, is given by [7]

$$\frac{\Delta\theta}{\theta_D} = \frac{a^2 + a \cos \alpha}{1 + a^2 + 2a \cos \alpha} \tag{9.3}$$

where a is the amplitude of the signal from the smaller scatterer relative to that from the larger, α is the phase angle difference between the signals from the scatterers, $\Delta\theta$ is the angle measured from radar boresight (aligned on the larger scatterer) to the apparent target radar center, and θ_D is the angular separation of the two scatterers. It is common to plot (9.3) as a function of phase angle α with parameter a [7, 13], and this shows that for target signal amplitude ratios in the neighborhood of one-half or greater, $\Delta\theta$ exhibits a very large change with phase angle α, for α in the region around 180°. Using more than one frequency, sequentially, to change phase angle α, and averaging the resulting angular equivalent positions has been considered for reducing target glint [14]. A similar technique has been considered using polarization variation [13].

A simple example will suffice to show that polarization variation, with averaging, can reduce the effect of target glint. Consider a target made up of two scatterers, as in Figure 9.2, with a constant aspect angle. Let the left-hand scatterer be a sphere, with constant cross section. Let the right-hand scatterer be a horizontal wire. The polarization variation to be considered will be particularly simple, with both transmitter and receiver linearly polarized at tilt angle δ ($0 \leq \delta \leq \pi/2$) with respect to the horizontal axis. Then the radar cross section of the horizontal wire, if it is chosen to have a maximum cross section of one, is

$$\sigma = \cos^2 \delta \tag{9.4}$$

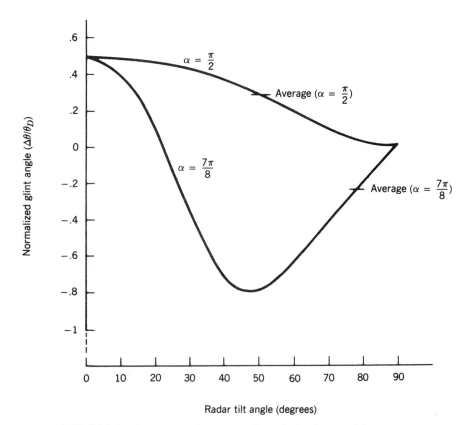

FIGURE 9.3 Apparent radar center of a polarization-sensitive target.

Using this in (9.3) gives for the equivalent angle measured by the tracking radar

$$\frac{\Delta \theta}{\theta_D} = \frac{\cos^2 \delta + \cos \delta \cos \alpha}{1 + \cos^2 \delta + 2 \cos \delta \cos \alpha} \tag{9.5}$$

Figure 9.3 is a plot of this equation as a function of the tilt angle of the radar for two values of the relative phase angle α. Shown also are the averages of the apparent radar center angles for a $\pi/2$ rotation of the radar tilt angle. For a relative phase angle of $7/8\,\pi$, a severe condition, the apparent phase center average lies outside the target. This case might benefit from a different polarization variation or more sophisticated processing than an unweighted average. Even the unweighted average is of benefit, however, in that it gives stability to the tracking system with target aspect changes. As phase angle α changes from 0 to 175°, the apparent radar center of the target varies from 0.5 to -5.2 if the amplitude of the right-hand scatterer of Figure 9.2 is such as to allow the maximum apparent center wander. The average of the apparent radar center angles, with polarization averaging, varies over the smaller range 0.36 to -1.8. It appears, then, that polarization variation with averaging is an effective way to reduce target glint in angle tracking radars.

9.6. A MATRIX TARGET – CLUTTER DISCRIMINANT

The methods suggested in the preceding sections for reducing the interfering effects of clutter do not fully utilize the polarization information available from the target and clutter. Instead, they take advantage of the fact that some polarizations of the radar antenna are better than others for reducing the effects of particular types of clutter. Such processes have been called *vector discriminants*. In more sophisticated techniques, the scattering matrix is measured by the radar and examined to see if it is characteristic of clutter alone or a target–clutter combination. This procedure is a *matrix discriminant* method. A key distinction is that the vector discrimination does not separate the effect of the radar antenna from that of the target (recall the comparison of radar cross sections, which include antenna effects, of raindrops for linear and circular polarizations), while the scattering matrix, obtained by a matrix discriminant method, does not include antenna effects [15].

One matrix discriminant will be introduced here by an example. Consider a target that is a wire or wire grid oriented at 45° from the radar horizontal, with scattering matrix

$$\mathbf{S}_t = \begin{bmatrix} \cos^2 \theta & \cos \theta \sin \theta \\ \cos \theta \sin \theta & \sin^2 \theta \end{bmatrix} = \frac{1}{2} \begin{bmatrix} 1 & 1 \\ 1 & 1 \end{bmatrix} \tag{9.6}$$

The target is in a resolution cell with rain, which is represented by a scattering matrix characteristic of a single spheroid with minor axis vertical, thus

$$\mathbf{S}_r = C \begin{bmatrix} 1 & 0 \\ 0 & a \end{bmatrix} \quad 0 < a \leq 1 \tag{9.7}$$

where C is a constant that relates the rain clutter cross section to that of the target.

The overall scattering matrix of the target in rain, if we make for simplifying purposes the assumption that the scattering matrices add in phase, is

$$\mathbf{S} = \frac{1}{2}\begin{bmatrix} 1 & 1 \\ 1 & 1 \end{bmatrix} + C\begin{bmatrix} 1 & 0 \\ 0 & a \end{bmatrix} \tag{9.8}$$

In essence the problem is, given a measured scattering matrix, to decide whether it falls into a class such as (9.7) for rain alone or one such as (9.8) for rain plus target.

This is a discrimination problem that under some circumstances might be solved by the simple use of circular polarization. For example, if $C = 1$ and $a = 0.9$ (nearly spherical raindrops), the radar cross section of target plus rain is about 14 dB greater than for rain alone, for circular polarization. For linear vertical polarization the signal-plus-clutter to clutter-alone ratio is 3.8 dB, but for both polarizations the cross section depends on the phase relationship of the matrices of (9.8). For a heavier rain, which increases the constant C in (9.8) and flattens the raindrops, the rain cross section increases. For $C = 3$ and $a = 0.5$ the ratio of signal-plus-rain clutter to rain clutter alone is only 1.6 dB for circular polarization, an amount inadequate for reliable detection.

In the approach used here, the measured return from a resolution cell being examined for the presence of a target is used to determine a null polarization of the scattering matrix, using the procedures of Chapter 6. This null polarization is then compared to the corresponding null polarization from a typical resolution cell containing only rain. If these null polarization states are plotted on the Poincaré sphere, there is reason to expect the plotted points to lie close to each other if the return being examined is from rain alone and to lie farther apart if the unknown return is from a cell containing a target alone, or a target plus rain.

For ease of plotting, one of the two-dimensional maps of the Poincaré sphere is used, although lengths are not preserved in these mappings. For the simple example considered here, the τ–ϵ (tilt-ellipticity) map is used, but it must be kept in mind that there is gross distortion for large values of the ellipticity angle. The copolarization nulls of the scattering matrices, corresponding to the polarization ratio P_3 of Chapter 6, will be plotted, although

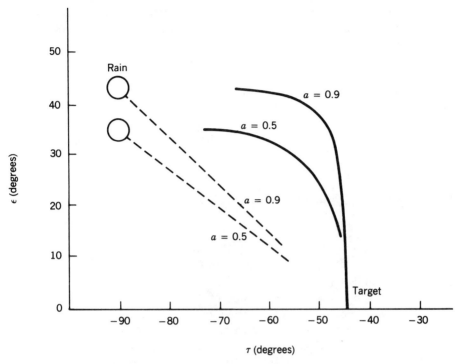

FIGURE 9.4 Null polarizations of a target in rain clutter.

other combinations of the characteristic polarization states might be as useful.

Figure 9.4 is a tilt–ellipticity angle diagram of the copolarization null of (9.8). It represents a 45°-rotated wire in rain with spheroidal raindrops. Two curves (solid lines) are shown, one for parameter a of (9.8) equal to 0.9 (nearly spherical raindrops characteristic of light rain), and for $a = 0.5$ (a flattened spheroidal raindrop more likely to be found with greater rainfall rates). The constant C of (9.8) is varied for each curve from a value of 0.1, representing a negligible amount of rain, to a value of 10, which virtually obscures the target (the maximum power density, with optimum polarization determined from the Graves matrix, is 1 for the target and C^2 for the rain clutter). The dashed curves of the figure are for the same values of a and range of C, but are for a 90° phase addition of the scattering matrices for target and rain. Also shown in the figure are the null polarizations for the target alone and for rain alone, with an assumed spheroidal shape. Since rain may contain drops with a range of ellipticities and wind may cant the drops so that the major axis is not horizontal, each point for the rain alone is surrounded by a circle of indeterminate size to suggest that a measured null for rain may fall within a region of nonzero size on the τ–ϵ map.

492 TARGET DETECTION

The radar detection decision is essentially a decision that the null polarization of a measured scattering matrix falls either into the category of null polarizations of target-plus-rain or the category of rain alone. The methodologies of target discrimination and recognition are beyond the scope of this book, but the reader may find in reference 1 an excellent bibliography on the subject. It must be noted again that for large values of the ellipticity angle, relatively large differences in the tilt angle τ correspond to only small distances on the Poincaré sphere, and the reader should guard against making a facile classification based on the location of a null polarization on this τ–ϵ map.

9.7. SEPARATION OF CONSTANT-NULL-POLARIZATION CLASSES

In the example of the previous section, the task of separating target and clutter was made easier by the choice of constant null polarizations for target and clutter. We shall examine here another case, using measured data, in which land and sea clutter are separated by a process similar to, but more sophisticated than, the simple method of the previous section.

The data to be considered were obtained by a National Aeronautics and Space Administration—Jet Propulsion Laboratory synthetic aperture radar operating at 1.225 GHz. The complete scattering matrix was obtained for each resolution cell, about 10 m × 10 m, of the San Francisco Bay area, containing buildings, streets, parks with vegetation, and ocean. Characteristic polarizations for each resolution cell can be found and used to classify the cell. Alternatively, the characteristic polarizations can be used to choose transmit and receive polarizations to enhance the contrast between the man-made and ocean areas, in effect to remove the ocean clutter and allow better identification of man-made structures. This discussion is based on an analysis by a group at the University of Illinois at Chicago [16]. Other useful analyses of the same data and a description of the radar used have been given [17, 18].

We may proceed with the analysis in various ways. The copolarization null polarizations for the resolution cells in an ocean patch can be obtained with the methods of Chapter 6 (the Kennaugh process) and examined for their statistics. This can be repeated for a known land area. If the statistics of the two areas are sufficiently distinct, each cell can be assigned (not always without ambiguity) to one of the classes. This same process may also be carried out using the cross-polarization nulls (polarizations for maximum received power). Alternatively, the co-pol nulls for an ocean area and the cross-pol nulls for land, or vice versa, might be used for classification. The three-stage optimization process of Chapter 7 may be advantageously used in certain circumstances. If the scattering matrices found from measurements are asymmetric, an optimum transmitting polarization can be found from the Graves matrix of the three-stage process, whereas the methods of Chapter 6 do not allow an overall optimum polarization (applicable to both transmitter and receiver) to be found. Also, if the cross-polarization nulls are found for each of the cells by the Kennaugh method (giving the same polarization for receiver and transmitter) and averaged, the result may not be the receiver polarization obtained by selecting an average transmitting polarization, maximizing the received power from each cell, and averaging the resulting receiving antenna polarizations.

Figure 9.5a is a histogram, on a tilt–ellipticity angle (τ–ϵ) plot, of the polarization vectors that give maximum power density at the receiver for 40,000 (200 × 200) cells in a selected land area. They are the eigenvectors

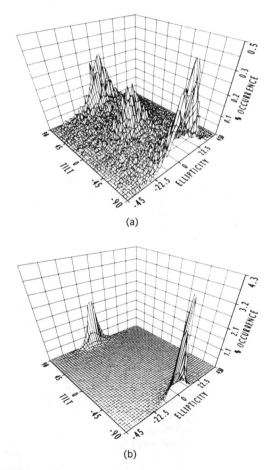

FIGURE 9.5 Polarization states for maximum and minimum power density: (a) maximum density for urban area; (b) minimum density for ocean area. (Reproduced with permission of the National Research Council of Canada, from the *Can. J. Phys.*, **66**, 871, 1988).

494 TARGET DETECTION

corresponding to the larger of the two eigenvalues of the Graves matrix

$$\sigma = S^{T*}S \qquad (7.326)$$

of each cell. The mode is near linear vertical, with substantial dispersion of the eigenvectors. Figure 9.5b shows a similar histogram for an ocean area of the same size, but with the eigenvectors determined from the smaller eigenvalue of the Graves matrix of each cell. This will be called for convenience the minimum power density eigenvector, although it was pointed out in Chapter 7 that it does not in general correspond to the polarization giving the smallest power density (only the smallest readily calculated power density). The minimum-power-density eigenvectors for the ocean are much more concentrated than the maximum-power-density eigenvectors for land, although a wavelength shorter than the 24.5 cm used here, or a higher sea state, would almost certainly alter the ocean area histogram toward a wider dispersion and might conceivably change the mode. By a fortunate coincidence the distribution mode for the minimum density ocean eigenvectors coincides approximately with that for the maximum density land eigenvectors. Therefore, transmission of a linear vertical wave will both maximize the received power density from land and minimize that from the ocean. Absent the fortuitous circumstance of eigenvector coincidence, an appropriate strategy is to choose the transmit polarization to maximize the land area return or to select one to minimize the ocean return. The relative sharpness of the two distributions is a factor in the decision; the sharper distribution should be favored. It should also be recalled that the smaller eigenvalue of the Graves matrix does not give a true minimum power density, and a choice that favors it may not be as effective as one that favors the greater eigenvalue.

After the polarization of the transmitting antenna is selected, the field at the receiver is found for each resolution area element from

$$E^s = CSh_t \qquad (9.9)$$

where h_t is the effective length selected for the transmitting antenna (normalized or otherwise) and C accounts for geometry, wavelength, and power transmitted.

The next step is to select a receiving-antenna polarization that maximizes power received from a land area cell and minimizes that from an ocean cell. If such a selection is impossible—and in the general case it is—a compromise choice must be made for the receiving-antenna polarization. The receiver polarization that maximizes the received power from land is found from the field intensity at the receiver, using

$$\hat{h}_r = \frac{E^{s*}}{|E^s|} \qquad (9.10)$$

where E^s is found from (9.9) for land area cells only.

SEPARATION OF CONSTANT-NULL-POLARIZATION CLASSES

To minimize received power from an ocean area the receiver polarization state is selected to meet the requirement

$$V = \hat{\mathbf{h}}_r^T \mathbf{E}^s = 0 \tag{9.11}$$

where $\mathbf{E}^s$ is the field from an ocean cell, and $\mathbf{h}_r$ and $\mathbf{E}^s$ are in the same coordinate system. Equation (9.11) is readily solved for $\hat{\mathbf{h}}_r$. Expansion gives

$$\hat{h}_{rx} E_x^s \begin{bmatrix} 1 & \dfrac{\hat{h}_{ry}}{\hat{h}_{rx}} \end{bmatrix} \begin{bmatrix} 1 \\ \dfrac{E_y^s}{E_x^s} \end{bmatrix} = 0$$

which gives

$$\frac{\hat{h}_{ry}}{\hat{h}_{rx}} = -\frac{E_x^s}{E_y^s} \tag{9.12}$$

This ratio can be used to form $\hat{\mathbf{h}}_r$ from

$$\hat{\mathbf{h}}_r = \hat{h}_{rx} \begin{bmatrix} 1 \\ \dfrac{\hat{h}_{ry}}{\hat{h}_{rx}} \end{bmatrix} = \hat{h}_{rx} \begin{bmatrix} 1 \\ -\dfrac{E_x^s}{E_y^s} \end{bmatrix} \tag{9.13}$$

Note that the absolute phase of $\hat{\mathbf{h}}_r$ is of no significance and since the normalized form of the antenna effective length is used,

$$|\hat{h}_{rx}|^2 + |\hat{h}_{ry}|^2 = 1 \tag{9.14}$$

from which it follows, using (9.12), that

$$|\hat{h}_{rx}| = \frac{1}{\sqrt{1 + |\hat{h}_{ry}/\hat{h}_{rx}|^2}} = \frac{1}{\sqrt{1 + |E_x^s/E_y^s|^2}} \tag{9.15}$$

Then the receiving antenna polarization state is

$$\hat{\mathbf{h}}_r = \frac{1}{\sqrt{1 + |E_x^s/E_y^s|^2}} \begin{bmatrix} 1 \\ -\dfrac{E_x^s}{E_y^s} \end{bmatrix} \tag{9.16}$$

The ratio of field components in the last equation is readily recognized as the inverse of the polarization ratio

$$P_s = -\frac{E_y^s}{E_x^s} \tag{9.17}$$

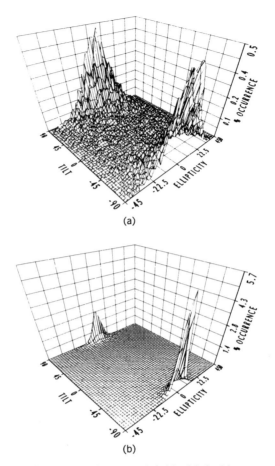

FIGURE 9.6 Polarization states of scattered field with incident wave linear vertical: (*a*) urban area; (*b*) ocean area. (Reproduced with permission of the National Research Council of Canada, from *Can. J. Phys.*, **66**, 871, 1988).

of the scattered wave, where the negative sign is used because the polarization ratio is defined in $x_2 y_2 z_2$ coordinates, with the x axis reversed from the coordinates we are using (see Section 6.4).

Figure 9.6*a* is a histogram of the normalized scattered field

$$\frac{\mathbf{E}^s}{|\mathbf{E}^s|}$$

from 200×200 urban area resolution cells, with the illuminating wave taken to be linear vertical (found previously to give maximum power density for urban area cells). A word of caution is in order here: $\mathbf{E}^s$ is given in $x_1 y_1 z_1$ coordinates of Section 6.4, even though the wave travels in the $-z_1$ direction.

Tilt and ellipticity angles must therefore be interpreted differently for $\mathbf{E}^s$ and $\mathbf{h}_t$. Confusion can be avoided in selecting $\hat{\mathbf{h}}_r$ by using (9.10) to determine $\hat{\mathbf{h}}_r$ for each cell and plotting the histogram for $\hat{\mathbf{h}}_r$. No confusion arises in this instance since the mode for the histogram of Figure 9.5a is approximately linear vertical. The dispersion shows that some repolarization exists for this built-up land area. It is readily seen that to maximize the power to a receiving antenna for the greatest number of land cells, the polarization should be linear vertical.

Figure 9.6b is a histogram of the scattered field from 200 × 200 ocean area resolution cells, again with a linear–vertical polarization assumed for the transmitting antenna. The same word of caution about tilt and ellipticity angles applies in this situation. If confusion exists, it can be removed by finding $\hat{\mathbf{h}}_r$ from (9.16) and plotting its histogram rather than that of $\mathbf{E}^s$. The mode for this distribution is again, by a fortunate coincidence, linear vertical. Now the field $\mathbf{E}^s$ used for this histogram was a minimum-power-density field. Then to minimize the power in the receiving antenna, $\hat{\mathbf{h}}_r$ should be selected to match the mode of $\mathbf{E}^s$; it should be linear vertical. We find, just as we did for the transmitter, that the same antenna both maximizes the urban area return and minimizes the ocean backscatter. If two different polarizations are found, one to maximize one type of return and a different one to minimize the second type of return, the reasonable choice is one of the two polarizations determined, selected with the guidelines used for the transmitter polarization.

The results of this antenna optimization may be used with the scattering matrix data for each resolution cell to increase the urban area–ocean contrast in an image (one of the purposes of the referenced study [16] of the San Francisco Bay area). It may also be used to select a polarization for other radar mapping of the same or similar area.

The reader may have noted that a conclusion of Section 9.4, that linear horizontal polarization is effective in reducing sea clutter, is in contrast to the results of this section showing ocean return to be reduced by linear vertical polarization. It should be kept in mind that sea clutter reduction with horizontal polarization is effective for low-grazing angles, whereas the San Francisco Bay area data were obtained at a relatively high grazing angle.

9.8. POLARIZATION DIVERSITY AND ADAPTATION

Figure 9.7 shows a lossless system that is able to transmit a wave of any desired polarization or to receive optimally a wave of any polarization when properly set for the polarization of the incident wave. A detailed analysis of the system was given in Chapter 5. The transmitted, or optimally received, polarization is controlled by two phase shifters. When they are set to receive optimally at one port of the lower 3-dB hybrid a wave from an antenna of fixed polarization, a transmitter then connected to the same port will radiate

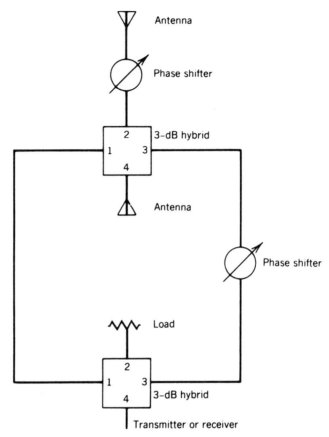

FIGURE 9.7 System for transmitting and receiving arbitrarily polarized waves.

back to the fixed-polarization antenna a wave that is optimally received by it. Further, a transmitter connected to the other port of the 3-dB hybrid will radiate a wave that is cross-polarized to the fixed-polarization antenna, and a receiver connected to the other port will receive optimally a wave orthogonally polarized to the wave from the fixed-polarization antenna. This property readily allows polarization coding of signals, using any desired orthogonal polarizations, or polarization diversity for other reasons, such as propagation path losses. Although the analysis of Chapter 5 was carried out for two linearly polarized antennas, it is obvious that two opposite-sense circular antennas, or any two orthogonally polarized antennas, can be used without affecting the ability of the system to achieve any desired polarization state.

With ordinary reception, using one receiver, it is normally desired to maximize the signal-to-interference-plus-noise ratio (SINR). This may be done by an operator, provided that the SINR can be determined in some fashion. On the other hand, this two-antenna configuration is a special case of an antenna array (with dissimilar elements) for which procedures exist that allow the array weights (phase shifters) to be set automatically to maximize the SINR, assuming appropriate information about the incoming signal is available. Adaptive antenna array theory can be used to develop weight-setting algorithms not only for one of these variable-polarization systems but for a $2N$-element array comprised of N such systems, allowing beam steering and polarization adaptation to be done simultaneously. An introduction to adaptive arrays was presented in Chapter 2, and Compton [19] covers the subject more thoroughly. The background required of the reader in antennas, random processes, analytic signals, and matrices is almost certainly possessed by the reader of this book, and that required in feedback control theory is supplied by an undergraduate degree in electrical engineering.

Figure 9.7 implies that the polarization adaptation is to be done at RF, but it can readily be done at IF on reception if the antenna elements are followed by mixers.

If the transmitter and receiver of Figure 9.7 are connected to the same port by a circulator and if the transmitter and receiver are appropriate to a radar, a polarization-adaptive radar can be constructed. This may not be the best way to configure one, however, since the adaptation criteria for a radar and a two-way communication system may not be the same. The polarization of the target return can vary quickly with time [3], and two targets illuminated by the same pulse at different ranges can have greatly different polarizations. The adaptivity requirements will be less stringent to minimize the clutter return (which generally has a slower time variation than the target return [3]) than to maximize the SINR. A second factor is important also. If two targets at different ranges are illuminated by the same pulse, the transmitter polarization cannot be optimum for both. A compromise transmitter polarization could be used, but even then it would be necessary to vary the polarization pulse-to-pulse, which is difficult because of adaptation time requirements [3]. Polarization switching for high-power radars in particular is slow.

The factors discussed above may well make it desirable to use polarization to minimize the clutter returns only, but the problem of transmitter polarization persists. Different clutter can exist in different range resolution cells (in the same beam direction), rain, for example, in one cell or group or cells, so no optimum transmitter polarization can be selected.

The *virtual polarization adaptation technique* of Poelman [20] offers an attractive solution to these problems. When it is used, no attempt is made to adapt the antennas, either on transmission or reception, to the target. Instead, two orthogonally polarized signals, for example, linear vertical and horizontal, are transmitted sequentially (feasible pulse-to-pulse). This can be done with the system of Figure 9.7 by switching the transmitter from port 2 to port 4 and back. For each transmitted polarization, a two-channel receiver measures the response at both ports 2 and 4. Information is then available to determine the scattering matrix of the resolution cell, and from this matrix the response to *any* polarization of transmitter and receiver can be calculated. The adaptation takes place in a processor that follows the two-channel receiver, and it can be done fast enough to accommodate each resolution cell. Some deficiencies of the method have been pointed out, namely that a detection loss is incurred relative to transmission of the optimum polarization, that coherence of the radar on reception is required, and that Doppler compensation is necessary [3], but nonetheless it is a useful technique for polarization-adaptive radar.

REFERENCES

1. N. F. Ezquerra, "Target Recognition Considerations," in *Principles of Modern Radar*, J. L. Eaves and E. K. Reedy, Eds., Van Nostrand Reinhold, New York, 1987.
2. M. I. Skolnik, *Introduction to Radar Systems*, McGraw-Hill, New York, 1962.
3. D. Giuli, "Polarization Diversity in Radars," *Proc. IEEE*, **79**(2), 245–269, February 1986.
4. D. E. Kerr and H. Goldstein, "Radar Targets and Echoes," in *Propagation of Short Radio Waves*, D. E. Kerr, Ed., McGraw-Hill, New York, 1951.
5. N. C. Currie, "Clutter Characteristics and Effects," in *Principles of Modern Radar*, J. L. Eaves and E. K. Reedy, Eds., Von Nostrand Reinhold, New York, 1987.
6. M. W. Long, *Radar Reflectivity of Land and Sea*, Lexington Books, Lexington, MA, 1975.
7. M. I. Skolnik, *Introduction to Radar Systems*, 2nd Ed., McGraw-Hill, New York, 1980.
8. B. R. Bean, E. J. Dutton, and B. D. Warner, "Weather Effects on Radar," in *Radar Handbook*, M. I. Skolnik, Ed., McGraw-Hill, New York, 1970.
9. A. P. Agrawal, "A Polarimetric Rain Backscatter Model Developed for Coherent Polarization Diversity Radar Applications," Ph.D. Thesis, University of Illinois, Chicago, IL, 1987.
10. F. E. Nathanson, *Radar Design Principles*, McGraw-Hill, New York, 1951.
11. M. I. Skolnik, "Sea Echo," in *Radar Handbook*, M. I. Skolnik, Ed., McGraw-Hill, New York, 1970.
12. J. Croney, "Civil Marine Radar," in *Radar Handbook*, M. I. Skolnik, Ed., McGraw-Hill, New York, 1970.
13. G. W. Ewell, N. T. Alexander, and E. L. Tomberlin, "Investigation of the Effects of Polarization Agility on Monopulse Radar Angle Tracking," Final Technical Report, Vol. 1, Contract DAAH01-70-C-0535, Project A-1225, Engineering Experiment Station, Georgia Institute of Technology, June 1971.
14. J. M. Loomis and E. R. Graf, "Frequency-Agility Processing to Reduce Radar Glint Pointing Error," *IEEE Tran. Aerospace Elec. Systems*, **AES-10**, 811–820, November 1974.

15. W. A. Holm, "Polarimetric Fundamentals and Techniques," in *Principles of Modern Radar*, J. L. Eaves and E. K. Reedy, Eds., Van Nostrand Reinhold, New York, 1987.
16. A. B. Kostinski, B. D. James, and W-M. Boerner, "Polarimetric Matched Filter for Coherent Imaging," *Canad. J. Phys.*, **66**, 871–877, 1988.
17. J. J. van Zyl and H. A. Zebker, "Imaging Radar Polarimetry," in *Polarimetric Remote Sensing*, PIER 3, J. A. Kong, Ed., Elsevier, New York, 1990.
18. J. A. Kong, S. H. Yuch, H. H. Lim, R. T. Shin, and J. J. van Zyl, "Classification of Earth Terrain Using Polarimetric Synthetic Aperture Radar Images," in *Polarimetric Remote Sensing*, PIER 3, J. A. Kong, Ed., Elsevier, New York, 1990.
19. R. T. Compton, Jr., *Adaptive Antennas*, Prentice-Hall, Englewood Cliffs, NJ, 1988.
20. A. J. Poelman and J. R. F. Guy, "Polarization Information Utilization in Primary Radar," in *Inverse Methods in Electromagnetic Imaging*, W-M. Boerner, Ed., D. Reidel, Dordrecht, Holland, 1985.

APPENDIX A

The Mueller Matrix

A.1. ELEMENTS

The Mueller matrix is given by

$$\mathbf{M} = \mathbf{Q}(\mathbf{T} \times \mathbf{T}^*)\mathbf{Q}^{-1} \qquad (7.199)$$

where

$$\mathbf{T} \times \mathbf{T}^* = \begin{bmatrix} |T_{xx}|^2 & T_{xx}T_{xy}^* & T_{xx}^*T_{xy} & |T_{xy}|^2 \\ T_{xx}T_{yx}^* & T_{xx}T_{yy}^* & T_{xy}T_{yx}^* & T_{xy}T_{yy}^* \\ T_{xx}^*T_{yx} & T_{xy}^*T_{yx} & T_{xx}^*T_{yy} & T_{xy}^*T_{yy} \\ |T_{yx}|^2 & T_{yx}T_{yy}^* & T_{yx}^*T_{yy} & |T_{yy}|^2 \end{bmatrix} \qquad (A.1)$$

is the direct or Kronecker product of the Jones matrix and its conjugate. Time averaging of the product may be appropriate in some circumstances but will not be explicitly considered here. In (7.199),

$$\mathbf{Q} = \begin{bmatrix} 1 & 0 & 0 & 1 \\ 1 & 0 & 0 & -1 \\ 0 & 1 & 1 & 0 \\ 0 & j & -j & 0 \end{bmatrix} \qquad (7.190)$$

with inverse

$$\mathbf{Q}^{-1} = \frac{1}{2}\begin{bmatrix} 1 & 1 & 0 & 0 \\ 0 & 0 & 1 & -j \\ 0 & 0 & 1 & j \\ 1 & -1 & 0 & 0 \end{bmatrix} \qquad (A.2)$$

ELEMENTS

If the multiplication in (7.199) is carried out, the elements of **M** in

$$\mathbf{M} = \mathbf{Q}(\mathbf{T} \times \mathbf{T}^*)\mathbf{Q}^{-1} = \begin{bmatrix} M_{11} & M_{12} & M_{13} & M_{14} \\ M_{21} & M_{22} & M_{23} & M_{24} \\ M_{31} & M_{32} & M_{33} & M_{34} \\ M_{41} & M_{42} & M_{43} & M_{44} \end{bmatrix} \quad (\text{A.3})$$

are given by

$$M_{11} = \tfrac{1}{2}\left(|T_{xx}|^2 + |T_{xy}|^2 + |T_{yx}|^2 + |T_{yy}|^2\right) \quad (\text{A.4a})$$

$$M_{12} = \tfrac{1}{2}\left(|T_{xx}|^2 - |T_{xy}|^2 + |T_{yx}|^2 - |T_{yy}|^2\right) \quad (\text{A.4b})$$

$$M_{13} = \operatorname{Re}(T_{xx}T_{xy}^* + T_{yx}T_{yy}^*) \quad (\text{A.4c})$$

$$M_{14} = \operatorname{Im}(T_{xx}T_{xy}^* + T_{yx}T_{yy}^*) \quad (\text{A.4d})$$

$$M_{21} = \tfrac{1}{2}\left(|T_{xx}|^2 + |T_{xy}|^2 - |T_{yx}|^2 - |T_{yy}|^2\right) \quad (\text{A.4e})$$

$$M_{22} = \tfrac{1}{2}\left(|T_{xx}|^2 - |T_{xy}|^2 - |T_{yx}|^2 + |T_{yy}|^2\right) \quad (\text{A.4f})$$

$$M_{23} = \operatorname{Re}(T_{xx}T_{xy}^* - T_{yx}T_{yy}^*) \quad (\text{A.4g})$$

$$M_{24} = \operatorname{Im}(T_{xx}T_{xy}^* - T_{yx}T_{yy}^*) \quad (\text{A.4h})$$

$$M_{31} = \operatorname{Re}(T_{xx}T_{yx}^* + T_{xy}T_{yy}^*) \quad (\text{A.4i})$$

$$M_{32} = \operatorname{Re}(T_{xx}T_{yx}^* - T_{xy}T_{yy}^*) \quad (\text{A.4j})$$

$$M_{33} = \operatorname{Re}(T_{xx}T_{yy}^* + T_{xy}T_{yx}^*) \quad (\text{A.4k})$$

$$M_{34} = \operatorname{Im}(T_{xx}T_{yy}^* - T_{xy}T_{yx}^*) \quad (\text{A.4l})$$

$$M_{41} = \operatorname{Im}(T_{xx}^*T_{yx} + T_{xy}^*T_{yy}) \quad (\text{A.4m})$$

$$M_{42} = \operatorname{Im}(T_{xx}^*T_{yx} - T_{xy}^*T_{yy}) \quad (\text{A.4n})$$

$$M_{43} = \operatorname{Im}(T_{xx}^*T_{yy} + T_{xy}^*T_{yx}) \quad (\text{A.4o})$$

$$M_{44} = \operatorname{Re}(T_{xx}T_{yy}^* - T_{xy}T_{yx}^*) \quad (\text{A.4p})$$

In order to find M_{ij} in terms of the Sinclair matrix **S**, which gives the scattered fields in $x_3 y_3 z_3$ coordinates, the elements of **T** are replaced according to

$$S_{xx} = -T_{xx} \quad (\text{A.5a})$$

$$S_{xy} = -T_{xy} \quad (\text{A.5b})$$

$$S_{yx} = T_{yx} \quad (\text{A.5c})$$

$$S_{yy} = T_{yy} \quad (\text{A.5d})$$

If the Sinclair matrix is symmetric, as it is for backscattering and some other cases as well,

$$S_{yx} = S_{xy} = -T_{xy} \tag{A.6}$$

If these substitutions are made, the Mueller elements become, for backscattering in terms of the Sinclair matrix,

$$M_{11} = \tfrac{1}{2}\left(|S_{xx}|^2 + 2|S_{xy}|^2 + |S_{yy}|^2\right) \tag{A.7a}$$

$$M_{12} = \tfrac{1}{2}\left(|S_{xx}|^2 - |S_{yy}|^2\right) \tag{A.7b}$$

$$M_{13} = \mathrm{Re}(S_{xx}S_{xy}^* + S_{xy}S_{yy}^*) \tag{A.7c}$$

$$M_{14} = \mathrm{Im}(S_{xx}S_{xy}^* + S_{xy}S_{yy}^*) \tag{A.7d}$$

$$M_{21} = M_{12} \tag{A.7e}$$

$$M_{22} = \tfrac{1}{2}\left(|S_{xx}|^2 - 2|S_{xy}|^2 + |S_{yy}|^2\right) \tag{A.7f}$$

$$M_{23} = \mathrm{Re}(S_{xx}S_{xy}^* - S_{xy}S_{yy}^*) \tag{A.7g}$$

$$M_{24} = \mathrm{Im}(S_{xx}S_{xy}^* - S_{xy}S_{yy}^*) \tag{A.7h}$$

$$M_{31} = -M_{13} \tag{A.7i}$$

$$M_{32} = -M_{23} \tag{A.7j}$$

$$M_{33} = -|S_{xy}|^2 - \mathrm{Re}(S_{xx}S_{yy}^*) \tag{A.7k}$$

$$M_{34} = -\mathrm{Im}(S_{xx}S_{yy}^*) \tag{A.7l}$$

$$M_{41} = M_{14} \tag{A.7m}$$

$$M_{42} = M_{24} \tag{A.7n}$$

$$M_{43} = -M_{34} \tag{A.7o}$$

$$M_{44} = |S_{xy}|^2 - \mathrm{Re}(S_{xx}S_{yy}^*) \tag{A.7p}$$

A.2. BACKSCATTERING MATRICES FOR COMMON SCATTERERS

Flat Plate

For a disk of radius R at normal incidence,

$$\mathbf{S} = \frac{2\sqrt{\pi}}{\lambda}(\pi R^2)\begin{bmatrix} -1 & 0 \\ 0 & -1 \end{bmatrix} \tag{6.249}$$

$$\mathbf{M} = \frac{4\pi^3 R^4}{\lambda^2}\begin{bmatrix} 1 & 0 & 0 & 0 \\ 0 & 1 & 0 & 0 \\ 0 & 0 & -1 & 0 \\ 0 & 0 & 0 & -1 \end{bmatrix} \tag{A.8}$$

Dihedral Corner

For the corner fold line tilted at angle θ from the y axis and incident wave at angle ϕ from one plane, the dihedral corner with dimensions a and b has

$$S = \frac{4\sqrt{\pi}\,ab\,\sin(\pi/4 + \phi)}{\lambda} \begin{bmatrix} -\cos 2\theta & \sin 2\theta \\ \sin 2\theta & \cos 2\theta \end{bmatrix} \quad (6.263)$$

$$M = \frac{16\pi a^2 b^2 \sin^2(\pi/4 + \phi)}{\lambda^2} \begin{bmatrix} 1 & 0 & 0 & 0 \\ 0 & \cos 4\theta & -\sin 4\theta & 0 \\ 0 & \sin 4\theta & \cos 4\theta & 0 \\ 0 & 0 & 0 & 1 \end{bmatrix} \quad (A.9)$$

Trihedral Corner

For a square trihedral corner with side L,

$$S = \frac{2\sqrt{3\pi}\,L^2}{\lambda} \begin{bmatrix} -1 & 0 \\ 0 & -1 \end{bmatrix} \quad (6.268a)$$

$$M = \frac{12\pi L^4}{\lambda^2} \begin{bmatrix} 1 & 0 & 0 & 0 \\ 0 & 1 & 0 & 0 \\ 0 & 0 & -1 & 0 \\ 0 & 0 & 0 & -1 \end{bmatrix} \quad (A.10)$$

Sphere

The Sinclair matrix is

$$S = S_{xx} \begin{bmatrix} 1 & 0 \\ 0 & 1 \end{bmatrix} \quad (6.251)$$

where S_{xx} is a function of frequency. Then

$$M = |S_{xx}|^2 \begin{bmatrix} 1 & 0 & 0 & 0 \\ 0 & 1 & 0 & 0 \\ 0 & 0 & -1 & 0 \\ 0 & 0 & 0 & -1 \end{bmatrix} \quad (A.11)$$

A.3. TRANSMISSION MATRICES FOR OPTICAL DEVICES

These devices are assumed here to operate without loss. The coordinate systems $x_1 y_1 z_1$ and $x_2 y_2 z_2$ coincide for transmission.

Rotator

A rotator changes the plane of polarization of a wave. Let it cause a clockwise ($x \rightarrow y$) rotation through angle α. Incident and transmitted waves

506 THE MUELLER MATRIX

are related by

$$E_x^t = E_x^i \cos \alpha - E_y^i \sin \alpha \tag{A.12a}$$

$$E_y^t = E_x^i \sin \alpha + E_y^i \cos \alpha \tag{A.12b}$$

giving a Jones matrix (or forward scattering matrix)

$$\mathbf{T} = \begin{bmatrix} \cos \alpha & -\sin \alpha \\ \sin \alpha & \cos \alpha \end{bmatrix} \tag{A.13}$$

and from (A.4) a Mueller matrix,

$$\mathbf{M} = \begin{bmatrix} 1 & 0 & 0 & 0 \\ 0 & \cos 2\alpha & -\sin 2\alpha & 0 \\ 0 & \sin 2\alpha & \cos 2\alpha & 0 \\ 0 & 0 & 0 & 1 \end{bmatrix} \tag{A.14}$$

Compensator

A compensator retards the phase of a wave component. If the retardation of E_y is 2δ, then

$$\mathbf{T} = \begin{bmatrix} e^{j\delta} & 0 \\ 0 & e^{-j\delta} \end{bmatrix} \tag{A.15}$$

and

$$\mathbf{M} = \begin{bmatrix} 1 & 0 & 0 & 0 \\ 0 & 1 & 0 & 0 \\ 0 & 0 & \cos 2\delta & \sin 2\delta \\ 0 & 0 & -\sin 2\delta & \cos 2\delta \end{bmatrix} \tag{A.16}$$

Polarizer

Let the polarizer transmit only the wave component at angle α to the x axis, as in Figure A.1. The incident wave is $\mathbf{E}^i$, and the wave $\mathbf{E}^t$ at angle α is transmitted. The direction of $\mathbf{E}^t$ is that of the unit vector

$$\mathbf{u} = \cos \alpha \mathbf{u}_x + \sin \alpha \mathbf{u}_y \tag{A.17}$$

and its length is $\mathbf{u} \cdot \mathbf{E}^i$. Then

$$\mathbf{E}^t = (\mathbf{u} \cdot \mathbf{E}^i)\mathbf{u} = \left(\cos \alpha E_x^i + \sin \alpha E_y^i\right)\left(\cos \alpha \mathbf{u}_x + \sin \alpha \mathbf{u}_y\right) \tag{A.18}$$

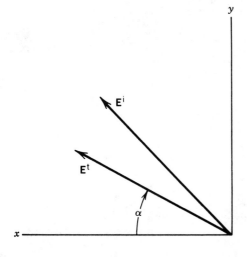

FIGURE A.1 Transmission through polarizer.

The transmission matrix is

$$T = \begin{bmatrix} \cos^2 \alpha & \cos \alpha \sin \alpha \\ \cos \alpha \sin \alpha & \sin^2 \alpha \end{bmatrix} \quad (A.19)$$

and the Mueller matrix is

$$M = \frac{1}{2} \begin{bmatrix} 1 & \cos 2\alpha & \sin 2\alpha & 0 \\ \cos 2\alpha & \cos^2 2\alpha & \sin 2\alpha \cos 2\alpha & 0 \\ \sin 2\alpha & \sin 2\alpha \cos 2\alpha & \sin^2 2\alpha & 0 \\ 0 & 0 & 0 & 0 \end{bmatrix} \quad (A.20)$$

Absorber

An absorber attenuates the x and y wave components differently (equal attenuation does not affect polarization). With attenuation constants α_x and α_y and distance d traveled, the transmission matrix is

$$T = \begin{bmatrix} e^{-\alpha_x d} & 0 \\ 0 & e^{-\alpha_y d} \end{bmatrix} \quad (A.21)$$

Let

$$\alpha = \tfrac{1}{2}(\alpha_x + \alpha_y) \quad (A.22a)$$

$$\epsilon = \tfrac{1}{2}(\alpha_x - \alpha_y) \quad (A.22b)$$

Then

$$\mathbf{T} = e^{-\alpha d}\begin{bmatrix} e^{-\epsilon d} & 0 \\ 0 & e^{\epsilon d} \end{bmatrix} \quad (A.23)$$

and

$$\mathbf{M} = e^{-2\alpha d}\begin{bmatrix} \cosh 2\epsilon d & \sinh 2\epsilon d & 0 & 0 \\ \sinh 2\epsilon d & \cosh 2\epsilon d & 0 & 0 \\ 0 & 0 & 1 & 0 \\ 0 & 0 & 0 & 1 \end{bmatrix} \quad (A.24)$$

APPENDIX B

The Kennaugh Matrix

B.1. ELEMENTS

The Kennaugh matrix used in radar scattering is

$$\mathbf{K} = \tfrac{1}{2}\mathbf{Q}^* \langle \mathbf{S} \times \mathbf{S}^* \rangle \mathbf{Q}^{T*} \qquad (7.247)$$

where

$$\mathbf{Q} = \begin{bmatrix} 1 & 0 & 0 & 1 \\ 1 & 0 & 0 & -1 \\ 0 & 1 & 1 & 0 \\ 0 & j & -j & 0 \end{bmatrix} \qquad (7.190)$$

and

$$\mathbf{S} \times \mathbf{S}^* = \begin{bmatrix} |S_{xx}|^2 & S_{xx}S_{xy}^* & S_{xx}^*S_{xy} & |S_{xy}|^2 \\ S_{xx}S_{yx}^* & S_{xx}S_{yy}^* & S_{xy}S_{yx}^* & S_{xy}S_{yy}^* \\ S_{xx}^*S_{yx} & S_{xy}^*S_{yx} & S_{xx}^*S_{yy} & S_{xy}^*S_{yy} \\ |S_{yx}|^2 & S_{yx}S_{yy}^* & S_{yx}^*S_{yy} & |S_{yy}|^2 \end{bmatrix} \qquad (B.1)$$

If the matrix multiplications are performed, the matrix elements are (with the time average implicit)

$$K_{11} = \tfrac{1}{2}\left(|S_{xx}|^2 + |S_{xy}|^2 + |S_{yx}|^2 + |S_{yy}|^2\right) \qquad (B.2a)$$

$$K_{12} = \tfrac{1}{2}\left(|S_{xx}|^2 - |S_{xy}|^2 + |S_{yx}|^2 - |S_{yy}|^2\right) \qquad (B.2b)$$

$$K_{13} = \mathrm{Re}\left(S_{xx}S_{xy}^* + S_{yx}S_{yy}^*\right) \qquad (B.2c)$$

$$K_{14} = \text{Im}(S_{xx}S_{xy}^* + S_{yx}S_{yy}^*) \qquad (B.2d)$$

$$K_{21} = \tfrac{1}{2}(|S_{xx}|^2 + |S_{xy}|^2 - |S_{yx}|^2 - |S_{yy}|^2) \qquad (B.2e)$$

$$K_{22} = \tfrac{1}{2}(|S_{xx}|^2 - |S_{xy}|^2 - |S_{yx}|^2 + |S_{yy}|^2) \qquad (B.2f)$$

$$K_{23} = \text{Re}(S_{xx}S_{xy}^* - S_{yx}S_{yy}^*) \qquad (B.2g)$$

$$K_{24} = \text{Im}(S_{xx}S_{xy}^* - S_{yx}S_{yy}^*) \qquad (B.2h)$$

$$K_{31} = \text{Re}(S_{xx}S_{yx}^* + S_{xy}S_{yy}^*) \qquad (B.2i)$$

$$K_{32} = \text{Re}(S_{xx}S_{yx}^* - S_{xy}S_{yy}^*) \qquad (B.2j)$$

$$K_{33} = \text{Re}(S_{xx}S_{yy}^* + S_{xy}^*S_{yx}) \qquad (B.2k)$$

$$K_{34} = \text{Im}(S_{xx}S_{yy}^* + S_{xy}^*S_{yx}) \qquad (B.2l)$$

$$K_{41} = \text{Im}(S_{xx}S_{yx}^* + S_{xy}S_{yy}^*) \qquad (B.2m)$$

$$K_{42} = \text{Im}(S_{xx}S_{yx}^* - S_{xy}S_{yy}^*) \qquad (B.2n)$$

$$K_{43} = \text{Im}(S_{xx}S_{yy}^* - S_{xy}^*S_{yx}) \qquad (B.2o)$$

$$K_{44} = \text{Re}(S_{xy}S_{yx}^* - S_{xx}S_{yy}^*) \qquad (B.2p)$$

For backscattering, $S_{yx} = S_{xy}$, and the matrix elements become

$$K_{11} = \tfrac{1}{2}(|S_{xx}|^2 + 2|S_{xy}|^2 + |S_{yy}|^2) \qquad (B.3a)$$

$$K_{12} = K_{21} = \tfrac{1}{2}(|S_{xx}|^2 - |S_{yy}|^2) \qquad (B.3b)$$

$$K_{13} = K_{31} = \text{Re}(S_{xx}S_{xy}^* + S_{xy}S_{yy}^*) \qquad (B.3c)$$

$$K_{14} = K_{41} = \text{Im}(S_{xx}S_{xy}^* + S_{xy}S_{yy}^*) \qquad (B.3d)$$

$$K_{22} = \tfrac{1}{2}(|S_{xx}|^2 - 2|S_{xy}|^2 + |S_{yy}|^2) \qquad (B.3e)$$

$$K_{23} = K_{32} = \text{Re}(S_{xx}S_{xy}^* - S_{xy}S_{yy}^*) \qquad (B.3f)$$

$$K_{24} = K_{42} = \text{Im}(S_{xx}S_{xy}^* - S_{xy}S_{yy}^*) \qquad (B.3g)$$

$$K_{33} = \text{Re}(S_{xx}S_{yy}^*) + |S_{xy}|^2 \qquad (B.3h)$$

$$K_{34} = K_{43} = \text{Im}(S_{xx}S_{yy}^*) \qquad (B.3i)$$

$$K_{44} = |S_{xy}|^2 - \text{Re}(S_{xx}S_{yy}^*) \qquad (B.3j)$$

B.2. BACKSCATTERING MATRICES FOR COMMON SCATTERERS*

Flat Plate

For a disk of radius R at normal incidence,

$$\mathbf{K} = \frac{4\pi^3 R^4}{\lambda^2} \begin{bmatrix} 1 & 0 & 0 & 0 \\ 0 & 1 & 0 & 0 \\ 0 & 0 & 1 & 0 \\ 0 & 0 & 0 & -1 \end{bmatrix} \quad (B.4)$$

Dihedral Corner

For the corner fold line tilted at angle θ from the y axis and incident wave at angle ϕ from one plane, the dihedral corner with dimensions a and b has matrix,

$$\mathbf{K} = \frac{16\pi a^2 b^2 \sin^2(\pi/4 + \phi)}{\lambda^2} \begin{bmatrix} 1 & 0 & 0 & 0 \\ 0 & \cos 4\theta & -\sin 4\theta & 0 \\ 0 & -\sin 4\theta & -\cos 4\theta & 0 \\ 0 & 0 & 0 & 1 \end{bmatrix} \quad (B.5)$$

Trihedral Corner

For a square trihedral corner with side L,

$$\mathbf{K} = \frac{12\pi L^4}{\lambda^2} \begin{bmatrix} 1 & 0 & 0 & 0 \\ 0 & 1 & 0 & 0 \\ 0 & 0 & 1 & 0 \\ 0 & 0 & 0 & -1 \end{bmatrix} \quad (B.6)$$

Sphere

$$\mathbf{K} = |S_{xx}|^2 \begin{bmatrix} 1 & 0 & 0 & 0 \\ 0 & 1 & 0 & 0 \\ 0 & 0 & 1 & 0 \\ 0 & 0 & 0 & -1 \end{bmatrix} \quad (B.7)$$

where S_{xx} is a function of frequency.

*Note: The Sinclair matrices for these scatterers appear in Appendix A.

Index

Absorber, 507–508
Adams, O. S., 179
Adaptive arrays, 79–88
  Applebaum array, 88
  LMS (Widrow) array, 80–87
  phase-lock-loop, 79–80
  Shor array, 88
Agrawal, A. P., 500
Alexander, N. T., 500
Allied integral, 370
Analytic signal, 368–372
Antenna(s):
  adaptive, see Adaptive arrays
  admittance, 38
  aperture, 38, 109
    tapered-feed, 109
  area, see Effective area
  arrays, see Adaptive arrays; Antenna arrays
  bandwidth, 2, 389, 424
  broadband, 95–98. See also Antenna(s), frequency-independent; Antenna(s), log-periodic
    biconical, 96–97
    bow-tie, 96–97
    ridged horn, 97–98
  circularly-polarized, 91
    axial-mode helix, 267–273
    logarithmic spiral, 90
    loop and dipole, 237–239
    traveling-wave loop, 264–267
  crossed-dipole, 232–233
    with ground plane, 233–235
  cross-polarized, 197
  dipole, 22, 25
    current distribution on, 22, 26, 56
    fields, 25–29
    half-wave, 37
    input impedance, 32–37
    short, 56, 229–231
  directivity, 21
  effective length, 182–183, 185–188, 323
  efficiency, 45. See also Beam efficiency; Radiation, efficiency
  elementary, 18, 22
  elliptically-polarized, 273–278. See also Elliptically-polarized antenna system
  equiangular spiral, 90
  equivalent circuit, 37–38, 44–45, 48, 56
  feed, 23, 38
    tapered, 109
  frequency-independent, 88–92
  gain, 21, 22, 25, 47
  helix, 267–273
  horn:
    E-plane sectoral, 245
    H-plane sectoral, 245
    pyramidal, 97, 242–246
  identical cross-polarized, 199
  identical polarization-matched, 198–199
  impedance, 37–38, 42
    input, 32–37
    self, 44
  infinitesimal, 18, 50
  isotropic, 51
  logarithmic spiral, 90
  log-periodic, 92–95
  loop, 235–236
  loop and dipole, 237–239
  losses, 23–24, 111
  loss resistance, 24, 37
  misaligned, 212–219

514 INDEX

Antenna(s) (*Continued*)
  noise temperature, 103, 109-111. *See also* Noise
  omnidirectional, 21
  open waveguide, 38-42, 239-242
  parabolic (paraboloidal) reflector, 57, 246-261
  pattern, 15-21. *See also* Antenna(s), receiving pattern
  polarization-matched, 196
  polarization, measurement of, 463-465
  polarization ratio, 192
  receiving, 42-50
    pattern, 45-47
  receiving polarization, 203, 211
  resistance, 22-25, 37, 45. *See also* Radiation, resistance
  ridged horn, 97-98
  slot, 91, 94
  spiral slot, 91
  temperature, 103, 109-111. *See also* Noise
  traveling-wave loop, 264-267
  turnstile, 232-233
  wire approximation:
    biconical, 96-97
    bow-tie, 96
    log-periodic, 94
Antenna arrays:
  adaptive, 79-88
  array factor, 69, 71
  beam cross section, 75-78
  beam scanning width, 73-78
  element pattern, 68, 71
  linear, 68-70
  narrow polarization beamwidth, 261-264
  planar, 70-78
Anterior scalar product, 67
Aperture antenna, 38
Aperture plane, 246, 251
Applebaum array, 88
Area. *See* Effective area
Area sweep rate, 128-129. *See also* Polarization ellipse
Arrays. *See* Antenna arrays
Atmospheric losses, 108, 110
Autocorrelation, 419
Axial ratio, 115, 123, 147. *See also* Polarization ellipse
Azzam, R. M. A., 459

Balanis, C. A., 55, 112, 287
Bandwidth, 369. *See also* Antenna(s), bandwidth
Barrick, D. E., 362
Bashara, N. M., 459

Basis functions (moment method), 60
Beam cross section, 75
Beam efficiency, 109
Beamwidth:
  half-power, 20
  overall, 229
  polarization, 229
  radiation intensity, 20, 228
Bean, B. R., 500
Beckmann, P., 362
Bellman, R., 459
Beran, M. J., 366, 459
Bistatic cross section, 52, 54
Blackbody, 100
Blacksmith, P., Jr., 477
Boerner, W-M., 362, 460, 501
Boltzmann's constant, 100
Born, M., 112, 179, 366, 459
Brewster angle, 299
Brightness, 98-100, 104
  average, 99
Brightness temperature, 100, 103, 104
Brillouin, L., 32
Broadband antennas, 95-98. *See also* Antenna(s), frequency-independent; Antenna(s), log-periodic
Brown, G. H., 112
Bushore, K. R., 287

Canonical problem, 65
Carter, P. S., 32, 55
Cauchy principal value, 371
Chaff, 484-485
Chan, C. Y., 362, 460, 478
Characteristic impedance:
  free space, 22
  transmission line, 140
Charge density:
  electric, 4
  magnetic, 3, 4
Circulator, 279
Clayton, L., Jr., 478
Cloude, S. R., 460
Clutter, 481-482
  chaff, 484-485
  rain, 482-484, 489-492
  sea, 485-486, 497
Coherence theory, 366
Coherency matrix, 373-378
  antenna, 387
  generalized, 427
  receiving, 388
Coherency vector, 404
Cohn, S. B., 112

INDEX   515

Collin, R. E., 55, 287
Communication system, polarization adaptive, 278-287
Compensator, 506
Complete polarization, 365, 376-378
Compton, R. T., Jr., 112, 459, 501
Conjugate functions, 370
Constitutive equations, 4
Copolarization maxima, 331
Copolarization nulls, 331
Copolarized waves, 331
Correlation functions, 419-421
Correlation matrix, 427
Correlation vector (LMS array), 83
Cosine integral, 36
Cosmic noise, 110
Critical angle:
  sea clutter, 485
  wave reflection, 300
Croney, J., 500
Crosscorrelation, 419
  measurement, 420-421
Cross-polarized antennas, 197
Cross-polarized waves, 331
Cross-polarization maxima, 361
Cross-polarization nulls, 331
Cross-polarization saddlepoint extrema, 361
Cross section:
  backscattering, 54
  bistatic, 52, 54
  monostatic, 52
  radar, 52-53, 333-336
  scattering, 52-53, 333-336
Crystal mixer, 111
Current:
  electric, 8
  magnetic, 9
Current density:
  electric, 4, 9, 40
  magnetic, 3, 4, 40
Current element, 7
Currie, N. C., 500

Deetz, C. H., 179
Degree of polarization, 379-381, 394-395
Depolarization, 366, 415
Diffracted field, 66
Diffraction, 65-67
Diffraction coefficient, 66
Dihedral corner, *see* Target, dihedral corner
Dipole, *see* Antenna(s), dipole
Dirac delta function, 34
Directive gain, *see* Directivity
Directivity, 21, 22, 57, 184

Direct product, 404
DuHamel, R. H., 112
Dutton, E. J., 500
Dyad, 67

Earth noise, 111
Eaves, J. L., 500, 501
Effective area, 47, 49, 57, 98, 184
Effective length, 182-183, 185-188, 323
Effective temperature, *see* Temperature
Efficiency, *see* Antenna(s), efficiency; Polarization, efficiency; Polarization, match factor; Radiation, efficiency
Eigenvalues:
  Graves matrix, 445
  scattering matrix, 327, 340-341
Eigenvectors:
  Graves matrix, 445-446
  scattering matrix, 329
Eikonal equation, 61-62
El-Arini, M. B., 362
Electric charge density, 4
Electric current, 8
Electric current density, 4, 9, 40
Electric source, *see* Source, electric
Elliott, R. S., 32, 55, 112, 287
Ellipse, *see* Polarization ellipse
Elliptically-polarized antenna system, 278-287
Elliptically-polarized waves, 115-125
  reflection of, 300-304
Ellipticity angle, 123. *See also* Polarization ellipse
E-plane, 19
Equivalence principle, 31, 40
Equivalent circuit of antennas, 37-38, 44-45, 48, 56
Equivalent current, 31, 40
Equivalent source, 3, 40
Euler angle matrix, 214
Euler's constant, 36
Ewell, G. W., 500
Ezquerra, N. F., 500

Faraday rotation, 324
Far zone, 10
Far-zone fields, 13
Fermat's principle, 64
Feynman, R. P., 179
Flat plate, *see* Target, flat plate
Focal point, 246, 248
Foo, B. Y., 362
Fourier integral, 370
Fourier transform, 370, 420

Fraunhofer zone, *see* Far zone
Freeny, C. C., 477
Frequency-independent antennas, 88-92
Fresnel coefficients, 294-297
Fresnel zone, 9-10
Friis transmission equation, 194

Gabor, D., 459
Gain, 21, 22, 25, 47, 183
Gain, directive, *see* Directivity
Geometrical theory of diffraction, 65-67
Geometric optics, 61-65
Ghose, R. N., 287
Giuli, D., 179, 500
Glint, 486-489
Goldstein, H., 219, 500
Goode, B. B., 112
Graf, E. R., 500
Grating lobes, 74
Graves, C. D., 362, 460
Graves matrix, 440-450
 eigenvalues, 445
 eigenvectors, 445-446
 measurement, 475-476
Greene, J. C., 112
Griffiths, L. J., 112
Guy, J. R. F., 179, 501

Harrington, R. F., 55, 112
Helix, 267-273
Helmholtz equation, 11
Hermitian matrix, 326
Hiatt, R. E., 477
Hickman, T. G., 478
Hilbert transform, 370-371
Hollis, J. S., 478
Holm, W. A., 362, 501
Horn, *see* Antenna(s), horn
H-plane, 19, 20
Huynen, J. R., 362, 459, 477
Huynen fork, 341-347, 349-360
Hybrid tee, 279

Illumination, tapered, 109
Image theory, 30-31
Impedance:
 antenna, *see* Antenna(s), impedance
 mode, 38
 transmission line characteristic, 140
Impedance match factor, 51
Impedance matrix, 43
Incidence plane, 62, 290
Index of refraction, 62
Induced emf, 32

Interface:
 reflection, 290-304
 transmission, 290-304
Interference region (sea clutter), 485
Interfering signal, 82
Invisible region, 74
Ip, W. S., 362
Isbell, D. E., 112
Isotropic radiation, 51

James, B. D., 501
Jammer, 82, 103
Jasik, H., 478
Jones matrix, 314-316, 368. *See also* Scattering matrix; Sinclair matrix
Jordan, E. C., 55

Kennaugh, E. M., 362, 460, 478
Kennaugh matrix, 415-419, 509-510. *See also* Mueller matrix, measurement, 476-477
Kennaugh vector, 418. *See also* Stokes vector
Kepler's law, 128
Kerr, D. E., 500
Kim, K., 460
Knittel, G. H., 478
Ko, H. C., 459
Kong, J. A., 501
Kostinski, A. B., 460, 501
Kouyoumjian, R. G., 477
Kraus, J. D., 55, 112, 287, 478
Krichbaum, C. K., 362
Kronecker product, 404

Lebenbaum, M. T., 112
Leighton, R. B., 179
Length, *see* Effective length
Lim, H. H., 501
LMS array, *see* Adaptive arrays
Lobe, radiation, 21
Log-periodic antennas, *see* Antenna(s), log-periodic
Long, M. W., 500
Loomis, J. M., 500
Lorentz reciprocity theorem, 30, 42
Losses, antenna, 23-24, 111
Loss resistance, 24, 37
Love equivalence principle, 31
Lyon, T. J., 478

Mack, R. B., 477
Magnetic charge density, 3, 4
Magnetic current, 9
Magnetic current density, 3, 4, 40
Magnetic source, *see* Source, magnetic

Mandel, L., 460
Mantey, P. E., 112
Map:
  Aitoff-Hammer equal-area, 115, 176–179
  authalic, see Map, equal-area
  conformal, 115, 168
  equal-area, 115, 168
  equidistant, 168
  homalographic, see Map, equal-area
  homolographic, see Map, equal-area
  Lambert azimuthal equal-area, 170–172
  Mercator, 173–174
  modified, 172–173
  Mollweide equal area, 115, 175–176
  orthogonal, 172–173
  scale factors, see individual maps
  stereographic, 159–166, 169–170
  tilt angle-ellipticity angle, 174–175
Marlow, H. C., 477
Maser, 111
Mastoris, P. M., 362
Match factor:
  impedance, 51
  polarization, see Polarization, efficiency; Polarization, match factor
Matching network, 38
Matrix, see Graves, Jones, Kennaugh, Mueller, Scattering, and Sinclair matrix
  impedance, 43
  unitary, 135, 384
Maxwell equations, 2
  complex time-invariant:
    negative sign convention, 2–3
    positive sign convention, 2–4
McQuiddy, D. N., 55, 287
Measurements, 463–478
Misaligned antennas, 212–219
Mixer, crystal, 111
Moment method, 58–61
Monochromatic wave, 155, 365, 393
Monostatic cross section, 52, 54
Mott, H., 55, 287
Mueller matrix, 403–407, 502–508. See also Kennaugh matrix
  average, 414–415
  measurement, 476–477
Multipath, 2
Mutual coherence, 381

Nash, R. T., 112
Nathanson, F. E., 500
Newton-Raphson iteration, 176
Noise, 98–112. See also Antenna(s), temperature; Brightness
  antenna loss, 111
  atmospheric loss, 110
  cosmic, 110
  earth, 111
  radome loss, 110
  sun, 104
Noise figure, 105
  of cascaded networks, 106
  of lossy medium, 107–108
North pole (Poincaré sphere), 157
Null:
  polarization, 331
  radiation pattern, 82
Nyquist, H., 102, 112

Omnidirectional antenna, 21
O'Neill, E. L., 460
Optical path length, 64
Optics, intensity law of, 65
Orthogonality, 117
Orthogonal vectors, 116, 134, 137

Panofsky, W. K. H., 5, 55
Parametric amplifier, 111
Parrent, G. B., 366, 459
Partial polarization, 365–366, 368–454. See also Wave(s), partially-polarized
Pattern:
  antenna, 15–21
  polarization, 465–467
  radiation, 15, 18, 45, 47
  receiving, 45, 47
Pattern multiplication, 69
Pease, M. C., III, 459
Peters, L., Jr., 477
Phase-lock-loop, 79–80
Phillips, M., 5, 55
Planck's constant, 100
Planck's radiation law, 100
Plane of incidence, 62, 290
Plane wave, 46
Plateau region (sea clutter), 485
Pocklington's equation, 59–61
Poelman, A. J., 179, 501
Poincaré sphere, 115, 156–157, 210, 396. See also Map
Point matching, 61
Poisson's equation, 8, 55
Polarizability, 407–408
Polarization:
  adaptation, 497–500
  chart, 140–147
  circular, 125, 378
  complete, 365, 376–378

Polarization (*Continued*)
  degree of, 379–381, 394–395
  efficiency:
    backscattering, 329
    of two-antenna system, 191, 194, 203, 210.
      See also Polarization, match factor
  ellipse, *see* Polarization ellipse
  elliptic, 117–125
  fork, 341–347, 349–360
  linear, 125, 378
  loss, *see* Polarization, efficiency, match factor
  match factor, 191, 196–200, 201–203, 203–210. *See also* Polarization, efficiency
  maximum power density, 445
  maximum received power, 322–329
  measurements, 463–478
  minimum power density, 449
  minimum received power, 327, 330
  optimum, 327, 447
  partial, 365–366, 368–454. *See also* Wave(s), partially-polarized
  pattern, 465–467
  receiving, 203, 211
  state, 323
Polarization-adaptive communication system, 88, 278–287
Polarization ellipse:
  area sweep rate, 128–129
  axial ratio, 115, 123, 147
  ellipticity angle, 123
  rotation rate, 126
  rotation sense, 123–125
  rotation with distance, 129
  tilt angle, 115, 120–123, 147
Polarization pattern, 465–467
Polarization power scattering matrix, *see* Graves matrix
Polarization ratio:
  antenna, 192
  circular, 133
  complex linear, 130–131, 133
  inverse circular, 132–133, 136–137
  modified, 135–136
  optimum, 196, 456
  partially-polarized waves, 399
  suboptimum, 456
Polarizer, 506–507
Polkinghorne, A. A., 112
Potential:
  electric scalar, 5
  electric vector, 6, 9
  integrals, 7–9, 13–14
  magnetic scalar, 6

  magnetic vector, 5, 8
Power, maximum received:
  monochromatic waves, 188–191
  partially-polarized waves, 400–403
Power density, 125
Power flux density, 98
Power gain, *see* Gain
Power spectral density, 419–440
Poynting vector, time average, 16, 98
P plane, 160–161
p plane, 141, 146, 160
Principal plane, 19–20
Projection, *see* Map
Propagation constant, 5
  transmission line, 140
Propagation vector, 3

q plane, 141, 147, 163
Quarter-wave plate, 273–274
Quasi-monochromatic wave, 369

Radar:
  bistatic, 52
  monostatic, 52
Radar cross section, 52–53, 333–336
Radar equation, 52–54, 333–336
Radar target, 52
Radiation:
  efficiency, 25, 45, 56, 57. *See also* Antenna(s), efficiency
  intensity, 16, 18
  lobe, 21
  pattern, 15–21
  resistance, 22, 37, 56, 184. *See also* Antenna(s), resistance
Radome, 108, 110–111
Raindrops, scattering from, 482–484, 489–492
Ramo, S., 55
Raven, R. S., 179
Rayleigh–Jeans radiation law, 100
Reaction, 42–44
Receiving pattern, 45–47
Receiving polarization, 203, 211
Reciprocity, 29, 30, 33, 42, 187
Reedy, E. K., 500, 501
Reflection:
  at interface, 290–304
  from conductor, 304–309
  of elliptically-polarized waves, 300–304
  total, 300
Reflection coefficients (Fresnel coefficients), 294–297
Reflection coefficients (transmission line):
  current, 140–141

INDEX 519

voltage, 140–141
Reflector, see Target
Refracted wave, 62
Refractive index, 62
Resistance, see also Antenna(s), resistance;
    Radiation, resistance
  high-frequency, 24
  loss, 24, 45, 56
Riblet, H. J., 287
Rotation rate of $\mathscr{E}$ vector:
  distance, 129
  time, 126
Rotation sense, 115, 123–125
Rotator, 505–506
Rubin, R., 464, 478
Ruck, G. T., 362
Rumsey, V. H., 55, 112, 179, 219

Saatchi, S., 362
Sands, M., 179
Satellite, 288
Scattering cross section, 52–53, 333–336
Scattering matrix, 311–322, 332–333, 336,
    347–348, 349–359. See also Jones matrix;
    Sinclair matrix
  absolute, 314
  diagonal, 324, 337–339
  with geometric variables, 336–348
  measurement, 472–474
  nonsymmetric, 455–459
  relative, 314, 326
Schiff, L. I., 460
Schwarz inequality, 375
Sea clutter, 485–486, 497
Shin, R. T., 501
Shor array, 88
Sidelobes, 20, 109
Signal vector (LMS array), 83
Silver, S., 38, 55, 287
Sinclair, G., 219
Sinclair matrix, 312–314, 318–319, 367. See
    also Jones matrix; Scattering matrix
  nonsymmetric, characteristic polarizations
    with, 361, 455–459
Sine integral, 36
Skin depth, 24
Skolnik, M. I., 112, 478, 500
Smith chart, 140–141
Snell's laws, 62, 293–294
Snyder, J. P., 179
Solid angle, 16
Sommerfeld, A., 5, 55
Source:
  electric, 4, 12, 13, 14

  equivalent, 3, 40
  magnetic, 4, 6, 11, 12, 13
South pole (Poincaré sphere), 157
Spectral coherency matrix, 427
Spectral density of signal, 434–435
Spectral flux density, 100
Spectral Stokes vector, 427
Specular region (sea clutter), 485
Sphere, 350, 505, 511
Stereographic projection, 159–166, 169–170.
    See also Map
Stokes correlation vector, 427
Stokes parameters:
  monochromatic waves, 155, 392
  partially-polarized waves, 392
Stokes vector, see also Kennaugh vector
  antenna, 397
  measurement of, 396–397, 470–472
  monochromatic wave, 393
  partially-polarized wave, 392
  receiving, 398
  unpolarized wave, 393
Stratton, J. A., 362
Stuart, W. D., 362
Sun:
  antenna temperature caused by, 104–105
  brightness, 104
  spectral flux density of, 104
  subtended angle, 104
Symmetric target, 357

Tai, C. T., 112
Tapered illumination, 109
Target:
  coherency matrix, 451–452
  covariance matrix, 452–453
  depolarizing, 366, 415
  detection, 479–480
  diagonal wires, 356–357
  dihedral corner:
    Huynen fork parameters, 354
    Kennaugh matrix, 511
    Mueller matrix, 505
    reflected polarization, 305–306
    Sinclair matrix, 350–354
  discrimination, 479–480
  disk:
    Kennaugh matrix, 511
    Mueller matrix, 504
  distributed, 365
  extended, 480
  flat plate:
    Huynen fork parameters, 349
    Kennaugh matrix, 511

Target, flat plate (*Continued*)
   Mueller matrix, 504
   reflected polarization, 304–305
   Sinclair matrix, 349
glint, 486–489
"helix", 359
nonsymmetric noise (N), 358–359
polarizability, 407–408
radar, 52
recognition, 479–480
skip angle, 341
sphere:
   Huynen fork parameters, 350
   Kennaugh matrix, 511
   Mueller matrix, 505
   Sinclair matrix, 350
symmetric, 357
trihedral corner:
   Huynen fork paramaters, 355
   Kennaugh matrix, 511
   Mueller matrix, 505
   reflected polarization, 307–309
   Sinclair matrix, 354–355
vector, 451
wire grid, 355–356
Teeter, W. L., 287
Temperature:
   antenna noise, 103, 111
   brightness, 100
   effective:
      of lossy medium, 107–108
      of network, 107
   example, 109–111
Thermal equilibrium, 102
Three-stage optimization, 455–459
Tilt angle, 115, 120–123, 147. *See also* Polarization ellipse
Titchmarsh, E. C., 459
Tomberlin, E. L., 500
Total reflection, 300
Total transmission, 299
Transition angle, 485
Transmission:
   coefficients, 294–297
   through interface, 290–300
   line, 1, 95, 140–141
   matrix, *see* Jones matrix
   total, 299
Tranter, W. H., 287, 460
Trihedral corner, *see* Target, trihedral corner

Unitary transformation, 324

Unpolarized wave, 99, 366, 376

Vaillancourt, R. M., 287
Van Duzer, T., 55
van Hoozer, C. H., 477
van Zyl, J. J., 460, 501
Vector length, *see* Effective length
Vectors, orthogonal, 116, 134, 137
Visible region, 73
von Aulock, W. H., 112

Warner, B. D., 500
Watson, D. C., 477
Wave(s):
   circular, 131–133
   completely polarized, 365, 376
   elliptically-polarized, 115–125
   harmonic, 115–118
   independent, 379
   linear horizontal, 136, 160
   linear vertical, 136, 160
   monochromatic, 155, 365, 393
   nonplanar, 115–118
   nonuniform, 300
   orthogonal, 133–134
   partially-polarized, 366
      reception of, 386–388, 397–399, 400–403, 417–418
      scattering from target, 389–391, 403–407, 415–419
   plane, 117, 118–125
   polarization measurement, 463–465
   polychromatic, 368, 371
   quasi-monochromatic, 369
   unpolarized, 99, 366, 376
Wave components:
   circular, 131–133, 316–318
   orthogonal, 133–134, 200
   rectangular, 118, 131
Wedge, 65
Weeks, W. L., 287
Weight(s):
   optimum LMS, 82–85
   setting, 85–87
   vector, 83
Weitenhagen, J. H., 112
Whinnery, J. R., 55
White noise, 102
Widrow, B., 112
Widrow LMS algorithm, 87
Wolf, E., 112, 179, 366, 459, 460

Woodward, O. M., 112
w plane, 145, 147, 163

Xi, A-Q., 362
X-pol nulls, *see* Cross-polarized nulls

Yuch, S. H., 501

Zebker, H. A., 501
Ziemer, R. E., 287, 460
Zucker, F. J., 55, 287